Lecture Notes in Computer Science 5957

Commenced Publication in 1973
Founding and Former Series Editors:
Gerhard Goos, Juris Hartmanis, and Jan van Leeuwen

T0216744

Gheorghe Păun Mario J. Pérez-Jiménez
Agustín Riscos-Núñez Grzegorz Rozenberg
Arto Salomaa (Eds.)

Membrane Computing

10th International Workshop, WMC 2009
Curtea de Arges, Romania, August 24-27, 2009
Revised Selected and Invited Papers

 Springer

Volume Editors

Gheorghe Păun
Institute of Mathematics of the Romanian Academy
Bucharest, Romania
E-mail: george.paun@imar.ro

Mario J. Pérez-Jiménez
Research Group on Natural Computing
University of Sevilla, Spain
E-mail: marper@us.es

Agustín Riscos-Núñez
Research Group on Natural Computing
University of Sevilla, Spain
E-mail: ariscosn@us.es

Grzegorz Rozenberg
Leiden Center of Advanced Computer Science (LIACS)
Leiden University, The Netherlands
E-mail: rozenber@liacs.nl

Arto Salomaa
Turku Centre for Computer Science (TUCS)
Turku, Finland
E-mail: asalomaa@cs.utu.fi

Library of Congress Control Number: 2009943037

CR Subject Classification (1998): F.1, F.4, I.6, J.3

LNCS Sublibrary: SL 1 – Theoretical Computer Science and General Issues

ISSN   0302-9743
ISBN-10   3-642-11466-0 Springer Berlin Heidelberg New York
ISBN-13   978-3-642-11466-3 Springer Berlin Heidelberg New York

Typesetting: Camera-ready by author, data conversion by Scientific Publishing Services, Chennai, India
Printed on acid-free paper   SPIN: 12839131   06/3180   5 4 3 2 1 0

Preface

This volume contains a selection of papers presented at the 10th Workshop on Membrane Computing, WMC 2009, which took place in Curtea de Argeş, Romania, during August 24–27, 2009.

The first three Workshops on Membrane Computing were organized in Curtea de Argeş, Romania – they took place in August 2000 (with the proceedings published in *Lecture Notes in Computer Science*, volume 2235), in August 2001 (with a selection of papers published as a special issue of *Fundamenta Informaticae*, volume 49, numbers 1–3, 2002), and in August 2002 (with the proceedings published in *Lecture Notes in Computer Science*, volume 2597). The next six workshops were organized in Tarragona, Spain (in July 2003), in Milan, Italy (in June 2004), in Vienna, Austria (in July 2005), in Leiden, The Netherlands (in July 2006), in Thessaloniki, Greece (in June 2007), and in Edinburgh, UK (in July 2008), with the proceedings published as volumes 2933, 3365, 3850, 4361, 4860, and 5391, respectively, of Springer's *Lecture Notes in Computer Science*.

The workshop changed its name in 2010, and the series will continue as the Conference on Membrane Computing, with the next edition, CMC11, to be held in Jena, Germany.

The tenth edition of WMC took place in Hotel Posada in Curtea de Argeş (http://www.posada.ro/) and it was organized by the National College "Vlaicu Vodă" of Curtea de Argeş, the University of Piteşti, Romania, and the Institute of Mathematics of the Romanian Academy, Bucharest, under the auspices of the European Molecular Computing Consortium (EMCC) and IEEE Computational Intelligence Society Emergent Technologies Technical Committee Molecular Computing Task Force, with the financial and organizational support of the Council of Argeş County and Seville University.

Being an anniversary edition of the workshop, ten researchers with fundamental contributions to membrane computing were invited to deliver talks covering important directions of research in this area. These invited speakers were: Erzsébet Csuhaj-Varjú, Budapest, Hungary; Rudolf Freund, Vienna, Austria; Pierluigi Frisco, Edinburgh, UK; Marian Gheorghe, Sheffield, UK; Oscar H. Ibarra, Santa Barbara, USA; Vincenzo Manca, Verona, Italy; Solomon Marcus, Bucharest, Romania; Giancarlo Mauri, Milan, Italy; Mario J. Pérez-Jiménez, Seville, Spain; Grzegorz Rozenberg, Leiden, The Netherlands.

Full papers associated with the invited talks or only extended abstract are included in the present volume.

The volume also contains 22 papers, most of them significantly rewritten according to the discussions held during WMC 2009. Each paper had three or four referee reports. The Program Committee consisted of Matteo Cavaliere (Trento, Italy), Erzsébet Csuhaj-Varjú (Budapest, Hungary), Rudolf Freund (Vienna, Austria), Pierluigi Frisco (Edinburgh, UK), Marian Gheorghe (Sheffield, UK), Thomas Hinze (Jena, Germany), Oscar H. Ibarra (Santa Barbara, USA),

Florentin Ipate (Piteşti, Romania), Shankara Narayanan Krishna (Mumbai, India), Vincenzo Manca (Verona, Italy), Giancarlo Mauri (Milan, Italy), Radu Nicolescu (Auckland, New Zealand), Linqiang Pan (Wuhan, China), Gheorghe Păun (Bucharest, Romania, and Seville, Spain) – Chair, Mario J. Pérez-Jiménez (Seville, Spain), and Claudio Zandron (Milan, Italy).

The program of WMC 2009 also included seven papers which were submitted after the deadline; they were allocated only 10 minutes for the presentation (and are not included in this volume). A pre-workshop proceedings volume, containing all papers, was available during the workshop.

During the workshop several prizes were awarded, some of them related to the 10th anniversary of the workshop (e.g., the youngest participant, the person who participated in most editions of WMC, the author of the largest number of papers in membrane computing, the author of the first PhD thesis in membrane computing, and so on), as well as the *best paper* award.

This award was shared by two papers:

1. Andrea Valsecchi, Antonio E. Porreca, Alberto Leporati, Giancarlo Mauri, Claudio Zandron: "An Efficient Simulation of Polynomial-Space Turing Machines by P Systems with Active Membranes"
2. Petr Sosík, Andrei Păun, Alfonso A. Rodriguez-Patón, David Pérez: "On the Power of Computing with Proteins on Membranes"

The Organizing Committee consisted of Gheorghe Păun – Chair, Costel Gheorghe – Co-chair, Gheorghe Barbu – Co-chair, Paul Radovici, Ştefana Florea, Ileana Popescu, Nicolae Lazăr, Marius Hirzoiu, Ştefana Dumitrache, Raluca Judeţ, Ana-Mariana Margarita, Vlad Bărbulescu, and Radu-Daniel Gheorghe.

Details about membrane computing can be found at: http://ppage. psystems.eu and its mirror page http://bmc.hust.edu.cn/ psystems. The workshop website, designed by Mihai Ionescu, is available at the address http:// wmc10.psystems.eu/.

The workshop was sponsored by the Council of Argeş County, Piteşti, Romania, and it was one of the events in the program of Argeş and Muscel Days, 2009. The pre-workshop proceedings volume was published (as TR 3/2009) by the Research Group on Natural Computing of Seville University, with the support of Proyecto de Excelencia con Investigador de Reconocida Valía, de la Junta de Andalucía, grant P08 – TIC 04220. Further local support by the City Hall of Curtea de Argeş and Hotel Posada is gratefully acknowledged.

The editors warmly thank the Program Committee, the invited speakers, the authors of the papers, the reviewers, and all the participants, as well as all who contributed to the success of WMC 2009. Special thanks are due to Springer for the pleasant cooperation in the timely production of this volume.

October 2009

Gheorghe Păun
Mario J. Pérez-Jiménez
Agustin Riscos-Núñez
Grzegorz Rozenberg
Arto Salomaa

Table of Contents

Invited Presentations

Regular Presentations

P Automata:
Concepts, Results, and New Aspects*

Erzsébet Csuhaj-Varjú

Computer and Automation Research Institute
Hungarian Academy of Sciences
Kende utca 13-17, 1111 Budapest, Hungary
and
Department of Algorithms and Their Applications
Faculty of Informatics, Eötvös Loránd University
Pázmány Péter sétány 1/c, 1117 Budapest, Hungary
csuhaj@sztaki.hu

Abstract. In this paper we discuss P automata, constructs combining properties of classical automata and P systems being in interaction with their environments. We describe the most important variants and their properties, and propose new topics and open problems for future research.

1 Introduction

Observing natural systems and processes, ideas for constructing unconventional computational models can be obtained. When a new computing device is introduced, its benefits usually are demonstrated through comparisons to its conceptual predecessors or to other classical computational models having features similar to the new one. This procedure takes place in the theory of P automata, a framework consisting of accepting variants of P systems which combine features of classical automata and membrane systems being in interaction with their environments.

Briefly, a P automaton is a P system receiving input in each computational step from its environment which influences its operation by changing its configuration and thus affecting its functioning. The sequences of inputs are distinguished as accepted or rejected input sequences. The input is given as a multiset of objects, where the objects can be elementary ones, i.e., without any structure (for example, symbols) or non-elementary, structured ones (for example, a P system).

Similarities between P automata and classical automata can immediately be observed, but the reader may easily notice differences between the two constructs as well: for example, conventional automata have separate state sets while in the case of P automata the actual state is represented by the actual configuration of the underlying P system. Another property which makes P automata different from

---

* Research supported in part by the Hungarian Scientific Research Fund (OTKA), Grant no. K75952.

G. Păun et al. (Eds.): WMC 2009, LNCS 5957, pp. 1–15, 2010.

classical automata is that the computational resource they can use is provided by the objects of the already consumed input multisets. In this way, the objects which enter the system become part of the description of the machine, that is, the object of the computation and the machine which performs the computation cannot be separated as it can be done in the case of customary automata.

The first variant of P automata, introduced in [14,15], was the so-called *one-way P automaton* where the underlying P system had only top-down symport rules with promoters (and implicitly inhibitors). Almost at the same time, a closely related notion, the *analyzing P system* was defined in [21] providing a slightly different concept of an automaton-like P system. Both models describe the class of recursively enumerable languages.

The property that purely communicating accepting P systems may represent computationally complete classes of computing devices, gave an impetus to the research in the theory of P automata, resulting in a detailed study of automaton-like P systems. Since that time, several variants have been introduced and investigated, which differ from each other in the main ingredients of these systems: the objects the P system operates with, the way of defining the acceptance, the way of communication with the environment, the types of the communication rules used by the regions, the types of the rules associated with the regions (whether or not evolution rules are allowed to be used), and whether or not the membrane structure changes in the course of the computation. Summaries on these constructs and their properties can be found in [32,10,13,41].

Due to the power of the underlying P system, several of the above variants determine the class of recursively enumerable languages, even with limited size parameters. Although these constructs offer alternatives for Turing machines, P automata with significantly less computational power are of special interest as well. For example, the generic variant which is based on antiport rules with promoters or inhibitors, and accepts with final states, if applies its rules sequentially and uses some appropriately chosen mappings for defining its language, determines a language class with sub-logarithmic space complexity. In this way, a "natural description" of this particular complexity class is provided.

In the following sections we describe the most important variants of P automata and their properties. We also discuss how some *variants of classical automata can be represented in terms of P automata*. Special emphasis is put on *non-standard features of P automata*, namely, that the same construct is able to operate over both finite and infinite alphabets, the underlying membrane structure may remain unchanged but it also may dynamically alter under functioning, and that to obtain large computational power they do not need workspace overhead.

We also propose new topics and problems for future research.

2 P Automaton - The Basic Model

2.1 The Formal Concept

In order to provide the reader with sufficient information to follow the discussion on P automata and its different variants, we present some formal details concerning the basic model, following mainly the terms and notations in [13]. For more information on the basics of membrane computing we refer to [37] and for more details on formal language and automata theory to [38].

Throughout the paper, we denote the class of context-sensitive and recursively enumerable languages by CS and RE. For a finite alphabet V, we designate by V^* the set of all strings over Σ; λ stands for the empty string. The set of finite multisets over a finite set V is denoted by V°, and the set of their sequences by $(V^\circ)^*$. If no confusion arises, the empty multiset is also denoted by λ; due to the representation of multisets by strings used in the literature of P systems.

The underlying membrane system of a P automaton is an antiport (symport) P system possibly having promoters and/or inhibitors. For details on symport/antiport the reader is referred to [35], for the use of promoters to [31].

Briefly, a symport rule is of the form (x, in) or $(x, out), x \in V^\circ$. When such a rule is applied in a region of a P system, then the objects of the multiset x enter the region from the parent region (in) or they leave to the parent region (out). An antiport rule is of the form $(x, out; y, in), x, y \in V^\circ$. In this case, the objects of y enter the region from the parent region and in the same step the objects of x leave to the parent region. The parent region of the skin region is the environment. All types of these rules might be associated with a promoter or an inhibitor multiset, denoted by $(x, in)|_z, (x, out)|_z$, or $(x, out; y, in)|_z, x, y \in V^\circ, Z \in \{z, \neg z \mid z \in V^\circ\}$. If $Z = z$, then the rule can only be applied if the region contains all objects of multiset z, and if $Z = \neg z$, then z must not be a submultiset of the multiset of objects present in the region. To simplify the notations, we denote symport and antiport rules with or without promoters/inhibitors by $(x, out; y, in)|_z, x, y \in V^\circ, Z \in \{z, \neg z \mid z \in V^\circ\}$ where we also allow x, y, z to be the empty multiset. If $y = \lambda$ or $x = \lambda$, then the notation above denotes the symport rule $(x, in)|_z$ or $(y, out)|_z$, respectively, if $Z = \lambda$, then the rules above are without promoters or inhibitors.

Definition 1. *A P automaton (with n membranes) is an $(n + 4)$-tuple, $n \geq 1$, $\Pi = (V, \mu, P_1, \ldots, P_n, c_0, \mathcal{F})$, where*

- *V is a finite alphabet of objects,*
- *μ is a membrane structure of n membranes with membrane 1 being the skin membrane,*
- *P_i is a finite set of antiport rules with promoters or inhibitors associated to membrane i for all $i, 1 \leq i \leq n$,*
- *$c_0 = (w_1, \ldots, w_n)$ is called the initial configuration (or the initial state) of Π where each $w_i \in V^\circ$ is called the initial contents of region i, $1 \leq i \leq n$,*
- *$\mathcal{F}$ is a computable set of n-tuples $(v_1, \ldots, v_n)$ where $v_i \subseteq V^\circ$, $1 \leq i \leq n$; it is called the set of accepting configurations of Π.*

An n-tuple $(u_1, \ldots, u_n)$ of finite multisets of objects over V present in the n regions of the P automaton Π is called a (possible) *configuration* of Π; u_i is the contents of region i in this configuration, $1 \leq i \leq n$.

A P automaton functions as a standard antiport (symport) P system (with promoters and/or inhibitors), it changes its configurations by applying rules according to a certain type of working mode. In the case of P automata, the two most commonly used variants are the sequential rule application, introduced in [14,15] (also called 1-restricted minimally parallel in [26]), and the maximally parallel rule application. In the case of sequential rule application, at any step of the computation the rule set to be applied is chosen in such a way that exactly one rule is applied in each region where the application of at least one rule is possible. When the the maximally parallel working mode is used, at every computational step as many rule application is performed simultaneously in each region as it is possible.

The set of the different types of working modes is denoted by $MODE$, we use *seq* and *maxpar* for the *sequential* and the *maximally parallel* rule application, respectively.

Definition 2. *Let $\Pi = (V, \mu, P_1, \ldots, P_n, c_0, \mathcal{F})$, $n \geq 1$, be a P automaton working in the X-mode of rule application, where $X \in MODE$. The transition mapping of Π is defined as a partial mapping $\delta_X : V^\circ \times (V^\circ)^n \to 2^{(V^\circ)^n}$ as follows:*

For two configurations $c, c' \in (V^\circ)^n$, we say that $c' \in \delta_X(u, c)$ if Π enters configuration c' from configuration c by applying its rules in the X-mode while reading the input $u \in V^\circ$, i.e., if u is the multiset of objects that enter the skin membrane from the environment while the underlying P system changes configuration c to c' by applying its rules in mode X.

The sequence of multisets of objects accepted by a P automaton is defined as the input sequence which is consumed by the skin membrane until the system reaches an accepting configuration.

Definition 3. *Let $\Pi = (V, \mu, P_1, \ldots, P_n, c_0, \mathcal{F})$, $n \geq 1$, be a P automaton. The set of input sequences accepted by Π with X-mode of rule application, $X \in MODE$, is defined as*

$$A_X(\Pi) = \{v_1 \ldots v_s \in (V^\circ)^* \mid \text{ there are } c_0, c_1, \ldots, c_s \in (V^\circ)^n, \text{ such that}$$
$$c_i \in \delta_X(v_i, c_{i-1}), 1 \leq i \leq s, \text{ and } c_s \in \mathcal{F}\}.$$

A P automaton Π, as above, is said to be accepting by final states if $\mathcal{F} = E_1 \times \ldots \times E_n$ for some $E_i \subseteq V^\circ$, $1 \leq i \leq n$, where E_i is either a finite set of finite multisets or $E_i = V^\circ$. Thus, a configuration $c = (u_1, \ldots, u_n)$ is final, if for all regions of Π, $u_i \in E_i$, $1 \leq i \leq n$.

If Π accepts by halting, then $\mathcal{F}$ contains all configurations c with no $c' \in (V^\circ)^n$ such that $c' \in \delta_X(v, c)$ for some $v \in V^\circ$, $X \in MODE$.

The accepted multiset sequences of a P automaton can be encoded to strings, thus making possible to assign languages to the P automaton. In the case of sequential rule application, the set of multisets that may enter the system is finite, thus the input multisets can obviously be encoded by a finite alphabet. This implies that any accepted input sequence can be considered as a string over a finite alphabet. In the case of parallel rule application, the number of objects which may enter the system in one step is not necessarily bounded by a constant. This implies that in this case the accepted input sequences correspond to strings over infinite alphabets.

In the following we consider languages over finite alphabets, therefore we apply a mapping to produce a finite set of symbols from a possibly infinite set of multisets.

Definition 4. *Let* $\Pi = (V, \mu, P_1, \ldots, P_n, c_0, \mathcal{F})$, $n \geq 1$, *be a P automaton,* Σ *be a finite alphabet, and let* $f : V^\circ \to \Sigma^*$ *be a mapping. The language accepted by* Π *with respect to* f *using the X-mode rule application, where* $X \in MODE$, *is defined as*

$$L_X(\Pi, f) = \{f(v_1) \ldots f(v_s) \in \Sigma^* \mid v_1 \ldots v_s \in A_X(\Pi)\}.$$

The class of languages accepted by P automata with respect to a class of computable mappings $\mathcal{C}$ with X-mode rule application, $X \in MODE$, is denoted by $\mathcal{L}_{X,\mathcal{C}}(PA)$.

We illustrate the notion of a P automaton by an example from [10].

Example 1. Let $\Pi = (\{S_1, S_2, S_3, a, b, c\}, [_1 \ [_2 \ [_3 \ ]_3 \ ]_2 \ ]_1 (S_1, P_1, \{d\}),$ $(S_2, P_2, \{S_1 S_2\}), (S_3, P_3, \emptyset)$, with

$$P_1 = \{(a, in)|_{S_1}, (a, in)|_a, (b, in)|_a, (b, in)|_b, (c, in)|_b, (c, in)|_c,$$
$$(d, in)|_c, (\lambda, in)|_d\},$$
$$P_2 = \{(S_1, in)|_{S_2}, (a, in)|_{S_1}, (b, in)|_{S_1}, (c, in)|_{S_1}, (\lambda, in)|_c\},$$
$$P_3 = \{(\lambda, in)|_{S_3}, (abc, in)|_{S_3}\},$$

where λ denotes the empty multiset. Then, for $f(x) = x$, where $x \in \{a, b, c, d\}$, Π accepts, with sequential application of rules and with only symport rules with promoters, words of the form $da^n b^n c^n$, $n \geq 1$. Thus, the language accepted by Π is a non-context-free context-sensitive language.

2.2   Computational Power of P Automata

Examining the concept of a language accepted by a P automaton, the reader can immediately notice that it strongly depends on the choice of the mapping f (see Definition 4). This implies that there might be cases where the power of the P automaton comes from f and not from the P automaton itself. Due to this property, the investigations on the accepting power of P automata have concentrated on the cases where f is of low complexity.

It can also easily be seen that P automata work with no workspace overhead, i.e., the computational resource the P automata can use for computation is provided by the objects of the already consumed input multisets. Although this property appears to significantly bound the computational power, since P automata may use maximally parallel working mode, i.e., may input an exponentially growing number of objects, the obtained computational power can be rather large.

We first recall some notations from [13]. Let **NSPACE**(S) designate the class of languages accepted by a non-deterministic Turing machine using a workspace which is bounded by a function $S : \mathbf{N} \to \mathbf{N}$ of the length of the input. We say that $L \in \mathbf{r1NSPACE}(S)$ if there is a Turing machine which accepts L by reading the input from a read-only input tape once from left to right, and for every accepted word of length n, there is an accepting computation during which the number of nonempty cells on the work-tape(s) is bounded in each step by $c \cdot S(d)$ where c is an integer constant, and $d \leq n$ is the number of input tape cells that have already been read, that is, the actual *distance* of the reading head from the left end of the one-way input tape.

Let $c = (u_1, \ldots, u_n)$ be a configuration of a P automaton. We designate by $card(c)$ the number of objects present inside the membrane system, that is, $card(c) = \Sigma_{i=1}^n card(u_i)$ where $card(u_i)$ denotes the number of objects of $u_i \in V°$.

The following statement describes the workspace of the P automaton used for computing and its language for a non-erasing mapping f [13]. (The mapping f is non-erasing if $f : V° \to \Sigma^*$ for some V, Σ with $f(u) = \lambda$ if and only if u is the empty multiset.)

Theorem 1. *Let Π be a P automaton, let $c_0, c_1, \ldots, c_m$ be a sequence of configurations during an accepting computation of Π, and let $S : \mathbf{N} \to \mathbf{N}$, such that $card(c_i) \leq S(d)$, $0 \leq d \leq i \leq m$, where $S(d)$ bounds the number of objects inside the system in the ith step of functioning and $d \leq i$ is the number of transitions in which a nonempty multiset entered the system from the environment.*

If f is non-erasing and $f \in \mathbf{NSPACE}(S_f)$, then for any $X \in MODE$, $L_X(\Pi, f) \in \mathbf{r1NSPACE}(\log(S) + S_f)$.

By applying the above theorem and its proof to three-counter machines, the following theorem was obtained (see [13]). (Three-counter machines are Turing machines with a one-way read only input tape and three work-tapes which can be used as three counters capable of storing any non-negative integer as the distance of the reading head from the only non-blank tape cell marked with the special symbol Z.)

The two results were first presented in [11,12].

Theorem 2

1. $\mathcal{L}_{seq,\mathcal{C}}(PA) = \mathbf{r1NSPACE}(\log(n))$ *for any class $\mathcal{C}$ of non-erasing mappings with a finite domain, and*
2. $\mathcal{L}_{maxpar,\mathcal{C}}(PA) = CS$ *for any class $\mathcal{C}$ of non-erasing linear space computable mappings.*

By the simulation of the three-counter machine which is used to prove the previous theorem, it follows that if we may use arbitrary linear space computable mappings for the input multisets of the P automaton to obtain the alphabet of the accepted language, then we yield a characterization of the class of recursively enumerable languages.

Corollary 1. $\mathcal{L}_{maxpar,\mathcal{C}}(PA) = RE$ *for any class $\mathcal{C}$ of linear space computable mappings.*

2.3   Discussion of the Basic Model

In the following we briefly discuss the *main ingredients* of P automata and propose topics for future research.

If we consider a P automaton as a *system being in interaction with its environment*, then not only input sequences but also output sequences and their relations are interesting for further examinations. An *input sequence* can be considered as a *representation of a sequence of impulses obtained from the environment*. A sequence of *outputs*, i.e., a sequence of multisets of objects that were sent to the environment at the steps of the computation, correspond to *reactions* to the effect of the previously obtained impulses and the change they caused in the behavior of the system. By obvious modifications of Definitions 2, 3, 4, we may associate a so-called *output language* with the P automaton. Output languages of P automata, supposing that the underlying P system issues at any computation step at least one object to the environment, would be a particularly interesting topic for investigations. The idea of input and output of a P automaton-like system was implemented in [9], where the so-called P transducers were introduced and examined.

The concept of an (accepted) output sequence of a P automaton opens *several topics to be examined*. For example, if u_i denotes the input and v_i the output of a P automaton Π at the ith computation step of a computation, then $d(i) =| card(u_i) - card(v_i) |$, i.e., the absolute value of the difference in the number of objects entering and leaving the system, describes the *volume of information exchange* at the given computation step and it is a characteristics of the P system. Based on this parameter, several descriptional complexity measures can be defined: for example, $maxd(\Pi)$, i.e., the supremum, or $mind(\Pi)$, i.e., the minimum of the difference of the volume of information exchange with respect to any accepting computation. We may consider the difference of these two measures as well. Especially *interesting topic for future research* would be the description of language classes of P automata classes where the value of measures $maxd$ and $mind$ regarding for any P automata in the class can be *bounded by linear, or polynomial, or exponential functions*, respectively.

The concepts of an input and an output of a P automaton raise *another issue*. As we have seen, unlike classical automata, the *whole input sequence* is not given at the beginning of the computation, but it will be *available step by step*, it is determined by the actual configuration (state) of the underlying P system. It is an obvious question, what happens if we *present the input sequence of the multisets of objects in advance* and we consider it as an accepted sequence if after consuming the elements of the sequence the underlying P system enters an accepting state. Obviously, *the details of this model should be elaborated*, since the multisets in the sequence need not to coincide with the multiset of objects the underlying P system is able to consume. Some steps, although in a bit different manner, have already been made in this direction, see, for example, [20]. We note that the existence of a designated input membrane does not necessarily alter the computational power.

2.4   Non-standard Features of P Automata

Examining the functioning of the basic model, it can be observed that in the case of maximally parallel rule application there may be P automata, where the number of objects entering the skin membrane during a successful computation can be arbitrarily large. This property makes possible to consider *P automata as tools for describing languages over infinite alphabets*, without any extension or additional component added to the underlying P systems.

An approach to this idea is the concept of a *P finite automaton*, introduced in [19].

This construct is a P automaton $\Pi = (V, \mu, P_1, \ldots, P_n, c_0, \mathcal{F})$ which applies the rules in the maximally parallel manner and accepts by final states. Its alphabet of objects, V, contains a distinguished element, a. The rules associated with the skin region, P_1, are of the form $(x, out; y, in)|_Z$ with $x \in V^\circ$, $y \in \{a\}^\circ$, $Z \in \{z, \neg z\}$, $z \in V^\circ$; and if $i \neq 1$, the rules of P_i are of the form $(x, out; y, in)|_Z$ with $Z \in \{z, \neg z\}$, $x, y, z \in V^\circ$. We also allow the use of rules of the form $(x, in)|_Z$ in the skin region in such a way that the application of any number of copies of the rule is considered in maximally parallel manner.

The domain of the mapping f is infinite and thus its range could also be defined to be infinite, as $f : \{a\}^\circ \to \Sigma \cup \{\lambda\}$ for an infinite alphabet $\Sigma = \{a_1, a_2, \ldots\}$ with $f(a^k) = a_k$ for any $k \geq 1$, and $f(\emptyset) = \lambda$.

The language accepted by a P finite automaton Π is defined as $L(\Pi) = L_{maxpar}(\Pi, f)$ for f as above.

In [19] it was shown that *for any $L \subseteq \Sigma^*$ over a finite alphabet Σ, L is regular if and only if $L = L(\Pi)$ for some P finite automaton Π*.

Due to this statement, the *languages which are defined over infinite alphabets and accepted by P finite automata* can be considered as extensions of the class of regular languages to infinite alphabets. The above construction significantly differs from other infinite alphabet extensions of regular languages defined by, for example, the finite memory automata from [29] or the infinite alphabet regular expressions introduced in [34], as it is shown in [19].

P automata models for *extensions of further language classes to infinite alphabets*, for example, to context-free languages, would also be an interesting research direction.

Accepting by final states, P automata provide possibilities of *describing* (possibly) *infinite runs* (sequences of configurations). This property is of crucial importance, since if we consider a P automaton as a system being in interaction with its environment, we also should consider communication processes not limited in time.

New variants of P automata, called *ω-P automata* were introduced in [25], inspired by the above considerations. These constructs (which have also so-called membrane channels) are counterparts of ω-Turing machines: In [25], it was shown that *for any well-known variant of acceptance mode of ω-Turing machines one can construct an ω-P automaton with two membranes which simulates the computations of the corresponding ω-Turing machine.*

2.5   Variants of P Automata

During the years, several types of automaton-like P systems were introduced with the aim of studying their properties and limits as computational devices.

A lot of efforts has been devoted to *describe the recursively enumerable language class in terms of P automata.* To be conform with formal language theoretic constructs, several variants have been introduced where input objects and auxiliary objects , i.e., terminal objects and nonterminal objects of the P automaton are distinguished. In this case, the accepted language is defined as the sequence of terminal strings of the input multisets during an accepting computation. Notice that *analyzing P systems* [21], one of the first P automata variants, which have only antiport rules, work with the maximally parallel rule application and accept by halting are *extended P automata.* As we mentioned in the Introduction, the authors of [21] proved that these systems are able to recognize any recursively enumerable language, furthermore, even with very small size parameters.

In [28] interesting results were obtained for automaton-like P systems, called *exponential-space symport/antiport acceptors,* working with other types of bounded resources, with a set of terminal objects Σ containing a distinguished symbol $, and four special types of rules of restricted forms. These systems work with maximally parallel application of rules and accept by final states; the language accepted by them is defined in a slightly different way from the one that is used in the case of an extended P automaton. The term "exponential-space symport/antiport acceptor" comes from the fact that due to the restricted form of the rules, the system contains no more than an exponential number of objects (up to some constant) at any time during the computation. Working with the maximally parallel rule application, these systems describe the class of context-sensitive languages [28].

The original motivation of introducing the concept of the P automaton was to study the power of purely communicating accepting P systems. For this reason, the question whether or not any change in the underlying communicating P system implies changes in the power and the size complexity of the respective new class of P automata has been intensively studied.

Given by a partial binary relation, priorities were associated with the application of the communication rules of the basic model in [6]. In the case of these *P automata with priorities*, the rules with the highest priority must be applied in configuration change. Two other P automata variants, with conditional symport/antiport rules, are *P automata with membrane channels* [32,22,23], motivated by certain natural processes taking place in cells, and *P automata with conditional communication rules associated with the membranes* [32,24]. All these models are computationally complete devices, in the latter two cases optimal results on their size parameters have also been obtained.

Although most of the variants of P automata are purely communicating, accepting P systems, the concept can be extended in a natural manner to be suitable for *describing complex evolving systems.* One approach in this directions is the *evolution-communication P automaton*, having both communication and evolution rules [1]. As expected, it provides a description of the class of recursively enumerable languages.

While most of the above variants of the basic model are based on changes in the use of communication rules, the following construct is given with a change in the main ingredients. As we observe, P automata have no separate internal state sets, the states are represented by the (possibly infinite) set of configurations. To make the basic concept more close to conventional automata, *P automata with states* were introduced and studied in [30], where both states and objects are considered, which together govern the communication. As expected, these constructs are computationally complete, moreover, any recursively enumerable language can be described by these systems with very restricted form [20].

3   Further Developments

3.1   P Automata Computing by Structure

The P automata variants we have discussed so far have *static membrane structure*, that is, the membrane structure is not altered during the functioning of the system. From modelling point of view, this condition is rather restrictive, since the architecture of natural systems may change in the course of their functioning.

A P automaton-like system working with a dynamically changing membrane structure is the *P automaton with marked membranes* ([16]), or a P_{pp} automaton for short. The concept was inspired by *the theory of P systems, brane calculi [5], and classical automata theory*. The underlying P system models the movements of proteins through the membranes in such a way that the moves may also imply changes in the membrane structure. As in the previous cases, the model is computationally complete. Its importance lies in the *bridge built between important research areas*.

Another variant of accepting P systems with dynamically changing membrane structure, is the *active P automaton*, that was proposed for parsing sentences of natural languages in [2,3]. It starts the computation with one membrane containing the string to be analyzed, together with some additional information assisting the computation. Then, it computes with the structure of the membrane system, using operations as membrane creation, division, and dissolution. There are also rules for extracting a symbol from the left-hand end of the input string and for processing assistant objects. The computation is successful (accepting) if all symbols from the string are consumed and all membranes are dissolved. It was shown that the model is suitable for recognizing any recursively enumerable language, and with restrictions in the types of rules, for determining other well-known language classes (the regular language class and the class of context-sensitive languages) as well. This special variant of accepting P systems resembles P automata since any symbol in the string can be considered as a multiset of objects with one element consumed from the environment. An important difference is that in this case the whole input is given at the beginning.

3.2   Classical Automata versus P Automata

Another important research area to study is how models and concepts of classical automata theory can be related to models and concepts in P automata theory. As we have seen above, finite automata can be represented in terms of P automata in a natural manner.

The property, that by using the maximally parallel working mode, during successful computations an object can appear in a region in an arbitrarily large number of copies, implies that strings (in the form of numbers which are values of numbers given in the k-ary notation) can be represented by contents of the regions of P automata. Based on this correspondence, contents of pushdown storages or stacks can be described, which natural observation is used for characterizing the context-free language class by a restricted variant of P automata, called *stack P automata* in [40]. Obviously, a pushdown storage can also be represented as a configuration of a P system with a linear structure, where there is only one object or one object of some distinguished type (representing a symbol that belongs to the pushdown alphabet) in each region [39]. If we allow changes in the linear membrane structure, i.e., the dissolution of the skin membrane and the creation of a new linear structure which embraces the remaining part of the original linear membrane structure, we can obtain a representation of a pushdown storage in some other manner. Both approaches are used in [17], where different languages classes, for example, the growing context-sensitive language class is determined by these special classes of P automata.

One can find counterparts of some other classical variants of automata in [7], where the so-called *Mealy multiset automata* and *elementary Mealy membrane automata*, inspired by concept of a Mealy automaton, are proposed and examined. As a further development, an augmented version of the elementary Mealy membrane automaton, with extended communication capabilities, called a *simple P machine* was investigated in [8].

The automata variants we have discussed so far process only input, but transducers, i.e., automata with input and output play outstanding role in classical automata theory. The concept of a *P transducer*, which is more or less a one-membrane P automaton working with input and output objects [9], realizes such a construction. Four types of these machines were investigated, two of them proved to be computationally complete, and the two other ones are incomparable to finite state sequential transducers. Iterating these latter classes of P transducers, new characterizations of the recursively enumerable language class were obtained.

3.3   P Automata and Words with Nested Data

Since membrane systems are *nested architectures*, investigations in relations of P automata theory to the theory of data languages, a theory mainly motivated by applications in XML databases and parametrized verification, are of particular importance. Research in this direction has started in [18].

In order to briefly report on the topic, we recall some notions on words with nested data, following the notations in [4]. Let V be a finite alphabet and Δ

an infinite set whose elements are called data values. For a natural number k, a word w with k layers of data is a string where every position, apart from a label in V, has k labels $d_1, \ldots, d_k \in \Delta$. The label d_i is called the ith data value of the position. Thus, $w \in (V \times \Delta^k)^*$. In w, the data values can be seen as inducing k equivalence relations $\sim_1, \ldots, \sim_k$ on the positions of w; two positions are related by $\sim_i$ if they agree on the ith data value. A word with k layers of data is said to have nested data if for each $i = 2, \ldots, k$ the relation $\sim_i$ is a refinement of $\sim_{i-1}$. Since P automata are able to operate over infinite alphabets, for representing sets of words with k layers of data or with k layers of nested data (over some alphabets V and Δ), P automata with dynamically changing linear structure and antiport rules can be constructed.

Unlike standard questions concerning the computational power of P automata, the main questions in this case are *how much change the input implies in the structure of the underlying P system* and in the contents of certain regions.

Another important research direction can be to *develop logic for these P automata (P systems)*, since certain properties of words with (k layers of) nested data, have been described in terms of a fragment of first order logic, thus these words were considered as models for logic, with logical quantifiers ranging over word positions.

The topic is closely related to the study of shuffle expressions, since connections between words with nested data and these expressions have been investigated, see, for example [4]. *Shuffle expressions* are regular expressions extended with intersections and the shuffle operation. Relations between shuffle expressions and so-called *high-order multicounter automata* was analyzed in [4], where it was shown that the class of languages defined by shuffle expressions, the class of languages defined by high-order multi-counter automata, and the recursively enumerable language class are the same. High-order multicounter automata are automata with several counters which can be incremented and decremented, but zero tests are only allowed at the end of the word. In [18] a new variant of P automata is defined with strong formal similarities to high-order multicounter automata. Based on the construction, results on P automata and shuffle expressions can be derived.

3.4   P Automata Expressions

One important research area of classical automata theory is the study of the closure with respect to certain operations, especially how to construct an automaton for languages obtained by certain operation among a given collection of automata. Questions related to *compositions of P automata* are of particular importance.

A step in this direction has been made in [27], where so-called *P automata with communication and active membrane rules working in the initial mode (CAIP)* have been introduced. The authors presented methods for constructing automata for accepting the union, the concatenation, the Kleene closure, or the ω closure of the given languages which are represented by some P automata. Starting from these results, and considering these and other operations and these and other (restricted) variants of P automata, it would be interesting to develop further descriptions of language classes in term of so-called *P-automata expressions*.

4 Conclusions

Investigations in the theory of P automata expected to be continued in several directions. Since P automata can be considered as constructs attempting to build a *bridge between classical automata theory and membrane systems theory*, similarities and differences between the two fields are certainly worth studying. But, as we mentioned in the Introduction, P automata are models of *dynamically changing systems which are in communication (interaction) with their environments* as well. According to this approach, investigations of P automata as *dynamical systems* form similarly important research directions. We hope to have new results in both directions in the future.

References

1. Alhazov, A.: Minimizing evolution-communication P systems and EC P automata. In: Cavaliere, M., et al. (eds.) Brainstorming Week on Membrane Computing. Technical Report 26/03 of the Research Group on Mathematical Linguistics, Rovira i Virgili University, Tarragona, Spain, pp. 23–31 (2003)
2. Bel-Enguix, G., Gramatovici, R.: Parsing with active P automata. In: Martín-Vide, C., Mauri, G., Păun, G., Rozenberg, G., Salomaa, A. (eds.) WMC 2003. LNCS, vol. 2933, pp. 31–42. Springer, Heidelberg (2004)
3. Bel-Enguix, G., Gramatovici, R.: Parsing with P automata. In: Ciobanu, G., Păun, Gh., Pérez-Jiménez, M.J. (eds.) Applications of Membrane Computing, pp. 389–410. Springer, Berlin (2006)
4. Björklund, H., Bojanczyk, M.: Shuffle expressions and words with nested data. In: Kučera, L., Kučera, A. (eds.) MFCS 2007. LNCS, vol. 4708, pp. 750–761. Springer, Heidelberg (2007)
5. Cardelli, L.: Brane calculi. Interactions of biological membranes. In: Danos, V., Schachter, V. (eds.) CMSB 2004. LNCS (LNBI), vol. 3082, pp. 257–280. Springer, Heidelberg (2005)
6. Cienciala, L., Ciencialova, L.: Membrane automata with priorities. Journal of Computer Science and Technology 19(1), 89–97 (2004)
7. Ciobanu, G., Gontineac, V.M.: Mealy multiset automata. International Journal of Foundations of Computer Science 17, 111–126 (2006)
8. Ciobanu, G., Gontineac, V.M.: P machines: An automata approach to membrane computing. In: Hoogeboom, H.-J., et al. (eds.) WMC 2006. LNCS, vol. 4361, pp. 314–329. Springer, Heidelberg (2006)
9. Ciobanu, G., Păun, Gh., Stefănescu, G.: P transducers. New Generation Computing 24(1), 1–28 (2006)
10. Csuhaj-Varjú, E.: P automata. In: Mauri, G., Păun, Gh., Jesús Pérez-Jímenez, M., Rozenberg, G., Salomaa, A. (eds.) WMC 2004. LNCS, vol. 3365, pp. 19–35. Springer, Heidelberg (2005)
11. Csuhaj-Varjú, E., Ibarra, O.H., Vaszil, Gy.: On the computational complexity of P automata. In: Ferretti, C., Mauri, G., Zandron, C. (eds.) DNA 2004. LNCS, vol. 3384, pp. 77–90. Springer, Heidelberg (2005)

12. Csuhaj-Varjú, E., Ibarra, O.H., Vaszil, Gy.: On the computational complexity of P automata. Natural Computing 5(2), 109–126 (2006)
13. Csuhaj-Varjú, E., Oswald, M., Vaszil, Gy.: P automata. In: Păun, G., Rozenberg, G., Salomaa, A. (eds.) Handbook of Membrane Computing. Oxford University Press, Oxford (to appear)
14. Csuhaj-Varjú, E., Vaszil, Gy.: P automata. In: Păun, G., Zandron, C. (eds.) Pre-Proceedings of the Workshop on Membrane Computing WMC-CdeA 2002, Curtea de Argeş, Romania, August 19-23, Pub. No. 1 of MolCoNet-IST-2001-32008, pp. 177–192 (2002)
15. Csuhaj-Varjú, E., Vaszil, Gy.: P automata or purely communicating accepting P systems. In: Păun, Gh., Rozenberg, G., Salomaa, A., Zandron, C. (eds.) WMC 2002. LNCS, vol. 2597, pp. 219–233. Springer, Heidelberg (2003)
16. Csuhaj-Varjú, E., Vaszil, Gy.: (Mem)brane automata. Theoretical Computer Science 404(1-2), 52–60 (2008)
17. Csuhaj-Varjú, E., Vaszil, Gy.: Representation of language classes in terms of P automata (manuscript, 2009)
18. Csuhaj-Varjú, E., Vaszil, Gy.: Logic for P automata (manuscript, 2009)
19. Dassow, J., Vaszil, Gy.: P finite automata and regular languages over countably infinite alphabets. In: Hoogeboom, H.J., Păun, G., Rozenberg, G., Salomaa, A. (eds.) WMC 2006. LNCS, vol. 4361, pp. 367–381. Springer, Heidelberg (2006)
20. Freund, R., Martín-Vide, C., Obtułowicz, A., Păun, Gh.: On three classes of automata-like P systems. In: Ésik, Z., Fülöp, Z. (eds.) DLT 2003. LNCS, vol. 2710, pp. 292–303. Springer, Heidelberg (2003)
21. Freund, R., Oswald, M.: A short note on analysing P systems. Bulletin of the EATCS 78, 231–236 (2002)
22. Freund, R., Oswald, M.: P automata with activated/prohibited membrane channels. In: Păun, Gh., Rozenberg, G., Salomaa, A., Zandron, C. (eds.) WMC 2002. LNCS, vol. 2597, pp. 261–269. Springer, Heidelberg (2003)
23. Freund, R., Oswald, M.: P automata with membrane channels. Artificial Life and Robotics 8, 186–189 (2004)
24. Freund, R., Oswald, M.: P systems with conditional communication rules assigned to membranes. Journal of Automata, Languages and Combinatorics 9(4), 387–397 (2004)
25. Freund, R., Oswald, M., Staiger, L.: ω-P automata with communication rules. In: Martín-Vide, C., Mauri, G., Păun, Gh., Rozenberg, G., Salomaa, A. (eds.) WMC 2003. LNCS, vol. 2933, pp. 203–217. Springer, Heidelberg (2004)
26. Freund, R., Verlan, S.: (Tissue) P systems working in the k-restricted minimally parallel derivation mode. In: Csuhaj-Varjú, E., et al. (eds.) International Workshop on Computing with Biomolecules, Wien, Austria, August 27, 2008, pp. 43–52. Österreichische Computer Gesellschaft (2008)
27. Long, H., Fu, Y.: A general approach for building combinational P automata. International Journal of Computer Mathematics 84(12), 1715–1730 (2007)
28. Ibarra, O.H., Păun, Gh.: Characterization of context-sensitive languages and other language classes in terms of symport/antiport P systems. Theoretical Computer Science 358(1), 88–103 (2006)
29. Kaminski, M., Francez, N.: Finite-memory automata. Theoretical Computer Science 134, 329–363 (1994)
30. Madhu, M., Krithivasan, K.: On a class of P automata. International Journal of Computer Mathematics 80(9), 1111–1120 (2003)
31. Martín-Vide, C., Păun, A., Păun, Gh.: On the power of P systems with symport rules. Journal of Universal Computer Science 8, 317–331 (2002)

32. Oswald, M.: P Automata. PhD dissertation, Vienna University of Technology (2003)
33. Oswald, M., Freund, R.: P Automata with membrane channels. In: Sugisaka, M., Tanaka, H. (eds.) Proc. of the Eights Int. Symp. on Artificial Life and Robotics, Beppu, Japan, pp. 275–278 (2003)
34. Otto, F.: Classes of regular and context-free languages over countably infinite alphabets. Discrete Applied Mathematics 12, 41–56 (1985)
35. Păun, A., Păun, Gh.: The power of communication: P systems with symport/antiport. New Generation Computing 20(3), 295–305 (2002)
36. Păun, Gh.: Computing with membranes. Journal of Computer and System Sciences 61(1), 108–143 (2000)
37. Păun, Gh.: Membrane Computing. An Introduction. Springer, Berlin (2002)
38. Rozenberg, G., Salomaa, A. (eds.): Handbook of Formal Languages. Springer, Berlin (1997)
39. Sburlan, D.: Private communication (2009)
40. Vaszil, Gy.: A class of P automata characterizing context-free languages. In: Gutiérrez-Naranjo, M.A., et al. (eds.) Proceedings of the Fourth Brainstorming Week on Membrane Computing, Sevilla, Spain, January 30-February 3, vol. II, RGNC Report, 03/2006, Fénix Editora, Sevilla, pp. 267–276 (2006)
41. Vaszil, Gy.: Automata-like membrane systems - A natural way to describe complex phenomena. In: Campeanu, C., Pighizzini, G. (eds.) Proceedings of 10th International Workshop on Descriptional Complexity of Formal Systems, Charlottetown, PE, Canada, July 16-18, pp. 26–37. University of Prince Edwards Island (2008)

Computational Nature of Processes Induced by Biochemical Reactions

Andrzej Ehrenfeucht[1] and Grzegorz Rozenberg[1,2]

[1] University of Colorado at Boulder, USA
[2] Leiden University, The Netherlands
rozenber@liacs.nl

Natural computing is concerned with human-designed computing inspired by nature as well as with computations taking place in nature, i.e., it investigates phenomena taking place in nature in terms of information processing.

Well-known examples of the first strand of research are evolutionary computing, neural computation, cellular automata, swarm intelligence, molecular computing, quantum computation, artificial immune systems, and membrane computing.

Examples of research themes from the second strand of research are computational nature of self-assembly, computational nature of developmental processes, computational nature of bacterial communication, computational nature of brain processes, computational nature of biochemical reactions, and system biology approach to bionetworks.

While progress in the first line of research often contributes to important progress in Information and Communication Technology (ITC), advances in the second line of research often remind the general scientific community that computer science is also the fundamental science of information processing, and as such a basic science for other scientific disciplines such as, e.g., biology.

The research we present is concerned with the computational nature of biochemical reactions in living cells. In particular we investigate the computational processes inspired (based on) biochemical reactions.

On the level of abstraction that we adopt, the functioning of a biochemical reaction is based on facilitation and inhibition: a reaction can take place if all of its reactants are present and none of its inhibitors is present. If a reaction takes place, then it produces its product. Therefore a *reaction* is defined as a triplet $a = (R, I, P)$, where R, I, P are finite sets called the *reactant set of a*, the *inhibitor set of a*, and the *product set of a*, and denoted by R_a, I_a, and P_a, respectively. If S is a set such that $R, I, P \subseteq S$, then we say that a is a *reaction in S*.

Then a reaction a takes place (in a given state – a given molecular soup) if all of its reactants are present and none of its inhibitors is present. Consequently, for a finite set (state) T, a is enabled by T if $R_a \subseteq T$ and $I_a \cap T = \emptyset$. The result of a on T, denoted by $res_a(T)$, is defined by: $res_a(T) = P_a$ if a is enabled on T, and $res_a(T) = \emptyset$ otherwise.

G. Păun et al. (Eds.): WMC 2009, LNCS 5957, pp. 16–17, 2010.

For a set A of reactions, the *result of A on T*, denoted $res_A(T)$, is defined by:

$$res_A(T) = \bigcup_{a \in A} res_a(T).$$

Finally, a *reaction system*, abbreviated rs, is an ordered pair $\mathcal{A} = (S, A)$ such that S is a finite set, called the *background set of $\mathcal{A}$*, and A is a set of reactions in S, called the *set of reactions of $\mathcal{A}$*. For a finite set (state) $T \subseteq S$, the *result of $\mathcal{A}$ on T*, denoted $res_{\mathcal{A}}(T)$, is defined by:

$$res_{\mathcal{A}}(T) = res_A(T).$$

The framework of reaction systems sketched above and motivated by organic chemistry of living organisms is based on assumptions that are very different from (and mostly orthogonal to) underlying assumptions of majority of models in theoretical computer science. We will discuss now some of these assumptions.

If a reaction a is enabled by a state T, then the result $res_a(T)$ is "locally determined" in the sense that it depends on R_a only. However, the effect of applying a to T is "dramatically global", because the whole set $T - P_a$ vanishes (to visualize this effect assume that the cardinalities of T, R_a, and P_a are 10000, 3, and 2 respectively; then 9998 elements of T will vanish while a has seen/used only 3 elements of T!!!). This is really orthogonal to models such as, e.g., Petri nets and membrane computing, and it reflects our assumption that there is no permanency of elements: an element of a global current state will vanish unless it is sustained by a reaction.

When a set of reactions A is applied to a state T, the result of application is cumulative: it is the union of the results of all individual reactions from A. Hence we do not have here a notion of conflict between reactions in A: even if $R_a \cap R_b \neq \emptyset$ for some $a, b \in A$, then still both a and b contribute to $res_A(T)$ – there is no conflict of resources here. Again this is in strong contrast to standard models in theoretical computer science such as, e.g., Petri nets and membrane computing. This reflects our assumption about the "threshold supply": either an element is present, and then there is "enough" of it, or an element is not present. Therefore, there is no counting in reaction systems, and consequently, reaction systems is a qualitative rather than a quantitative model.

Finally, we note that in reaction systems reactions are primary while structures are secondary. We do not have permanency of elements, and consequently, in transitions between states, reaction systems *create* states (rather than they transform states). Therefore, reaction systems do not work in an environment, but rather they create an environment.

Transition and Halting Modes
in (Tissue) P Systems

Rudolf Freund

Faculty of Informatics, Vienna University of Technology
Favoritenstr. 9, 1040 Vienna, Austria
rudi@emcc.at

Abstract. A variety of different transition modes for (tissue) P systems
as well as several halting modes currently are used in the area of membrane computing. In this paper, the definitions of the most important
transition modes and halting modes are explained based on networks of
cells, a general model for tissue P systems. Moreover, some results for
specific variants of (tissue) P systems working on multisets of objects are
recalled.

1 Introduction

Membrane systems were introduced by Gheorghe Păun one decade ago as distributed parallel computing devices, based on inspiration from biochemistry,
especially with respect to the structure and the functioning of a living cell,
which is considered as a set of compartments enclosed by membranes containing objects and evolution rules. In the original model of membrane systems,
the objects evolve in a hierarchical membrane structure (see [8], [16]); in tissue P systems (e.g., see [20], [21], and [11]), the cells communicate within an
arbitrary graph topology. In the original model of membrane systems, the *maximally parallel transition mode* was used, yet later on also other new transition
modes for P systems and tissue P systems have been introduced and investigated, for example, the *sequential* and the *asynchronous transition mode* as well
as the *minimally parallel transition mode* (see [6]). In [12], a formal framework
for (tissue) P systems capturing the formal features of these transition modes
was developed, based on a general model of membrane systems as a collection
of interacting cells containing multisets of objects (compare with the models of
networks of cells as discussed in [5] and networks of language processors as considered in [7]). Continuing the formal approach started in [12], the *k-bounded
minimally parallel transition mode* (see [13]) was introduced, where at most k
rules can be taken from each of the sets covering the whole set of rules into a
multiset of rules used in the minimally parallel transition mode.

In most models of (tissue) P systems, a computation continues as long as
still a (multiset of) rule(s) can be applied; the result of a computation then is
taken at the end of a halting computation *(total halting)*. Recently, various other
halting conditions have been investigated; for example, when using *partial halting*

G. Păun et al. (Eds.): WMC 2009, LNCS 5957, pp. 18–29, 2010.

(see [2], [3], [10]), a computation may only continue as long as from each set of a covering of the whole rule set at least one rule can still be applied. The result of a computation may also be extracted at each step of a (halting or non-halting) computation, e.g., see [4].

The main parts of notions, definitions, and results presented in the following are taken from [12] and [13] as well as from [3] and [10]. For an introduction to the area of membrane computing we refer the interested reader to the monograph [17], the actual state of the art can be seen in the web [22].

2   Preliminaries

We recall some of the notions and the notations we use (for further details see [8] and [19]). Let V be a (finite) alphabet; then V^* is the set of all strings (a language) over V, and $V^+ = V^* - \{\lambda\}$ where λ denotes the empty string. RE, REG ($RE(T), REG(T)$) denote the families of recursively enumerable and regular languages (over the alphabet T), respectively; MAT^λ denotes the family of languages generated by context-free matrix grammars. For any family of string languages F, PsF denotes the family of Parikh sets of languages from F and NF the family of Parikh sets of languages from F over a one-letter alphabet. By $\mathbb{N}$ we denote the set of all non-negative integers, by $\mathbb{N}^k$ the set of all vectors of non-negative integers; $[k..m]$ for $k \le m$ denotes the set of natural numbers n with $k \le n \le m$. In the following, we will not distinguish between NRE, which coincides with $PsRE(\{a\})$, and $RE(\{a\})$.

Let V be a (finite) set, $V = \{a_1, ..., a_k\}$. A *finite multiset* M over V is a mapping $M : V \longrightarrow \mathbb{N}$, i.e., for each $a \in V$, $M(a)$ specifies the number of occurrences of a in M. The size of the multiset M is $|M| = \sum_{a \in V} M(a)$. A multiset M over V can also be represented by any string x that contains exactly $M(a_i)$ symbols a_i for all $1 \le i \le k$, e.g., by $a_1^{M(a_1)}...a_k^{M(a_k)}$. The set of all finite multisets over the set V is denoted by $\langle V, \mathbb{N} \rangle$. Throughout the rest of the paper, we will not distinguish between a multiset from $\langle V, \mathbb{N} \rangle$ and its representation by a string over V containing the corresponding number of each symbol. We also consider mappings M of the form $M : V \longrightarrow \mathbb{N}_\infty$ where $\mathbb{N}_\infty = \mathbb{N} \cup \{\infty\}$, i.e., elements of M may have an infinite multiplicity; we shall call such multisets where $M(a_i) = \infty$ for at least one i, $1 \le i \le k$, *infinite multisets*. The set of all such multisets M over V with $M : V \longrightarrow \mathbb{N}_\infty$ is denoted by $\langle V, \mathbb{N}_\infty \rangle$.

3   Networks of Cells

In this section we consider membrane systems as a collection of interacting cells containing multisets of objects like in [5] and [12].

Definition 1. *A* network of cells *– we shall also use the notion* tissue P system *– with checking sets, of degree $n \ge 1$, is a construct*

$$\Pi = (n, V, w, R)$$

where

1. n is the number of cells;
2. V is a finite alphabet;
3. $w = (w_1, \dots, w_n)$ where $w_i \in \langle V, \mathbb{N}_\infty \rangle$, for all $1 \le i \le n$, is the multiset initially associated to cell i *(in most of the cases, at most one cell, then being called the environment, will contain symbols occurring with infinite multiplicity)*;
4. R is a finite set of rules of the form

$$(E : X \to Y)$$

where E is a recursive condition for configurations of Π *(see definition below)* as well as $X = (x_1, \dots, x_n)$, $Y = (y_1, \dots, y_n)$, with $x_i, y_i \in \langle V, \mathbb{N} \rangle$, $1 \le i \le n$, are vectors of multisets over V. We will also use the notation

$$(E : (x_1, 1) \dots (x_n, n) \to (y_1, 1) \dots (y_n, n))$$

for a rule $(E : X \to Y)$. If no conditions E are used, we use the simpler notations $X \to Y$ etc.

A network of cells consists of n cells, numbered from 1 to n, that contain (possibly infinite) multisets of objects over V; initially cell i contains w_i. A *configuration* C of Π is an n-tuple of multisets over V $(u_1, \dots, u_n)$; the *initial configuration* of Π, C_0, is described by w, i.e., $C_0 = w = (w_1, \dots, w_n)$. Cells can interact with each other by means of the rules in R. An interaction rule

$$(E : (x_1, 1) \dots (x_n, n) \to (y_1, 1) \dots (y_n, n))$$

is applicable to a configuration C if and only if C fulfills condition E and every cell i contains the multiset x_i; its application means rewriting objects x_i from cells i into objects y_j in cells j, $1 \le i, j \le n$.

The set of all multisets of rules *applicable* to C is denoted by *Appl* (Π, C) (a procedural algorithm how to obtain *Appl* (Π, C) is described in [12]).

For the specific *transition modes* to be defined in the following, the selection of multisets of rules applicable to a configuration C has to be a specific subset of *Appl* (Π, C); for the transition mode ϑ, the selection of multisets of rules applicable to a configuration C is denoted by *Appl* (Π, C, ϑ).

Definition 2. *For the* asynchronous *transition mode (asyn)*,

$$\textit{Appl} \, (\Pi, C, asyn) = \textit{Appl} \, (\Pi, C) \, ,$$

i.e., *there are no particular restrictions on the multisets of rules applicable to* C.

Definition 3. *For the* sequential *transition mode (sequ)*,

$$\textit{Appl} \, (\Pi, C, sequ) = \{ R' \mid R' \in \textit{Appl} \, (\Pi, C) \ \textit{and} \ |R'| = 1 \} \, ,$$

i.e., *any multiset of rules* $R' \in \textit{Appl} \, (\Pi, C, sequ)$ *has size* 1.

The most important transition mode considered in the area of P systems from the beginning is the *maximally parallel* transition mode where we only select multisets of rules R' that are not extensible, i.e., there is no other multiset of rules $R'' \supsetneq R'$ applicable to C.

Definition 4. *For the* maximally parallel *transition mode (max),*

$$Appl\,(\Pi, C, max) = \{R' \mid R' \in Appl\,(\Pi, C) \ \text{and there is}$$
$$\text{no } R'' \in Appl\,(\Pi, C) \ \text{with } R'' \supsetneq R'\}.$$

For the *minimally parallel* transition mode, we need an additional feature for the set of rules R, i.e., we consider a covering of R by subsets R_1 to R_h. Usually, this covering of R may coincide with a specific assignment of the rules to the cells.

There are several possible interpretations of this minimally parallel transition mode which informally can be described as applying multisets in such a way that from every set R_j, $1 \le j \le h$, at least one rule – if possible – has to be used (e.g., see [6]). For the basic variant as defined in the following, in each transition step we choose a multiset of rules R' from $Appl\,(\Pi, C, asyn)$ that cannot be extended to $R'' \in Appl\,(\Pi, C, asyn)$ with $R'' \supsetneq R'$ as well as $(R'' - R') \cap R_j \ne \emptyset$ and $R' \cap R_j = \emptyset$ for some j, $1 \le j \le h$, i.e., extended by a rule from a set of rules R_j from which no rule has been taken into R'.

Definition 5. *For the* minimally parallel *transition mode (min),*

$$Appl\,(\Pi, C, min) = \{R' \mid R' \in Appl\,(\Pi, C, asyn) \ \text{and}$$
$$\text{there is no } R'' \in Appl\,(\Pi, C, asyn)$$
$$\text{with } R'' \supsetneq R', \ (R'' - R') \cap R_j \ne \emptyset$$
$$\text{and } R' \cap R_j = \emptyset \text{ for some } j, \ 1 \le j \le h\}.$$

In [12], further restricting conditions on the four basic modes defined above, especially interesting for the minimally parallel transition mode, were considered. The following variant $all_{aset}min$ requires that from every applicable partition at least one rule has to be applied:

Definition 6. *For the* using all applicable sets minimally parallel *transition mode ($all_{aset}min$),*

$$Appl\,(\Pi, C, all_{aset}min) = \{R' \mid R' \in Appl\,(\Pi, C, min) \ \text{and}$$
$$\text{for all } j, \ 1 \le j \le h,$$
$$R_j \cap Appl\,(\Pi, C) \ne \emptyset$$
$$\text{implies } R_j \cap R' \ne \emptyset\}.$$

We now consider a restricted variant of the minimally parallel transition mode allowing only a bounded number of at most k rules to be taken from each set R_j, $1 \le j \le h$, of the covering into a multiset of rules applicable in the minimally parallel transition mode.

Definition 7. *For the k-restricted minimally parallel transition mode (min$_k$),*

$$Appl\,(\Pi, C, min_k) = \{R' \mid R' \in Appl\,(\Pi, C, min) \text{ and } \\ |R' \cap R_j| \le k \text{ for all } j,\ 1 \le j \le h\}.$$

For all the transition modes defined above, we now can define how to obtain a next configuration from a given one by applying an applicable multiset of rules according to the constraints of the underlying transition mode:

Definition 8. *Given a configuration C of Π and a transition mode ϑ, we may choose a multiset of rules $R' \in Appl\,(\Pi, C, \vartheta)$ in a non-deterministic way and apply it to C. The result of this transition step from the configuration C with applying R' is the configuration $Apply\,(\Pi, C, R')$, and we also write $C \Longrightarrow_{(\Pi, \vartheta)} C'$. The reflexive and transitive closure of the transition relation $\Longrightarrow_{(\Pi, \vartheta)}$ is denoted by $\Longrightarrow^*_{(\Pi, \vartheta)}$.*

Definition 9. *A configuration C is said to be accessible in Π with respect to the derivation mode ϑ if and only if $C_0 \Longrightarrow^*_{(\Pi, \vartheta)} C$ (C_0 is the initial configuration of Π). The set of all accessible configurations in Π is denoted by $Acc\,(\Pi)$.*

Definition 10. *A derivation mode ϑ is said to be deterministic (det-ϑ) if $|Appl\,(\Pi, C, \vartheta)| \le 1$ for any accessible configuration C.*

Definition 11. *A computation in a tissue P system Π, $\Pi = (n, V, w, R)$, starts with the initial configuration $C_0 = w$ and continues with transition steps according to the chosen transition mode ϑ.*

3.1 Halting Conditions

A halting condition is a predicate applied to an accessible configuration. The system halts according to the halting condition if this predicate is true for the current configuration. In such a general way, the notion halting with final state or signal halting can be defined as follows:

Definition 12. *An accessible configuration C is said to fulfill the signal halting condition or final state halting condition (S) if and only if*

$$S\,(\Pi, \vartheta) = \{C' \mid C' \in Acc\,(\Pi) \text{ and } State\,(\Pi, C', \vartheta)\}.$$

Here $State\,(\Pi, C', \vartheta)$ means a decidable feature of the underlying configuration C', e.g., the occurrence of a specific symbol (signal) in a specific cell.

The most important halting condition used from the beginning in the P systems area is the *total halting*, usually simply considered as *halting*:

Definition 13. *An accessible configuration C is said to fulfill the total halting condition (H) if and only if no multiset of rules can be applied to C with respect to the derivation mode anymore, i.e.,*

$$H\,(\Pi, \vartheta) = \{C' \mid C' \in Acc\,(\Pi) \text{ and } Appl\,(\Pi, C', \vartheta) = \emptyset\}.$$

The adult halting condition guarantees that we still can apply a multiset of rules to the underlying configuration, yet without changing it anymore:

Definition 14. *An accessible configuration C is said to fulfill the adult halting condition (A) if and only if*

$$A(\Pi, \vartheta) = \{C' \mid C' \in Acc(\Pi), \; Appl(\Pi, C', \vartheta) \neq \emptyset \text{ and}$$
$$Apply(\Pi, C', R') = C' \text{ for every } R' \in Appl(\Pi, C', \vartheta)\}.$$

We should like to mention that we could also consider $A(\Pi, \vartheta) \cup H(\Pi, \vartheta)$ instead of $A(\Pi, \vartheta)$.

For introducing the notion of partial halting, we have to consider a covering of R by subsets R_1 to R_h as for the minimally parallel transition mode. We then say that we are not halting only if there still is a multiset of rules R' from $Appl(\Pi, C)$ with $R' \cap R_j \neq \emptyset$ for all j, $1 \leq j \leq h$:

Definition 15. *An accessible configuration C is said to fulfill the partial halting condition (h) if and only if*

$$h(\Pi, \vartheta) = \{C' \mid C' \in Acc(\Pi) \text{ and there is}$$
$$no \; R' \in Appl(\Pi, C') \text{ with}$$
$$R' \cap R_j \neq \emptyset \text{ for all } j, \; 1 \leq j \leq h\}.$$

3.2 Goal and Result of a Computation

The computations with a tissue P system may have different goals, e.g., to generate (*gen*) a (vector of) non-negative integers in a specific output cell (membrane) or to accept (*acc*) a (vector of) non-negative integers placed in a specific input cell at the beginning of a computation. Moreover, the goal can also be to compute (*com*) an output from a given input or to output **yes** or **no** to decide (*dec*) a specific property of a given input.

The results not only can be taken as the number (N) of objects in a specified output cell, but, for example, also be taken modulo a terminal alphabet (T) or by subtracting a constant from the result ($-k$).

Such different tasks of a tissue P system may require additional parameters when specifying its functioning, e.g., we may have to specify the output/input cell(s) or the terminal alphabet.

We shall not go into the details of such definitions here, we just mention that the goal of the computations $\gamma \in \{gen, acc, com, dec\}$ and the way to extract the results ρ are two other parameters to be specified and clearly defined when defining the functioning of a tissue P system.

3.3 Taxonomy of (Tissue) P Systems

For a particular variant of networks of cells or tissue P systems we have to specify the transition mode, the halting condition as well as the procedure how to get the result of a computation, but also the specific kind of rules that are used, especially some complexity parameters.

For tissue P systems, we shall use the notation

$$O_m t C_n \left(\vartheta, \phi, \gamma, \rho \right) \left[\text{parameters for rules} \right]$$

to denote the family of sets of vectors of non-negative integers obtained by tissue P systems $\Pi = (n, V, w, R)$ of degree n with $m = |V|$, as well as $\vartheta, \phi, \gamma, \rho$ indicating the transition mode, the halting condition, the goal of the computations, and the way how to get results, respectively; the *parameters for rules* describe the specific features of the rules in R. If any of the parameters m and n is unbounded, we replace it by $*$.

If the communication structure in the tissue P system is a tree as in the original model of membrane systems, then we omit the t and use the notation $O_m C_n \left(\vartheta, \phi, \gamma, \rho \right)$ [parameters for rules].

4   Examples and Results

In this section, we give some examples how several well-known models of (tissue) P systems can be expressed within the general framework presented in the preceding section.

4.1   P Systems with Symport/Antiport Rules

For definitions and results concerning P systems with symport/antiport rules, we refer to the original paper [15] as well as to the overview given in [18]. An *antiport rule* is a rule of the form $(x, i) (u, j) \rightarrow (x, j) (u, i)$, in membrane systems usually written as $(x, out; u, in)$, $xu \neq \lambda$, where j is the region outside the membrane i in the underlying tree structure. A *symport rule* is of the form $(x, i) \rightarrow (x, j)$ or $(u, j) \rightarrow (u, i)$, usually written as (x, out) and (u, in), respectively.

The weight of the antiport rule $(x, i) (u, j) \rightarrow (x, j) (u, i)$ is defined as $\max \left\{ |x|, |u| \right\}$. Using only antiport rules with weight k induces the type of rules α usually written as $anti_k$. The weight of a symport rule $(x, i) \rightarrow (x, j)$ or $(u, j) \rightarrow (u, i)$ is defined as $|x|$ or $|u|$, respectively. Using only symport rules with weight k induces the type of rules α usually written as sym_k. If only antiport rules $(x, i) (u, j) \rightarrow (x, j) (u, i)$ of weight ≤ 2 and with $|x| + |u| \leq 3$ as well as symport rules of weight 1 are used, we shall write $anti_{2'}$. The following result is well known:

Theorem 1. $O_* C_1 \left(max, H, gen, N \right) \left[anti_{2'} \right] = NRE$.

Observe that, within the normal framework of membrane systems, we only need one membrane separating the environment and the skin region, but this means that two regions corresponding to two cells are involved, i.e.,

$$O_* t C_2 \left(max, H, gen, N \right) \left[anti_{2'} \right] = NRE.$$

4.2 Purely Catalytic P Systems

Already in the original paper of Gheorghe Păun (see [16]), membrane systems with catalytic rules were defined, but used together with other noncooperative rules. In [9] it was shown that only three catalysts are sufficient in one membrane, using only catalytic rules with the maximally parallel transition mode, to generate any recursively enumerable set of natural numbers.

A *noncooperative rule* is of the form $(I : (a, i) \to (y_1, 1) \dots (y_n, n))$ where a is a single symbol and I denotes the condition that is always fulfilled. A *catalytic rule* is of the form $(I : (c, i)(a, i) \to (c, i)(y_1, 1) \dots (y_n, n))$ where c is from a distinguished subset $C \subset V$ such that in all rules (noncooperative evolution rules, catalytic rules) of the whole system the y_i are from $(V - C)^*$ and the symbols a are from $(V - C)$. Imposing the restriction that the noncooperative rules and the catalytic rules in a tissue P system allow for finding a hierarchical tree structure of membranes such that symbols either stay in their membrane region or are sent out to the surrounding membrane region or sent into an inner membrane, then we get the classical catalytic P systems without priorities. Allowing regular sets checking for the non-appearance of specific symbols instead of I, we even get the original P systems with priorities. Catalytic P systems using only catalytic rules are called purely catalytic P systems. As we know from [9], only two (three) catalysts in one membrane are needed to obtain NRE with (purely) catalytic P systems without priorities working in the maximally parallel transition mode, i.e., we can write these results as follows:

Theorem 2. $NRE = O_* C_1 (max, H, gen, -2) [cat_2]$
$$= O_* C_1 (max, H, gen, -3) [pcat_3].$$

In a purely catalytic P system, for each catalyst present in a membrane we may take the corresponding set of noncooperative rules thus defining a covering of the rule set; the sets of rules of this covering, when working in the 1-restricted minimally parallel transition mode, then replace the use of the catalysts, because by definition from each of these sets – if possible – exactly one rule (as with the use of the corresponding catalyst) is chosen: from the set of purely catalytic rules R we obtain the corresponding set of noncooperative rules R' as

$$R' = \{(a, i) \to (y_1, 1) \dots (y_n, n) \mid \\ (c, i)(a, i) \to (c, i)(y_1, 1) \dots (y_n, n) \in R\}$$

as well as the corresponding covering of R' by the sets

$$R'_{i,c} = \{(a, i) \to (y_1, 1) \dots (y_n, n) \mid \\ (c, i)(a, i) \to (c, i)(y_1, 1) \dots (y_n, n) \in R\}.$$

Considering purely catalytic P systems in one membrane, we therefore infer the following quite astonishing result that when using the 1-restricted minimally parallel transition mode for a suitable covering of rules we only need noncooperative rules:

Theorem 3. $NRE = O_*C_1\,(min_1, H, gen, N)\,[noncoop]$.

When using the asynchronous or the sequential transition mode, we only obtain regular sets:

Theorem 4. *For every* $\vartheta \in \{asyn, sequ\}$ *and* $\phi \in \{H, h\}$,

$$NREG = O_*tC_*\,(\vartheta, \phi, gen, N)\,[noncoop].$$

4.3 Extended Spiking Neural P Systems

In extended spiking neural P systems (without delays, see [1]), the rules are applied in a sequential way in each neuron, but on the level of the whole system, the maximally parallel transition mode is applied (every neuron which may use a spiking rule has to spike, i.e., to apply a rule, see the original paper [14]). When partitioning the rule set according to the set of neurons, the application of the 1-restricted minimally parallel transition mode exactly models the original transition mode defined for spiking neural P systems.

An *extended spiking neural P system* (of degree $m \geq 1$) (in the following we shall simply speak of an *ESNP system*) is a construct $\Pi = (m, S, R)$ where

- m is the number of *neurons*; the neurons are uniquely identified by a number between 1 and m;
- S describes the *initial configuration* by assigning an initial value (of spikes) to each neuron;
- R is a finite set of *rules* of the form $\left(i, E/a^k \to P\right)$ such that $i \in [1..m]$ (specifying that this rule is assigned to neuron i), $E \subseteq REG\left(\{a\}\right)$ is the *checking set* (the current number of spikes in the neuron has to be from E if this rule shall be executed), $k \in \mathbb{N}$ is the "number of spikes" (the energy) consumed by this rule, and P is a (possibly empty) set of *productions* of the form (l, a^w) where $l \in [1..m]$ (thus specifying the target neuron), $w \in \mathbb{N}$ is the *weight* of the energy sent along the axon from neuron i to neuron l.

A *configuration* of the ESNP system is described by specifying the actual number of spikes in every neuron. A *transition* from one configuration to another one is executed as follows: for each neuron i, we non-deterministically choose a rule $\left(i, E/a^k \to P\right)$ that can be applied, i.e., if the current value of spikes in neuron i is in E, neuron i "spikes", i.e., for every production (l, w) occurring in the set P we send w spikes along the axon from neuron i to neuron l. A *computation* is a sequence of configurations starting with the initial configuration given by S. An ESNP system can be used to generate sets from NRE (we do not distinguish between NRE and $RE\left(\{a\}\right)$) taking the contents, i.e., the number of spikes, of a specific neuron called *output neuron* in halting computations.

We now consider the ESNP system $\Pi = (m, S, R)$ as a tissue P system $\Pi' = (m, \{a\}, S, R')$ working in the 1-restricted minimally parallel transition mode, with

$$R' = \left\{ \begin{array}{l} \left(E : \left(a^k, i\right) \to (a^{w_1}, l_1) \ldots (a^{w_n}, l_n)\right) \mid \\ \left(i, E/a^k \to (l_1, a^{w_1}) \ldots (l_n, a^{w_n})\right) \in R \end{array} \right\}$$

and the partitioning R_i', $1 \leq i \leq m$, of the rule set R' according to the set of neurons, i.e.,

$$R_i' = \left\{ \left(E : \left(a^k, i \right) \rightarrow \left(a^{w_1}, l_1 \right) \ldots \left(a^{w_n}, l_n \right) \right) \mid \right.$$
$$\left. \left(E : \left(a^k, i \right) \rightarrow \left(a^{w_1}, l_1 \right) \ldots \left(a^{w_n}, l_n \right) \right) \in R' \right\}.$$

The 1-restricted minimally parallel transition mode chooses one rule – if possible – from every set R_i and then applies such a multiset of rules in parallel, which directly corresponds to applying one spiking rule in every neuron where a rule can be applied. Hence, it is easy to see that Π' and Π generate the same set from $RE\{a\}$ if in both systems we take the same cell/neuron for extracting the output. Due to the results valid for ESNP systems, see [1], we obtain the following result:

Theorem 5. $NRE = O_1 t C_3 \left(min_1, H, gen, N \right) [ESNP]$.

4.4 A General Result

For any tissue P system using rules of type α, with a transition mode ϑ, $\vartheta \in \{all_{aset}min, asyn, sequ\}$, and partial halting, we only get Parikh sets of matrix languages (regular sets of non-negative integers), provided the checking set for each rule can be simulated by checking the (independent) applicability of a finite set of rules (fixed for each rule):

Theorem 6. *For every* $\vartheta \in \{all_{aset}min, asyn, sequ\}$,

$$O_* t C_* \left(\vartheta, h, gen, T \right) [\alpha] \subseteq PsMAT \text{ and }$$
$$O_* t C_* \left(\vartheta, h, gen, N \right) [\alpha] \subseteq NREG.$$

The proof follows the ideas of a similar result proved for a general variant of P systems with permitting contexts in [3] and therefore is omitted. We do not know whether a similar result also holds true for the transition mode min itself instead of $all_{aset}min$.

5 Conclusions

In the general framework considered in this paper, many variants of static tissue P systems (and P systems as well) can be represented. Although during the last decade, a great variety of such systems working in different transition mode has been considered, many specific models of (tissue) P systems still wait for being considered with other transition modes, for example, with the k-restricted minimally parallel transition mode. Moreover, different variants of halting, especially partial halting, should be considered for a lot more models of (tissue) P systems in the future.

Acknowledgements

I am very grateful to the coauthors of those papers from which most of the definitions and results have been taken for this paper, especially to Artiom Alhazov, Marion Oswald, and Sergey Verlan, yet most of all to Gheorghe Păun, whose great ideas have inspired myself to develop many new variants of P systems.

References

1. Alhazov, A., Freund, R., Oswald, M., Slavkovik, M.: Extended spiking neural P systems generating strings and vectors of non-negative integers. In: Hoogeboom, H.J., Paun, Gh., Rozenberg, G. (eds.) Pre-proceedings of Membrane Computing, International Workshop, WMC7, Leiden, The Netherlands, pp. 88–101 (2006)
2. Alhazov, A., Freund, R., Oswald, M., Verlan, S.: Partial versus total halting in P systems. In: Gutiérrez-Naranjo, M.A., Păun, Gh., Pérez-Jiménez, M.J. (eds.) Cellular Computing (Complexity Aspects), ESF PESC Exploratory Workshop, Fénix Editorial, Sevilla, pp. 1–20 (2005)
3. Alhazov, A., Freund, R., Oswald, M., Verlan, S.: Partial halting in P systems using membrane rules with permitting contexts. In: Durand-Lose, J., Margenstern, M. (eds.) MCU 2007. LNCS, vol. 4664, pp. 110–121. Springer, Heidelberg (2007)
4. Beyreder, M., Freund, R.: (Tissue) P systems using noncooperative rules without halting conditions. In: Frisco, P., et al. (eds.) Pre-Proc. Ninth Workshop on Membrane Computing (WMC9), Edinburgh, pp. 85–94 (2008)
5. Bernardini, F., Gheorghe, M., Margenstern, M., Verlan, S.: Networks of Cells and Petri Nets. In: Gutiérrez-Naranjo, M.A., et al. (eds.) Proc. Fifth Brainstorming Week on Membrane Computing, Sevilla, pp. 33–62 (2007)
6. Ciobanu, G., Pan, L., Păun, Gh., Pérez-Jim énez, M.J.: P systems with minimal parallelism. Theoretical Computer Science 378(1), 117–130 (2007)
7. Csuhaj-Varjú, E.: Networks of language processors. Current Trends in Theoretical Computer Science, 771–790 (2001)
8. Dassow, J., Păun, Gh.: On the power of membrane computing. Journal of Universal Computer Science 5(2), 33–49 (1999)
9. Freund, R., Kari, L., Oswald, M., Sosík, P.: Computationally universal P systems without priorities: two catalysts are sufficient. Theoretical Computer Science 330, 251–266 (2005)
10. Freund, R., Oswald, M.: Partial halting in P systems. Intern. J. Foundations of Computer Sci. 18, 1215–1225 (2007)
11. Freund, R., Păun, Gh., Pérez-Jiménez, M.J.: Tissue-like P systems with channel states. Theoretical Computer Science 330, 101–116 (2005)
12. Freund, R., Verlan, S.: A formal framework for P systems. In: Eleftherakis, G., Kefalas, P., Paun, Gh. (eds.) Pre-proceedings of Membrane Computing, International Workshop – WMC8, Thessaloniki, Greece, pp. 317–330 (2007)
13. Freund, R., Verlan, S.: (Tissue) P systems working in the k-restricted minimally parallel derivation mode. In: Csuhaj-Varjú, E., et al. (eds.) Proceedings of the International Workshop on Computing with Biomolecules, Österreichische Computer Gesellschaft, pp. 43–52 (2008)
14. Ionescu, M., Păun, Gh., Yokomori, T.: Spiking neural P systems. Fundamenta Informaticae 71(2-3), 279–308 (2006)

15. Păun, A., Păun, Gh.: The power of communication: P systems with symport/antiport. New Generation Computing 20(3), 295–306 (2002)
16. Păun, Gh.: Computing with membranes. J. of Computer and System Sciences 61(1), 108–143 (2000); TUCS Research Report 208 (1998), http://www.tucs.fi
17. Păun, Gh.: Membrane Computing. An Introduction. Springer, Berlin (2002)
18. Rogozhin, Y., Alhazov, A., Freund, R.: Computational power of symport/antiport: history, advances, and open problems. In: Freund, R., Păun, Gh., Rozenberg, G., Salomaa, A. (eds.) WMC 2005. LNCS, vol. 3850, pp. 1–30. Springer, Heidelberg (2006)
19. Rozenberg, G., Salomaa, A. (eds.): Handbook of Formal Languages, 3 vols. Springer, Berlin (1997)
20. Păun, Gh., Sakakibara, Y., Yokomori, T.: P systems on graphs of restricted forms. Publicationes Matimaticae 60, 635–660 (2002)
21. Păun, Gh., Yokomori, T.: Membrane computing based on splicing. In: Winfree, E., Gifford, D.K. (eds.) DNA Based Computers V. DIMACS Series in Discrete Mathematics and Theoretical Computer Science, vol. 54, pp. 217–232. American Mathematical Society, Providence (1999)
22. The P Systems, http://ppage.psystems.eu

Conformon P Systems and
Topology of Information Flow

Pierluigi Frisco

School of Mathematical and Computer Sciences
Heriot-Watt University
EH14 4AS Edinburgh, UK
pier@macs.hw.ac.uk

Abstract. We survey some of the results about conformon P systems
and link these results to the topology of information flow. This topology
is studied with models of Petri nets. Several directions of research and
open problems are given.

1 Introduction

The ten years young field of Membrane Computing saw, in between other
things, the definition of a number of formal models of computation all sharing a
well defined topological structure, locality of interaction and parallel processing
[30,10,28]. These models allowed us to broaden our understanding of computation. Now, for instance, we know that the simple passage of symbols from
one compartment to another in P systems with symport/antiport is sufficient to
compute [27], that conformon P systems with either positive or negative values
have similar computational power [7], that dissolution can play an important
role in the computing power of P systems with active membranes [17], etc.

All these results told us a lot about *how* to perform computation. One important question that often went unanswered is *why* a certain model of P system
could or could not compute a specific set of numbers. The answer to this *why*
would:

link all the different models of P systems (and any formal system) based on
multiset rewriting now regarded as different because of their definition;
generalise results to any other formal system based on multiset rewriting;
allow us to understand more fundamental features that have to be present in a
formal system in order to compute;
classify formal systems of computation in different ways;
give new tools to prove the computational power and other properties of these
formal systems.

Recently a way to answer this *why*, using the topology of information flow, has
been suggested [6,9,10,13]. In this paper we survey the known results linking the
topology of information flow to computational power and we show how these
results can be applied to conformon P systems. No new result is introduced.

G. Păun et al. (Eds.): WMC 2009, LNCS 5957, pp. 30–53, 2010.

We refer readers to the original papers for the proofs of the used theorems. The description is kept rather informal even if formal definitions are provided. The given directions of research and open problems are meant to stimulate and inspire further developments in this line of research.

2   Preliminaries

We assume the reader to have familiarity with basic concepts of formal language theory [19], and in particular with the topic of membrane computing [10,29]. In this section we recall particular aspects relevant to our presentation. We denote with $\mathbb{N}_0$ the set of natural numbers $\{0, 1, 2, \ldots\}$ and $\mathbb{N} = \mathbb{N}_0 \setminus \{0\}$.

3   About Simulations

In order to study the topology of information flow of cP systems with P/T systems we have to relate configurations of the former with configuration of the latter. We do this through a definition of *simulation* (different than the ones in [24,25,33]). Here we generalise the definitions of simulation given in [10,6].

A *multiset* over a set A is a total function $M : A \to \mathbb{N}$. For every $a \in A, M(a)$ denotes the multiplicity of a in the multiset M. The *support* of a multiset M is the set $supp(M) = \{a \in A \mid M(a) > 0\}$. The set of multisets over a set A is denoted by $\mathbb{M}_A = \{M \mid M : A \to \mathbb{N}\}$. If A and B are sets, $\alpha \subseteq A \times B$ is a relation and if $(a, b) \in \alpha, a \in A, b \in B$, then we say that b *is returned* by α (on a).

The *formal systems* we consider have *configurations*, that is, *snapshots* of relevant elements defining a formal system while it computes. Different formal systems have different relevant elements defining configurations. Even if we are not able to formally define *configuration* for a generic formal system, it is possible to formally define this concepts for specific formal systems. Formal systems can have an infinite set of configurations.

Given an initial configuration, a formal system can pass from one configuration to another. This process is called *computation* and the passage from one configuration to another is called *transition*. A transition occurs because some *operations* (rules, instructions, transitions, etc.) are applied to one configuration. If a computation is finite, then the last configuration is called *final*. We assume that final configurations have certain properties that make them such (that is, different from the other configurations). There is no transition from a final configuration. If a final configuration meets certain criteria, then it is said that the the systems *halts*, otherwise it is said that the system *stops*.

Let S be a formal systems with O set of operations and $C = \{c_1, c_2, \ldots\}$ set of configurations.

We denote with $\overset{\sigma}{\Rightarrow}$, σ multiset over O, the transition from one configuration to another in a computation of S according to the application of the operations in σ. With $\overset{\sigma_1, \ldots, \sigma_n}{\Rightarrow}{}^+$ we denote non-empty sequences of transitions from one configuration to another in a computation of S according to the application of the operations in $\sigma_1, \ldots, \sigma_n$ in sequence. So, for example, if $c_1 \overset{\sigma_1}{\Rightarrow} c_2 \overset{\sigma_2}{\Rightarrow} c_3$, then we can write $c_1 \overset{\sigma_1,\sigma_2}{\Rightarrow}{}^+ c_3$ and we say that c_3 is *reachable* from c_1.

Depending on the operational mode of S, the multiset σ can be a multiset of a specific kind. For instance, σ can be such that it returns at most 1.

Given a set A we denote with $\mathcal{P}(A)$ its *power set*, that is the set of its subsets. If S is a formal system with C set of configurations, then the function $reach_S : C \to \mathcal{P}(C)$ returns the set of configurations reachable from a given configuration. The returned set includes the given configuration. In the previous example $reach_S(c_1) = \{c_1, c_2, c_3\}$, $reach_S(c_2) = \{c_2, c_3\}$ and $reach_S(c_3) = \{c_3\}$.

Moreover, if O is the set of operations of S, then $oper_S : C \to \mathcal{P}(O)$ returns the set of operations applied by the system to reach any configuration reachable from a given configuration. In the previous example, $oper_S(c_1) = supp(\sigma_1) \cup supp(\sigma_2)$, $oper_S(c_2) = supp(\sigma_2)$ and $oper_S(c_3) = \emptyset$.

Definition 1. *Let S and S' be two formal systems with:*

O *and* O' *their respective sets of operations,*
C *and* C' *their respective sets of configurations,*
$F \subset C$ *and* $F' \subset C'$ *sets of final configurations,*
$c_{init} \in C$ *and* $c'_{init} \in C'$ *respective initial configurations.*

We say that S' $\alpha\beta$ simulates S if for each c_{init} there are two relations $\alpha \subseteq C \times \mathcal{P}(C')$ and $\beta \subseteq O \times \mathbb{M}_{O'}$ such that:

i) $(c_{init}, \{c'_{init}\}) \in \alpha$;
ii) *for each $c_1, c_2 \in reach(c_{init})$, $\bar{C} \subseteq C'$ and $\sigma \in oper_S(c_{init})$, if $c_1 \overset{\sigma}{\Rightarrow} c_2$ and $(c_1, C') \in \alpha$, then there is $\bar{\bar{C}} \subseteq C'$ such that $\bar{c} \overset{\sigma'}{\Rightarrow} \bar{\bar{c}}$ with $\bar{c} \in \bar{C}, \bar{\bar{c}} \in \bar{\bar{C}}, (c_2, \bar{C}) \in \alpha$ and $(\sigma, \Sigma') \in \beta, \sigma' \in \Sigma'$. Moreover, for each $c \in \bar{C} \cup \bar{\bar{C}}, c \in reach_{S'}(c'_{init})$;*
iii) *for each $\bar{C}, \bar{\bar{C}} \subseteq C', \bar{c} \in \bar{C}, \bar{\bar{c}} \in \bar{\bar{C}}, c_1 \in reach_S(c_{init}), \Sigma' \in \mathcal{P}(O'), \sigma' \in \Sigma'$ such that for each $c \in \bar{C} \cup \bar{\bar{C}}, c \in reach_{S'}(c'_{init})$, if $\bar{c} \overset{\sigma'}{\Rightarrow} \bar{\bar{c}}$ and $(c_1, \bar{C}) \in \alpha$, then there is $c_2 \in C$ such that $c_1 \overset{\sigma}{\Rightarrow} c_2$ with $(c_2, \bar{\bar{C}}) \in \alpha$ and $(\sigma, \Sigma') \in \beta$;*
iv) *for each $f \in F$, then $(f, \Omega') \in \alpha$ only if there is $f' \in F', f' \in \Omega'$ and there are not $c' \in C', c' \in reach_{S'}(c'_{init})$ and $\sigma' : O' \to \mathbb{N}$ such that $f' \overset{\sigma'}{\Rightarrow} c'$.*

In the previous definition:

i) means that both systems start from corresponding configurations;
ii) means that if both systems are in corresponding configurations and S applies one operations σ, then S can simulate that operation applying the operations in σ'. When this happens, then S and S' are in corresponding configurations. Moreover, all configurations in $\bar{C} \cup \bar{\bar{C}}$ are not isolated, that is, they can be reached by the initial configuration of S';
iii) means that S' does not do anything more than S does;
iv) means that when S is in a final configuration, then S' can also be in a final configuration.

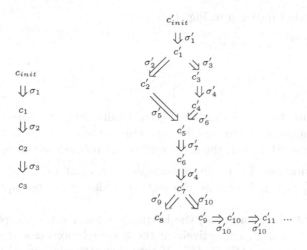

Fig. 1. Transition graphs related to Definition 1

In the rest of this paper we say that a system S' *simulates* a system S if there are relations α and β as in Definition 1 such that S' $\alpha\beta$ simulates S. Moreover, for brevity, we do not provide definitions for these relations.

In order to further clarify the previous definition we give an example. In Fig. 1 two *transitions graphs* (graphs indicating transitions in formal systems) are depicted. The dots at the bottom-right of this figure indicate that there is an infinite number of transitions each different than the previous one (indicated by the increasing subscript of the c') but obtained applying the same multiset of operations σ'_{10}.

In this figure the graph depicted on the left side refers to a system S, while the one depicted on the right side refers to a system S'. These systems are such that:

$O = \{o_i \mid 1 \le i \le 50\}$;
$O' = \{o'_i \mid 1 \le i \le 30\}$;
$C = \{c_{init}\} \cup \{c_1 \mid 1 \le i \le 40\}$;
$C' = \{c'_{init}\} \cup \{c'_i \mid i \ge 1\}$;
$F = \{c_3\}$;
$F' = \{c'_8\}$.

The multisets of applied operations are such that $\sigma_i(o_i) = 1$ while $\sigma_i(o_j) = 0$ for $i \ne j$. Similarly for the multisets σ'. It is important to notice that not all elements in the sets of operations and sets of configurations are present in Fig. 1. In this figure only the configurations reachable from c_{init} and c'_{init} and the multisets of operations applied in order to reach these configurations are present.

The two relations α and β are:

$\alpha = \{(c_{init}, \{c'_{init}\}), (c_1, \{c'_1, c'_2, c'_3, c'_4\}), (c_2, \{c'_5, c'_6\}), (c_3, \{c'_8\})\}$;
$\beta = \{(\sigma_1, \sigma'_1), (\sigma_2, \{\sigma'_5, \sigma'_6\}), (\sigma_3, \sigma'_4)\}$.

Considering what depicted in Fig. 1 we have:

$reach_S(c_{init}) = \{c_{init}, c_1, c_2, c_3\};$
$reach_{S'}(c'_{init}) = C';$
$oper_S(c_{init}) = \{\sigma_1, \sigma_2, \sigma_3\};$
$oper_{S'}(c'_{init}) = \{\sigma'_i \mid 1 \leq i \leq 10, i \neq 8\}.$

It is important to notice that the set of configurations reachable from an initial configuration can be an infinite set. This is the case for $reach_{S'}(c'_{init})$ in the given example. Moreover, the same multiset of operations can be applied to different configurations. In the given example $c'_3 \overset{\sigma'_4}{\Rightarrow} c'_4$ and $c'_6 \overset{\sigma'_6}{\Rightarrow} c'_7$. Clearly, in this example, c'_3 and c'_6 share some elements that allow σ'_4 to be applied to both configurations.

We denote with $L(S)$ and $L(S')$ the language generated (or accepted) by the formal systems S and S', respectively. If there are relations α and β such that S' $\alpha\beta$ simulates S, then $L(S) \subseteq L(S')$. If there are also relations α' and β' such that S $\alpha'\beta'$ simulates S', then $L(S') = L(S)$.

Given a set A a partition of A is a collection of subsets of A, $A_i \subseteq A, 1 \leq i \leq n$ such that $\bigcup_{i=1}^{n} A_i = A$ and $A_p \cap A_q = \emptyset, p \neq q, 1 \leq p, q \leq n$.

Lemma 1. *Let S and S' be formal systems and let α and β be relations such that S' $\alpha\beta$ simulates S. Then the relation α induces a partition in the set of configurations of S'.*

Proof. Let us assume that $c_1, c_3, c_3 \in C$, C being the set of configurations of S, $c_1, c_2, c_3 \in reach_s(c_{init})$, c_{init} initial configuration of S, that c_1, c_2, c_3 are different from eachother and that $c_1 \overset{\sigma_1}{\Rightarrow} c_2 \overset{\sigma_2}{\Rightarrow} c_3$ where σ_1 and σ_2 are multisets of configurations over C. There is not σ_3, multiset of configurations over C, such that $c_1 \overset{\sigma_3}{\Rightarrow} c_3$.

As S' $\alpha\beta$ simulates S, there are $P', Q', R' \in \mathcal{P}(C')$, C' set of configurations of S', such that $(c_1, P'), (c_2, Q'), (c_3, R') \in \alpha$ and $p' \in P', q' \in Q', r' \in R'$ such that $p' \overset{\sigma'_1}{\Rightarrow} q' \overset{\sigma'_2}{\Rightarrow} r'$ with σ'_1, σ'_2 multisets of configurations over C', $(\sigma_1, \Sigma'_1), (\sigma_2, \Sigma'_2) \in \beta$, $\sigma'_1 \in \Sigma'_1, \sigma'_2 \in \Sigma'_2$.

Moreover, there is not σ'_3, multiset of configurations over C', such that $p'' \overset{\sigma'_3}{\Rightarrow} r''$ with $p'' \in P', r'' \in R'$. If α would not induce a partition over C', then it could be that $Q' \cap R' \neq \emptyset, t' \in Q' \cap R'$. But then it could also be that there is $p''' \in P'$ and σ''' multiset of operations over C', such that $p''' \overset{\sigma'''}{\Rightarrow} t'$. A contradiction.

The partition induced by α is:

$\bar{C}_j = \{c' \in C'_j \mid (c_j, C'_j) \in \alpha, c_j \in reach(c_{init})\} \ 1 \leq j \leq |reach(c_{init})|,$
$\bar{C} = C' \setminus \bigcup_{j=1}^{|reach(c_{init})|} \bar{C}_j.$

Here we used $|A|$ to denote the *cardinality* of a set, that is, the number of elements in a set A. $\qquad\square$

Similar definitions of configurations have been given in the literature [24,33]. These definitions have not being used to study the computational power of formal systems, these definition (sometimes augmented with a labelling of the operations) aimed to study the equivalence of specific formal systems and their processes. The definition given by us generalises similar definitions in [24,33]. In the following we state in terms of Definition 1 the different definitions of simulations present in [24,33]. We will see that different kinds of simulations can be defined on how many elements are returned by α and β and by what kind of relations α and β are (bijections, functions, etc.).

Definition 2. *Let S and S' formal systems and let $\alpha, \alpha', \beta, \beta'$ relations such that S $\alpha\beta$ simulates S and S $\alpha'\beta'$ simulates S', then S and S' are:*

isomorphic *if α and α' always return 1 configurations, β and β' always return 1 multiset of operations having support 1 and all four relations are bijections;*
configuration equivalent *if α and α' always return 1 configurations, β and β' always return 1 multiset of operations and all four relations are bijections;*
weakly configuration equivalent *if α and α' always return 1 configurations, β and β' always return 1 multiset of operations and β and β' are bijections. The relations α and α' are called bisimulations [24].*

Definition 1 can be generalised even further with, for instance, $\alpha \subseteq \mathcal{P}(C) \times \mathcal{P}(C')$. In this case it is possible that n consecutive configurations of the simulated systems S can be simulated by $m < n$ configurations of the simulating systems S'.

3.1 Conformon P Systems

In [14] a model of membrane systems called *conformon P systems* (cP systems) was introduced. This model, later studied also in [3,6,4,7,12,15], is based on simple and basic concepts inspired by a theoretical model of the living cell centred around *conformon* [21,22].

We will consider a few models of cP systems. Here we introduce one of these models.

Definition 3. *An accepting conformon P system is a construct*

$$\Pi = (V, \mu, input_z, ack, L_1, \ldots, L_m, R_1, \ldots, R_m)$$

where:

V *is an alphabet;*
$\mu = (Q, E, lr)$ *is an edge-labelled directed multigraph (a cell-graph) underlying Π where:*

  $Q \subset \mathbb{N}$ *contains vertices. For simplicity we define $Q = \{1, \ldots, m\}$. Each vertex in Q defines a compartment of Π;*
  $E \subseteq Q \times Q$ *defines directed labelled edges between vertices, denoted by (i, j), $i, j \in Q$, $i \neq j$;*

$lr : Q \times Q \rightarrow pred(\mathbb{N}_0)$ *is the labelling relation where for each* $n \in \mathbb{N}_0$ *we consider* $pred(n) \in \{\geq n, \leq n\}$ *set of* predicates;

$input_z$ *with* $z \in Q$. *The compartment* z *contains the input;*

$ack \in Q$ *denotes the* acknowledge *compartment;*

$L_i : (V \times \mathbb{N}_0) \rightarrow \mathbb{N}_0 \cup \{+\infty\}$, $i \in Q$, *are multisets of conformons initially associated with the vertices in* Q;

R_i, $i \in Q$, *are finite sets of interaction rules associated with the vertices in* Q.

If $(i, j) \in E$ and $lr(i, j) = pred(n)$, then we write $(i, j, pred(n))$, a *labelled edge*. Let M_i and R_i be the multiset of conformons and the set of interaction rules, respectively, associated with the compartment $i, i \in Q$. Two conformons present in i can interact according to a rule also present in i such that the multiset of conformons M_i changes into M_i'. If, for instance, $[\Phi, a], [\Upsilon, b] \in M_i$, $\Phi \xrightarrow{e} \Upsilon \in R_i$, and $a \geq e$, then $M_i' = M_i \backslash \{[\Phi, a], [\Upsilon, b]\} \cup \{[\Phi, a - e], [\Upsilon, b + e]\}$.

A conformon $[\Phi, a]$ present in i can *pass* to compartment j if $(i, j, pred(n)) \in E$ and $pred(a)$ holds on a. That is, if $pred(n)$ is $\leq n$, then $a \leq n$; if $pred(n)$ is $\geq n$, then $a \geq n$. This passage changes the multisets of conformons M_i and M_j into M_i' and M_j', respectively. In this case $M_i' = M_i \backslash \{[\Phi, a]\}$ and $M_j' = M_j \cup \{[\Phi, a]\}$.

At the moment we do not assume any requirement (as maximal parallelism, priorities, etc.) on the application of operations. That is, cP systems operate in an *asynchronous* way. If a conformon can pass to another compartment or interact with another conformon according to a rule, then one of the two operations or none of them is non-deterministically chosen.

The possibility to carry out one of the two allowed operations in a same compartment or none of them lets cP systems to be non-deterministic. Non-determinism can also arise from the configurations of a cP system if in a compartment a conformon can interact with more than one conformon and also from the cell-graph underlying Π if a compartment has edges with the same predicate going to different compartments.

A *configuration* of Π is an m-tuple $(M_1, \ldots, M_m)$ of multisets over $V \times \mathbb{N}_0$. The m-tuple $(L_1, \ldots, L_m)$, $supp(L_{ack}) = \emptyset$, is called *initial configuration* (so in the initial configuration the acknowledge compartment does not contain any conformon) while any configuration having $supp(M_{ack}) \neq \emptyset$ is called *final configuration*. In a final configuration no operation is performed even if it could. If in a configuration with no conformon in M_{ack} no operation can be performed, then we say that the system *stops*.

For two configurations $(M_1, \ldots, M_m)$, $(M_1', \ldots, M_m')$ of Π we write $(M_1, \ldots, M_m) \Rightarrow (M_1', \ldots, M_m')$ to denote a *transition* from $(M_1, \ldots, M_m)$ to $(M_1', \ldots, M_m')$, that is, the application of one operation to at least one conformon. In other words, in any configuration in which $supp(M_{ack}) \neq \emptyset$ any conformon present in a compartment can either interact with another conformon present in the same compartment or pass to another compartment or remain in the same compartment unchanged. If no operation is applied to a multiset M_i, then $M_i' = M_i$. The reflexive and transitive closure of $\Rightarrow$ is denoted by $\Rightarrow^*$.

A *computation* is a sequence of transitions between configurations of a system Π starting from $(L_1, \ldots, L_m)$. If a computation is finite, then the last configuration is called *final*.

The input of Π is given by the number of conformons (counted with their multiplicity) present in L_z. The input is accepted by Π if it reaches a final configuration in which a (any) conformon is present in ack. When this happens the computation *halts*, that is, no transition takes place even if it could.

Formally:

$$N(\Pi) = \{|L_z| \mid (L_1, \cdots, L_m) \Rightarrow^* (M'_1, \cdots, M'_m) \Rightarrow^* (M_1, \cdots, M_m),$$
$$supp(M'_{ack}) = \emptyset, supp(M_{ack}) \neq \emptyset\}.$$

Other models of cP systems are given in the following sections.

Figures and Modules. We do not provide formal definitions for all the cP systems considered in this paper. Figures depicting them are provided instead. These figures have compartments represented by rectangles having their label written in **bold** on their right-upper corner. Ovals with a label in them refer to the compartment having that label. Conformons and interaction rules related to a compartment are written inside a rectangle.

Conformons present in the initial configuration of a system are written in **bold** inside a rectangle. If m conformons $[\Phi, a]$ can be present in a compartment, then $([\Phi, a], m)$ is indicated. If an infinite number of conformons $[\Phi, a]$ is present in a compartment, then $([\Phi, a], +\infty)$ is indicated.

If in a compartment the interaction rules $\Phi \xrightarrow{e} \Upsilon$ and $\Upsilon \xrightarrow{e} \Phi$ are present, then $\Phi \xleftrightarrow{e} \Upsilon$ is indicated. The interaction rules $\Phi \xrightarrow{e} \Upsilon$ and $\Upsilon \xrightarrow{e} \Phi$ are one the *reverse* of the other.

Directed edges between compartments are represented as arrows with their predicate indicated close to them. The figures representing cP systems can contain also shorthands for *modules*. Modules are groups of compartments with conformons and interaction rules in a cP system able to perform a specific task. Here the list of module we will use and their representation:

separator: it can select conformons depending on their name. Separators are represented with conformons close to an edge. This denotes that only conformons as the one indicated can pass through the edge. Separators alternated with a slash (/) indicate the presence of more than one edge between the same compartments. For instance, an edge with labes $[A, 3]/[A, 5]$ denotes two edges, one with separator $[A, 3]$ and another with separator $[A, 5]$.

increaser/decreaser: it can increase/decrease the value of conformons until a specific amount. Increasers are represented with rectangles with a thicker line having **[x]inc** as label. This denotes that the value of conformons entering that module is increased until it reaches x and then the conformons can leave the module. Decreasers are represented with rectangles with a thicker line having **[x]dec** as label. This denotes that the value of conformons entering that module is decreased until it reaches x and then the conformon can leave the module.

More details about modules can be found in [10].

3.2   Register Machines

The devices we define in this section use numbers to perform computations. They have *registers* (also called *counters*) each of unbounded capacity recording a natural number or zero. Simple operations can be performed on the registers: addition of one unit and conditional subtraction of one unit. After each of these operations the machine can change state.

Formally a *register machine* with n registers ($n \in \mathbb{N}$), each register able to store any element in $\mathbb{N}_0$, is defined as $M = (S, I, s_1, s_f)$, where $S = \{s_1, \ldots, s_f\}$ is a finite set of *states*, $s_1, s_f \in S$ are respectively called the *initial* and *final* states, I is the finite set of *instructions* of the form (s, γ_i^-, v, w) or (s, γ_i^+, v) with $s, v, w \in S$, $s \neq s_f, 1 \leq i \leq n$.

A *configuration* of a register machine M with n registers is given by an element in the $n + 1$-tuples $S \times \mathbb{N}_0^n$. Given two configurations $(s, val(\gamma_1), \ldots, val(\gamma_n))$, $(s', \gamma_1', \ldots, \gamma_n')$ (where $val : \{\gamma_1, \ldots, \gamma_n\} \to \mathbb{N}_0$ is the function returning the content of a register) we define a *computational step* as $(s, val(\gamma_1), \ldots, val(\gamma_n)) \vdash (s', \gamma_1', \ldots, \gamma_n')$ and:

if $(s, \gamma_i^-, v, w) \in I$ and $val(\gamma_i) \neq 0$, then $s' = v$, $\gamma_i' = val(\gamma_i) - 1$, $\gamma_j' = val(\gamma_j)$, $j \neq i$, $1 \leq j \leq n$;
if $val(\gamma_i) = 0$, then $s' = w$, $\gamma_j' = val(\gamma_j)$, $1 \leq j \leq n$;
(informally: in state s if the content of register γ_i is greater than 0, then subtract 1 from that register and change state into v, otherwise change state into w)
if $(s, \gamma_i^+, v) \in I$, then $s' = v$, $\gamma_i' = val(\gamma_i) + 1$, $\gamma_j' = val(\gamma_j)$, $j \neq i$, $1 \leq j \leq n$;
(informally: in state s add 1 to register γ_i and change state into v).

The reflexive and transitive closure of $\vdash$ is denoted by $\vdash^*$.

A *computation* is a sequence of computational steps of a register machine M starting from the *initial configuration* $(s_1, val(\gamma_1), 0, \ldots, 0)$. If a computation is finite, then the last configuration is called *final*. If a final configuration has s_f as state, then we say that M *halts* and it *accepts* the input $val(\gamma_1)$. For this reason γ_1 is called the *input register* and M is called an *accepting register machine*. Starting from an initial configuration $(s_1, val(\gamma_1), 0, \ldots, 0)$ a register machine M could have a finite sequence of computational steps in which the last one does not have s_f as state. In this case we say that M *stops* and $val(\gamma_1)$ is not accepted.

The set of numbers *accepted* by M is defined as $\mathsf{N}(M) = \{val(\gamma_1) \mid (s_1, val(\gamma_1), 0, \ldots, 0) \vdash^* (s_f, \gamma_1', \ldots, \gamma_n'), \gamma_1', \ldots, \gamma_n' \in \mathbb{N}_0\}$.

The set of numbers accepted by register machines is $\mathsf{N} \cdot \mathsf{RE}$.

It has been proved that the acceptance of $\mathsf{N} \cdot \mathsf{RE}$ can be obtained by a register machine with three registers, two registers are necessary and sufficient if one uses a specific input format (for example, 2^x instead of x).

Some authors considered register machines equipped with instructions of the kind: $(s, \gamma_i^-, v), (s, \gamma_i^{=0}, w)$ and $(s, \gamma_i^+, v), 1 \leq i \leq n$. Informally, instructions of the kind (s, γ_i^-, v) let a register machine in state s and with $val(\gamma_i) > 0$ to change state into v and decrease by 1 the content of γ_i. If instead $val(\gamma_i) = 0$, then the register machine stops. Instructions of the kind $(s, \gamma_i^{=0}, w)$ let a register machine in state s and with $val(\gamma_i) = 0$ change state into w. If instead $val(\gamma_i) > 0$, then the register machine stops. Instructions of the kind (s, γ_i^+, v) perform what indicated earlier.

Instructions of the kind (s, γ_i^-, v, w) and $(s, \gamma_i^{=0}, w)$ are also called *test on 0* or *0-test*. This is because they allow the machine to detect if a register stores zero and perform some operations as a consequence of this. The name given to instructions of these kinds is misleading as these instructions perform more than just a test.

Partially blind register machines are defined as register machines without test on zero. The only allowed operations are (s, γ^+, v) and (s, γ^-, v) where γ is a register. In case the machine tries to subtract from a register having value zero it stops. They are strictly less powerful from a computational point of view than register machines.

Restricted register machines are defined as register machines restricted in their operations: they can increase the value of a register, say β, only if they decrease the value of another register, say γ at the same time.

So, restricted register machines have only one kind of instruction: $(s, \gamma^-, \beta^+, v, w)$ with s, v, w states and γ, β different registers of the restricted register machine. If when in state s the content of register γ can be decreased by 1, then the one of register β is increased by 1 and the machine goes into state v, otherwise no operation is performed on the registers and the machine goes into state w.

Here is a result proved in [20]:

Theorem 1. *Restricted register machines with* n+1 *registers are more powerful from a computational point of view than the ones with* n *registers.*

A consequence of this theorem is that an infinite hierarchy is induced, by means of the number of registers, among families of computed sets of numbers.

3.3 P/T Systems

The study of the topology of information flow we referred to in Section 1 is done through *P/T systems* a model of Petri nets.

Definition 4. *A place/transition system (P/T system) is a tuple* $N = (P, T, F, W, K, C_{in})$, *where:*

i) (P, T, F) *is a net:*
    1. P *and* T *are sets with* $P \cap T = \emptyset$;
    2. $F \subseteq (P \times T) \cup (T \times P)$;
    3. *for every* $t \in T$ *there exist* $p, q \in P$ *such that* $(p, t), (t, q) \in F$;
ii) $W : F \to \mathbb{N}$ *is a* weight function;
iii) $K : P \to \mathbb{N} \cup \{+\infty\}$ *is a* capacity function;
iv) $C_{in} : P \to \mathbb{N}_0$ *is the* initial configuration *(or initial marking).*

We consider P/T systems in which the weight function returns always 1 and the capacity function returns always $+\infty$. We introduced these functions in the previous definition for consistency with the (for us) standard definition of P/T systems and for consistency with the definition in [6,9,10]. We follow the very well established notations (places are represented by empty circles, transitions by full rectangles, tokens by bullets, etc.), concepts and terminology (configuration, input set, output set, sequential configuration graph, etc.) relative to P/T systems [10,32,31].

Here we informally give a few details on the concepts and terminology we use. A P/T system can *run* in different ways, that is, its transitions can fire following different rules. Depending on these rules different *configuration graphs* can be associated to a P/T system with a given initial configuration:

Sequential Configuration Graph (SCG): obtained by a P/T system in which in each configuration at most (any) one transition fires;

Maximal Strategy Configuration Graph (MSCG): obtained by a P/T system in which in each configuration all transitions that can fire do so only once. This means that if in a configuration a transition can fire more than one once, then it only fires once;

Maximal Parallelism Configuration Graph (MPCG): obtained by a P/T system in which in each configuration all transitions that can fire do so as many times as they can. This means that if in a configuration a transition can fire more than once, then it does fire the maximum number of times it can fire;

Priority Configuration Graph (PrCG): only possible for P/T system in which a priority between transitions is defined. This graph is obtained by a P/T system in which for each configuration the set of firing transitions U is such that no other set of firing transitions U' for the same configuration has elements with an higher priority of the elements in U.

Despite their different definitions $MSCG$ and $MPCG$ are actually equivalent as this result from [10] states:

Lemma 2. *Let N be a P/T system. It is possible to define a P/T system N' and two bijections α and β such that:*

$MPCG(N')$ *simulates* $MSCG(N)$ *according to α and β;*
$MSCS(N) = MSCG(N')$.

The definition of simulation if given in Section 3. In this paper we consider P/T systems as accepting computing devices. The definition of accepting P/T systems includes the indication of a set $P_{in} \subset P$ of *input places*, one *initial place*, $p_{init} \in P \setminus P_{in}$, and one *final place*, $p_{fin} \in P \setminus P_{in}$. The places in $P \setminus P_{in}$ are called *work places*.

An *accepting P/T system* N with input C_{in} is denoted by $N(C_{in}) = (P, T, F, W, K, P_{in}, p_{init}, p_{fin})$ where $C_{in} : (P_{in} \cup \{p_{init}\}) \to \mathbb{N}_0$, $C_{in}(p_{init}) = 1$, is the initial configuration of the input places. So, in the initial configuration some input places can have tokens and the work place p_{init} has one token. All the remaining places are empty in the initial configuration. A configuration $C_{fin} \in \mathbb{C}_N$, the set of all reachable configurations of N, is said to be *final* (or *dead state*) if no firing is possible from C_{fin}.

We say that a P/T system $N(C_{in}) = (P, T, F, W, K, P_{in}, p_{init}, p_{fin})$ with $P_{in} = \{p_{in,1}, \ldots, p_{in,k}\}$, $k \in \mathbb{N}$, *accepts* the vector $(C_{in}(p_{in,1}), \ldots, C_{in}(p_{in,k}))$ if in the sequential configuration graph of $N(C_{in})$ there is a final configuration C_{fin} such that:

$C_{fin}(p_{fin}) > 0$;
there is at least one path from C_{in} to C_{fin};
no other configuration D in the paths from C_{in} to C_{fin} is such that $D(p_{fin}) > 0$.

The *set of vectors accepted* by N is denoted by $\mathsf{N}^k(N)$ and it is composed by the vectors $(C_{in}(p_{in,1}), \ldots, C_{in}(p_{in,k}))$ accepted by N. The just given definition of (vector) acceptance for P/T systems is new in Petri nets. Normally, Petri nets are generating devices having labels associated to the transitions. A generated word is given by the concatenations of the labels associated to the transitions in a firing sequence.

As in [9] we call the nets *join* and *fork building blocks*, see Figure 2, where the places in each building block are distinct.

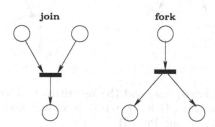

Fig. 2. Building blocks: *join* and *fork*

From [9] we also take:

Definition 5. *Let $x, y \in \{join, fork\}$ be building blocks and let $\bar{t}_x$ and $\hat{t}_y$ be the transitions present in x and y respectively.*

We say that y comes after x (or x is followed by y, or x comes before y or x and y are in sequence) if $\bar{t}_x^\bullet \cap {}^\bullet\hat{t}_y \neq \emptyset$ and ${}^\bullet\bar{t}_x \cap \hat{t}_y = \emptyset$. We say that x and y are in parallel if ${}^\bullet\bar{t}_x \cap \hat{t}_y \neq \emptyset$ and $\bar{t}_x^\bullet \cap {}^\bullet\hat{t}_y = \emptyset$.

We say that a net is composed of building blocks (it is composed of x) if it can be defined by building blocks (it is defined by x) sharing places but not transitions. So, for instance, to say that a net is composed of joins means that the only building blocks present in the net are join.

Figure 4.a, Fig. 4.c and Fig. 3 building blocks are in sequence and in parallel.

Sequences of building blocks can be *re-written* in a 'compressed' form as depicted in Fig. 4.

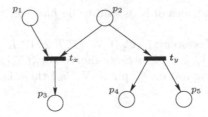

Fig. 3. A *join* and a *fork* in parallel. © Used with permission from Oxford University Press [10]

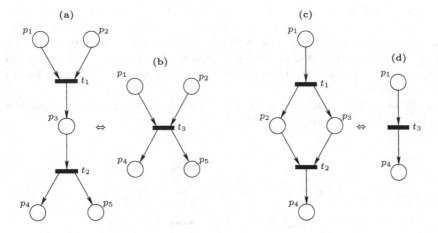

Fig. 4. (a) *join* and *fork* in sequence and (b) re-written in a 'compressed' form; (c) *fork* and *join* in sequence and (d) re-written in a 'compressed' form. © Used with permission from Oxford University Press [10]

In this paper we consider accepting P/T systems (in which the weight functions returns always 1 and the capacity function returns always $+\infty$) with different nets. All the nets considered by us can be obtained by compositions of *join* and *fork*.

4   Conformon P Systems and P/T Systems

Let us call *total value* the sum of all the values of the conformons present in a cP system.

A cP systems can be simulated by a P/T systems in the following way. A place identifies a conformon present in a compartment. If, for instance, the conformon $[A, 5]$ is present in compartment 1, then the relative place is $[A, 5]_{[1]}$. The number of tokens present in a place identifies the number of occurrences of that conformon in that compartment and the transitions identify passage and interaction rules.

For instance, the passage of $[A, 5]$ from compartment 1 to compartment 2 can be simulated by the net depicted in Fig. 5.a. The interaction of $[A, 5]$ and $[B, 0]$ according to the rule $A \xrightarrow{3} B$ in compartment 1 can be simulated by the net depicted in Fig. 5.b.

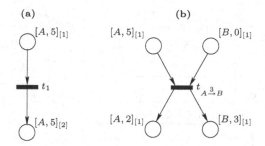

Fig. 5. (a) Nets simulating the passage and **(b)** the interaction of conformons. © Used with permission from Oxford University Press [10]

We know from Section 3.3, see Fig. 4, that the building blocks can be arranged so to obtain the nets depicted in Fig. 5.

5 Measures of Infinity

In [3] the following is proved:

Theorem 2. *The family of sets of numbers accepted by cP systems coincides with the one accepted by partially blind register machines.*

The cP system simulating a register machine is depicted in Fig. 6.

If we consider what in said in Section 4, then the simulation of instructions of the kind (s_i, γ^-, s_j) can be performed by the P/T system depicted in Fig. 7.a. This net is composed by the sub-nets depicted in Fig. 5.

The net in Fig. 7.a can be simulated by the one depicted in Fig. 7.b. Here the conformons $[\gamma, 0]$ in compartment 5 are disregarded as they are present in an infinite amount (that is, they are invariant). Notice that what depicted in Fig. 7.b is just a *join* building block.

The remaining nets present in this paper are not composed by the sub-nets depicted in Fig. 5, for simplicity only re-writings as the one in Fig. 7.b are depicted.

One can draw the net underlying the P/T system simulating cP system in Fig. 6 simulating instructions of the kind (s_i, γ^+, s_j). The rewriting of such net would result in the one depicted in Fig. 7.c (where, as before, invariant conformons are disregarded).

The cP system of Theorem 2 has initially an infinite number of $[\gamma, 0]$ conformons in compartment 5 (see [3]). This can be simulated by a P/T system as the

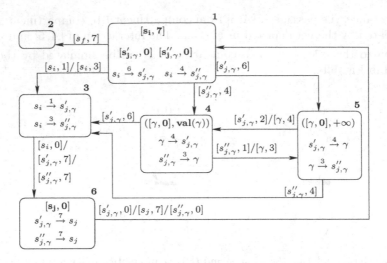

Fig. 6. The cP system related to Theorem 2

one depicted in Fig. 8. What depicted in this figure allows the P/T system to have a finite initial configuration with only one token in p_0, to 'load' a random number of tokens in p_γ and then to put one token in p_{s_1} starting in this way the simulation of the cP system.

It is important to notice that:

the P/T system is composed of *join* and *fork* arranged in any way;
the SCG of the P/T system can simulate partially blind register machines. This is due to the fact that the cP system in Theorem 2 operates in an asynchronous way: the cP system can simulate the partially blind register machines if in each configuration at least one operation is performed. This translates in at least one transition firing in the P/T system, so a SCG is obtained.

The following then holds [10]:

Theorem 3. *For each partially blind register machine M there is an accepting P/T systems N whose underlying net is composed by building blocks (and using an unbounded number of tokens) such that $SCG(N)$ simulates M. Moreover, for each accepting P/T systems N whose underlying net is composed by building blocks (and using an unbounded number of tokens) there is a partially blind register machine simulating $SCG(N)$.*

In the following we recall a result partially answer the questions: *How does the way to run a P/T system relate to the sets of vectors accepted by it?*

We say that a net (P, T, F) in a P/T system $N = (P, T, F, W, K, C_{in})$ is *connected* if when all elements of the flow relation F are replaced by bi-directional connections, then for each $x_1, x_2 \in P \cup T$ there is a path from x_1 to x_2.

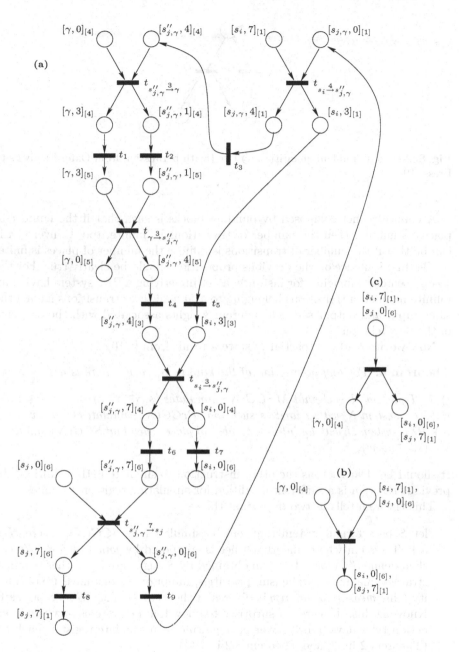

Fig. 7. (a) Net for the simulation of (s_i, γ^-, s_j) related to Theorem 2, (b) its rewriting and the re-writing of the net simulating (s_i, γ^+, s_j). © Used with permission from Oxford University Press [10]

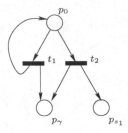

Fig. 8. Net to fix random quantities. © Used with permission from Oxford University Press [10]

A connected net composed by building blocks is such that if the number of places is infinite, then the number of transitions is infinite, too. Conversely, it can be that if the number of transitions is infinite the number of places is finite.

The first sentence on the previous proposition should be clearly true. For the second sentence: imagine, for instance, a net underlying a P/T system having an infinite number of transitions belonging to a *join* all these transitions having the same input and output sets but different weights associated with the elements of the flow relation.

Now we are ready to (partially) state a result from [8,10]:

Theorem 4. *For any instruction of the kind* (s, γ^-, v, w) *there is a*

i) P/T system N such that $MSCG(N)$ *simulates* (s, γ^-, v, w);
ii) P/T system N with priorities such that $PrCG(N)$ *simulates* (s, γ^-, v, w);
iv) P/T system N with an infinite number of places such that $SCG(N)$ *simulates* (s, γ^-, v, w).

It should be clear that, as the other instruction to simulate is the addition, the previous theorem is a sufficient condition for simulating register machines.

This theorem tells us two important things:

1. let S be a formal system that can be simulated by $SCG(N)$, where N is a P/T system whose underlying net is composed by *join* and *fork*. We can then define S': a formal system obtained by S augmented by either maximal strategy, maximal parallelism, priorities, inhibitors or one measure of infinity (this *measure of infinity* is discuss in the item 2). Then S' accepts NRE. Knowing this, then are not surprised to know that cP systems in which interaction rules have priority over passage rules can simulate register machines (Theorem 2 in [3] and Theorem 8.2 in [10]).

2. let Π be the cP system in Theorem 2. This system has an infinite amount of $[\gamma, 0]$. Moreover, let N be the P/T systems simulating Π and putting a random number of tokens in a specific place. We call this infinite number of conformons, or random number of tokens the *first measure of infinity*. From Theorem 4 we know that there is another measure of infinity (the number of places) such that if N' is similar to N but it has an infinite number of places,

then $SCG(N')$ simulates register machine. We call this infinite number of places the *second measure of infinity.*

These two measures of infinity are different: the SCG of a P/T system with a finite number of places and an infinite number of tokens cannot simulate the SCG of a P/T system with an infinite number of places and a finite number of tokens. If this was not the case, then a partially blind program machine could simulate a program machine.

Using cP system it is possible to clearly separate these two measures of infinity. Let us consider the following result from [3]:

Theorem 5. *The family of sets numbers accepted by cP systems with infinite total value coincides with the one accepted by register machines.*

The cP system associated to Theorem 5, depicted in Fig. 9, has an infinite total value. This is given by the conformons in the module **[5]inc** allowing the value of the incoming conformons to be increased to 5 (see [10]).

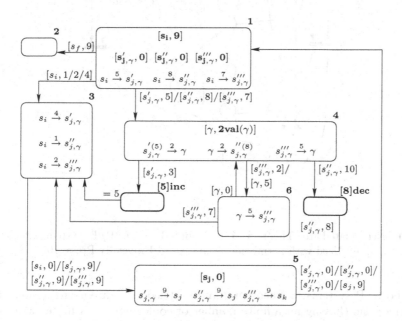

Fig. 9. The cP system with infinite total value related to Theorem 5

In the proof of Theorem 5 the simulation of instructions of the kind $(s_i, \gamma^-, s_j, s_k)$ is performed 'gambling': the system randomly gambles if the counter γ is empty or not. The computation goes on only if the gamble was correct, otherwise it will never halt. A net performing such simulation is depicted in Fig. 10, where the dotted lines suggest the presence of an infinite number of places, transitions and elements of the flow relation. We are not detailed in the name of the places in this net. Instead of indicating the full configuration for each place in this net (as done in Fig. 7b and Fig. 7.c) we only indicate the conformon and compartment that are essential in the configuration to performed some operations. We are also not precise in the indication of the transitions: we indicate only one transition instead of several.

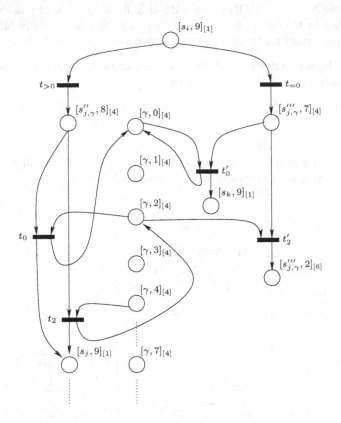

Fig. 10. Net related to Theorem 5 and underlying a P/T system simulating $(s_i, \gamma^-, s_j, s_k)$. © Used with permission from Oxford University Press [10]

So, both cP systems in Theorem 2 and in Theorem 5 are asynchronous, but the former one (having an infinite number of conformons and finite value) can simulate partially blind register machines, while the latter one (having an infinite total value) can simulate register machines. The P/T systems simulating the cP systems in these two theorem have an underlying net composed of *join* and *fork*, but the P/T system related to Theorem 2 has a finite number of places and uses a random (so, unbounded) number of tokens (see row 1 Table 1), while the P/T system related to Theorem 5 has an infinite number of places and a finite number of tokens (see row 3 Table 1).

6   Infinite Hierarchies

In this section we describe how restricted models of cP system can induce infinite hierarchies on the computation they can perform and how this related to features of the P/T systems simulating them.

Conformon-restricted cP systems can have more than one input compartment and they have only one conformon with a distinguished name, let us say l, encoding the input. The formal definition of such cP systems changes then into:

$$\Pi = (V, \mu, input_{z_1}, \ldots, input_{z_n}, L_1, \ldots, L_m, R_1, \ldots, R_m)$$

where $z_1, \ldots, z_n \in Q$ denote the input compartments. In the initial configuration the input compartments contain only l conformons and no other compartment contains l conformons. The definitions of configuration, transition, computation, halt, stop and set of numbers accepted follow from the ones given in Section 3.1.

The following result is from [5]:

Theorem 6. *The family of sets numbers accepted by conformon-restricted cP systems with* n *input compartments coincides with the one accepted by restricted register machines with* n *registers.*

A conformon-restricted cP system with 2 input compartments, 4 and 7, simulating a restricted register machines with 2 registers is depicted in Fig. 11.

The cP system depicted in Fig. 6 is very similar to the one depicted in Fig. 9. The only difference between the two figures is that the increaser and decreaser

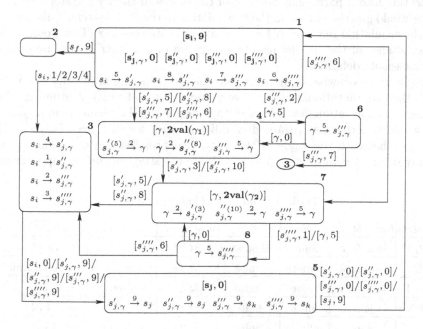

Fig. 11. The conformon-restricted cP system related to Theorem 6

modules present in Fig. 9 have been replaced by compartment 7 in Fig. 11. This means that it is possible to define a P/T system N having as underlying net composed by *join* and *fork* and having a finite number of tokens such that $SCG(N)$ simulates restricted register machines. This is summarised in row 2 of Table 1. As a matter of fact, a further restriction can be imposed to the net underlying N: it can composed by only *join* and *fork* in sequence in a 'compressed' form (see Fig. 4.b).

7   Final Remarks

It should be clear from the previous sections that the sole study of models of Petri nets suffices to link topology (and other features as the number of measures of infinity, the way to run, etc.) to the computation power of these nets. In turn, the definition of simulation we gave allows these links to be valid to many formal systems based on multiset rewriting (as the cP systems considered in the present paper).

One could then ask why to study the computational power of such formal systems. If from the one hand the analysis of the topology of information flow allows to obtain results that are very general, from the other hand this analysis does not say much on the details of the implementation in a specific formal system. In some models of membrane systems, for instance, the topological structure of the compartments is a cell-tree and symbols can pass from one compartment to another only following the topological structure. A simulation could translate this feature into a particular labelling of the places in the P/T system, this labelling would put then limits in the flow relation of the P/T system. This implies that the simulation relations α and β would be quite complex. This would limit further studies of the specific membrane system or resulting in very unnatural (and inelegant) definitions.

Despite these drawbacks we think that the study linking properties of a Petri net to their computational power is very important and worth continuing. The topology of information flow allows us to regard as similar elements in a system that would be otherwise regarded as different (see, for instance, Theorem 4). This, together with the study of the dynamical properties of a system, allows to unify different computing models, facilitate the study of their computational

Table 1. Summary of known results

n.	system	build. blocks	n. places	n. tokens	way to run	acc./gen.	class
1	P/T	join, fork	finite	unbounded	SCG	acc.	= part. blind r.m.
2	P/T	join, fork	infinite	finite	SCG	acc.	= restricted r.m.
3	P/T	join, fork	finite	finite	$MSCG$	acc./gen.	= $\mathbb{N}\cdot$RE
4	EN	join	finite	finite	SCG	acc.	= $\mathbb{N}\cdot$FIN
5	P/T'	join	infinite	finite	SCG	acc./gen.	= $\mathbb{N}\cdot$RE
6	P/T	join	finite	finite	SCG	acc.	J languages

power and introduces new measured of complexity. In [10] it is shown how the computational power of models of P systems can be analysed through the study of the topology of their information flow. If it is not needed to know the details of the considered formal system, then this kind of study is very useful as it avoids tedious proofs. Table 1 summarises the known results [6,9,10,13].

This table is not complete: more rows and also more columns can be added. Clearly, the rows would emerge from different values and their combinations. More columns could result from some of the following suggestions for research.

Suggestion for research 1. *In Section 3.3 we said that our definition of acceptance for a P/T system is unusual. For this reason it would be interesting to study how our definition relates to the standards one (concatenations of transition labels). Are the results in Table 1 dependent on the definition of acceptance?*

Suggestion for research 2. *Are join and fork the only building blocks that lead to the result in Table 1? Are there other building blocks leading to the same or different results?*

Suggestion for research 3. *The Petri nets considered in this study are built ad hoc. What about the computational power of Petri nets built in a pseudo random way, that is, with different percentages of join and fork? Would such Petri nets have a different computational power depending on these percentages?*

This last suggestion for research could have strong links with biology. Several studies show that the topological structure of a network, that is the way the different agents interact, is a key element in the dynamics (and other properties as response to signals, robustness, etc.) showed by the network. Some examples are: [23] where the model "not only fits the data but its behaviour is also robust to parameter changes", and [16] where "the evolved network has robustness against parameter variation under fixed network topology". Moreover, it has been shown [26,1] that specific *motifs* are at the base of complex, biological and not, networks.

Suggestion for research 4. *Is there any link between some properties of biological networks (response to signals, robustness, kinds of motifs present in the network, etc.) and concepts in formal language theory?*

More suggestions for research can be found in [10,11].

Acknowledgements. Frisco's attendance at WMC10 has been fully funded by The Royal Society, International Travel Grant Scheme 2009R2, application number TG090724.

References

1. Alon, U.: An introduction to Systems Biology. Chapman and Hall, Boca Raton (2006)
2. Freund, R., Lojka, G., Oswald, M., Păun, G.: WMC 2005. LNCS, vol. 3850. Springer, Heidelberg (2006)

3. Frisco, P.: The conformon-P system: A molecular and cell biology-inspired computability model. Theoretical Computer Science 312(2-3), 295–319 (2004)
4. Frisco, P.: Infinite hierarchies of conformon-P systems. In: Hoogeboom, et al. (eds.) [18], pp. 395–408
5. Frisco, P.: Infinite hierarchies of conformon-P systems. In: Hoogeboom, et al. (eds.) [18], pp. 395–408
6. Frisco, P.: P systems, Petri nets, and program machines. In: Freund, et al. (eds.) [2], pp. 209–223
7. Frisco, P.: Conformon-P systems with negative values. In: Eleftherakis, G., Kefalas, P., Păun, G., Rozenberg, G., Salomaa, A. (eds.) WMC 2007. LNCS, vol. 4860, pp. 331–344. Springer, Heidelberg (2007)
8. Frisco, P.: An hierarchy of recognising computational processes. Technical report, Heriot-Watt University. HW-MACS-TR-0047 (2007)
9. Frisco, P.: A hierarchy of computational processes. Technical report, Heriot-Watt University, HW-MACS-TR-0059 (2008), http://www.macs.hw.ac.uk:8080/techreps/index.html
10. Frisco, P.: Computing with Cells. In: Advances in Membrane Computing. Oxford University Press, Oxford (2009)
11. Frisco, P.: P systems and topology: some suggestions for research. In: Seventh Brainstorming Week on Membrane Computing (2009)
12. Frisco, P., Gibson, R.T.: A simulator and an evolution program for conformon-P systems. In: SYNASC 2005, 7th International Symposium on Simbolic and Numeric Algorithms for Scientific Computing. Workshop on Theory and Applications of P Systems, TAPS, Timisoara, Romania, September 26-27, pp. 427–430. IEEE Computer Society, Los Alamitos (2005)
13. Frisco, P., Ibarra, O.H.: On languages accepted by P/T systems composed of *joins*. In: DCFS 2009, 11th workshop on Descriptional Complexity of Formal Systems (2009) (to appear in EPTCS)
14. Frisco, P., Ji, S.: Conformons-P systems. In: Hagiya, M., Ohuchi, A. (eds.) DNA 2002. LNCS, vol. 2568, pp. 291–301. Springer, Heidelberg (2003)
15. Frisco, P., Ji, S.: Towards a hierarchy of info-energy P systems. In: Păun, G., Rozenberg, G., Salomaa, A., Zandron, C. (eds.) WMC 2002. LNCS, vol. 2597, pp. 302–318. Springer, Heidelberg (2003)
16. Fujimoto, K., Ishihara, S., Kaneko, K.: Network evolution of body plants. PLoS ONE 3(7), 1–13 (2008)
17. Gutírrez-Naranjo, M.A., Pérez-Jiménez, M.J., Riscos-Núñez, A., Romero-Campero, F.J.: On the power of dissolution in P systems with active membranes. In: Freund, et al. (eds.) [2], pp. 226–242
18. Hoogeboom, H.J., Păun, G., Rozenberg, G., Salomaa, A. (eds.): WMC 2006. LNCS, vol. 4361. Springer, Heidelberg (2006)
19. Hopcroft, J.E., Ullman, D.: Introduction to Automata Theory, Languages, and Computation. Addison-Wesley, Reading (1979)
20. Ibarra, O.H.: On membrane hierarchy in P systems. Theoretical Computer Science 334, 115–129 (2005)
21. Ji, S.: The Bhopalator: a molecular model of the living cell based on the concepts of conformons and dissipative structures. Journal of Theoretical Biology 116, 395–426 (1985)
22. Ji, S.: The Bhopalator: an information/energy dual model of the living cell (II). Fundamenta Informaticae 49(1-3), 147–165 (2002)

23. Locke, J.C.W., Southern, M.M., Kozma-Bognár, L., Hibberd, V., Brown, P.E., Turner, M.S., Millar, A.J.: Extension of a genetic network model by iterative experimentation and mathematical analysis. Molecular Systems Biology 1, 1–9 (2005)
24. Milner, R.: Communication and Concurrency. Prentice-Hall, Englewood Cliffs (1989)
25. Milner, R.: Communicating and Mobile Systems: the π-calculus. Cambridge University Press, Cambridge (1999)
26. Milo, R., Shen-Orr, S., Itzkovitz, S., Kashtan, N., Chklovskii, D., Alon, U.: Network motifs: simple building blocks of complex networks. Science 298(5594), 824–827 (2002)
27. Păun, A., Păun, G.: The power of communication: P systems with symport/antiport. New Generation Computing 20(3), 295–306 (2002)
28. Păun, G.: Computing with membranes. Journal of Computer and System Science 1(61), 108–143 (2000)
29. Păun, G.: Membrane Computing. An Introduction. Springer, Berlin (2002)
30. Păun, G., Rozenberg, G., Salomaa, A. (eds.): The Oxford Handbook of Membrane Computing. Oxford University Press, Oxford (in press, 2010)
31. Reisig, W.: Petri Nets: An Introduction. Monographs in Theoretical Computer Science. An EATCS Series, vol. 4. Springer, Berlin (1985)
32. Reisig, W., Rozenberg, G. (eds.): APN 1998. LNCS, vol. 1491. Springer, Heidelberg (1998)
33. Rozenberg, G., Engelfriet, J.: Elementary net systems. In: [32], pp. 12–121

Formal Verification and Testing
Based on P Systems

Marian Gheorghe[1,2], Florentin Ipate[2], and Ciprian Dragomir[1]

[1] Department of Computer Science, The University of Sheffield
Regent Court, Portobello Street, Sheffield S1 4DP, UK
M.Gheorghe@dcs.shef.ac.uk
[2] Department of Computer Science
Faculty of Mathematics and Computer Science
The University of Piteşti
Str. Târgu din Vale 1, 110040 Piteşti
florentin.ipate@ifsoft.ro

Abstract. In this paper it is surveyed the set of formal verification methods and testing approaches used so far for applications based on P systems.

1 Introduction

P systems (also called membrane systems) represent a class of parallel and distributed computing devices which are inspired by the structure and functioning of the living cells [18], [19]. The model has been intensively investigated from a theoretical perspective, and studies related to computational power, complexity aspects, hierarchies of different mechanisms and connections with other similar models have been undertaken. Another important line of research has considered P systems as a vehicle to represent different problems from various domains [20]. A rich set of software tools has been produced, implementing simulators for various classes of P systems or translators to formal verification tools [11].

As a consequence of using membrane systems to specify, model and simulate various systems, certain methods and techniques have been employed to verify they produce the expected results.

Formal methods have been used for various types of membrane systems and using different formalisms. Petri nets have been used to express the semantics of certain classes of P systems and methods to translate these systems into Petri nets have been developed. Tools and techniques produced and utilised for Petri nets become available for the description, analysis, and verification of membrane systems [17]. They also allow to study specific properties of such systems, like causality and (a)synchrony.

Structural operational semantic for certain some classes of P systems has been systematically investigated and their translation into specific rewriting logic formalisms provided by Maude [9], [5], has been defined. This approach allows to formally verify properties of such systems by using linear temporal logic model checking procedures [3].

G. Păun et al. (Eds.): WMC 2009, LNCS 5957, pp. 54–65, 2010.
© Springer-Verlag Berlin Heidelberg 2010

For probabilistic and stochastic P systems special relationships with classes of stochastic process algebras and Petri nets have been investigated and a special purpose model checking approach based on PRISM, which will be dicussed in the next section, has been studied [6].

A complementary approach to formal verification is usually based on testing. In the case of P systems this route has been recently open to research by considering some classical test coverage criteria [14]. More specifically, model based testing has been investigated for simple classes of P systems [14], [16] and ways to devise adequate test sets have been proposed. These techniques are somehow similar to studies investigating the role of the so called observers [7], [8] for certain classes of P systems where the behaviour is just filtered through some mechanisms for a well-defined purpose. In the case of testing the behaviour which is selected needs to obey some testing rules. The problem of testing P systems will be surveyed in section 3.

2   P Systems Verification

In many research areas, like, cellular biology, ecology, social insects study, the accurate simulations play a very important role as they reveal new properties that can be difficult or impossible to discover through direct experiments. One key question is what one can do with a model, other than just simulate trajectories. The problem of formal verification of the behaviour of such systems represents an alternative to simulations for verifying certain properties and for revealing new behaviour. Formal verification methods have been studied for various classes of P systems. This research can be classified according to certain criteria that reveal various formal aspects of such systems.

Due to a rapid proliferation of various variants of P systems, the research on various formal semantics started quite early and consists of a wealth of different approaches. Structural operational semantics for basic classes of P systems [3] together with rewriting logic semantics [5] offer the appropriate framework to develop not only rigorous formal definitions of certain classes of P systems, but also the opportunity to define adequate translations of such systems into a well-known model checker, Maude [5]. Using this framework a large variety of queries can be formulated in Linear Temporal Logic (LTL for short) and then verified using suitable Maude tools [3]. Executable semantics based on an extension to Maude, called K, is provided for certain classes of P systems [22]. Another class of semantics is based on Petri nets [17] and it brings also the wealth of tool support that comes together with these models. Finally, semantics aiming to include compositionality aspects using a process algebra style, is studied for very basic classes of P systems [4].

There have been studies on mapping various models of P systems into other computational models that benefit from well-established formal verification and testing methods. A class of bounded symport/antiport is coded as brane calculi [24] and basic membrane systems are matched against X-machines [1].

Relationships between membrane systems and other formalisms, like, ambients [2], cellular automata [13], Petri nets [12], and X-machines [23], have been also considered. Specific model-checking decidability results for P systems have been studied [10]. More precisely, decidability results for queries formulated in Computation Tree Logic (CTL for short) augmented with atomic predicates in REG and LIN, are investigated for the class of bounded P systems – only rewriting rules with the left hand side bigger than the right hand side are utilised.

The problems described above have been considered for various classes of deterministic or nondeterministic P systems. Stochastic systems behave in an unintuitive way and consequently are harder to conceive and verify. The field is widely open for theoretical investigations and applications in various areas. A way of formally verifying stochastic behaviour is to use specific model checking tools to analyse in an automatic way various properties of the model. The class of P systems which is suitable for such formal verification consists of stochastic P systems.

In most of the variants of P systems the rules are applied in a maximally parallel way. This mechanism allows an efficient execution of the systems and proves to be very effective for theoretical investigations where it plays a major role in many circumstances. However in applications where the number of molecules is not that big, a different variant seems to be more suitable. This variant has been introduced in [21] and will be discussed in some detail below.

Definition 1. *A stochastic P system is a construct*

$$\Pi = (O, L, \mu, M_1, M_2, \ldots, M_n, R_1, \ldots, R_n)$$

where:

- *O is a finite alphabet of symbols, called* objects*;*
- *L is a finite alphabet of* labels *associated with compartments;*
- *μ is a membrane structure containing $n \geq 1$ membranes labelled by elements from L;*
- *$M_i = (l_i, w_i, s_i)$, for each $1 \leq i \leq n$, is the initial configuration of membrane i, with $l_i \in L$, the label of this membrane, $w_i \in O^*$, a finite multiset of objects and s_i, a finite set of strings over O;*
- *R_i, for each $1 \leq i \leq n$, is a finite set of rewriting rules associated with membrane labelled i, having one of the following two forms:*
  - *Multiset rewriting rules:*

$$obj_1 \, [\, obj_2 \,]_l \xrightarrow{\ k\ } obj_1' \, [\, obj_2' \,]_l$$

with $obj_1, obj_2, obj_1', obj_2' \in O^*$ some finite multisets of objects and l a label from L. A multiset of objects, obj is represented as $obj = o_1 + \ldots + o_m$ with $o_1, \ldots, o_m \in O$.

  These multiset rewriting rules are applicable on both sides of each membrane; a multiset obj_1 which is outside a membrane l and a multiset obj_2 placed inside the same membrane can simultaneously be rewritten by a multiset obj_1' and a multiset obj_2', respectively.

- *String rewriting rules:*

$$[\, obj_1 + str_1; \ldots ; obj_p + str_p \,]_l \xrightarrow{k}$$

$$[\, obj'_1 + str'_{1,1} + \ldots str'_{1,i_1}; \ldots ; obj'_p + str'_{p,1} + \ldots str'_{p,i_p} \,]_l$$

A string str is represented as follows $str = \langle s_1.s_2.\cdots.s_i \rangle$ where $s_1, \ldots, s_i \in O$. In this case each multiset of objects obj_j and string str_j, $1 \le j \le p$, are replaced by a multiset of objects obj'_j and strings $str'_{j,1} \ldots str'_{j,i_j}$.

The stochastic constant k is used to compute the propensity of the rule by multiplying it by the number of distinct possible combinations of the objects and substrings that occur on the left-side of the rule with respect to the current contents of membranes involved in the rule. The propensity associated with each rule is further utilised in generating the probability of the rule and time necessary to execute it.

Stochastic P systems are utilised to specify cellular systems consisting of molecular interactions taking place in different locations of living cells. Different regions and compartments are represented by membranes. Each molecular species is an object in the multiset associated with the region or compartment where the molecule is located. Strings are utilised to specify the genetic information encoded by DNA and RNA. Molecular interactions, compartment translocation and gene expression are specified using rewriting rules on multisets of objects and strings.

In stochastic P systems [21] constants are associated with rules in order to compute their probabilities and time needed to be applied according to Gillespie algorithm. This approach is based on a Monte Carlo algorithm for stochastic simulation of molecular interactions taking place inside a single volume or across multiple compartments [6].

In order to construct and analyse a stochastic P system model this can be translated into an adequate model checker like PRISM. This has its own language, a simple, high level, state-based language. The fundamental components of the PRISM language are modules, variables and commands. Each model is composed of a number of modules which can interact with each other. A module contains a number of local variables and commands utilised to specify certain behaviour.

The variables are utilised to keep verious values that constitute the states of the module. The space of reachable states is computed using the range of each variable and its initial value. The global state of the whole model is determined by the local state of all modules.

A command defining some behaviour within a module has the following form:

$$[\ \ \texttt{action}\ \ ]\ g \to \lambda_1 : u_1 + \cdots + \lambda_n : u_n;$$

The guard g is a predicate over all the variables of the model. Each update u_i describes the new values of the variables in the module specifying a transition of the module. The expressions λ_i are used to compute probabilities associated to transitions.

The label **action** placed inside the square brackets are used to synchronise different commands spread across the system. The rate of the transition resulting in this case is equal to the product of the individual rates, since the processes involved are assumed to be independent events.

There is a straightforward way of mapping stochastic P systems into PRISM representation. Compartments are mapped into modules, objects are translated as local variables, with some initial values, and rules are transcribed as transitions. For instance, a rule like

$$obj_1\,[\,obj_2\,]_l \xrightarrow{\ k\ } obj_1'\,[\,obj_2'\,]_l$$

where $obj_1 = obj_1' = \lambda$ and $obj_2 = a_1 + \cdots + a_p$, $obj_2' = b_1 + \cdots + b_q$, is translated into the following PRISM code given that all symbols are distinct

$$[]\ \mathbf{a_1 > 0 \&\cdots\&a_p > 0 \to k:}$$
$$\mathbf{(a_1' = a_1 - 1)\&\cdots\&(a_p' = a_p - 1)\&(b_1' = b_1 + 1)\&\cdots\&(b_q' = b_q - 1)}.$$

After an entire translation of the P system into PRISM specification language is obtained, various properties of the system can be formulated into adequate logics and simulations and verifications of certain properties can be checked.

3   P Systems Testing

All software applications, irrespective of their use and purpose, are tested before being released, installed and used. Testing is not a replacement for formal verification, it is a necessary mechanism to increase the confidence in software correctness and to make sure it works properly. Although formal verification, as previously discussed, has been applied for different P system models, testing has been neglected until [14], [16]. In the sequel it is presented a testing framework and its underpinning theory which based on formal grammars and finite state machines. We develop this testing theory based on formal grammars and finite state machines because these models of computation are the closest formalisms to P systems and testing approaches for them have been very well developed.

We will consider basic P systems in this section.

Definition 2. *A P system is a tuple* $\Pi = (V, \mu, w_1, ..., w_n, R_1, ..., R_n)$, *where*

- *V is a finite set, called* alphabet;
- *μ defines the membrane structure; a hierarchical arrangement of n compartments called* regions *delimited by* membranes; *these membranes and regions are identified by integers 1 to n;*
- *w_i, $1 \leq i \leq n$, represents the initial multiset occurring in region i;*
- *R_i, $1 \leq i \leq n$, denotes the set of rules applied in region i.*

The rules in each region have the form $a \rightarrow (a_1, t_1)...(a_m, t_m)$, where $a, a_i \in V$, $t_i \in \{in, out, here\}$, $1 \le i \le m$. When such a rule is applied to a symbol a in the current region, the symbol a is replaced by the symbols a_i which stays in this region if $t_i = here$; symbols a_i are sent to the outer region, when $t_i = out$, and symbols a_i, with $t_i = in$, are sent into one of the regions contained in the current one, arbitrarily chosen. In the following definitions and examples all the symbols $(a_i, here)$ are used as a_i, i.e., $here$ destination will be removed. The rules are applied in the maximally parallel mode which means that they are used in all the regions in the same time and in each region all symbols that may be processed, must be.

A configuration of the P system Π is a tuple $c = (u_1, ..., u_n)$, $u_i \in V^*$, $1 \le i \le n$. A derivation of a configuration c_1 to c_2 using the maximal parallelism mode is denoted by $c_1 \Longrightarrow c_2$. We will distinguish terminal configurations, $c = (u_1, ..., u_n)$, as being configurations where no u_i can be further processed.

The set of all halting configurations is denoted by $L(\Pi)$, whereas the set of all configurations reachable from the initial one (including the initial configuration) is denoted by $S(\Pi)$.

Definition 3. *A* deterministic finite automaton *(abbreviated* DFA*), M, is a tuple (A, Q, q_0, F, h), where:*

- *A is the finite* input *alphabet;*
- *Q is the finite set of states;*
- *$q_0 \in Q$ is the* initial *state;*
- *$F \subseteq Q$ is the set of final states;*
- *$h : Q \times A \longrightarrow Q$ is the* next-state *function.*

The next-state function h can be extended to a function $h : Q \times A^* \longrightarrow Q$ defined by:

- $h(q, \epsilon) = q$, $q \in Q$;
- $h(q, sa) = h(h(q, s), a)$, $q \in Q$, $s \in A^*$, $a \in A$.

For simplicity the same name h is used for the next-state function and for the extended function.

Given $q \in Q$, a sequence of input symbols $s \in A^*$ is said to be accepted by M in q if $h(q, s) \in F$. The set of all input sequences accepted by M in q_0 is called the *language defined (accepted) by* M, denoted $L(M)$.

3.1 Grammar-Like Testing

In *grammar engineering*, formal grammars are used to specify complex software systems, like compilers, debuggers, documentation tools, code pre-processing tools etc. One of the areas of grammar engineering is *grammar testing* which covers the development of various testing strategies for software based on grammar specifications. One of the main testing methods developed in this context refers to rule coverage, i.e., the testing procedure tries to cover all the rules of a specification [14].

In the context of grammar testing it is assumed that for a given specification defined as a grammar, an implementation of it exists and this will be tested. In order to test the implementation, a test set is built, as a finite set of sequences, that reveals potential errors. As opposed to testing based on finite state machines, where it is possible to prove that the specification and implementation, have the same behaviour, in the case of general context-free grammars this is no longer possible as it reduces to the equivalence of two such devices, which is not decidable. Of course, for specific restricted classes of context-free grammars there are decidability procedures regarding the equivalence problem and these may be considered for testing purposes as well. The best we can get is to cover as much as possible from the languages associated to the two mechanisms, and this is the role of a test set.

Although there are similarities between context-free grammars utilised in grammar testing and basic P systems, like those considered in this section, there are also major differences that pose new problems in defining testing methods and strategies. Some of the difficulties that we encounter in introducing some grammar-like testing procedures are related to: the hierarchical compartmentalisation of the entire model, parallel behaviour, communication mechanisms, the lack of a non-terminal alphabet and the use of multisets of objects instead of sets of strings.

We define some rule coverage criteria by firstly starting with one compartment P system, i.e., $\Pi = (V, \mu, w, R)$, where $\mu = [_1]_1$. In the sequel, if not otherwise stated, we will consider that the specification and the implementation are given by the P systems Π and Π', respectively. For such a P system Π, we define the following concepts.

Definition 4. *A multiset denoted by $u \in V^*$, covers a rule $r : a \to v \in R$, if there is a derivation $w \Longrightarrow^* xay \Longrightarrow x'vy' \Longrightarrow^* u$; $w, x, y, v, u \in V^*$, $a \in V$.*

Definition 5. *A set $T \subseteq V^*$, is called a* test set *that satisfies the* rule coverage *(RC) criterion if for each rule $r \in R$ there is $u \in T$ which covers r.*

The above criterion can be defined for terminal derivations as well and we get a new type of test set.

The following one compartment P systems are considered, $\Pi_i, 1 \leq i \leq 4$, having the same alphabet and initial multiset [14]:

$$\Pi_i = (V_i, \mu_i, w_i, R_i)$$

where

- $V_1 = V_2 = V_3 = V_4 = \{s, a, b, c\}$;
- $\mu_1 = \mu_2 = \mu_3 = \mu_4 = [_1]_1$ - i.e., one compartment, denoted by 1;
- $w_1 = w_2 = w_3 = w_4 = s$;
- $R_1 = \{r_1 : s \to ab, r_2 : a \to c, r_3 : b \to bc, r_4 : b \to c\}$;
- $R_2 = \{r_1 : s \to ab, r_2 : a \to \lambda, r_3 : b \to c\}$;
- $R_3 = \{r_1 : s \to ab, r_2 : a \to bcc, r_3 : b \to \lambda\}$;
- $R_4 = \{r_1 : s \to ab, r_2 : a \to bc, r_3 : a \to c, r_4 : b \to c\}$.

In the sequel for each multiset w, we will use the following vector of non-negative integer numbers $(|w|_s, |w|_a, |w|_b, |w|_c)$.

The sets of all configurations expressed as vectors of non-negative integer numbers, computed by the P systems Π_i, $1 \leq i \leq 4$ are:

- $S(\Pi_1) = \{(1,0,0,0), (0,1,1,0)\} \cup \{(0,0,k,n)|k = 0,1; n \geq 2\}$;
- $S(\Pi_2) = \{(1,0,0,0), (0,1,1,0), (0,0,0,1)\}$;
- $S(\Pi_3) = \{(1,0,0,0), (0,1,1,0), (0,0,1,2), (0,0,0,2)\}$;
- $S(\Pi_4) = \{(1,0,0,0), (0,1,1,0), (0,0,1,2), (0,0,0,2), (0,0,0,3)\}$.

Test sets for Π_1 satisfying the RC criterion are

- $T_{1,1} = \{(0,1,1,0), (0,0,1,2), (0,0,0,2)\}$ and
- $T_{1,2} = \{(0,1,1,0), (0,0,1,2), (0,0,0,3)\}$,

whereas $T'_{1,1} = \{(0,1,1,0), (0,0,0,2)\}$ and $T'_{1,2} = \{(0,1,1,0), (0,0,1,2)\}$ are not, as they do not cover the rules r_3 and r_4, respectively.

If we consider Π_1 a specification with test sets $T_{1,1}$ and $T_{1,2}$, then we observe that Π_2 fail to pass $T_{1,1}$ and $T_{1,2}$, and Π_3 fails on and $T1, 2$. Hence, these are faulty implementations and the errors are revealed by the test sets above. Π_4 instead, although is not correct, passes both tests. In this case a more powerful critrion is needed [14].

3.2   Finite State Machine Based Testing

We first present the process of constructing a DFA for one compartment P system. Let $\Pi = (V, \mu, w, R)$, where $\mu = [_1]_1$ be such a system. In this case, the configuration of Π can change as a result of the application of some rule in R or of a number of rules, in parallel. In order to guarantee the finiteness of this process, for a given integer k, only computations of maximum k steps will be considered. For example, for $k = 4$, the tree in Figure 1 depicts all derivations in Π_1 of length less than or equal to k. The terminal nodes are in bold.

As only sequences of maximum k steps are considered, for every rule $r_i \in R$ there will be some N_i such that, in any step, r_i can be applied at most N_i times. Thus, the tree that depicts all the derivations of a P system Π with rules $R = \{r_1, \ldots, r_m\}$ can be described by a DFA Dt over the alphabet $A = \{r_1^{i_1} \ldots r_m^{i_m} \mid 0 \leq i_1 \leq N_1, \ldots, 0 \leq i_m \leq N_m\}$, where $r_1^{i_1} \ldots r_m^{i_m}$ describes the multiset with i_j occurrences of r_j, $1 \leq j \leq m$.

As Dt is a DFA over A, one can construct the minimal DFA that accepts *precisely* the language $L(Dt)$ defined by Dt. However, as only sequences of at most k transitions are considered, it is irrelevant how the constructed automaton will behave for longer sequences. Thus, a finite cover automaton can be constructed instead.

A *deterministic finite cover automaton (DFCA)* of a finite language U is a DFA that accepts all sequences in U and possibly other sequences that are longer than any sequence in U.

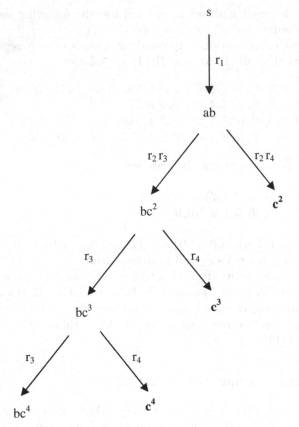

Fig. 1. Derivation tree for Π_1 and $k = 4$

Definition 6. *Let $M = (A, Q, q_0, F, h)$ be a DFA, $U \subseteq A^*$ a finite language and l the length of the longest sequence(s) in U. Then M is called a* deterministic finite cover automaton *(DFCA) of U if $L(A) \cap A[l] = U$, where $A[l] = \bigcup_{0 \leq i \leq l} U^i$ denotes the sets of sequences of length less than or equal to l with members in the alphabet A.*

A *minimal* DFCA of U is a DFCA of U having the least number of states. Unlike the case in which the acceptance of the precise language is required, the minimal DFCA is not necessarily unique (up to a renaming of the state space).

Any DFA that accepts U is also a DFCA of U and so the size (number of states) of a minimal DFCA of U cannot exceed the size of the minimal DFA that accepts U. On the other hand, as shown by examples in this paper, a minimal DFCA of U may have considerably fewer states than the minimal DFA that accepts U.

A minimal DFCA of the language $L(Dt)$ defined by the previous derivation tree is represented in Figure 2; q_3 in Figure 2 is final state. It is implicitly assumed that a non-final "sink" state, denoted q_S, also exists, that receives all

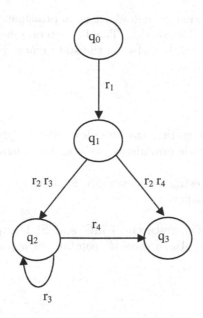

Fig. 2. Minimal DFCA for Π_1 and $k = 4$

"rejected" transitions. For testing purposes we will consider all the states as final. For details see [14].

Once the minimal DFCA $M = (A, Q, q_0, F, h)$ has been constructed, various specific coverage levels can be used to measure the effectiveness of a test set. In this paper we use two of the most widely known coverage levels for finite automata: *state coverage* and *transition coverage*.

Definition 7. *A set* $T \subseteq V^*$, *is called a* test set *that satisfies the* state coverage *(SC) criterion if for each state q of M there exists $u \in T$ and a path $s \in A^*$ that reaches q $(h(q_0, s) = q)$ such that u is derived from w through the computation defined by s.*

Definition 8. *A set* $T \subseteq V^*$, *is called a* test set *that satisfies the* transition coverage *(TC) criterion if for each state q of M and each $a \in A$ such that a labels a valid transition from q $(h(q, a) \neq q_S)$, there exist $u, u' \in T$ and a path $s \in A^*$ that reaches q such that u and u' are derived from w through the computation defined by s and sa, respectively.*

Clearly, if a test set satisfies TC, it also satisfies SC. A test set for Π_1 satisfying the SC criterion is

$$T_{1,1} = \{(1,0,0,0), (0,1,1,0), (0,0,1,2), (0,0,0,2)\},$$

whereas a test set satisfying the TC criterion is

$$T_{1,s} = \{(1,0,0,0), (0,1,1,0), (0,0,1,2), (0,0,0,2), (0,0,1,3), (0,0,0,3)\}.$$

The TC coverage criterion defined above is, in principle, analogous to the RC criterion given in the previous section. The TC criterion, however, does not only depend on the rules applied, but also on the state reached by the system when a given rule has been applied.

4   Conclusions

In this paper are reviewed certain aspects of formally verifying properties of some classes of P systems. Non-deterministic and stochastic classes are presented and discussed.

Testing is another investigation developed for basic classes of P systems and briefly analysed in this paper.

Acknowledgements. The research of MG and FI is supported by CNCSIS grant IDEI no.643/2009, *An integrated evolutionary approach to formal modelling and testing.*

References

1. Aguado, J., Bălănescu, T., Cowling, A., Gheorghe, M., Holcombe, M., Ipate, F.: P systems with replicated rewriting and stream X-machines (Eilenberg machines). Fundamenta Informaticae 49, 17–33 (2002)
2. Aman, B., Ciobanu, G.: Translating mobile ambients into P systems. Electronic Notes in Theoretical Computer Science 171, 11–23 (2007)
3. Andrei, O., Ciobanu, G., Lucanu, D.: Executable specifications of P systems. In: Mauri, G., Păun, Gh., Jesús Pérez-Jímenez, M., Rozenberg, G., Salomaa, A. (eds.) WMC 2004. LNCS, vol. 3365, pp. 126–145. Springer, Heidelberg (2005)
4. Barbuti, R., Maggiolo-Schettini, A., Milazzo, P., Tini, S.: Compositional semantics and behavioral equivalences for P systems. Theoretical Computer Science 395, 77–100 (2008)
5. Andrei, O., Ciobanu, G., Lucanu, D.: A rewriting logic framework for operational semantics of membrane systems. Theoretical Computer Science 373, 163–181 (2007)
6. Bernardini, F., Gheorghe, M., Romero-Campero, R., Walkinshaw, N.: Hybrid approach to modeling biological systems. In: Eleftherakis, G., Kefalas, P., Păun, Gh., Rozenberg, G., Salomaa, A. (eds.) WMC 2007. LNCS, vol. 4860, pp. 138–159. Springer, Heidelberg (2007)
7. Cavaliere, M.: Computing by observing: A brief survey. In: Beckmann, A., Dimitracopoulos, C., Löwe, B. (eds.) CiE 2008. LNCS, vol. 5028, pp. 110–119. Springer, Heidelberg (2008)
8. Cavaliere, M., Mardare, R.: Partial knowledge in membrane systems: A logical approach. In: Hoogeboom, H.J., Păun, Gh., Rozenberg, G., Salomaa, A. (eds.) WMC 2006. LNCS, vol. 4361, pp. 279–297. Springer, Heidelberg (2006)
9. Ciobanu, G.: Semantics of P Systems. In: Păun, Gh., Rozenberg, G., Salomaa, A. (eds.) Handbook of membrane computing, ch. 16, pp. 413–436. Oxford University Press, Oxford (to appear)
10. Dang, Z., Ibarra, O.H., Li, C., Xie, G.: Decidability of model-checking P systems. Journal of Automata, Languages and Combinatorics 11, 179–198 (2006)

11. Díaz-Pernil, D., Graciani, C., Gutiérrez-Naranjo, M.A., Pérez-Hurtado, I., Pérez-Jiménez, M.J.: Software for P systems. In: Păun, Gh., Rozenberg, G., Salomaa, A. (eds.) Handbook of membrane computing, ch. 17, pp. 437–454. Oxford University Press, Oxford (to appear)
12. Frisco, P.: P systems, Petri nets, and program machines. In: Freund, R., Păun, G., Rozenberg, G., Salomaa, A. (eds.) WMC 2005. LNCS, vol. 3850, pp. 209–223. Springer, Heidelberg (2006)
13. Frisco, P., Corne, D.W.: Dynamics of HIV infection studied with cellular automata and conformon-P systems. BioSystems 91, 531–544 (2008)
14. Gheorghe, M., Ipate, F.: On testing P systems. In: Corne, D.W., Frisco, P., Paun, G., Rozenberg, G., Salomaa, A. (eds.) WMC 2008. LNCS, vol. 5391, Springer, Heidelberg (2009)
15. Hinton, A., Kwiatkowska, M., Norman, G.: PRISM – A tool for automatic verification of probabilistic systems. In: Hermanns, H., Palsberg, J. (eds.) TACAS 2006. LNCS, vol. 3920, pp. 441–444. Springer, Heidelberg (2006)
16. Ipate, F., Gheorghe, M.: Testing non-deterministic stream X-machine models and P systems. Electronic Notes in Theoretical Computer Science 227, 113–226 (2008)
17. Kleijn, J., Koutny, M.: Petri nets and membrane computing. In: Păun, G., Rozenberg, G., Salomaa, A. (eds.) Handbook of membrane computing, ch. 15, pp. 389–412. Oxford University Press, Oxford (to appear)
18. Păun, Gh.: Computing with membranes. Journal of Computer and System Sciences 61, 108–143 (2000)
19. Păun, Gh., Rozenberg, G.: A guide to membrane computing. Theoretical Computer Science 287, 73–100 (2002)
20. Păun, Gh.: Membrane Computing. An Introduction. Springer, Berlin (2002)
21. Pérez-Jiménez, M.J., Romero-Campero, F.: P systems, a new computational modelling tool for systems biology. In: Priami, C., Plotkin, G. (eds.) Transactions on Computational Systems Biology VI. LNCS (LNBI), vol. 4220, pp. 176–197. Springer, Heidelberg (2006)
22. Şerbănuţă, T., Ştefănescu, Gh., Roşu, G.: Defining and executing P systems with structured data in K. In: Corne, D.W., Frisco, P., Paun, G., Rozenberg, G., Salomaa, A. (eds.) WMC 2008. LNCS, vol. 5391, pp. 374–393. Springer, Heidelberg (2009)
23. Stamatopoulou, I., Kefalas, P., Gheorghe, M.: Transforming state-based models to P systems models in practice. In: Corne, D.W., Frisco, P., Paun, G., Rozenberg, G., Salomaa, A. (eds.) WMC 2008. LNCS, vol. 5391, pp. 260–273. Springer, Heidelberg (2009)
24. Vitale, A., Mauri, G., Zandron, C.: Simulation of a bounded symport antiport P system with brane calculi. Biosytems 91, 558–571 (2008)

A Look Back at Some Early Results in Membrane Computing*

Oscar H. Ibarra

Department of Computer Science
University of California
ibarra@cs.ucsb.edu

Abstract. On this tenth anniversary of the Workshop on Membrane Computing, it seems appropriate and fitting to look back at some early basic contributions in the area. We give a brief summary of results, some of which answered fundamental open questions in the field. These concern complexity issues such as universality versus non-universality, determinism versus nondeterminism, various notions of parallelism, membrane and alphabet-size hierarchies, and characterizations of some classes of P systems.

1 Introduction

There have been tremendous research activities in the area of membrane computing initiated by Gheorghe Păun in a seminal paper [13] ten years ago (see also [14]). Membrane computing identifies an unconventional computing model, namely a P system, from natural phenomena of cell evolutions and chemical reactions. Due to the built-in nature of maximal parallelism inherent in the model, P systems have a great potential for implementing massively concurrent systems in an efficient way that would allow us to solve currently intractable problems in much the same way as the promise of quantum and DNA computing, once future bio-technology (or silicon-technology) gives way to a practical bio-realization (or chip-realization).

A P system is a computing model, which abstracts from the way the living cells process chemical compounds in their compartmental structure. The regions defined by a membrane structure contain objects that evolve according to specified rules. The objects can be described by symbols or by strings of symbols, and multisets of these objects are placed in the regions of the membrane structure. The membranes themselves are organized as a Venn diagram or a tree structure where one membrane may contain other membranes. By using the rules in a nondeterministic, maximally parallel manner, transitions between the system configurations can be obtained. A sequence of transitions shows how the system is evolving. Various ways of controlling the transfer of objects from a region to another and applying the rules, as well as possibilities to dissolve, divide or create membranes have been studied. P systems were introduced with the goal

---

* This research was supported in part by NSF Grant CCF-0524136.

G. Păun et al. (Eds.): WMC 2009, LNCS 5957, pp. 66–73, 2010.

to abstract a new computing model from the structure and the functioning of the living cell (as a branch of the general effort of Natural Computing – to explore new models, ideas, paradigms from the way nature computes). Membrane computing has been very successful: many models have been introduced, most of them Turing complete and/or able to solve computationally intractable problems (NP-complete, PSPACE-complete) in a feasible time, by trading space for time; development of software and simulations; proposals for various potential applications. See the P system website at http://ppage.psystems.eu/ for a large collection of papers in the area, and in particular the monograph [15].

On this tenth anniversary of the Workshop on Membrane Computing, it seems appropriate and fitting to look back at some early basic contributions in the area. In this talk, we will give a brief summary of results (mostly by the author and his collaborators: Zhe Dang, Andrei Păun, Gheorghe Păun, Hsu-Chun Yen, Sara Woodworth), some of which answered fundamental open questions in the field. These concern complexity issues such as universality versus non-universality, determinism versus nondeterminism, various notions of parallelism, membrane and alphabet-size hierarchies, and characterizations of some classes of P systems. These investigations into complexity issues in membrane computing are natural and interesting from the points of view of foundations and applications, e.g., in modeling and simulating of cells.

2   Hierarchies in P Systems

2.1   The First Membrane Hierarchy Result

An important open problem that was raised early on in the membrane computing research community (see [15]) was whether one can exhibit a non-universal (i.e., non-Turing-complete) model of a membrane system for which the number of membranes induces an infinite hierarchy on the computations that can be performed by such systems. This question was affirmatively answered in [5]. The basic model investigated in [5] is a restricted model of a communicating P system (CPS)[17], called an RCPS. The environment of an RCPS does not contain any object initially. The system can expel objects into the environment but only expelled objects can be retrieved from the environment. Such a system is initially given an input $a_1^{i_1}...a_n^{i_n}$ (with each i_j representing the multiplicity of distinguished object a_i, $1 \leq i \leq n$) and is used as an acceptor. An RCPS is equivalent to a two-way multihead finite automaton operating on bounded languages (i.e., the inputs, with left and right end markers ¢ and \$, come from $a_1^*...a_n^*$ for some distinct symbols $a_1,...,a_n$). It was shown in [5] that there is an infinite hierarchy of RCPS's in terms of the number of membranes: For every r, there is an $s > r$ and a unary language L accepted by an RCPS with s membranes that cannot be accepted by an RCPS with r membranes. Partial solutions to the membrane hierarchy problem were previously given in [2,11] which, however, were based on definitions that were considered too restrictive (hence not "completely innocent"), so [15] considered the hierarchy problem still open. The solution presented in [5] (the RCPS model and the hierarchy proof) is considered as convincingly answering the open problem. Some variants/generalizations of RCPS's that also form an infinite hierarchy with respect to the number of membranes were considered in [5]. In particular, a model of a CPS was proposed that can be used as an acceptor of languages (sets of strings), called CPSA. A CPSA can have abundant (i.e., infinite) supply of some objects in the environment. CPSA's accept precisely the recursively enumerable languages. A characterization of a special case when the CPSA is restricted to use only a polynomial (on the length n of the input string) amount of objects from the environment was shown in [5]: For any positive integers k and r, there is an $s > r$ and a language L that can be accepted by an n^k-CPSA with s membranes that cannot be accepted by any n^k-CPSA with r membranes.

2.2   Hierarchies for Symport/Antiport Systems

A restricted model of a one-membrane symport/antiport system [12] called bounded S/A system was studied in [8]. The rules are of the form: $(u, out; v, in)$, where u, v are strings representing multisets of objects (i.e., symbols) with the restriction that $|u| = |v| \geq 1$. (Note that only the multiplicities of the objects are of interest.) An input $z = a_1^{n_1} \ldots a_k^{n_k}$ (each n_i a nonnegative integer) is accepted if the system when started with wz, where w is a fixed string independent of z and not containing a_i $(1 \leq i \leq k)$ eventually halts. The following results were shown in [8]:

1. A language $L \subseteq a_1^* \ldots a_k^*$ is accepted by a bounded S/A system if and only if it is accepted by a *log n* space-bounded Turing machine. This holds for both deterministic and nondeterministic versions.
2. For every positive integer r, there is an $s > r$ and a unary language L that is accepted by a bounded S/A system with s objects that cannot be accepted by any bounded S/A system with only r objects. This holds for both deterministic and nondeterministic versions.
3. Deterministic and nondeterministic bounded S/A systems over a unary input alphabet are equivalent if and only if deterministic and nondeterministic linear-bounded automata (over an arbitrary input alphabet) are equivalent.

Also studied were multi-membrane S/A systems, called special S/A systems. They are restricted in that only rules of the form $(u, out; v, in)$, where $|u| = |v| \geq 1$, can appear in the skin membrane. Thus, the number of objects in the system during the computation remains the same. Let E be the alphabet of symbols in the environment (note that there may be other symbols in the system that are not transported into the environment and, therefore, not included in E). It was shown that for every nonnegative integer t, special S/A systems with environment alphabet E of t symbols has an infinite hierarchy in terms of the number of membranes. Again, this holds for both deterministic and nondeterministic versions. Also introduced in [8] is a model of a one-membrane bounded S/A system that accepts string languages and show that the deterministic version is strictly weaker than the nondeterministic version.

The results above answer some important open questions in the field (see, e.g., [16]).

3 Determinism versus Nondeterminism in P Systems

3.1 Nonuniversality of Deterministic Catalytic Systems

In the standard semantics of P systems [14,15], each evolution step of a system $\mathcal{P}$ is a result of applying all the rules in $\mathcal{P}$ in a maximally parallel manner. More precisely, starting from the initial configuration, w, the system goes through a sequence of configurations, where each configuration is derived from the directly preceding configuration in one step by the application of a multiset of rules, which are chosen nondeterministically. For example, a catalytic rule $Ca \rightarrow Cv$ in membrane m is applicable if there is a catalyst C and an object (symbol) a in the preceding configuration in membrane m. The result of applying this rule is the evolution of v from a. If there is another occurrence of C and another occurrence of a, then the same rule or another rule with Ca on the left hand side can be applied. Thus, in general, the number of times a particular rule is applied at anyone step can be unbounded. We require that the application of the rules is maximal: all objects, from all membranes, which *can be* the subject of local evolution rules *have to* evolve simultaneously. Configuration z is reachable (from the starting configuration) if it appears in some execution sequence; z is halting if no rule is applicable on z.

Two popular models of P systems are the catalytic system [14] and the symport/antiport system [12]. An interesting subclass of the latter was studied in [3] – each system is *deterministic* in the sense that the computation path of the system is unique, i.e., at each step of the computation, the maximal multiset of rules that is applicable is unique. It was shown in [3] that any recursively enumerable unary language $L \subseteq o^*$ can be accepted by a deterministic 1-membrane symport/antiport system. Thus, for symport/antiport systems, the deterministic and nondeterministic versions are equivalent and they are universal. It also follows from the construction in [17] that for another model of P systems, called communicating P systems, the deterministic and nondeterministic versions are equivalent as both can accept any unary recursively enumerable language. However, the deterministic-versus-nondeterministic question was left open in [3] for the class of catalytic systems (these systems have rules of the form $Ca \rightarrow Cv$ or $a \rightarrow v$), where the proofs of universality involved a high degree of parallelism [17,4]. For a discussion of this open question and its importance, see [1,16]. This question was later resolved in the negative in [9]. Since nondeterministic catalytic systems are universal, this result also gave the first example of a P system for which the nondeterministic version is universal, but the deterministic version is not.

For a catalytic system serving as a *language acceptor*, the system starts with an initial configuration wz, where w is a fixed string of catalysts and noncatalysts not containing any symbol in z, and $z = a_1^{n_1} \ldots a_k^{n_k}$ for some nonnegative integers $n_1, \ldots, n_k$, with $\{a_1, \ldots, a_k\}$ a distinguished subset of noncatalyst symbols (the input alphabet). At each step, a maximal multiset of rules are nondeterministically selected and applied in parallel to the current configuration to derive the next configuration (note that the next configuration is not unique, in general).

The string z is accepted if the system eventually halts. Unlike nondeterministic 1-membrane catalytic system acceptors (with 2 catalysts) which are universal, we are able to show using a graph-theoretic approach that the Parikh map of the language $\subseteq a_1^* \dots a_k^*$ accepted by any deterministic catalytic system is a simple semilinear set which can also be effectively constructed. Our result gives the first example of a P system for which the nondeterministic version is universal, but the deterministic version is not. For deterministic 1-membrane catalytic systems using only rules of type $Ca \rightarrow Cv$, we show the set of reachable configurations from a given initial configuration to be effective semilinear. In contrast, the reachability set is no longer semilinear in general if rules of type $a \rightarrow v$ are also used. Our result generalizes to multi-membrane catalytic systems. Determinisic catalytic systems which allow rules to be prioritized were also considered in [9]. Three such systems, namely, *totally prioritized*, *strongly prioritized* and *weakly prioritized* catalytic systems, were investigated, and the question of whether such systems are universal were answered in [9].

3.2   Determinism versus Nondeterminism for Nonuniversal Systems

We also mention that [6] exhibited two interesting restricted classes of communicating P systems where the following results were shown:

1. For the first class, the deterministic and nondeterministic versions are equivalent if and only if deterministic and nondeterministic linear bounded automata are equivalent. The latter problem is a long-standing open question in complexity theory.
2. For the second class, the deterministic version is strictly weaker than the nondeterministic version.

Both classes are nonuniversal, but can accept fairly complex languages. Similar results were later obtained for restricted classes of symport/antiport P systems [8]

4   Characterizations of Context-Sensitive and Other Families of Languages

A natural and fundamental question was considered in [7]: Can we characterize families of languages (other than the recursively enumerable languages) by means of of classes of symport/antiport P systems? The problem is trivial for regular languages, but challenging for other families from Chomsky hierarchy, in particular, for the families of context-free and context-sensitive languages. In [7], "syntactic" characterizations of context-sensitive languages (CSLs) in terms of some restricted models of symport/antiport P systems were given. These were the first such characterizations of CSLs in terms of P systems. In particular, the following were shown for any language L over a binary alphabet:

1. Let m be any integer ≥ 1. Then L is a CSL if and only if it can be accepted by a restricted symport/antiport P system with m membranes and multiple number of symbols (objects). Moreover, holding the number of membranes at m, there is an infinite hierarchy in computational power (within the class of binary CSLs) with respect to the number of symbols.
2. Let s be any integer ≥ 14. Then L is a CSL if and only if it can be accepted by a restricted symport/antiport P system with s symbols and multiple number of membranes. Moreover, holding the number of symbols at s, there is an infinite hierarchy in computational power with respect to the number of membranes.

(Similar results hold for languages over an alphabet of $k \geq 2$ symbols.) Thus (1) and (2) say that in order for the restricted symport/antiport P systems to accept all binary CSLs, at least one parameter (either the number of symbols or the number of membranes) must grow. These were the first results of their kind in the P systems area. They contrast a known result that (unrestricted) symport/antiport P systems with $s \geq 2$ symbols and $m \geq 1$ membranes accept (or generate) exactly the recursively enumerable sets of numbers even for $s+m = 6$. We also note that previous characterizations of formal languages in the membrane computing literature are mostly for the Parikh images of languages. Variations of the model yield characterizations of regular languages, languages accepted by one-way $\log n$ space-bounded Turing machines, and recursively enumerable languages.

5   Three Notions of Parallelism in P Systems

Let G be a P systems, and consider the following definition (different from the standard definition in the literature) of "maximal parallelism" in the application of evolution rules: Let $R = \{r_1, \ldots r_k\}$ be the set of (distinct) rules in the system. G operates in maximally parallel mode if at each step of the computation, a maximal subset of R is applied, and at most one instance of any rule is used at every step (thus at most k rules are applicable at any step). We refer to this system as a maximally parallel system. In [10], the computing power of P systems under three semantics of parallelism were investigated. For a positive integer $n \leq k$, define:

n-**Max-Parallel:** At each step, nondeterministically select a maximal subset of at most n rules in R to apply (this implies that no larger subset is applicable).

$\leq n$-**Parallel:** At each step, nondeterministically select any subset of at most n rules in R to apply.

n-**Parallel:** At each step, nondeterministically select any subset of exactly n rules in R to apply.

In all three cases, if any rule in the subset selected is not applicable, then the whole subset is not applicable. When $n = 1$, the three semantics reduce to the **Sequential** mode.

The focus in [10] were two models of P systems: multi-membrane catalytic systems and communicating P systems (though the results hold for other systems, like the symport/antiport systems). It was shown that for these systems, n-**Max-Parallel** mode is strictly more powerful than any of the following three modes: **Sequential**, $\leq n$-**Parallel**, or n-**Parallel**. As a corollary, a maximally parallel communicating P system is universal for $n = 2$. However, under the three limited modes of parallelism, the system is equivalent to a vector addition system, which is known to only define a recursive set. This shows that "maximal parallelism" is key for the model to be universal. Some of the results are rather surprising. For example, it was shown in [10] that a **Sequential** 1-membrane communicating P system can only generate a semilinear set, whereas with k membranes, it is equivalent to a vector addition system for any $k \geq 2$ (thus the hierarchy collapses at 2 membranes - a rare collapsing result for nonuniversal P systems). Another proof (using vector addition systems) of the known result [4] that a 1-membrane catalytic system with only 3 catalysts and (non-prioritized) catalytic rules operating under 3-**Max-Parallel** mode can simulate any 2-counter machine M. Unlike in [4], the catalytic system needs only a *fixed* number of noncatalysts, independent of M.

A simple cooperative system (SCO) is a P system where the only rules allowed are of the form $a \rightarrow v$ or of the form $aa \rightarrow v$, where a is a symbol and v is a (possibly null) string of symbols not containing a. It show that a 9-**Max-Parallel** 1-membrane SCO is universal.

6   Conclusion

We gave a short summary of early results in membrane computing, some of which answered fundamental open questions in the field These concerned complexity issues such as universality versus non-universality, determinism versus nondeterminism, various notions of parallelism, membrane and alphabet-size hierarchies, and characterizations of some classes of P systems.

References

1. Calude, C.S., Păun, G.: Computing with Cells and Atoms: After Five Years. Taylor and Francis, London (2001); Pushchino Publishing House (2004)
2. Freund, R.: Special variants of P systems inducing an infinite hierarchy with respect to the number of membranes. Bulletin of the EATCS (75), 209–219 (2001)
3. Freund, R., Păun, Gh.: On deterministic P systems (2003), http://ppage.psystems.eu
4. Freund, R., Kari, L., Oswald, M., Sosik, P.: Computationally universal P systems without priorities: two catalysts are sufficient. Theoretical Computer Science 330(2), 251–266 (2005)
5. Ibarra, O.H.: On membrane hierarchy in P systems. Theor. Comput. Sci. 334(1-3), 115–129 (2005)
6. Ibarra, O.H.: On determinism versus nondeterminism in P systems. Theoretical Computer Science 344(2-3), 120–133 (2005)

7. Ibarra, O.H., Păun, G.: Characterizations of context-sensitive languages and other language classes in terms of symport/antiport P systems. Theor. Comput. Sci. 358(1), 88–103 (2006)
8. Ibarra, O.H., Woodworth, S.: On bounded symport/antiport P systems. In: Carbone, A., Pierce, N.A. (eds.) DNA 2005. LNCS, vol. 3892, pp. 129–143. Springer, Heidelberg (2006)
9. Ibarra, O.H., Yen, H.-C.: Deterministic catalytic systems are not universal. Theor. Comput. Sci. 363(2), 149–161 (2006)
10. Ibarra, O.H., Yen, H.-C., Dang, Z.: On various notions of parallelism in P Systems. Int. J. Found. Comput. Sci. 16(4), 683–705 (2005)
11. Krishna, S.: Infinite hierarchies on some variants of P systems (2002) (submitted); See also Languages of P systems: computability and complexity, PhD thesis, Indian Institute of Technology Madras, India (2001)
12. Păun, A., Păun, Gh.: The power of communication: P systems with symport/antiport. New Generation Computing 20(3), 295–306 (2002)
13. Păun, Gh.: Computing with membranes. Turku University Computer Science Research Report No. 208 (1998)
14. Păun, Gh.: Computing with membranes. Journal of Computer and System Sciences 61(1), 108–143 (2000)
15. Păun, Gh.: Membrane Computing: An Introduction. Springer, Heidelberg (2002)
16. Păun, Gh.: Further twenty six open problems in membrane computing. In: Written for the Third Brainstorming Week on Membrane Computing, Sevilla, Spain (February 2005), P Systems Web Page, http://ppage.psystems.eu
17. Sosik, P.: P systems versus register machines: two universality proofs. In: Pre-Proceedings of the 2nd Workshop on Membrane Computing (WMC-CdeA 2002), Curtea de Arges, Romania, pp. 371–382 (2002)

From P to MP Systems

Vincenzo Manca

Department of Computer Science
University of Verona
vincenzo.manca@univr.it

Abstract. Metabolic P systems (MP systems) represent metabolic processes in a discrete mathematical framework based on P systems. MP systems are presented, with a special emphasis to their roots and to their relationship with P systems, which provided the right conceptual framework for their development. A synthetic algebraic formulation of MP system is given, and the log-gain theory of MP systems is outlined, by discussing the research perspectives and the methodological aspects of this approach.

1   Introduction

Metabolism is one of the basic phenomenon on which life is based. Any living organism has to maintain processes which introduce matter of some kind from the external environment, transform internal matter by changing its distribution in a number of biochemical species, and expel outside matter which is not useful or dangerous for the organism. Of course life cannot be reduced to this basic cycle of matter transformation, but no life can exist without such a kind of basic mechanism. To be more realistic, metabolism is not a unique process, but a network of strictly related processes, usually indicated as metabolic pathways. They differ for the involved substances, for the reactions and the enzymes performing them, for the shapes of the dynamical curves they determine (the amount of substances during time). The main questions on the origin of life involve the essence of metabolic processes, their reliability, their integration and their relationship with other essential life functionalities which need metabolism as a basic energetic fuel.

A (finite) multiset is a collection of elements where the same kind of element may occur many times, therefore a chemical reaction is representable by a multiset of rewriting rules.

A metabolic P system, shortly an MP system, is essentially a multiset grammar with rules *regulated* by functions. As it will results evident from the next section, the letter P of MP systems comes from the theoretical framework of P systems introduced by Păun, in the context of membrane computing [40]. In fact, MP systems are a special class of P systems introduced for expressing metabolism in a discrete mathematical setting.

A peculiar aspect of MP systems is given by the Log-gain theory, specifically devised for them [29]. This theory, provides tools for solving the inverse dynamical problem for real metabolic processes. This means that, given a time series of

G. Păun et al. (Eds.): WMC 2009, LNCS 5957, pp. 74–94, 2010.

the states of an observed metabolic system (at a specified time interval τ), then it is possible to deduce, by suitable algebraic manipulations, the functions regulating the rules which represent the metabolic transformations in terms of multiset rewriting. In this manner, an MP system can be defined which coincides, within a certain approximation, with the observed real system. This coincidence is, in many cases, an evidence of adequacy between the systemic logic of the observed real system and the mathematical structure of the deduced MP system.

Many phenomena were reconstructed in terms of MP systems (*e. g.*, Goldbeter's mitotic oscillator, Belousov-Zhabotinski reaction in the Nicolis and Prigogine's formulation, and Lotka-Volterra's Prey-Predator model [15,31,16]). In all these cases a complete concordance with the classical models was found. Moreover, some synthetic oscillators with interesting behaviors were easily discovered [28,29,34], and some MP models were directly deduced by using the Log-gain theory (a part of the photosynthetic NPQ phenomenon of NonPhotochemical Quenching, for which no standard reliable model is known) [37]. A specific software was developed for MP systems, starting from a prototypal version developed by Luca Bianco (Psim, MPsim, MetaPlab) [9,11,36,32] which is downloadable from http://mplab.scienze.univr.it.

In this paper we give a quick presentation of the theory of MP systems, with a special emphasis to its roots and to its relationship with P systems, which provided the right conceptual framework for its development.

2 Historical Backgrounds

The occasion for writing this paper, the decennial anniversary of Membrane Computing, suggested me to briefly reconstruct the initial ideas underlying the MP systems, aimed at developing a discrete theory of metabolic processes based on P systems. Along the line of this historical reconstruction it is possible to grasp in a deeper way the link between P systems and MP system, which rather than of a technical nature is based on the essential assimilation of P perspective in the context of symbolic analysis of metabolism.

My interests in this direction date around the late years 1990. The initial intuition of such a kind of research was the apparent similarity between processes of symbol transformation, typical of logic or formal language theory, with the processes of matter transformations typical of chemistry and biochemistry. If we represent atoms and molecules by suitable symbols, then any chemical reaction is directly translated by a rule of symbol manipulation.

Let me report an example which was a sort of initial formalization exercise. It describes a famous process known as Daniell's cell, a variant of Volta's pile. I presented this example during my invited talk in a meeting organized in 1997 by Gheorghe Păun in Mangalia (not so far from Curtea de Argeş) [22].

Daniell' cell is constituted by two rods of two different metals, zinc and copper (Zn, Cu) which are partially immersed in two solutions where the respective salts in ionic state $ZnSO_4, CuSO_4$ are present (see Fig 1). The two salt solutions are separated, but a salt bridge allows ions to pass through the two compartments. In the zinc compartment, the Zn metal molecules prefer to pass from the metal state to the ion state Zn^{++}, therefore some electrons are in abundance on the zinc rod. If a conductor wire connects the two metal rods, these electrons, according to the greater electron affinity (electronegativity) of copper with respect to the zinc, flow from the zinc rod to the copper rod. After that, the copper ions in the copper solutions, after attracting these exceeding electrons, pass from the ion state to the metal state. At this point, a different electrical charge is present in the two solutions, because in the zinc compartment is present a quantity of SO_4^{--} ions which are not balanced by Zn^{++}, while in the copper solution, the opposite phenomenon happens, because a quantity of Cu^{++} is not balanced by the corresponding SO_4^{--} ions. In this situation, a passage happens of SO_4^{--} ions from the copper to the zinc compartment, in order to restore the electrical equilibrium. In conclusion, an electrical flow along the conductor wire between the rods is coupled with the ion flows through the salt bridge. This provides a cycle which persists, consuming the metal zinc, producing metal copper, and moving ions. In principle the cycle continues until zinc is available, and both kinds of ions are present in both compartments. The membrane perspective of this example is apparent. According to Păun's terminology, in this case a neuron-like membranes system represents the process, which is essentially based on transformation and passage of object symbols through membranes.

In my formalization the concept of membrane was explicit, but the symbol manipulation was based on a special kind of Post rules, which I was very familiar with, and which are a powerful formalism for symbol manipulation. But this is exactly the crucial point which made my formalization unsatisfactory in many aspects. Post rules are *too powerful*, and moreover, in this context strings are not the right data structure for expressing the chemical reactions.

Maybe Gheorghe Păun got some suggestions from my conference in Mangalia in August of 1997 (the paper [35]) including Daniell's cell example was published in 1999). However, Gheorghe Păun (informally, George) sent me a preliminary version of his seminal paper on Membrane Computing [39] in the October of 1998. In his paper, membranes were acutely conjugated with multiset rewriting,

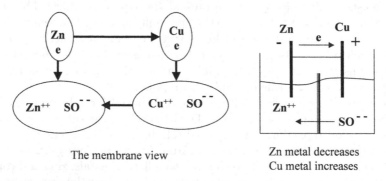

The membrane view

Zn metal decreases
Cu metal increases

Fig. 1. A Daniell's cell (on the right) and its membrane representation (on the left)

and from it I surely got the idea of using multisets in the representation of biochemical reactions.

This perspective emerged to me quite slowly, because I spent almost one year by searching the right form of a combinatorial mechanism for molecule manipulations, by essentially considering special forms of Post rules (with string variables and suitable constraints) [23]. In any case, in 2001, I realized what now seems to me almost obvious: that molecule populations and their tranformations are the essence of metabolism and that multiset rewriting is the natural way to mathematically express this reality. However, an aspect of Paun's P systems was not the exact ingredient to use. The original way of applying rules in P systems was the nondeterministic maximal parallel approach. This perspective is mathematically clear and elegant, moreover allows the proof of computational universality for many variants of P systems. But it is not realistic to assume that biochemical reactions work in this way. For example, if a so efficient approach were applied to the ATP → ADP molecule transformation in our cells, then our bodies would almost instantaneously burned. Therefore, the next step, for a P system perspective to metabolism was the *molar perspective* and the *mass partition principle* which we will briefly recall in the next section. Another aspect deserves to be preliminarily remarked. Biological processes are subjected to noise, fluctuations, external influxes, but at large, they are essentially deterministic. This determinism is of statistical nature. In fact, the individual behavior is strongly variable, but populations obey to strict laws. This introduces a second level of considering multiset. A rule $W + 6C \rightarrow Z + 6O$ (we use multiset polynomial notation) has to be read not only as one molecule occurrence of W (water) and six of C (carbon) to be replaced by one of Z (sugar) and six of O (oxygen), but rather, as a replacement of populations of N and $6N$ objects. The size N is the *(molar) reaction unit*, depending, in general, on the state of the system. This perspective of multiset rewriting changed completely the discrete mathematical point of view about metabolism, providing the right conceptual framework for quantitative analysis of metabolic processes.

In 2004, I started to apply this idea during the supervision of Luca Bianco's Phd thesis [4] (in the meantime I moved from Pisa to Verona). Luca was asked to model some biological phenomena where differential models were available, by trying to find the same dynamics given by these known models, by using a P system perspective (a similar attempt, more devoted to aspects of biological localization, was afforded in [18]). Finding the rules was generally a simple task, but the definition of the strategy for rewriting rules was very hard. Finally, we found a procedure, later called "Metabolic P Algorithm" (MPA), which was adequate for the example we considered, and which was based on a multiset representation of chemical transformations (I realized in [28] that they were an abstract formulation of Avogadro and Dalton principles in chemistry). The "official" appearance of MP system was in 2004 [31], but initially, their focus was on a new rewriting strategy for P systems [5,6,7]. Later it was clear that this was only an aspect of the MP approach, because other radical changes were necessary, and MPA was a particular case of a regulation mechanism based on the notion of population mole. In fact, the name of MP systems was introduced in 2006, when this awareness emerged [24,25]. In the membrane computing community, rewriting strategies different from maximal parallel rewriting were proposed, especially according to probabilistic approaches [41,42], however neither of them adopted the molar perspective, which is peculiar to the development of the log-gain theory of MP systems. The interest in metabolism was a specific aspect of a more general interest in a dynamical, rather than computational, perspective in the study of P systems, addressed in [2], and more recently in [34]. The paper [43] was particularly influential in drawing my attention toward oscillatory phenomena.

3   The Molar Perspective in Multiset Rewriting

Let us give a first intuition of the molar perspective in the multiset representation of biochemical reactions. A reaction $2a + b \to c$ identifies a transformation such that, when it is applied to a population of objects where types a and b occur in more than 20000 and 10000 elements respectively, and when its flux regulation map specifies a reaction unit of, say 10000 elements, then, in the passage from two time instants at a given time distance τ, these 30000 elements are replaced by 10000 new objects of type c. For example, 20000 molecules of Hydrogen, plus 10000 molecules of Oxygen, are transformed into 10000 molecules of water. Time interval between consecutive instants depends on the macroscopic level chosen for considering the dynamics in question. The state, on which reaction units depend, is given by the value of some magnitudes, called parameters, which can influence the reactions (*e.g.*, temperature and pressure) and on the sizes of the different populations inside the system, in correspondence to the different kinds of objects.

A metabolic P system is a discrete representation of a metabolic system. It is essentially given by a set of *reactions* (*reactions* and *rules* are synonymously used). Each reaction is equipped with a corresponding *regulator* (or *flux regulation map*). Such a regulator provides, for any state of the system, a *reaction flux* (*reaction fluxes* and *reaction units* are synonymously used).

The notion of MP system was explicitly defined, as a special class of P systems, during the Brainstorming Week on Membrane Computing, held in Sevilla in 2006 [24]. The initial formulations of MP systems were based on the usual string notation of P systems (sometimes using the additive notation). in Table 1 is given

Table 1. The rules of a mitotic oscillator

$$
\begin{array}{l}
r_1 : \lambda \to C \\
r_2 : C \to \lambda \\
r_3 : C + M_p \to C + M \\
r_4 : C + X \to X \\
r_5 : M \to Mp \\
r_6 : X_p + M \to X + M \\
r_7 : X \to X_p
\end{array}
$$

an example of this notation for Golbeter's model of mitotic oscillator, which we will consider later on. In this case, the rules are based on five substances $\{C, M, M_p, X, X_p\}$ (Cyclin, M-active kinase, M-inactive kinase, X-protease, X–inactive protease).

However, the same multiset grammar can be easily expressed in algebraic notation. In fact any multiset over $\{C, M, M_p, X, X_p\}$ is easily denoted by a vector of $\mathbb{N}$ having as its first component the multiplicity of C, as second component the multiplicity of M, and so forth (in tis context, an implicit order is assumed over substances). In this manner a multiset rewriting rule $\alpha_r \to \beta_r$ becomes representable by a pair of vector (r^-, r^+) (left and right vector), where r^- is the vector expressing the multiset α_r, and r^+ is the vector expressing the multiset β_r. For example the rule $r_3 : C + M_p \to C + M$ is denoted by the pair of vectors

$$\begin{pmatrix} 1 \\ 0 \\ 1 \\ 0 \\ 0 \end{pmatrix} \begin{pmatrix} 1 \\ 1 \\ 0 \\ 0 \\ 0 \end{pmatrix}$$

The algebraic sum of the right component minus the left one provides the *stoichiometric balance* of the rule. It is important to distinguish in a rule its left part, its right part, and its stoichiometric balance. The left part (left vector) expresses the reactants necessary for activating the rule, the right part expresses the products replacing the reactants, while the stoichiometric balance expresses the effective variation performed by the application of the rule. Even if two rules have the same stoichiometric balance, they can be different in the amount of matter they need for their activation. For example the rule $2C + M_p \to 2C + M$ has the same stoichiometric balance of the rule $C + M_p \to C + M$, but the latter needs half of the quantity of C necessary for its activation.

$$\begin{pmatrix} 1 \\ 1 \\ 0 \\ 0 \\ 0 \end{pmatrix} - \begin{pmatrix} 1 \\ 0 \\ 1 \\ 0 \\ 0 \end{pmatrix} = \begin{pmatrix} 0 \\ 1 \\ -1 \\ 0 \\ 0 \end{pmatrix}$$

This algebraic representation of rules remarkably simplifies the definition of MP system. The reader is advised to compare the next definition with the previous definitions of MP system [28,29,33]. However, it is not only matter of notation simplification. In fact, important properties of reactions need to be expressed by usual linear algebra concepts. For example, as it will be explained, the linear independence of some reactions is an essential requirement for discovering the fluxes responsible of a given dynamics.

An MP system is a discrete dynamical system where dynamics is considered at steps indexed in the set $\mathbb{N}$ of natural numbers.

Definition 1. *Let $\mathbb{R}^n$ be the vector (phase) space of dimension n over real numbers. An MP system M of type (n, m, k), that is, of n substances, m reactions and k parameters is specified by:*

$$M = (S, R, H, \Phi, \tau, \nu, \mu)$$

where:

• *S is a set of substances, considered in a conventional order, determining, for any metabolic state of the system, a vector X of substance quantities which varies on the real vectors of $\mathbb{R}^n$;*

• *R is a set of reactions, which are given by m pairs $(r_1^-, r_1^+) \ldots, (r_m^-, r_m^+) \in \mathbb{N}^n \times \mathbb{N}^n$, composed by the left and right vectors of the reactions (relative to the reactants and to the products respectively). For any $r \in R$ its stochiometric balance $r^\#$ is given by $r^+ - r^-$. The matrix $A = (r_1^\#, \ldots, r_m^\#)$ is the stoichiometric matrix associated to the reactions having as columns the stoichiometry balances of the rules;*

• *H is a function $H : \mathbb{N} \to \mathbb{R}^k$ providing, at each step $i \in \mathbb{N}$, the vector $H[i]$ of parameters;*

• *$\Phi = (\varphi_1, \ldots, \varphi_m)$ is a vector of regulators (or flux regulation functions) where $\Phi : \mathbb{R}^n \times \mathbb{R}^k \to \mathbb{R}^m$ provides the fluxes of reactions corresponding to any global state of the system, that is, a pair in $\mathbb{R}^n \times \mathbb{R}^k$ constituted by the metabolic state and by the parameter vector. However, given a reaction r, only some of the substances or parameters, which constitute the tuners of the reaction, may occur as arguments of the corresponding regulator φ_r. If the flux $\varphi_r(a, b, \ldots)$ corresponding to the values $a, b, \ldots$ of the tuners of r is greater than the amount of some reactant of r, then the corresponding reaction do not apply (that is $\varphi_r(a, b, \ldots)$ is forced to be null);*

• *$\tau \in \mathbb{R}$ is a time interval between two consecutive steps;*

• *$\nu \in \mathbb{R}$ is a conventional mole;*

• *$\mu \in \mathbb{R}^n$ is a vector of the mole masses of substances.*

Given an initial state vector $X[0] \in \mathbb{R}^n$, the dynamics of M is provided by the following vector recurrent equation, called $EMA[i]$ (Equational Metabolic Algorithm), where $U[i] = \Phi(X[i], H[i])$:

$$X[i + 1] = A \times U[i] + X[i] \tag{1}$$

which computes the state $X[i + 1]$, for each step $i \in \mathbb{N}$, ($\times$ is the usual matrix product, and, in dependence on the context, $+$ is the usual sum or the componentwise vector sum). □

The intuition behind the previous definition is that of a system defined by: *substances, reactions, parameters, regulators,* and *scale factors* (time, population, and substance mass units). Reactions transform substances, while regulators establish the amount of matter (expressed in moles) transformed by each reaction at each step. Parameters are not directly involved in reactions, but together with the substance quantities, may enter as arguments of regulators (which are maps).

Table 2. MP formulation of Goldbeter's mitotic oscillator

$K_1 = 0.005 \, \nu$	$K_2 = 0.005 \, \nu$	$K_3 = 0.005 \, \nu$
$K_4 = 0.005 \, \nu$	$V_{M1} = 3 \, \nu$	$V_i = 0.025 \cdot 10^{-6} \, \nu$
$V_2 = 1.5 \, \nu$	$V_4 = 0.5 \, \nu$	$Q_d = 0.02 \cdot 10^{-6} \, \nu$
$V_d = 0.25$	$K_c = 0.5 \cdot 10^{-6} \, \nu$	$\tau = 0.001 \, min$
$K_d = 0.01$	$S = 0.001$	$\nu = 6.02 \times 10^{23}$

$$C = 0.01 \cdot 10^{-6} \, \nu \quad M = 0.01 \, \nu \quad M_p = 0.99 \, \nu \quad X = 0.01 \, \nu \quad X_p = 0.99 \, \nu$$

$r_1 : \lambda \to C$	$\varphi_1 = S \cdot V_i$
$r_2 : C \to \lambda$	$\varphi_2 = S \cdot K_d \cdot C$
$r_3 : C + M_p \to C + M$	$\varphi_3 = (S \cdot V_1 \cdot M_p)/(K_1 + M_p)$
$r_4 : C + X \to X$	$\varphi_4 = (S \cdot V_d \cdot X \cdot C)/(Q_d + C)$
$r_5 : M \to M_p$	$\varphi_5 = (S \cdot V_2 \cdot M)/(K_2 + M)$
$r_6 : X_p + M \to X + M$	$\varphi_6 = (S \cdot M \cdot X_p)/(K_3 + X_p)$
$r_7 : X \to X_p$	$\varphi_7 = (S \cdot V4 \cdot X)/(K_4 + X)$

$$V_1[i] = (C[i] \cdot V_{M1})/(K_c + C[i])$$

Scale factors do not enter in the mathematical description of the dynamics, but they define its physical interpretation, according to an adequate time/mass scale of the phenomena under investigation.

MP systems can be depicted by means of MP graphs [30,19] with five kinds of nodes and four kinds of edges (see Fig. 3). Nodes are: substance nodes, reaction nodes, regulation nodes, parameter nodes, and gate nodes denoting matter fluxes from/to the external environment (lambda rules). Edges are: transformation edges (consumption and production), regulation edges and dependency edges.

Table 2 specifies, an MP model, of type $(5, 7, 1)$, for a famous oscillator occurring in the mitosis of early amphibian embryos, established by Goldbeter in terms of differential equations [21]. In the order, are indicated: i) the constants (used for e better reading of formulae and including the temporal interval τ and the population unit ν, but leaving unspecified the molar weights), ii) the initial values of substance quantities, iii) the rules with the corresponding flux regulation maps, and iv) the parameters with their evolution functions. This MP formulation is obtained by extending a procedure introduced in [17] and provides the same dynamics of the original differential model (see [28,29] for Goldbeter's differential equations, for other MP models, and for discussions concerning their identification).

4   The Log-Gain Theory of MP Systems

The main question, at beginning of the log-gain theory for MP systems, is the following inverse dynamic problem. Given a time series $(X[i], H[i]) \in \mathbb{R}^{n+k}$ (for

$i = 0, 1, 2, \ldots, t$) of some consecutive states and parameters of a metabolic system (at a time interval τ), is it possible to deduce a corresponding time series of vectors $U[i] \in \mathbb{R}^m$ which put in the equation (1) provide the time series of substance quantities? This is the discrete dynamical problem of reaction flux discovery. The deduction of time series $U[i]$ implies the knowledge, at the time granularity τ, of the systemic logic governing the matter transformations underlying the observed metabolic states. When vectors $U[i]$ are known, the discovery of maps Φ which provide $U[i]$, in correspondence to the vectors $(X[i], H[i])$, is a typical problem of approximation which can be solved with standard techniques of mathematical regression. Fig. 2 expresses graphically the two procedures, going in the opposite verses, of generation of a dynamics from a given MP system, and of providing an MP system fitting with an observed dynamics. The equation linear systems EMA provides the dynamics of an MP system, while the equation linear system OLGA, allows us to perform the opposite task. In the following, we will outline the log-gain theory, which determines the methods for constructing the OLGA systems.

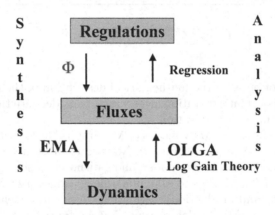

Fig. 2. Synthesis and analysis of dynamics by means of MP systems: direct and inverse dynamical problems

An important remark is due in this context (which will be more extensively reconsidered, in the final section). The approach of flux discovery is essentially observational, macroscopic, and global, in a sense which is opposite to the perspective of differential models, which is infinitesimal, microscopic and local. In fact, we do not intend to discover the real kinetic responsible, at a microscopic level, of the biochemical dynamics of each reaction, but we only try to capture the global pattern of reaction ratios of an observed dynamics. In other words, leaving unknown the *real* local internal dynamics, we decide to consider the system at an abstraction level which is sufficient to reveal the logic of the behavior we observe. This more abstract approach can be less informative, with respect to specific important details, but such a more generic information could be very useful in discriminating important aspects of the reality, and often, especially in the case of very complex systems, is the only way for grasping a kind of comprehension of the reality under investigation.

We call it EMA (Equational Metabolic Algorithm) when it is used for calculating the substance quantities, from the knowledge of flux regulation maps, while we call ADA (Avogadro and Dalton Action) the system (1) where we search to determine $U[i]$ from the knowledge of substance quantities (Avogadro refers to the integer stoichiometric coefficients, and Dalton to the summation of the effects of reactions).

Unfortunately, often, ADA is not sufficient to provide the solutions because the number m of reactions is greater than the number n of substances. Therefore, we need to extend ADA by adding new equations.

The log-gain principle assists us in the search of further equations for identifying the fluxes. This principle derives from a general biological principle called *allometry*, according to which, in a living organism, the global variation of its typical magnitudes follow a sort of *harmonic rule* according to which their relative variations are proportional to the relative variations of the magnitudes related to them. In differential terms the relative variation in time of a magnitude coincides with the variation of its logarithm, therefore we used the term "log-gain" for any law grounded on this assumption. In the specific context of our problem, we assume that the relative variations of a reaction flux is a linear combination of the relative variations of substance quantities and parameters affecting the reaction, and in a more restrict case, it is the sum of the relative variations of its reactants. We refer to the papers [29] for a detailed account on the log-gain theory of MP systems. The principle was initially formulated starting from its general form. Then, in three subsequent transformations, it provided an equation system COLG (Covering Offset Log-Gain), involving fluxes, with a number of equations equal to the number of reactions, but with additional unknown variables, called *offset log-gain*, equal to the number of substances. This means that the whole system constituted by ADA and COLG has $2m + n$ variables. Moreover, if we consider the two systems, at the same observation step i, then it results a nonlinear system.

Here, an induction argument helps us to obtain a further reduction of variables, in order to get a *square equation linear system*. In fact, if we consider $ADA[i + 1]$ and $COLG[i]$, assuming to know the fluxes at step i, we contemporarily reduce the variables to $n + m$ and remove the nonlinearity of the system.

Now we report the final form of a system of equations called OLGA (Offset Log Gain Adaptation) which solves our initial problem of flux discovery ($\times$ is the usual matrix product, while $+, \cdot, -, /$ are the componen-twise vector operations of sum, product, difference and division, respectively).

$$X[i + 2] = \mathbb{A} \times U[i + 1] + X[i + 1] \tag{2}$$
$$(U[i + 1] - U[i])/U[i] = B \times (W[i + 1] - W[i])/W[i] + C \cdot P \tag{3}$$

where W is the (n+k) dimensional vector of substances and parameters, B is a boolean matrix choosing, for any reaction, its *tuners*, that is, the magnitudes affecting its flux, and P is an m-dimensional vector of reals, expressing the reaction *offsets*, that is, the errors introduced in the log-gain approximations of fluxes, while C is a boolean m-dimensional vector, such that $\sum C = n$, that is, the sum of its components is equal to n.

We assume that the stoichiometric matrix $\mathbb{A}$ has maximum rank. This assumption is not restrictive because it implies that no substance variation is a linear combination of the variations of other substance. If this is the case we can remove the substance variation which is combination of other variations, without loss of information, by obtaining a stoichiometric matrix of maximum rank.

We say that a rule is *linearly dependent* on other rules if its stoichiometric balance is a vector which is linearly dependent on the stoichiometric balance of other rules. A set of rules are linearly independent if no rule of this set is dependent on other rules of the set. We say that a subset R_0 of n rules is a *covering* of the set R of rules, if any substance is reactant or product of some rule in R_0.

The following theorems are a natural consequence of the algebraic formulation of rules (see [20] for proofs).

Theorem 1. *Given a set of rules with stoichiometric matrix of maximum rank, then there exits a covering of linearly independent rules.*

Theorem 2. *Let R_0 be a subset of rules of R which are linearly independent. Let OLGA be a system with a covering vector C corresponding to R_0 ($C(i) = 1$ iff $r_i \in R_0$). Then, OLGA has one and only one solution.*

The previous theorems show that the problem of finding fluxes of a metabolic system is solvable under very general assumptions.

However, given the inductive nature of our method, in order to generate the time series of $U[i]$, for $i > 0$, we need the knowledge of $U[0]$. An algorithm for achieving this task was recently found [38], which was tested in many cases with a good success. This problem is essentially an optimum problem based on the notion of activation matrix. This matrix has as columns the left components of the rules of R. If we multiply it with the flux vector $U[i]$, then we get, for each component, the amount of a substance necessary, at step i, to activate all the rules which need that substance. Other constraints regard the positivity of fluxes and a sort of Lavoisier principle (the absolute variation of matter between two consecutive states has to equate the absolute difference between the sums of in-coming and out-coming fluxes).

The determination of the covering vector C is another important aspect in the construction of the OLGA system. Some investigations are in progress for the search of an *optimal* covering, or for showing that, under suitable conditions, the goodness of solutions can be independent on the choice of a specific covering. However, in the study of this aspect it seems useful to consider the Galois connection arising between substances and reactions. Given a substance x, we denote by $R(x)$ the set of reactions where x occurs (as product or reactant), but symmetrically, given a reaction r, we can define $S(r)$ as the set of substances involved in the reaction r. If we extend R, S as functions from set of substances to set of reactions, and *viceversa*, we get a Galois connection, which is a very general and powerful algebraic concept. It seems possible that, rule covering, and other metabolic concepts, are related to properties which can be analyzed in this algebraic setting.

The following theorem shows a relevant aspect of the notion of covering. In fact, for the application of the log-gain principle, the flux log-gain of a rule should consider non only its reactants, but its tuners, that is, all magnitudes (substances and parameters) which influence the rule. Unfortunately, the knowledge of tuners of reactions is very often not available. The following theorem (see [20] for a proof) ensures that fluxes can be deduced even with this lack of knowledge. Therefore, the analysis about tuners, for determining fluxes, could be focused on the uncovered reactions.

Theorem 3. *Consider an OLGA system based on a linearly independent covering R_0. The fluxes which are solutions of this system do not depend on the tuners which are chosen for the rules of R_0 in the flux log-gains of these rules.*

In conclusion, tuners of rules of R_0 can be reduced only to the reactants of there rules, and the solutions of OLGA systems, one for each step, provide the time series $U[i]$ that solve the flux discovery problem, posed at the beginning of our discourse.

Results of equivalence of MP systems with other formalisms were developed [17,13,14].

5   Fluxes, Reactivity, Inertia, and Differential Models

The analysis process which provides an MP system from an observed dynamics is directly related to the notion of reaction fluxes. However, in the process of synthesizing dynamics is more natural to associate to every reaction a reactivity parameter determining a sort of score in the competition for getting the reactants necessary for the activation of the reaction. This competition concerns the part of matter available in a given state, therefore another parameter is necessary, for each substance, which provides the amount of substance that, in a given state, can be partitioned among all reactions competing for it, or equivalently, the amount of substance that is not transformed, called the *inertia* of the substance (at a given step). These systems were the first kind of MP systems formally defined [28], and correspond to the special class of *reactive* MP systems. In a reactive MP system there is a parameter for each substance q_x ($x \in X$), providing its *inertia* and a parameter for each reaction f_r ($r \in R$), providing its *reactivity*. Let us set by $R^-(x)$ the set reactions consuming the substance x, and by $S^-(r)$ the reactants of reaction r. Then, another parameter p_x of (metabolic) *reactance* can be associated to any substance $x \in X$:

$$p_x = \frac{x}{q_x + \sum_{r \in S^-(x)} f_r}. \tag{4}$$

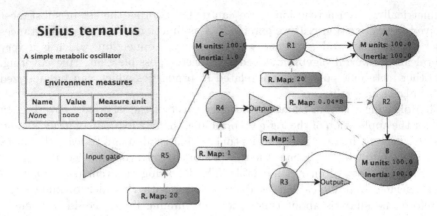

Fig. 3. The MP system Sirius ternarius. Big circles are substances, small circles are reactions, rectangles are reactivity parameters, and triangles indicate matter flows from/to the external environment. Fluxes are not indicated because deduced from inertias and reactivities by means of formulae (4) and (5) (their values, expressed in conventional moles of arbitrary size, are $q_A = 100, q_B = 100, q_C = 1$).

Let $minA$ be the minimum over a finite set A of numbers, conventionally extended to the empty set by $\min \emptyset = 1$. With these notations, the flux regulation maps of a reactive MP system are given by:

$$\varphi_r = f_r \min\{p_x \mid x \in S^-(r)\}. \tag{5}$$

Therefore, in reactive MP systems, flux regulation maps φ_r ($r \in R$) are completely determined by reactivities and inertias, thus may not be explicitly mentioned. In Fig. 3, an MP graph is given, which describes the simple metabolic oscillator *Sirius ternarius*, a variant of an oscillator widely studied in the context of MP systems [28,29,34]. The core of this oscillations is the reaction from $A \rightarrow B$, with a flux which linearly depends on the amount of B. In fact, when this quantity increases too much, then the reactant of $A \rightarrow B$ is greatly consumed, and consequently also the reaction flux diminishes. In such a way A, which is produced by $C \rightarrow A$ can increase and consequently also the reaction $A \rightarrow B$ returns again to work actively, so that the condition for a new cycle is restored.

The following theorem (see [29] for a proof) states the dynamical equivalence between any MP system and a suitable reactive reactive MP system (starting by the same state they provide the same sequence of states).

Theorem 4. *For any MP system there exists a reactive MP system which is dynamically equivalent to it.*

In Fig. 4 is given the oscillatory dinamics of the MP system of Fig. 3, computed by Psim software.

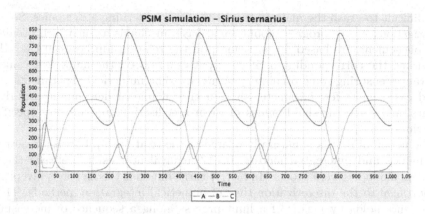

Fig. 4. Sirius ternarius' dynamics where EMA of Definition 1 is computed by Psim software (see: http://www.cbmc.it and http://mplab.sci.univr.it)

If we avoid the rule consuming C, the dynamics changes dramatically, even if we reduce sensibly the value of rule introducing C. This show that the analysis of metabolic processes is very complex and very often the behavior of a system is hardly deducible by the MP graph, without a direct inspection of its dynamics. The form of trajectories are related to the graph structure, but very often their shape is very robust for big changes of regulation maps and initial values, but very fragile with respect to some parameters. This kind of investigations applied to real metabolic oscillators are very important for establishing the key features responsible for maintain some dynamical regimes of interest.

A notion of *abstraction order* can be defined for MP systems, which result useful in the determination of models. A system M is more abstract than a system M' if the substance of M are a subset of those of M' and the dynamics of M coincide with the dynamics of M' on their common substances. In many cases a right abstraction level could be more informative of a too detailed system where

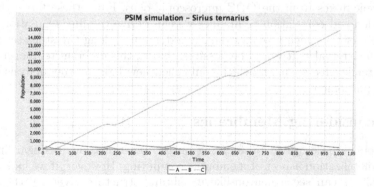

Fig. 5. Sirius ternarius' dynamics where the reaction $\lambda \to C$ is removed

it is difficult to grasp the main feature of the logic governing a dynamics. Some investigations are in progress about some basic mechanisms on which oscillatory phenomena are based, in particular, on the relationship between the MP graph and the corresponding oscillatory pattern, and on the numerical values and ranges ensuring some oscillatory forms. In some numerical experiments we found cases where few parameters have a crucial role in determining the dynamics, and some threshold values of them are discriminant for very specific behaviors.

The notion of inertia is naturally related to the relationship between reactive MP models and differential models. In [17] equivalence results between these two kind of models were proven. In fact, it turns out that the inertia is inversely proportional to the discretization time of numerical integration methods. This equivalence holds by means of a limit process along a sequence of increasing values of inertia, which is supposed to be equal for all substances.

A general theorem can be easily proved stating an equivalence between the dynamics of a differential model, computed by the Euler method of numerical integration, and the dynamics computed by EMA for an MP model which is deduced by means of a straightforward "rule-driven translation" of the right members of differential equations (the procedure used in Sect. 3 for the MP formulation of Goldbeter's mitotic oscillator). In this case, the MP time interval coincides with the discretization time of the numerical integration.

However, a deeper relationship can be established between differential and MP models. In fact, let us suppose, to have an ODE (Ordinary Differential Equation) model of a metabolic process. According to it, any derivative of substance quantity is the sum of some additive terms relative to the infinitesimal fluxes of the rules consuming and producing that substance. Assume to use a numerical integration method, and to solve the differential equations with a discretization time Δt. Now, if we consider a time interval τ and perform $\tau/\Delta t$ numerical integration steps (the natural number rounding this value), then we can deduce the fluxes of all the reactions involved in the system in the time interval τ. This means that we get exactly what the log-gain theory provides by solving the OLGA systems along a number of observation steps. In other words, we get the macroscopic fluxes from the ODE microscopic ones. From these fluxes, by approximation and correlation techniques we can derive the flux regulation maps of an MP system which provides the same dynamics along the steps separated at the time interval τ. It would be possible, that at this different temporal grain, some systemic effects emerge which could shed new light on the analysis of the modeled phenomenon.

6   Reconsidering Membranes

MP systems are described by focusing on the reactions, but disregarding the compartmetalization aspect of membrane computing. However, if we look at the MP graph we can see a neuron-like membrane structure given by the nodes along which the matter flows. This means that if we model substances as different membranes, and we fill them of a unique kind of substance (e. g. water) we are in a perfect membrane setting. This is a general aspect which it would be interesting to analyze in general terms. Objects and membrane are dual concepts which can be reciprocally reduced (an analogous situation arises in set theory). This duality is a special case of the space/matter duality formulated in the context of a discrete framework. In fact a physical object, having a spatial extension comprises a portion of space, the internal space occupied by it, that can be separated by an implicit membrane delimiting its internal region. Conversely, a membrane is an object with an internal region which can include other objects. Therefore, we may consider an object of type a as equivalent to an empty membrane $[\ ]_a$. Analogously an object a inside the membrane of label j, $[a]_j$, is represented by as an object a_j with the index denoting the localization of a. In general, we may reverse the relationship of containment of membranes and objects, by expressing the localization of an object by putting its *membrane address* (for example, a string of membrane labels). Here we do not enter into further details. However, many aspects deserve a careful analysis. Namely, a sharp examination of the notion of object distinguishability could show some subtle implicit pitfalls. For example, different occurrences in time, of the same kind of object, may not coincide with different occurrences in space of the same object (in the two cases different masses could be involved).

According to the perspective of addressed objects, moving an object from a membrane to another one results to be a transformation acting on the index part of the object. In many modeling context this is the natural approach adopted for expressing localization changes. For example a protein p which can be localized in two places A, B is modeled by two species p_A and p_B and its displacement is assimilated to a transformation of matter. This discussion shows that the more appropriate way to model a reality depends on the specific aspects we are interested to model, but in principle "membranization" or "demembranization", or a mixing of the two strategies, are related to different perspectives of investigation.

In [1,2] the boundary notation for membrane rules was introduced in order to cope with more general membrane rules (an application and extension of boundary notation was developed in [8]). In fact, in Păun's original formulation, rules are inside membranes and everything is outside the membrane where a rule is located remains unknown to the rule. But, in many cases a wider visibility is required. The essential point of boundary representation is the idea of rules with a greater level of localization knowledge about the objects which they apply to. An important case of this situation is present in the anti-port rules which postulate to consider objets inside and outside membranes. How can be this idea further generalized for coping with different contexts of application?

Another natural generalization of P rules concerns the possibility of high-order multisets. This is not a mathematical generalization, but expresses a natural necessity for representing biochemical transformations. In fact, in many reactions two or three level multisets occur. Even in the simple case of water formation, the usual chemical notation is $2H_2 + O_2 \rightarrow 2H_2O$. Here we have multiplicative numeric coefficients and numerical indexes, that we could express, by using parentheses, as $2(2H) + (2O) \rightarrow 2((2H)O)$. In this case, parentheses are not membrane parentheses, but express a two level multiset. In fact, the rule transforms a multiset of objects which are multisets too, that is, a second order (finite) multiset into in another one of the same kind.

In many phenomena the localization aspect is predominant, but in a way that membranes are not adequate. It is the case of *gradients* in morphogenesis. In this case, what is important, rather than containment relations, are the distances with respect to some coordination points, therefore indices memorizing these values are the natural way for handling this aspect.

In a discrete setting, *loci* could be represented by (localization) *binders* attached to the objects, which become relevant in relocation rules, while they are dummy when internal transformations are performed. Binders are useful for encoding physical features or parameters (*polarization* and *thickness* are examples of them, already investigated in membrane computing).

In conclusion, a very synthetic way for expressing the original P-system perspective could be: grammars of "parenthesized multisets" (possible with different types of parentheses). The passage from *boundaries* to *binders* and all the aspects mentioned above could enlarge the spectrum of modeling possibilities of P and MP systems toward the study of dynamics of high level discrete spatial complexity.

7   Open Problems and Methodological Issues

Many lines of development emerged, in the context of MP systems. Some of them, as it was argued in the previous section, are related to the theory of P systems. Other research lines are specifically focused on the log-gain theory. The hot points in this direction are: i) the determination of the initial fluxes, ii) the determination of the more appropriate covering for the OLGA systems, iii) the determination of the tuners of reactions, and iv) the determination of the flux regulation maps associated to the fluxes and to their tuners. Some investigations are in progress and some partial results are available. It is interesting that in the search of solutions a variety of methods naturally occurs, going from vector algebra and vector optimization to artificial neural networks [12,38]. The next kinds of modeling applications which we intend to realize are phenomena related to gene regulation networks and to signal transduction mechanisms. From the computational side, many plugins are under development for extending the MetaPlab software, according to specific needs of the experiments which could orientate the theoretical and applicative research. Presently, a plug-in is available for computing MP dynamics by means of EMA, moreover a plug-in is also available for the flux discovery by means of OLGA, other visualizations and format translation plug-ins are available, and prototypal plug-ins for polynomial regression and artificial neural network correlation plug-ins were developed [11,12,32].

Other research lines of MP systems theory are more specifically related to the metabolism and to the population perspective of biological phenomena. Many aspects of metabolic dynamics can be expressed and abstractly studied on MP systems [34]. In particular, a general study of metabolic oscillators seems to be especially adapt to be investigated by using reactive MP systems. This class of systems are especially suited for synthesizing specific behaviors, in order to identify the specific structural features related to some dynamical properties. For example, a catalog of basic MP metabolic oscillators is under investigation, which is aimed to instantiate experiments of computational synthetic biology.

I want to conclude by stressing an important methodological aspect which is very often source of misunderstanding, because it remarkably differs from the usual modeling approaches in computational biology.

When we design an MP model by using the log-gain theory we start with time series of observations. The model we get at end of the process is a model of what we observed. We adopt a perspective which could be described as the *Boltzmann's analogy*. According to Boltzmann's mechanical statistics, the macroscopic state of a thermodynamic system (a gas inside a volume at a given pressure and temperature) is given by the distribution function $f(z)$ providing the number of molecules in the ensemble z (a kind of energetic level). In our case, we claim that in a biochemical system, with a number of chemical species, its macroscopic state depends on the number of molecules which are present for each species. The passage from a state to another one is completely due to the change of molecule distribution per species.

We do non know and we do not claim to describe what happens at the microscopic reaction level. We observe that some species are related by some reasonable transformations and we assume that the variations are due to the action of these transformations. These transformations could be executed in many ways and maybe they involve other underlying very complex transformations, at different sublevels. However, this is outside the objective of the model. It tries to find the logic underlying the specified species and the chosen transformations. In other words, we explain what is observed in terms of the species and the transformations under investigation. If the choice of the species and of the transformations is not the right one, this means that the model was not adequate, but this is independent from the methodology, it is only a matter of the specific modeling design. In conclusion, MP modeling, according to the log-gain analysis, is deliberately at a different, more abstract, level with respect to ODE models. This does not means that it is less adherent to the reality, but simply that it is focused on a different level of reality.

A model is either good or bad only to the extent it helps us in predicting and explaining what we can observe. No other criterion can be discriminant, and it is ingenuous to adopt a mirror analogy with an absolute character. In fact, many mirrors could be available, and some could be more useful than others in certain contexts. Reality is different when it is considered at different levels of observation. When the level of phenomena under investigation is very different (too small or too big, or too complex) with respect to the observer level, the true scientific ability concerns the right theoretical and experimental choices about what has to be observed and about how the observation results have to be related. A priori is very hard to chose the "pertinent aspects" of a phenomenon and to disregard what is not relevant.

What is the reality adherence of the physical theories at quantum levels or at cosmological levels? What is the reality of the probability wave in Shrödinger equation? We trust them because they work. No mirror principle can assist us for their evaluation. Models are creations of the human invention. Modeling is an art, and it cannot follow easy prefixed procedures. This art is based on the right choice of what has to be observed, what relationships are relevant among the observed features, how translate them in a chosen conceptual universe, and how to interpret the findings which result from this translation.

References

1. Bernardini, F., Manca, V.: P Systems with boundary rules. In: Păun, G., Rozenberg, G., Salomaa, A., Zandron, C. (eds.) WMC 2002. LNCS, vol. 2597, pp. 107–118. Springer, Heidelberg (2003)
2. Bernardini, F., Manca, V.: Dynamical aspects of P systems. Biosystems 70, 85–93 (2003)
3. von Bertalanffy, L.: General Systems Theory: Foundations, Developments, Applications. George Braziller Inc., New York (1967)
4. Bianco, L.: Membrane models of biological systems. Ph.D. Thesis, University of Verona (April 2007)
5. Bianco, L., Fontana, F., Franco, G., Manca, V.: P systems for biological dynamics. In: [10], pp. 81–126
6. Bianco, L., Fontana, F., Manca, V.: Reaction-driven membrane systems. In: Wang, L., Chen, K., Ong, Y.S. (eds.) ICNC 2005. LNCS, vol. 3611, pp. 1155–1158. Springer, Heidelberg (2005)
7. Bianco, L., Fontana, F., Manca, V.: P systems with reaction maps. Intern. J. Found. Computer Sci. 17, 27–48 (2006)
8. Bianco, L., Manca, V.: Encoding-decoding transitional systems for classes of P systems. In: Freund, R., Păun, G., Rozenberg, G., Salomaa, A. (eds.) WMC 2005. LNCS, vol. 3850, pp. 134–143. Springer, Heidelberg (2006)
9. Bianco, L., Manca, V., Marchetti, L., Petterlini, M.: Psim: a simulator for biochemical dynamics based on P systems. In: 2007 IEEE Congress on Evolutionary Computation, Singapore (September 2007)
10. Ciobanu, G., Păun, Gh., Pérez-Jiménez, M.J. (eds.): Applications of Membrane Computing. Springer, Heidelberg (2006)
11. Castellini, A., Manca, V.: MetaPlab: A Computational Framework for Metabolic P Systems. In: Corne, D.W., Frisco, P., Paun, G., Rozenberg, G., Salomaa, A. (eds.) WMC 2008. LNCS, vol. 5391, pp. 157–168. Springer, Heidelberg (2009)
12. Castellini, A., Manca, V.: Learning regulation functions of metabolic systems by artificial neural networks. Electronic Notes in Theoretical Computer Science (2009), www.elsevier.nl/locate/entcs
13. Castellini, A., Franco, G., Manca, V.: Hybrid functional Petri nets as MP systems. Natural Computing (2009), doi:10.1007/s11047-009-9121-4
14. Castellini, A., Manca, V., Marchetti, L.: MP systems and hybrid Petri nets. Studies in Computational Intelligence 129, 53–62 (2008)

15. Fontana, F., Bianco, L., Manca, V.: P systems and the modeling of biochemical oscillations. In: Freund, R., Păun, G., Rozenberg, G., Salomaa, A. (eds.) WMC 2005. LNCS, vol. 3850, pp. 199–208. Springer, Heidelberg (2006)
16. Fontana, F., Manca, V.: Predator-prey dynamics in P systems ruled by metabolic algorithm. BioSystems 91, 545–557 (2008)
17. Fontana, F., Manca, V.: Discrete solutions to differential equations by metabolic P systems. Theoretical Computer Sci 372, 165–182 (2007)
18. Franco, G., Manca, V.: A membrane system for the leukocyte selective recruitment. In: Martín-Vide, C., Mauri, G., Păun, G., Rozenberg, G., Salomaa, A. (eds.) WMC 2003. LNCS, vol. 2933, pp. 181–190. Springer, Heidelberg (2004)
19. Franco, G., Guzzi, P.H., Mazza, T., Manca, V.: Mitotic oscillators as MP graphs. In: Hoogeboom, H.J., Păun, G., Rozenberg, G., Salomaa, A. (eds.) WMC 2006. LNCS, vol. 4361, pp. 382–394. Springer, Heidelberg (2006)
20. Franco, G., Manca, V., Pagliarini, R.: Regulation and covering problems in MP systems. In: WMC 2009. LNCS, vol. 5957. Springer, Heidelberg (2009)
21. Goldbeter, A.: A minimal cascade model for the mitotic oscillator involving cyclin and cdc2 kinase. PNAS 88, 9107–9111 (1991)
22. Manca, V.: String rewriting and metabolism: A logical perspective. In: Păun, Gh. (ed.) Computing with Bio-Molecules, pp. 36–60. Springer, Heidelberg (1998)
23. Manca, V.: Monoidal systems and membrane systems. In: Calude, C.S., Dinneen, M.J., Păun, Gh. (eds.) WMC-CdeA 2000, Workshop on Multiset Processing, August 2000. CDMTCS Research Report Series, pp. 176–190 (2000)
24. Manca, V.: Topics and problems in metabolic P systems. In: Gutiérrez-Naranjo, M.A., et al. (eds.) Proc. of the Fourth Brainstorming Week on Membrane Computing, Sevilla, Spain, January 30 - February 3, pp. 173–183 (2006)
25. Manca, V.: MP systems approaches to biochemical dynamics: Biological rhythms and oscillations. In: Hoogeboom, H.J., Păun, G., Rozenberg, G., Salomaa, A. (eds.) WMC 2006. LNCS, vol. 4361, pp. 86–99. Springer, Heidelberg (2006)
26. Manca, V.: Metabolic P systems for biochemical dynamics. Progress in Natural Sciences 17, 384–391 (2007)
27. Manca, V.: Discrete simulations of biochemical dynamics. In: Garzon, M.H., Yan, H. (eds.) DNA 2007. LNCS, vol. 4848, pp. 231–235. Springer, Heidelberg (2008)
28. Manca, V.: The metabolic algorithm for P systems: Principles and applications. Theoretical Computer Sci. 2, 142–157 (2008)
29. Manca, V.: Log-gain principles for metabolic P systems. In: Condon, A., et al. (eds.) Algorithmic Bioprocesses. Natural Computing Series, vol. 28. Springer, Heidelberg (2009)
30. Manca, V., Bianco, L.: Biological networks in metabolic P systems. BioSystems 91, 489–498 (2008)
31. Manca, V., Bianco, L., Fontana, F.: Evolutions and oscillations of P systems: Applications to biological phenomena. In: Mauri, G., Păun, G., Jesús Pérez-Jímenez, M., Rozenberg, G., Salomaa, A. (eds.) WMC 2004. LNCS, vol. 3365, pp. 63–84. Springer, Heidelberg (2005)
32. Manca, V., et al.: MetaPlab 1.1 Official Guide, Tutorials (2009), http://mplab.scienze.univr.it
33. Manca, V.: Fundamentals of metabolic P systems. In: Păun, G., Rozenberg, G., Salomaa, A. (eds.) Handbook of Membrane Computing, ch. 19, pp. 475–498. Oxford University Press, Oxford (2009)
34. Manca, V.: Metabolic P Dynamics. In: Păun, G., Rozenberg, G., Salomaa, A. (eds.) Handbook of Membrane Computing, ch. 20, pp. 499–528. Oxford University Press, Oxford (2009)

35. Manca, V., Martino, M.D.: From string rewriting to logical metabolic systems. In: Păun, G., Salomaa, A. (eds.) Grammatical Models of Multiagent Systems, pp. 297–315. Gordon and Breach Science Publishers, London (1999)
36. Manca, V., Marchetti, L.: XML representation of MP systems. In: 2009 IEEE Congress on Evolutionary Computation (CEC 2009), Trondheim, Norway, May 18-21, pp. 3103–3110 (2009)
37. Manca, V., Pagliarini, R., Zorzan, S.: A photosynthetic process modelled by a metabolic P system. Natural Computing (2009), doi:10.1007/s11047-008-9104-x
38. Pagliarini, R., Franco, G., Manca, V.: An Algorithm for Initial Fluxes of Metabolic P Systems. Int. J. of Computers, Communications & Control IV(3), 263–272 (2009)
39. Păun, Gh.: Computing with membranes. J. Comput. System Sci. 61, 108–143 (2000)
40. Păun, Gh.: Membrane Computing. An Introduction. Springer, Heidelberg (2002)
41. Pérez-Jiménez, M.J., Romero-Campero, F.J.: A study of the robustness of the EGFR signalling cascade using continuous membrane systems. In: Mira, J., Álvarez, J.R. (eds.) IWINAC 2005. LNCS, vol. 3561, pp. 268–278. Springer, Heidelberg (2005)
42. Pescini, D., Besozzi, D., Mauri, G., Zandron, C.: Dynamical probabilistic P systems. Intern. J. Found. Computer Sci. 17, 183–204 (2006)
43. Suzuki, Y., Tanaka, H.: A symbolic chemical system based on an abstract rewriting system and its behavior pattern. J. of Artificial Life and Robotics 6, 129–132 (2002)

The Biological Cell in Spectacle

Solomon Marcus

Romanian Academy, Bucharest, Romania
solomon.marcus@imar.ro

1 Introduction

It seems that, for some hot topics, the today means aiming to monitor and to survey the published literature in the respective fields are no longer able to accomplish their task. One of these topics is the biological cell.

Obviously, for biology the aim to investigate its functioning is for long time a basic task. It happens, however, that beginning with the middle of the past century, a lot of other disciplines became interested in the problems raised by the biological cell: physics, linguistics, mathematics, computer science, semiotics, philosophy, sociology are only some of them. Having the opportunity to attend some of their meetings and to read some of the respective papers, I was struck by the fact that, despite their different profile, they all start with statements like this: "Our aim is to understand the functioning of the biological cell". But in their next steps you hardly recognize that they have a common aim. Each of them adopts a specific terminology, a specific jargon, and has specific bibliographic references; moreover, to a large extent, each of them has its specific journals. Under these conditions, you expect that these different directions of research need to interact, but these expectations are not satisfied. They almost ignore each other.

It is scandalous! It is incredible that this may happen! It is a symptom of a grave disease of human communication. I call it a spectacle, but it is rather a spectacle belonging to the absurd; maybe we should describe it as a mixture of absurd and schizophrenia.

In the following, I will sketch some of these directions of research.

2 A First Controversy: How Old Is the Biological Cell?

We read in the *Encyclopedia of Computer Science* (Eds. Anthony Ralston et al., Oxford, Nature Publ. Group, 2000), that the first biological cell appeared 90 million years ago, but with the indication: address unknown. On the other hand, in the literature devoted to ciliates, we read that these most complex unicellular organisms on the earth appeared 2.5 billion years ago. We learn also that 200 million years ago cells were assembled into the first memory and attached to a rudimentary central processing unit.

G. Păun et al. (Eds.): WMC 2009, LNCS 5957, pp. 95–103, 2010.
© Springer-Verlag Berlin Heidelberg 2010

3 Dreams about Life, by Wolfram and Chaitin

Life is not unusual. No essential difference exists between a living being (and even a human being) and a universal Turing machine (UTM) (Stephen Wolfram, *A New Kind of Science*, Wolfram Media, October 2001). DNA is essentially a digital software. Human beings have much more DNA than viruses and bacteria. We are UTMs and we are surrounded by UTMs, but these UTMs differ in their program-size complexity. Life is a collection of UTMs whose software evolves in complexity (Gregory Chaitin, in *Bull. of EATCS*, 2002). So, Chaitin agrees with Wolfram in considering life anUTM, but he adds: life is a UTM of high program size complexity.

4 Bridging Membrane Computing and Biosemiotics

The syntagma "membrane computing" was invented in 1998, by Gheorghe Păun (G.P.) at a moment when G.P. already accumulated a considerable work in the field of formal languages and their applications to economics, linguistics and mainly to biology, related to DNA computing; see his joint monograph, with G. Rozenberg and A. Salomaa, on DNA computing, at Springer. It happened that, in the same year 1998, when membrane computing emerged, the biological membrane became a more important actor in the field of second order cybernetics and in biosemiotics. This fact stimulated us to try to bridge these two lines of development Here are some of our publications in this respect:

1. Membrane versus DNA. *Fundamenta Informaticae*, 49, 1/3 (2002), 223–227.
2. An emergent triangle: semiotics, genomics, computation. *Proc. of the International Congress of German Semiotic Society*, Kassel, 2002. CD-ROM 2003.
3. Bridging P systems and genomics. In *Membrane Computing* (eds. G. Păun, G. Rozenberg, A. Salomaa, C. Zandron), LNCS 2597, Springer, Berlin, 2003, 371–378.
4. The duality of patterning in molecular genetics. In *Aspects of Molecular Computing* (eds. N. Jonoska, G. Păun, G. Rozenberg), LNCS 2950, Springer, Berlin, 2004, 318–321.
5. The semiotics of the infinitely small: molecular computing and quantum computing. In *Semiotic Systems and Communication-Action-Interaction-Situation and Change. Proceedings of the 6th National Congress of the Hellenic Semiotic Society* (eds. K. Tsoukala et al.), Thessaloniki, 2004, 15–32.
6. Semiotic perspectives in the study of cell. In *Proceedings of the Workshop on Computatonal Models for Cell Processes* (eds. R.-J. Back, I. Petre). TUCS General Publication No. 47, 2008, Turku, Finland, 2008, 63–68.

5 Life Is DNA Software + Membrane Software

Here are some basic ideas related to the biosemiotic line of development related to membranes. Bridging this line with Păun's membrane computing seems to be an attractive, if not also a necessary investigation. Everything will argue here in favor of the slogan announced in the title of this section.

Let us first refer to Jesper Hoffmeyer ("Surfaces inside surfaces", in *Cybernetics and Human Knowing*, 5, 1 (1998), 33–42, and, the same author, "The biology of signification", in *Perspectives in Biology and Medicine*, 43, 2(2000), 252–268), claiming that "life is a surface activity", "life is fundamentally about insides and outsides". Hoffmeyer has in view the membrane and he quotes in this respect Heinz von Foerster, one of the pioneers of the second order cybernetics (differing from Norbert Wiener's cybernetics by the involvement of the subject), who proposed the Möbius strip as a topological representation of the kind of logic pertaining to self-referential cybernetic systems. Living systems may be seen as consisting essentially of surfaces inside of the surfaces. Within this framework, we can speak of an outside interior and of an inside exterior. These categories are realized through semiotic loops.

Relevant parts of the environment are internalized as an inside exterior/inner outside (the so-called Uexküll's Umwelt; see J. Uexküll, "The theory of meaning", *Semiotica*, 42,1 (1982)[1940], 25–82). The representation of certain environmental features is realized inside an organism by various means, while the interior becomes externalized as an outside interior/outer inside, in the form of the "semiotic niche" (Hoffmeyer 1998), as changed by the inside needs of the organism pertaining to that niche; see C. Emmeche, K. Kull, F. Stjernfelt, *Reading Hoffmeyer, Rethinking Biology*, Tartu Semiotic Library 3, Tartu University Press, 2002. This inside/outside interplay is made possible by the membrane strictly governing the traffic between them. Now we can claim that P systems (G. Păun, *Membrane Computing: An Introduction*. Springer, Berlin, 2002) find their starting point in this biological reality, to which a computational dimension is added. The slogan in the title of this section is just in the order of ideas of the previous slogans by Wolfram and Chaitin. There is a need to bridge genomics with membrane computing, via biosemiotics.

6 Is the Cell a Semiotic System? Modern Biology and Biosemiotics in Controversy

We will take as a guide Marcello Barbieri (*Introduction to Biosemiotics*, Walter de Gruyter, Berlin, 2007). The main question in respect to our interest here seems to be: Is the cell a semiotic system? Modern biology says: No! Biosemiotics argue in favor of an affirmative answer.

First argument for a negative answer: 'Sequences' and 'codes' are only metaphors and not fundamental entities. They can be reduced to physical entities. Semiotics' answer: Only formation of spontaneous molecules can be described by physical quantities; molecular artifacts like genes, proteins, require new basic entities, such as 'organic information', 'organic sign', 'organic meaning' defined by operative definitions, as objective and reproducible as physical quantities.

Second argument for a negative answer: Semiosis is a result of interpretation, whereas the genetic code does not depend on interpretation. Reply by biosemiotics: The qualifying feature of semiosis is not interpretation, it is 'coding', which requires only adaptors and codemakers.

Third argument for a negative answer: Cell is a duality of genotype and phenotype. Reply by biosemiotics: A semiotic system is not a couple, but a triad: sign, object, interpretant. The duality claimed by modern semiotics would be valid only if the cell had evolved from spontaneous genes and proteins; but spontaneous processes do not produce specific sequences; only manufactured molecules have biological specificity during cellular evolution and that gave origin to more complex types of cells. That means that semiosis was instrumental not only to the origin but also to the subsequent evolution of life, all the way up to the origin of language and culture. The "agents" that gave origin to the cell, therefore, were the molecular machines that manufactured genes and proteins by copying and coding and these machines form the ribotype the "third party" that still exists in every cell and that represent its codemaker, the seat of the genetic code.

Fourth argument for a negative answer: The cell does not have all the essential components of a semiotic system, because signs and meanings do not exist at the molecular level. Reply by biosemiotics: Signs and meanings are codemaker-dependent entities, whereas sequences of RNA and proteins that appear in protein synthesis have precisely the codemaking-dependence characteristics defining signs and meanings.

Fifth argument for a negative answer: There are only two types of codes in the living world: the genetic code and the cultural code. Reply by biosemiotics: If this were true, we would have to conclude that the cell did not produce any other code for most four billion years, virtually the entire history of life on earth, and it would be legitimate to doubt that it is a true semiotic system. But the genetic code was only the first of a long series of organic codes, such as the splicing code, the signal transduction codes, the compartment codes, the sequence code, the adhesive code, the sugar code.

7   The Origin of Genes

In the history of life, molecular copying came into being when the first copymakers appeared on the primitive earth and started making copies of nucleic acids. This means that natural nucleic acids had already been formed by spontaneous reactions on our planet, but that was no sign of evolution. Only the copying of genes could ensure their survival and have long-term effects, so it was really the arrival of copymaking that set in motion the chain of processes that we call evolution. The first major transition of the history of life (Maynard Smith & Szathmary, 1995) is described as the origin of genes, i.e., of the first molecular machines making copies of nucleic acids.

8   The Origin of Proteins

Unlike genes, proteins cannot be their own templates: one cannot get proteins by copying other proteins. The information for manufacturing proteins comes from genes. The important feature of the protein-makers was the ability to ensure a one-to-one correspondence between genes and proteins, otherwise it would be no biological specificity, i.e., no heredity and no reproduction; life would not exist. Protein synthesis arose from the integration of two different processes and the final machine was a code-and-template-dependent-peptide-maker, i.e., a codemaker. The second major transition of the history of life is described as the origin of proteins, i.e., of codemaking and of codemakers, the first molecular machines that discovered molecular coding and populate the earth with codified proteins.

9   Three Different Semiotic Models of the Biological Cell

Chronologically, the first model starts with Ferdinand de Saussure (1916); according to him, a semiotic system is a duality of "signifier and signified" (sign and meaning). Starting from this representation, Marcel Florkin (1974) defined the biological system as a duality of genotype and phenotype, because it is entirely accounted for by genes and proteins.

A second viewpoint was proposed by Thomas A. Sebeok, the founder of biosemiotics. He adopts the triadic scheme of Charles Sanders Peirce, according to which "there can be no semiosis in absence of interpretation"; "interpretation is a necessary and sufficient condition for something to be a semiosis" (Sebeok, 2001). So, the cell is a semiotic system only if the genetic code is based on some kind of interpretation.

As we have seen, modern biology used Sebeok's viewpoint in order to reject the claim that the biological cell is a semiotic system. But Marcello Barbieri, as we have pointed out, rejects Sebeok's viewpoint and considers that he necessary and sufficient condition to have a semiosis is to have a coding process. According to this attitude, the cell is a semiotic system, defined by the triad: genotype, phenotype, ribotype, where the ribotype is the ribo-nucleo protein system of the cell and represents its codemaker.

10   How Many Organic Codes Exist?

As we have already pointed out, modern biology claims that the genetic code is the only one existing in the organic world. Barbieri argues against this claim. He starts from the difference between copying and coding, visible in transcription and translation. In transcription, an RNA sequence is assembled from the linear information of a DNA sequence and a normal biological catalyst (an RNA polymerase) is sufficient, because each step requires a single recognition process. In translation instead, two independent recognition processes must be performed at each step, and the system that performs the reactions (the ribosome) needs special molecules, first called adaptors and then transfer RNAs, in order to associate codons to amino acids, by means of the genetic code; in its absence, we would loose biological specificity. Barbieri observes that these things can be generalized: WE are used to think that biochemical processes are all catalyzed reactions, but the difference that exists between copying and coding tells us that we should distinguish between catalyzed and codified reactions. In the first case, only one recognition process appears at each step, while in the codified reactions two independent recognition processes at each step are necessary and we need a set of coding rules. The catalyzed reactions require catalysts, the codified reactions need adaptors, i.e., catalysts plus a code. An organic code is just a set of rules of correspondence between two independent worlds, such that there is a system of molecular adaptors and a set of rules leading to biological specificity. The key molecules of the organic codes are the adaptors.

11   Organic Information

'Information' in molecular biology is a very controversial topic. When dealing with individual DNA strings, Shannon information is powerless, because Shannon's theory concerns global aspects only, i.e, the behavior of a whole system. On the other hand, Kolmogorov-Chaitin algorithmic information bridges this gap, but it leads to non-computable quantities, so it is again useless. Barbieri makes the story of organic information. In 1953, Watson and Crick proposed that the linear sequence of nucleotides represents the information carried by a gene. A few years later, the mechanism of the protein synthesis was discovered with a transfer of linear information from genes to proteins. In both types of molecules, biological information was identified and defined by the specific sequence of their subunits. All these things were expressed by saying that heredity is transmission of information. As a matter of fact, already in the 19th century the biologist Augustus von Weissmann observed that heredity cannot be described in terms of matter and energy only, a third term is needed and he proposed to call it information.

According to the physicalist thesis, heredity and organic information are ultimately physical. This would imply no essential difference between life and inanimate matter. But organic information is not only a specific sequence, it is the process of producing this sequence by copying. So, organic information is not quantitative and both Shannon's and Kolmogorov's theories are irrelevant here. Information in molecular biology and its relation to the sign remain hot topics.

12   Linguistics and Genetics

Our guide here is Wolfgang Raible, with his chapter "Linguistics and genetics: systematic parallels" in (eds.) Martin Haspelmath et al., *Language Typology and Language Universals. An International Handbook*, vol.1, Walter de Gruyter, Berlin, New York, 2001, p. 103–123. The main claim of this approach is that the biological cell follows a linguistic structure, a deep isomorphism with the natural language. The written language as a metaphor starts with the early atomists Leucippus and Democritus and continues with Aristotle's *Metaphysics*: the variety of the visible world follows from the differences between atoms and from their combinatorial capacity in the same way in which the potentially infinite variety of the Greek texts follows from the difference between the 20 letters of the Greek alphabet and from their combinatorial capacity. The same metaphor works in biology. In 1869, Friedrich Miescher discovered the existence of nucleic acid in the center of living cells; in 1893, he claims that the relation holding between letters of the alphabet and the huge number of words obtained from them could explain the relationship between the information contained in the nuclei of or cells and the variety of life forms. In 1943, Erwin Schrödinger (*What is Life?*) suggests a genetic alphabet similar to the Morse code. Mistakes in the process of reading and copying the code lead to mutations. Already in 1944, Oswald Avery observes that the nucleic acids - and not the proteins - contain the genetic information necessary to unfold the functions of a pneumonic bacterium. This fact makes Erwin Chargaff to propose the idea of a "grammar of biology". (Let us recall that the first grammar of proteins, in the sense the word 'grammar' is used in the field of dependency grammars, was proposed by Z. Pawlak in the sixties of the past century.)

All these proposals were confirmed in 1953 (F. Crick, J.Watson): the long strands of DNA (Schrödinger's punched Morse tapes) have the structure of a double helix. The metaphor of language in the form of alphabetic script is generally accepted in molecular biology.

13 The Power of the Language Metaphor in Molecular Biology

After a long period in which Darwinian biology was a source of ideas for historical linguistics, beginning with the middle of the past century the metaphorical transfer took the opposite direction. A lot of linguistic ideas invaded the field of molecular biology, as it is well illustrated by the bibliographic references in our article "Linguistic structures and generative devices in molecular genetics" (*Cahiers de Linguistique Théorique et Appliquée*, 11 (1974) 2, 77–104). The size of this transfer is now indicated by the U.S. The National Center for Biotechnology Information (NCBI), whose data base called MEDLINE, until 1997 had a subset in molecular biology (genetics). By means of this data base, permitting searching for any expression, it was possible to demonstrate the strong presence of the language and the script metaphor in the whole range of texts in molecular biology. In the recent ten years, a new metaphor emerged: the genome is like an encyclopedia.

14 Structural Similarities Between Genetics and Linguistics

In view of their common linearity, both linguistics and genetics are faced with the same problem: How can a one-dimensional medium transmit the information required for the construction of three (or even more) dimensional entities? The complexity of the information to be transmitted appears in the fact that all human beings originated from one single egg-cell whose genome was replicated billion of times by successive divisions of cells, so-called mitoses. The final result of this evolution is the three-dimensional human body. Something similar takes place in human speech, by means of which we express complex ideas and representations, involving internal structures such as sections, subsections, paragraphs, sentences, propositions, clauses, phrases, words, letters. It is rue that the breaking down of the complex representations in successive layers resulting eventually in a sequence of basic units does not have its counterpart in genetics, one can assume that genetic processes rely on principles strikingly similar to those holding in language. In both language and molecular genetics the principles allowing the reconstruction of multi-dimensional wholes from linear sequences of basic elements ar identical: double articulation; different classes of 'signs'; hierarchy; combinatorial rules on the different levels of hierarchy; linking the principles of hierarchy and combinatorial rules: wholes are always more than the sum of their parts. In linguistics, the double articulation comes from André Martinet; in the domain of life, a similar principle comes from Emile Boutroux (1875, 1991), *De la contingence des lois de la nature*(Paris, PUF), bridging in this way the seemingly antinomic Cartesian gap between mind and matter, by introducing the idea of a continuum. As have signs in language, DNA and proteins have two different functions: one of them, represented by the coding sequence of genes, is coding for proteins; the other – totally different one – represented by binding-sites, is to passively make possible a functional marking by other proteins, whose task is to activate (or to block) the process of reading specific genes. This implies the existence of two totally different kinds of genes and, as a consequence, of proteins translated from these genes: proteins that give the cell its specific shape and its typical metabolic functions and proteins whose task is to regulate the reading of other genes. The same kind of regulatory processes are necessary for the functioning of language.

Raible goes further, deeper and deeper, in the analysis of similarities and differences between linguistic systems and genetic systems. We cannot describe them here. He shows how some authors detect regularities in DNA by linguistic methods. The so-called hidden Markov models (HMMs), algorithms shaped according to the models of context-free grammars, turned out to be powerful tools for the detection of the so-called homologs: chains of amino acids sharing common function and evolutionary ancestry without being entirely identical since the function of divergent proteins may be conserved through evolution, even though sequence elements are free to change in some areas. The concept of HMMs based on the similarity of protein families(or of the underlying DNA) is used to statistically describe the so-called consensus sequence of a protein family and to detect new members belonging to the same family.

15   Other Approaches to the Biological Cell

I will make a short list of the fields I had in view in the initial section of this article: The study of biological cell, by biologists; DNA computing; Membrane computing; The study of ciliates (as it was started by G. Rozenberg); Computational molecular biology, as it has been developed within the framework of the Human Genome Project (see, for instance, its presentation by R.M. Karp (*Bul. of EATCS*, 71 (2000), 151–159 and *Notices of the Amer. Math. Soc.*, 49, 5 (2002), 544–553); The biological cell within the framework of biosemiotics; The biological cell within the framework of Linguistics and Genetics (see its synthetic presentation by Wolfgang Raible, in the Handbook quoted above); Matrix genetics (the algebraic study of genetics, by means of matrix algebras, investigating noise immunity and efficiency of discrete information transfer; see, for instance, the articles of Sergey V. Petoukhov in *Symmetry: Culture and Science*, in the last decade); Various metaphorical uses of genetics: Genetic algorithms; Richard Dawkins: Genes, beyond biology (see his books: *The Selfish Gene*, Oxford University Press, 1989-1999); *The Extended Phenotype*, Freeman, 1982; Oxford Univ. Press 1989). Books by Daniel Dennett; From genes to memes; Self organization and selection in the evolution of matter, molecules and life (topics proposed by people working in engineering fields); Approaches by physicists and engineering approaches.

All these approaches are concerned, in various ways, with the biological cell, as they were described in the first section of this article

Energy-Based Models of P Systems

Giancarlo Mauri, Alberto Leporati, and Claudio Zandron

Dipartimento di Informatica, Sistemistica e Comunicazione
Università degli Studi di Milano – Bicocca
Viale Sarca 336/14, 20126 Milano, Italy
{mauri,leporati,zandron}@disco.unimib.it

Abstract. Energy plays an important role in many theoretical computational models. In this paper we review some results we have obtained in the last few years concerning the computational power of two variants of P systems that manipulate energy while performing their computations: energy-based and UREM P systems. In the former, a fixed amount of energy is associated to each object, and the rules transform objects by manipulating their energy. We show that if we assign local priorities to the rules, then energy–based P systems are as powerful as Turing machines, otherwise they can be simulated by vector addition systems and hence are not universal. We also discuss the simulation of conservative and reversible circuits of Fredkin gates by means of (self)–reversible energy–based P systems. On the other side, UREM P systems are membrane systems in which a given amount of energy is associated to each membrane. The rules transform and move single objects among the regions. When an object crosses a membrane, it may modify the associated energy value. Also in this case, we show that UREM P systems reach the power of Turing machines if we assign a sort of local priorities to the rules, whereas without priorities they characterize the class $PsMAT^\lambda$, and hence are not universal.

1   Introduction

Membrane systems (also known as *P systems*) have been introduced in [13] as a parallel, nondeterministic, synchronous and distributed model of computation inspired by the structure and functioning of living cells. The basic model consists of a hierarchical structure composed by several membranes, embedded into a main membrane called the *skin*. Membranes divide the Euclidean space into *regions*, that contain multisets of *objects* (represented by symbols of an alphabet) and *evolution rules*. Using these rules, the objects may evolve and/or move from a region to a neighboring one. Usually, the rules are applied in a nondeterministic and maximally parallel way. A *computation* starts from an initial configuration of the system and terminates when no evolution rule can be applied. The result of a computation is the multiset of objects contained into an *output membrane*, or emitted from the skin of the system. For a systematic introduction to P systems we refer the reader to [14], whereas the latest information can be found in [17].

G. Păun et al. (Eds.): WMC 2009, LNCS 5957, pp. 104–124, 2010.

Since the introduction of P systems, many investigations have been performed on their computational properties: in particular, many variants have been proposed in order to study the contribution of various ingredients (associated with the membranes and/or with the rules of the system) to the achievement of the computational power of these systems. In this paper we review some computational features of two models of membrane systems that manipulate *energy* while performing their computations: energy-based P systems and UREM P systems.

In *energy–based P systems*, a given amount of energy is associated to each object. Moreover, instances of a special symbol are used to denote free energy units occurring inside the system. These energy units can be used to transform objects, through appropriate rules that manipulate energy, while satisfying the principle of energy conservation. In particular, if the object to which the rule is applied contains less (resp., more) energy than the one which has to be produced, then the necessary free energy units can be taken from (resp., released to) the region where the rule is applied. We assume that the application of rules consumes no energy: in particular, objects can be moved between adjacent regions of the system without energy consumption. Rules are applied in a sequential manner: at each computation step, one of the enabled rules is nondeterministically selected and applied. We show that, if a potentially infinite amount of free energy units is available, then energy–based P systems are able to simulate register machines (hence, the model is universal). This is done by assigning a form of local priorities to the rules: if two or more rules can be applied in a given region, then the one which consumes or releases the largest amount of free energy units is applied (if two or more of the enabled rules manipulate exactly the same maximal amount of free energy, then one of them is nondeterministically chosen). Instead, if we disregard priorities, then energy–based P systems can be simulated by vector addition systems, and hence are not universal. On the other hand, if we do not allow the presence of an infinite amount of energy, then the power of energy–based P systems reduces to that of finite state automata, both when considering priorities associated with the rules and when disregarding them. We also show that energy–based P systems can be used to simulate reversible and conservative (that is, energy–preserving) boolean circuits composed of Fredkin gates; the simulating P systems are themselves reversible and logically complete, and so we have the possibility to compute *any* boolean function by energy–based P systems in a reversible way.

The second model of membrane systems we consider are *P systems with unit rules and energy assigned to membranes* (UREM P systems, for short). In these systems, the rules are directly assigned to membranes (and not to the regions, as it is usually done in membrane computing). Every membrane carries an energy value that can be changed during a computation by objects passing through the membrane. Also in this case, rules are applied in the sequential way. The input, and the result of a successful computation, are considered to be the distributions of energy values carried by the membranes in the initial and in the halting configuration, respectively. We show that UREM P systems using a sort of local priority relation on the rules are Turing–complete. On the contrary, by omitting the priority relation we obtain a characterization of $PsMAT^\lambda$, the family of Parikh sets generated by context–free matrix grammars (with λ-rules and without occurrence checking). Alternatively, we can obtain Turing–completeness without using priorities, by applying rules in the maximally parallel mode.

The paper is organized as follows. In Section 2 we recall the definition of three computational models that will be used throughout the paper, to study the computational power of energy–based and UREM P systems: register machines, vector addition systems, and Fredkin circuits. In Sections 3 and 4 we review the computational power of energy–based and of UREM P systems, respectively. Section 5 concludes the paper and gives some directions for further research.

2   Preliminaries

In the following subsections we briefly recall the definition of three computational models that will be used in the rest of the paper to study the computational power of UREM and energy–based P systems.

2.1   Deterministic Register Machines

A *deterministic n–register machine* is a construct $M = (n, P, m)$, where $n > 0$ is the number of registers, P is a finite sequence of instructions (programs) bijectively labelled with the elements of the set $\{1, 2, \ldots, m\}$, 1 is the label of the first instruction to be executed, and m is the label of the last instruction of P. Registers contain non–negative integer values. The instructions of P have the following forms:

- $j : (INC(r), k)$, with $j, k \in \{1, 2, \ldots, m\}$ and $r \in \{1, 2, \ldots, n\}$
  This instruction, labelled with j, increments (by 1) the value contained in register r, and then jumps to instruction k.
- $j : (DEC(r), k, l)$, with $j, k, l \in \{1, 2, \ldots, m\}$ and $r \in \{1, 2, \ldots, n\}$
  If the value contained in register r is positive, then decrement it (by 1) and jump to instruction k. If the value of r is zero, then jump to instruction l (without altering the contents of the register).
- $m : HALT$
  Stop the execution of the program. Note that, without loss of generality, we may assume that this instruction always appears exactly once in P, with label m.

Computations start by executing the first instruction of P (labelled with 1), and terminate when they reach instruction m. Register machines provide a simple universal computational model [12]. In particular, the results proved in [5] immediately lead to the following proposition.

Proposition 1. *For any partial recursive function $f : \mathbb{N}^\alpha \to \mathbb{N}^\beta$ there exists a deterministic $(\max\{\alpha, \beta\} + 2)$–register machine M computing f in such a way that, when starting with $(n_1, \ldots, n_\alpha) \in \mathbb{N}^\alpha$ in registers 1 to α, M has computed $f(n_1, \ldots, n_\alpha) = (r_1, \ldots, r_\beta)$ if it halts in the final label m with registers 1 to β containing r_1 to r_β, and all other registers being empty. If the final label cannot be reached, then $f(n_1, \ldots, n_\alpha)$ remains undefined.*

2.2   Vector Addition Systems

Vector addition systems were introduced in [7] as a mathematical tool for analyzing systems of parallel processes. It is known that they are not Turing–complete, as they are equivalent to self–loop–free Petri nets [16]. Formally, a vector addition system (VAS, for short) is a pair $V = (B, s)$, where $B = \{b_1, b_2, \ldots, b_m\}$ is a set of m vectors, called *basis* or *displacement* vectors, and s is the *start* vector. All vectors consist of n integer values. The elements of s are non–negative (in what follows, we denote this as $s \geq 0$). The *reachability set* $R(V)$ for a VAS V is the smallest set of vectors such that: (1) $s \in R(V)$, and (2) if $x \in R(V)$, $b_j \in B$ and $x + b_j \geq 0$, then $x + b_j \in R(V)$. By considering a subset of $\beta \geq 1$ components as the output places, we can generate a set of vectors of β components by means of a VAS as follows. The VAS is started in the initial configuration. At each computation step the VAS, being in a configuration described by a vector $x \in R(V)$, chooses in a nondeterministic way a basis vector $b_j \in B$ such that $x + b_j \geq 0$ and goes to the resulting configuration $x + b_j$. The computation halts when no basis vector b_j satisfies the condition $x + b_j \geq 0$, for the current configuration x. In such a case, the values occurring at the output places of x constitute the output of the computation. Non–halting computations produce no output.

2.3   Fredkin Gates and Circuits

The *Fredkin gate* is a three–input/three–output boolean gate, whose input/output map $FG : \{0,1\}^3 \rightarrow \{0,1\}^3$ is logically reversible (that is, its inputs can always be deduced from its outputs) and preserves the number of 1's given as input. The map FG associates any input triple $(\alpha_i, \beta_i, \gamma_i)$ with its corresponding output triple $(\alpha_o, \beta_o, \gamma_o)$ according to the following relations: $\alpha_o = \alpha_i, \beta_o = (\neg\alpha_i \wedge \beta_i) \vee (\alpha_i \wedge \gamma_i)$, $\gamma_o = (\alpha_i \wedge \beta_i) \vee (\neg\alpha_i \wedge \gamma_i)$ (see the truth table in Figure 1). It is worth noting that the Fredkin gate behaves as a conditional switch, since α_i can be considered as a control line whose value determines whether the input values β_i and γ_i have to be exchanged or not: $FG(1, \beta_i, \gamma_i) = (1, \gamma_i, \beta_i)$ and $FG(0, \beta_i, \gamma_i) = (0, \beta_i, \gamma_i)$ for all $\beta_i, \gamma_i \in \{0, 1\}$.

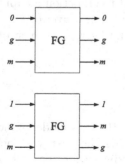

α_i β_i γ_i	$\mapsto$	α_o β_o γ_o
0 0 0		0 0 0
0 0 1		0 0 1
0 1 0		0 1 0
0 1 1		0 1 1
1 0 0		1 0 0
1 0 1		1 1 0
1 1 0		1 0 1
1 1 1		1 1 1

Fig. 1. The Fredkin gate: its behavior as a conditional switch (left) and its truth table (right)

The Fredkin gate is functionally complete for boolean logic: by fixing $\gamma_i = 0$ we obtain $\gamma_o = \alpha_i \wedge \beta_i$, whereas by fixing $\beta_i = 1$ and $\gamma_i = 0$ we obtain $\beta_o = \neg\alpha_i$. By inspecting the truth table, we can see that the Fredkin gate is also logically reversible, since the map FG is a bijection on $\{0,1\}^3$. Moreover, it is conservative: for every input/output pair the number of 1's in the input triple is the same as the number of 1's in the output triple. In other words, the output triple is obtained by applying an appropriate (input–dependent) permutation to the input triple.

The Fredkin gate is the basis of the model of conservative logic introduced in [2], which describes computations by considering some notable properties of microdynamical laws of physics, such as reversibility and the conservation of the internal energy of the physical system by which computations are performed. Within that model, computations are performed by reversible *Fredkin circuits*, which are acyclic and connected directed graphs made up of *layers* of Fredkin gates. Figure 2 depicts an example of Fredkin circuit having three gates arranged in two layers. The evaluation of a Fredkin circuit in topological order (i.e. layer by layer) defines the boolean function computed by the circuit, which is obtained as the composition of the functions computed by each layer. The conservativeness of the circuit (preservation of the number of 1's) is equivalent to the requirement that the output n-tuple is obtained by applying an appropriate (input–dependent) permutation to the corresponding input n-tuple.

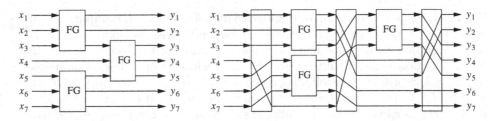

Fig. 2. A reversible Fredkin circuit (on the left) and its normalized version

A *reversible* n–input Fredkin circuit is a Fredkin circuit FC_n which computes a bijective map $f_{FC_n} : \{0,1\}^n \to \{0,1\}^n$. Note that the function computed by a reversible Fredkin circuit is also conservative: in fact, every layer of FC_n is composed by Fredkin gates, which are conservative, and by wires, which obviously preserve the number of 1's given as input.

3   Energy–Based P Systems

In this section we consider *energy–based P systems* [10,11], a model of membrane systems whose computations occur by manipulating the energy associated to the objects, as well as the free energy units occurring inside the regions of the system. These energy units can be used to transform objects, using appropriate rules, which are defined according to conservativeness considerations. Formally, an energy–based P system of degree $m \geq 1$, as defined in [11], is a construct $\Pi = (A, \varepsilon, \mu, e, w_1, \ldots, w_m, R_1, \ldots, R_m, i_{\text{in}}, i_{\text{out}})$ where:

- A is an alphabet; its elements are called *objects*;
- $\varepsilon : A \to \mathbb{N}$ is a mapping that associates to each object $a \in A$ the value $\varepsilon(a)$ (also denoted by ε_a), which can be viewed as the "energy value of a". If $\varepsilon(a) = \ell$, we also say that object a *embeds* ℓ units of energy;
- μ is a hierarchical membrane structure consisting of m membranes, each labelled in a unique way with a number in the set $\{1, \ldots, m\}$;
- $e \notin A$ is a special symbol that denotes one *free energy* unit, that is, one unit of energy which is not embedded into any object;
- w_i, with $i \in \{1, \ldots, m\}$, specifies the multiset (over $A \cup \{e\}$) of objects initially present in region i. In what follows we will sometimes assume that the number of e's in some regions of the system is unbounded. In any case, the number of objects from A will always be bounded;
- R_i, with $i \in \{1, \ldots, m\}$, is a finite set of multiset rewriting rules over $A \cup \{e\}$ associated with region i. Rules can be of the following types:

$$ae^k \to (b, p) \, , \qquad a \to (b, p)e^k \, , \qquad e \to (e, p) \, , \qquad a \to (b, p)$$

where $a, b \in A$, $p \in \{\text{here}, \text{in}(name), \text{out}\}$ and k is a non negative integer. Rules satisfy the *conservativeness condition*, whereby the sum of all (free and embedded) energy values appearing in the left hand side of each rule equals the sum of all (free and embedded) energy values in the corresponding right hand side;

- i_{in} is an integer between 1 and m and specifies the input membrane of Π;
- i_{out} is an integer between 0 and m and specifies the output membrane of Π. If $i_{\text{out}} = 0$ then the environment is used for the output, that is, the output value is the multiset of objects over $A \cup \{e\}$ ejected from the skin.

When a rule of the type $ae^k \to (b, p)$ is applied, the object a, in presence of k free energy units, is allowed to be transformed into object b (note that $\varepsilon_a + k = \varepsilon_b$, for the conservativeness condition). If $p = $ here, then the new object b remains in the same region; if $p = $ out, then b exits from the current membrane. Finally, if $p = \text{in}(name)$, then b enters into the membrane labelled with $name$, which must be directly contained inside the current membrane in the membrane hierarchy. The meaning of rule $a \to (b, p)e^k$, where k is a positive integer number, is similar: the object a is allowed to be transformed into object b by releasing k units of free energy (here, $\varepsilon_a = \varepsilon_b + k$). As above, the new object b may optionally move one level up or down into the membrane structure. The k free energy units might then be used by another rule to produce "more energetic" objects from "less energetic" ones. When $k = 0$ the rule $ae^k \to (b, p)$, also written as $a \to (b, p)$, transforms the object a into the object b (note that in this case $\varepsilon_b = \varepsilon_a$) and moves it (if $p \neq $ here) upward or downward into the membrane hierarchy, without acquiring or releasing any free energy unit. Analogously, rules $e \to (e, p)$ simply move (if $p \neq $ here) one unit of free energy upward or downward into the membrane structure.

An important observation concerns the application of rules. In the original definition of energy–based P systems, given in [10], the rules were applied in the maximally parallel way, as it usually happens in membrane systems. In the next section we will assume instead that the rules are applied in the *sequential* manner: at each computation step (a global clock is assumed), exactly one among the enabled rules is nondeterministically chosen and applied in the system. We will return to the maximally parallel mode of application in the subsequent section, where we will simulate Fredkin gates and circuits.

A configuration of Π is the tuple $(M_1, \ldots, M_m)$ of multisets (over $A \cup \{e\}$) of objects contained in each region of the system; $(w_1, \ldots, w_m)$ is the *initial* configuration. A configuration where no rule can be further applied is said to be *final*. A computation is a sequence of transitions between configurations of Π, starting from the initial one. A computation is *successful* if and only if it reaches a final configuration or, in other words, it *halts*. The multiset $w_{i_{\text{in}}}$ of objects occurring inside the input membrane is the *input* for the computation, whereas the multiset of objects occurring inside the output membrane (or ejected from the skin, if $i_{\text{out}} = 0$) in the final configuration is the *output* of the computation. A non–halting computation produces no output. As an alternative, we can consider the Parikh vectors associated with the multisets, and see energy–based P systems as computing devices that transform (input) Parikh vectors to (output) Parikh vectors. Optionally, we can disregard the number of free energy units that occur in the input and in the output region of the system, when defining the input and the output multisets (or Parikh vectors).

Since energy is an additive quantity, it is natural to define the *energy of a multiset* as the sum of the amounts of energy associated to each instance of the objects which occur into the multiset. Similarly, the energy of a configuration is the sum of the amounts of energy associated to each multiset which occurs into the configuration. A *conservative computation* is a computation where each configuration has the same amount of energy. A *conservative energy–based P system* is an energy–based P system that performs only conservative computations.

In what follows we will sometimes consider a slightly modified version of energy–based P systems as defined above, in which there are $\alpha \geq 1$ input membranes and $\beta \geq 1$ output membranes. As it will become clear in the following, this modification does not increase the computational power of energy–based P systems; this is due to the fact that, for any fixed value of $\alpha \geq 1$ (resp., $\beta \geq 1$), the set $\mathbb{N}^\alpha$ (resp., $\mathbb{N}^\beta$) is isomorphic to $\mathbb{N}$, as it is easily shown by using the Cantor mapping. Sometimes we will also use energy–based P systems as generating devices: we will disregard the input membrane, and will consider the multisets (or Parikh vectors) produced in the output membrane at the end of the (halting) nondeterministic computations of the system.

3.1 Computational Power

In this section we recall some results, taken from [8], concerning the computational power of energy–based P systems.

Let Π be an energy–based P system as formally defined above. First of all we observe that if we assume that the number of free energy units is *bounded* in each region of Π, then only a finite number of distinct configurations can be obtained, starting from the initial configuration. In fact, each object of Π can only be transformed into another object (it can never be created or destroyed), and possibly moved to another region, according to the rules listed in the definition of the system. In the "worst" case, every object can be transformed into any other object, and can be sent to any region of Π; however, also in this case the number of possible combinations is finite, and thus we obtain a finite number of configurations. By associating a state to each possible configuration of Π, it is not difficult to see that bounded energy–based P systems can be simulated by finite state automata: an arc of the state diagram connects two vertices u and v if and only if the configuration of Π that corresponds to v can be obtained in one step (that is, by applying one rule) from the configuration that corresponds to u.

In order to compare the computational power of energy–based P systems with that of Turing machines, from now on we assume that, in the initial configuration, some regions of the system contain an unlimited number of free energy units. Moreover, we define the following *local* priorities associated to the rules of the system: in each region, if two or more rules can be applied at a given computation step, then one of the rules that manipulate the *maximum* amount of free energy units is nondeterministically chosen and applied. Clearly, even if we impose this policy on energy–based P systems that have a bounded amount of free energy units in each region, we cannot go beyond the computational power of finite state automata.

Assuming an infinite amount of free energy units in the initial configuration, energy–based P systems with priorities assigned to the rules are universal, as stated in the following theorem.

Theorem 1. *Every partial recursive function $f : \mathbb{N}^\alpha \to \mathbb{N}^\beta$ can be computed by an energy–based P system with an infinite supply of free energy units and priorities assigned to rules, with (at most) $\max\{\alpha, \beta\} + 3$ membranes.*

Proof. We prove this proposition by simulating deterministic register machines. Let $M = (n, P, m)$ be a deterministic n–register machine that computes f. Observe that, according to Proposition 1, $n = \max\{\alpha, \beta\} + 2$ is enough.

The input values $x_1, \ldots, x_\alpha$ are expected to be in the first α registers of M, and the output values are expected to be in registers 1 to β at the end of a successful computation. Moreover, without loss of generality, we may assume that at the beginning of a computation all registers except (possibly) registers 1 to α contain zero. We construct the energy–based P system $\Pi = (A, \varepsilon, \mu, e, w_s, w_1, \ldots, w_n, R_s, R_1, \ldots, R_n)$ where:

- $A = \{p_j : j \in \{1, 2, \ldots, m\}\} \cup \{\widetilde{p}_j : j \in \{1, 2, \ldots, m-1\}$ and j is the label of an INC instruction$\} \cup \{p'_j : j \in \{1, 2, \ldots, m-1\}$ and j is the label of a DEC instruction$\}$;

- $\varepsilon : A \to \mathbb{N}$ is defined as follows:
  - $\varepsilon(p_j) = 2$ for all $j \in \{1, 2, \ldots, m\}$;
  - $\varepsilon(\widetilde{p}_j) = 1$ for all $j \in \{1, 2, \ldots, m-1\}$ such that j is the label of an INC instruction;
  - $\varepsilon(p'_j) = 3$ for all $j \in \{1, 2, \ldots, m-1\}$ such that j is the label of a DEC instruction;
- $\mu = [_s[_1]_1 \cdots [_\alpha]_\alpha \cdots [_n]_n]_s$ (note that label s denotes the skin membrane);
- $w_s = \{p_1\}$, plus an infinite supply of free energy units;
- $w_i = \begin{cases} \{e^{x_i}\} & \text{if } 1 \le i \le \alpha \\ \emptyset & \text{if } \alpha+1 \le i \le n \end{cases}$
- $R_s = \{p_j \to (p_j, in(r)) : j \in \{1, 2, \ldots, m-1\}$ and the j-th instruction of P operates on register $r\} \cup \{\widetilde{p}_j e \to (p_\ell, here) : j \in \{1, 2, \ldots, m-1\}$ and j is the label of an INC instruction that jumps to label $\ell\} \cup \{p'_j \to (p_{\ell_1}, here)e :$ $j \in \{1, 2, \ldots, m-1\}$ and j is the label of a DEC instruction whose first jump label is $\ell_1\}$;
- $R_i = \{p_j \to (\widetilde{p}_j, out)e : j \in \{1, 2, \ldots, m-1\}$ and j is the label of an INC instruction that affects register $i\} \cup \{p_j e \to (p'_j, out) : j \in \{1, 2, \ldots, m-1\}$ and j is the label of a DEC instruction that affects register $i\} \cup \{p_j \to (p_{\ell_2}, out) : j \in \{1, 2, \ldots, m-1\}$ and j is the label of a DEC instruction that affects register i and whose second jump label is $\ell_2\}$, for all $i \in \{1, 2, \ldots, n\}$.

Informally, the system is composed of the skin membrane, that contains one elementary membrane for each register of M. At each moment during the computation, the value r_i contained in register i, $1 \le i \le n$, is represented by the number of free energy units contained in the i-th elementary membrane. Hence, the elementary membranes from 1 to α contain the input at the beginning of the computation, whereas the elementary membranes from 1 to β contain the output if and when the computation halts. The region enclosed by the skin contains one object of the kind p_j, $j \in \{1, 2, \ldots, m\}$, which represents the value j (that is, the instruction labelled with j) of the program counter of M. To simulate the instruction $j : (INC(r), \ell)$, the object p_j enters into the region r thanks to the rule $p_j \to (p_j, in(r))$. In this region, p_j is transformed into $\widetilde{p}_j$ by means of the rule $p_j \to (\widetilde{p}_j, out)e$, thus releasing one free energy unit, while the resulting object $\widetilde{p}_j$ is sent back to the region enclosed by the skin. There, a rule of the kind $\widetilde{p}_j e \to (p_\ell, here)$ produces the object which represents the label of the next instruction to be executed. As we can see, the application of this rule requires the presence of a free energy unit in the region enclosed by the skin.

To simulate the instruction $j : (DEC(r), \ell_1, \ell_2)$, the object p_j, which occurs in the region enclosed by the skin, enters into region r by means of the rule $p_j \to (p_j, in(r))$. Assuming that there is at least one free energy unit inside region r, the object p_j can be transformed into p'_j thanks to the rule $p_j e \to (p'_j, out)$. One free energy unit is thus consumed in region r, and the resulting object is sent back to the region enclosed by the skin. There, it is transformed into p_{ℓ_1} thanks to the rule $p'_j \to (p_{\ell_1}, here)e$, by releasing one unit of free energy. On the other hand, if membrane r does not contain free energy units (and *only* in this case) then object p_j – just arrived from the region enclosed by the skin – is transformed into p_{ℓ_2} by means of the rule: $p_j \rightarrow (p_{\ell_2}, out)$. In this case no free energy units are involved in the transformation, and the resulting object is immediately sent to the region enclosed by the skin. Note that the correct simulation of the DEC instruction is guaranteed by the priorities associated with the rules: when object p_j enters into membrane r, then the rule $p_j e \rightarrow (p'_j, out)$ has priority over the rule $p_j \rightarrow (p_{\ell_2}, out)$, since it manipulates more free energy units than the other.

The halt instruction is simply simulated by doing nothing with the object p_m when it appears in region s. It is apparent from the description given above that, after the simulation of each instruction, the number of free energy units contained into membrane i equals the value contained in register i, with $1 \leq i \leq n$. Hence, when the halting symbol p_m appears in region s, the contents of membranes 1 to β equal the output of the program P.   □

The following corollary is an immediate consequence of Theorem 1, by taking $\beta = 0$.

Corollary 1. *Let $L \subseteq \mathbb{N}^\alpha$, $\alpha \geq 1$, be a recursively enumerable set of (vectors of) non–negative integers. Then L can be accepted by an energy–based P system with an infinite supply of free energy units and priorities assigned to rules, with (at most) $\alpha + 3$ membranes.*

For the generating case we have to simulate *nondeterministic* register machines, which are defined exactly as the deterministic version, the only difference being in the INC instruction, that now has the form $j : (INC(r), k, \ell)$; when executing this instruction, after incrementing register r, the computation continues nondeterministically either with the instruction labelled by k or with the instruction labelled by ℓ. The necessary changes in the above simulation are obvious, and hence are here omitted. Under this setting, the following corollary is also an immediate consequence of Theorem 1, by taking $\alpha = 0$.

Corollary 2. *Let $L \subseteq \mathbb{N}^\beta$, $\beta \geq 1$, be a recursively enumerable set of (vectors of) non–negative integers. Then L can be generated by an energy–based P system with an infinite supply of free energy units and priorities assigned to rules, with (at most) $\beta + 3$ membranes.*

On the other hand, if we assume that an infinite amount of free energy units occurs in the initial configuration but no priorities are assigned to the rules, then energy–based P systems are *not* universal, as proved in the following theorem.

Theorem 2. *Energy–based P systems with an infinite supply of free energy units, and without priorities assigned to the rules, can be simulated by vector addition systems.*

Proof. Let Π be an energy–based P system that contains an infinite supply of free energy units in its initial configuration. Denoted by m the degree of Π, by n the cardinality of the alphabet A, and by R the total number of rules in Π, we define a vector addition system $V = (B, s)$, with $B = \{b_1, b_2, \ldots, b_R\}$, as follows. The vectors $s, b_1, b_2, \ldots, b_R$ have one component for each possible object/region pair (a, i) of Π, that is, for all $a \in A \cup \{e\}$ and $i \in \{1, 2, \ldots, m\}$

(note that here we treat e just like the objects of A). The start vector s reflects the initial configuration of Π: for all $a \in A \cup \{e\}$ and for all $i \in \{1, 2, \ldots, m\}$, the component of s associated with the pair (a, i) is set to the number of copies of a in the i-th region of Π. The only exception is given for those regions of Π where an infinite number of free energy units occur: the corresponding components of s are initialized with E, which is defined as the maximum number of free energy units which are necessary to execute *any* rule of Π (formally, $E = \max\{k \,|\, ae^k \rightarrow (b, p)$ is a rule of $\Pi\}$). So doing, we are able to initialize every component of s with a finite value.

Each rule of the kind $ae^k \rightarrow (b, p) \in R_i$ is translated into a basis vector $b_l \in B$, $l \in \{1, \ldots, R\}$, as follows: since one copy of a and k copies of e are removed from region i, the component of b_l that corresponds to the pair (a, i) will be equal to -1, and the component that corresponds to (e, i) will be equal to $-k$. Similarly, denoted by j the region determined by the target p, since one copy of b will be sent to region j, the corresponding component of b_l will be equal to 1. Rules of the kind $a \rightarrow (b, p)e^k$, as well as rules of the kind $a \rightarrow (b, p)$ and $e \rightarrow (e, p)$, are translated into appropriate basis vectors in a similar way. An important observation is that each component of the basis vectors that corresponds to a pair (e, i), such that region i of Π contains an infinite supply of free energy units in its initial configuration, is set equal to E. So doing, at each computation step E copies of e are added to those components of the VAS which correspond to the regions of Π that contain an infinite amount of e. Thus, at the beginning of the next computation step, such components have a value which is finite but sufficiently high to simulate any rule of Π.

It is clear that any feasible sequential computation of Π corresponds to a sequence of applications of basis vectors of V, and that for each pair (a, i), with $a \in A \cup \{e\}$ and $i \in \{1, 2, \ldots, m\}$, the number of copies of object a in the region i of Π after the application of a rule matches the value of the component of the state vector that corresponds to (a, i), with the exception of the pairs (e, i) for those regions i of Π that contain an infinite number of free energy units in the initial configuration. However, any multiset (or its corresponding Parikh set) generated by Π can also be generated by V by means of the above simulation.   □

3.2   Simulating the Fredkin Gate

Let us now describe an energy–based P system which simulates the Fredkin gate. The results contained in this section are taken from [10,11]; as stated above, we switch to the maximally parallel mode of applying the rules.

The system, illustrated in Figure 3, is defined as follows. The alphabet contains 12 kinds of objects. For the sake of clarity, we denote these objects by $[b, j]$ and $[c, j]$, with $b \in \{0, 1\}$, $c \in \{0', 1'\}$ and $j \in \{1, 2, 3\}$. Intuitively, $[b, j]$ and $[c, j]$ indicate the boolean value which occurs in the j-th line of the Fredkin gate. It will be clear from the simulation that we need two different symbols to represent each of these boolean values. Every object of the kind $[b, j]$, with $b \in \{0, 1\}$ and $j \in \{1, 2, 3\}$, has energy equal to 3, whereas the objects $[c, 1]$ have energy equal to 1 and the objects $[c, 2]$ and $[c, 3]$ (with $c \in \{0', 1'\}$) have energies equal to 4.

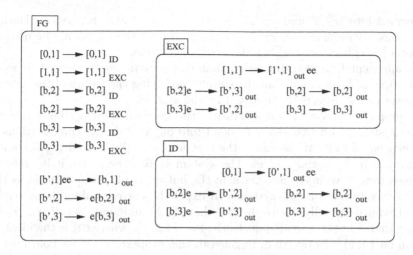

Fig. 3. An energy–based P system which simulates the Fredkin gate

The simulation works as follows. The input values $[x_1, 1], [x_2, 2], [x_3, 3]$, with $x_1, x_2, x_3 \in \{0, 1\}$, are injected into the skin. If $x_1 = 0$ then the object $[0, 1]$ enters into membrane ID, where it is transformed to the object $[0', 1]$ by releasing 2 units of energy. The object $[0', 1]$ leaves membrane ID and waits for 2 energy units to transform back to $[0, 1]$ and leave the system. The objects $[x_2, 2]$ and $[x_3, 3]$, with $x_2, x_3 \in \{0, 1\}$, may enter nondeterministically either into membrane ID or into membrane EXC; however, if they enter into EXC they cannot be transformed to $[x_2', 3]$ and $[x_3', 2]$ since in EXC there are no free energy units. Thus the only possibility for objects $[x_2, 2]$ and $[x_3, 3]$ is to leave EXC and choose again between membranes ID and EXC in a nondeterministic way. Eventually, after some time they enter (one at the time or simultaneously) into membrane ID. Here they have the possibility to be transformed into $[x_2', 2]$ and $[x_3', 3]$ respectively, using the 2 units of free energy which occur into the region enclosed by ID (alternatively, they have the possibility to leave ID and choose nondeterministically between membranes ID and EXC once again). When the objects $[x_2', 2]$ and $[x_3', 3]$ are produced they immediately leave ID, and are only allowed to transform back to $[x_2, 2]$ and $[x_3, 3]$ respectively, releasing 2 units of energy. The objects $[x_2, 2]$ and $[x_3, 3]$ just produced leave the system, and the 2 units of energy can only be used to transform $[0', 1]$ back to $[0, 1]$ and expel it from the skin.

On the other hand, if $x_1 = 1$ then the object $[1, 1]$ enters into membrane EXC where it is transformed into the object $[1', 1]$ by releasing 2 units of energy. The object $[1', 1]$ leaves membrane EXC and waits for 2 energy units to transform back to $[1, 1]$ and leave the system. Once again the objects $[x_2, 2]$ and $[x_3, 3]$, with $x_2, x_3 \in \{0, 1\}$, may choose nondeterministically to enter either into membrane ID or into membrane EXC. If they enter into ID they can only exit again since in ID there are no free energy units. When they enter into EXC they can be transformed to $[x_2', 3]$ and $[x_3', 2]$ respectively, using the 2 free energy units which occur into the region, and leave EXC. Now objects $[x_2', 3]$ and $[x_3', 2]$ can only be transformed into $[x_2, 3]$ and $[x_3, 2]$ respectively, and leave the system. During this transformation 2 free energy units are produced; these can only be used to transform $[1', 1]$ back to $[1, 1]$, which leaves the system.

It is apparent from the simulation that the system can be defined to work on any triple of lines of a circuit, by simply modifying the values of the second component of the objects manipulated by the system.

The proposed P system is conservative: the number of energy units present into the system (both free and embedded into objects) during computations is constantly equal to 9. At the end of the computation, all these energy units are embedded into the output values. The system is also reversible: it is immediately seen that if we inject into the skin the output triple just produced as the result of a computation, the system will expel the corresponding input triple. This behavior is trivially due to the fact that the Fredkin gate is *self–reversible*, meaning that FG ∘ FG = ID$_3$ (equivalently, FG = FG^{-1}), where ID$_3$ is the identity function on $\{0, 1\}^3$. Notice that, in general, this property does not hold for the functions $f : \{0, 1\}^n \to \{0, 1\}^n$ computed by n–input reversible Fredkin circuits. This means that in general the P system that simulates a given Fredkin circuit must be appropriately designed in order to be self–reversible.

3.3   Simulation of Reversible Fredkin Circuits

Basing upon the simulation of the Fredkin gate we have exposed in the previous section, in [11] we have shown that any reversible Fredkin circuit can be simulated by an appropriate energy–based P system. Since the construction is quite involved, in what follows we just give a few details.

Let FC_n be an n–input reversible Fredkin circuit of depth d, and let $L_1, L_2, \ldots, L_d$ denote the layers of FC_n. As we can see on the left side of Figure 2, each layer is composed by some number of Fredkin gates and some non–intersecting wires. Let k_j, with $j \in \{1, 2, \ldots, d\}$, be the number of Fredkin gates occurring in layer L_j. First of all we define the P systems $G_{j,i}$, for $j \in \{1, 2, \ldots, d\}$ and $i \in \{1, 2, \ldots, k_j\}$, by modifying the P system FG exposed in the previous section as follows. The objects of $G_{j,i}$ are denoted by $[b, \ell, j]$ and $[c, \ell, j]$, with $b \in \{0, 1\}$, $c \in \{0', 1'\}$, $\ell \in \{\ell_1, \ell_2, \ell_3\} \subseteq \{1, 2, \ldots, n\}$ such that $\ell_1 \neq \ell_2 \neq \ell_3$, and $j \in \{1, 2, \ldots, d\}$. Intuitively, $G_{j,i}$ simulates the i-th Fredkin gate occurring in layer L_j of FC_n, and $[b, \ell, j]$, $[c, \ell, j]$ indicate the boolean value which occurs in the ℓ-th line of L_j. The values ℓ_1, ℓ_2 and ℓ_3 correspond to the three lines of the circuit upon which the Fredkin gate operates. The objects $[b, \ell, j]$ have energy equal to 3, whereas the energy of objects $[c, \ell_1, j]$ is 1 and the energy of objects $[c, \ell_2, j]$ and $[c, \ell_3, j]$ is equal to 4. The system $G_{j,i}$ processes the objects $[b, \ell, j]$ given as input exactly as FC would process the corresponding objects $[b, \ell]$, with the only difference that, when it expels the results of the computation in its environment, it changes objects $[b, \ell, j]$ to $[b, \ell, j + 1]$. This is done in order to indicate that the simulation of FC_n can continue with the next layer.

We can now build an energy–based P system P_n which simulates FC_n as follows. To simplify the exposition, we will consider the P systems $G_{j,i}$ defined above as black boxes that, when fed with input values (represented as appropriate

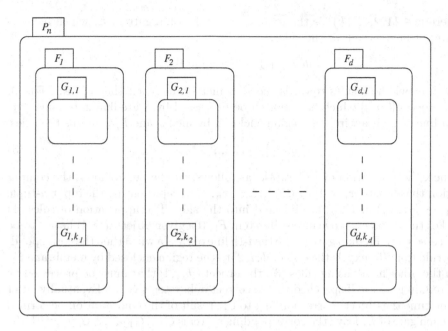

Fig. 4. Structure of the P system P_n which simulates an n–input reversible Fredkin circuit FC_n. Every subsystem F_j simulates the corresponding layer L_j of FC_n, whereas the subsystems $G_{j,i}$ simulate the Fredkin gates occurring in L_j.

objects), after some time produce their results. The objects of P_n are denoted by $[b, i, j]$, with $b \in \{0, 1\}$, $i \in \{1, 2, \ldots, n\}$ and $j \in \{1, 2, \ldots, d+1\}$. The energy of all these objects is equal to 3. As before, $[b, i, j]$ indicates the presence of the boolean value b on the i-th input line of the j-th layer of FC_n. Note that some of these objects are also used in subsystems $G_{j,i}$. The system P_n, illustrated in Figure 4, is composed by a main membrane (the skin) that contains a subsystem F_j for each layer L_j of FC_n. Every subsystem F_j simulates the corresponding layer L_j of the circuit, using the subsystems $G_{j,1}, G_{j,2}, \ldots, G_{j,k_j}$ to simulate the Fredkin gates which occur in L_j. The region associated to the skin membrane contains the rules:

$$[b, i, j] \to [b, i, j]_{F_j} \tag{1}$$

and the rules:

$$[b, i, d+1] \to [b, i, d+1]_{out} \tag{2}$$

for every $b \in \{0, 1\}$, $i \in \{1, \ldots, n\}$ and $j \in \{1, \ldots, d\}$. The application of rules (1) makes the objects representing the boolean values occurring in the i-th input line of layer L_j move into subsystem F_j, whereas rules (2) expel the result of the simulation to the environment. The region associated to membrane F_j, for $j \in \{1, 2, \ldots, d\}$, contains the rules:

$$[b, i, j] \to [b, i, j]_{G_{j, r_i}} \tag{3}$$

where $r_i \in \{1, 2, \ldots, k_j\}$ is the number of the Fredkin gate in L_j which has i as an input line, as well as the rules:

$$[b, i, j + 1] \rightarrow [b, i, j + 1]_{out} \tag{4}$$

which expel the results towards the skin membrane when they appear. For all the objects $[b, i, j]$ which have not to be processed by a Fredkin gate (since the i-th line of L_j is a wire) the region enclosed by membrane F_j contains the rules:

$$[b, i, j] \rightarrow [b, i, j + 1]_{out} \tag{5}$$

Hence, the simulation of FC_n works as follows. At the beginning of the computation the objects $[x_1, 1, 1], [x_2, 2, 1], \ldots, [x_n, n, 1]$, representing the input n-tuple $(x_1, x_2, \ldots, x_n)$ of FC_n, are injected into the skin. The application of rules (1) makes these objects move into subsystem F_1. If a given object $[b, i, 1]$ hasn't to be processed by a Fredkin gate (since the i-th line of L_1 is a wire) then the corresponding rule from (5) expels the object $[b, i, 2]$ to the region enclosed by membrane F_1. On the other hand, using rules (3), the objects $[b, i, 1]$ that must be processed by a Fredkin gate are dispatched to the correct subsystems G_{1,r_i}. Eventually, after some time the objects corresponding to the result of the computation performed by each gate of L_1 leave the corresponding systems $G_{1,1}, G_{1,2}, \ldots, G_{1,k_1}$, with the third component incremented by 1. These objects are expelled from F_1 using rules (4). As objects $[b, i, 2]$ are expelled from F_1, rules (1) dispatch them to subsystem F_2. The simulation of FC_n continues in this way until the objects $[b, i, d+1]$ leave the subsystem F_d. Here they activate rules (2), that expel them into the environment as the result of the computation performed by P_n.

The formal definition of P_n can be found in [11]. Let us note that the system is conservative, since the amount of energy units present into the system (both free and embedded into objects) during computations is constantly equal to $3n$. The number of rules and the number of membranes in the system are directly proportional to the number of gates in FC_n. Differently from the other approaches seen in literature, the depth of hierarchy μ in system P_n is constant; in particular, it does not depend upon the number of gates occurring in FC_n.

Reverse computations. If a Fredkin circuit FC_n is reversible, then there exists a Fredkin circuit FC'_n which computes the inverse function $f_{FC_n}^{-1} : \{0,1\}^n \rightarrow \{0,1\}^n$. This circuit can be easily obtained from FC_n by reversing the order of all layers. Actually, in [11] we have shown that the P system P_n that simulates FC_n can be modified in order to become *self–reversible*, that is, able to compute both f_{FC_n} and $f_{FC_n}^{-1}$. To this aim, we add a further component $k \in \{0, 1\}$ to the objects of P_n, which is used to distinguish between "forward" and "backward" computations. Precisely, the objects which are used to compute f_{FC_n} have $k = 0$, and those used to compute $f_{FC_n}^{-1}$ have $k = 1$. A forward computation starts by injecting the objects $[x_1, 1, 1, 0], [x_2, 2, 1, 0], \ldots, [x_n, n, 1, 0]$ into the skin of P_n. The computation proceeds as described above, with the rules modified in order to consider the presence of the new component $k = 0$. The objects produced in output are $[y_1, 1, d+1, 0], \ldots, [y_n, n, d+1, 0]$, where $(y_1, \ldots, y_n) = f_{FC_n}(x_1, \ldots, x_n)$.

Analogously, a "backward" computation should start by injecting the objects $[y_1, 1, 1, 1], [y_2, 2, 1, 1], \ldots, [y_n, n, 1, 1]$ into the skin. The computation of $f_{FC_n}^{-1}$ can be accomplished by incorporating the rules of the region enclosed by the skin and the subsystems of P_n' (both modified in order to take into account the presence of the new component $k = 1$) into P_n. Interferences between the rules concerning forward and backward computations do not occur since they act on different kinds of objects.

A further improvement is obtained by observing that each layer of FC_n is self–reversible, and that the layers of FC_n' are the same as the layers of FC_n, in reverse order. Hence we can merge each subsystem F_j, which simulates layer L_j of FC_n, with the subsystem F_{d-j+1}', which simulates layer L_{d-j+1}' of FC_n'. The merge operation consists in putting the rules and the subsystems of F_{d-j+1}' into F_j. Of course we have also to modify the rules in the region enclosed by the skin so that the objects that were previously moved to F_{d-j+1}' are now dispatched to F_j. Recursively, since each Fredkin gate is self–reversible, we can merge also subsystems $G_{j,1}, \ldots, G_{j,k_j}$ occurring into F_j with the corresponding subsystems $G_{d-j+1,1}', \ldots, G_{d-j+1,k_j}'$ which occur into F_{d-j+1}'. In this way, we obtain a self–reversible P system which is able to compute both f_{FC_n} and $f_{FC_n}^{-1}$. The new system has the same number of membranes as P_n, and the double of rules.

Reducing the Number of Subsystems. As we have seen in the previous sections, the number of membranes and the number of rules of the P system P_n that simulates the reversible Fredkin circuit FC_n grow linearly *with respect to the number of gates occurring in the circuit*. Actually, the number of membranes in P_n can be made linear with respect to n, *independently of the number of gates* occurring in the simulated Fredkin circuit FC_n. To compensate the reduced number of membranes, the number of rules in the system will grow accordingly. For the sake of simplicity, let us consider only forward computations, involving objects of the kind $[b, i, j]$, with $b \in \{0, 1\}$, $i \in \{1, \ldots, n\}$ and $j \in \{1, \ldots, d+1\}$.

First of all, every n–input reversible Fredkin circuit FC_n can be "normalized" by moving the Fredkin gates contained into each layer as upward as possible, as illustrated on the right side of Figure 2. The resulting layers are called *normalized* layers. In order to keep track of which input value goes into which gate, we precede each normalized layer by a fixed (that is, non input–dependent) permutation, which is realized by rearranging the wires as required. A final fixed permutation, occurring after the last normalized layer, allows the output values of FC_n to appear on the correct output lines. Observe that the number of possible n–input normalized layers of Fredkin gates is $\lfloor \frac{n}{3} \rfloor$. We can thus number all possible normalized layers with an index $\ell \in \{1, \ldots, \lfloor \frac{n}{3} \rfloor\}$, and describe a normalized Fredkin circuit by a sequence of indexes $\ell_1, \ell_2, \ldots, \ell_d$ together with a corresponding sequence of fixed permutations $\pi_1, \pi_2, \ldots, \pi_{d+1}$.

The normalization of every layer L_j of FC_n can be performed in linear time with respect to n, as described in [11]. The time needed to normalize the entire circuit is thus bounded by $O(n \cdot d)$, the size of the circuit.

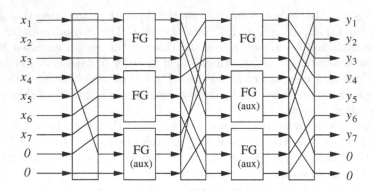

Fig. 5. A normalized Fredkin circuit with auxiliary lines and gates. The number of gates is the same in each layer.

An energy–based P system that simulates a normalized Fredkin circuit can be built by composing (at most) the $\lfloor \frac{n}{3} \rfloor$ subsystems $F_1, \ldots, F_{\lfloor n/3 \rfloor}$, each one capable to simulate a fixed normalized layer of Fredkin gates. The region enclosed by the skin contains the rules $[b, i, j] \rightarrow [b, \pi_j(i), j]_{F_{\ell_j}}$ for all $b \in \{0, 1\}$, $i \in \{1, \ldots, n\}$ and $j \in \{1, \ldots, d\}$, as well as the rules $[b, i, d+1] \rightarrow [b, \pi_{d+1}(i), d+1]_{out}$. These rules implement the fixed permutations, move the objects to the subsystem that simulates the next normalized layer, and expel the results of the computation into the environment. The simulation of each normalized layer is analogous to the simulation of the layers of a non–normalized Fredkin circuit, as described above. Note that the objects emerge from subsystems $F_1, \ldots, F_{\lfloor n/3 \rfloor}$ with the j component incremented by 1, so that they are ready for the next computation step. If the same normalized layer occurs in two or more positions in the normalized Fredkin circuit, then the corresponding subsystem must contain the rules which allow to process all the objects which appear in these positions.

A further transformation of the Fredkin circuit allows to perform the simulation with just *one* subsystem. Starting from a normalized n–input Fredkin circuit NFC_n, we transform each normalized layer so that in the resulting circuit every layer contains the same number of gates. Figure 5 shows the result of this transformation, applied to the normalized Fredkin circuit illustrated in Figure 2. Informally, the transformation is performed as follows. Considering one normalized layer at a time, we first add a number of auxiliary lines, fed with the boolean constant 0. The number of auxiliary lines added depends upon the number of *free* lines (that is, lines not affected by any gate) in the given layer. As a result, the total number of lines is a multiple of 3. We can thus add an appropriate number of auxiliary Fredkin gates (denoted by "FG (aux)" in Figure 5) to the layer, each one taking an auxiliary line as its first input, so that every auxiliary gate computes the identity function. At the end of this process, we add (if needed) to each layer further auxiliary lines, in order to obtain the same number of input/output lines for all the layers. Since the auxiliary lines have been added at the bottom of the circuit, we have to permute them together with the original free lines to feed them correctly to the transformed layer. The details can be found in [11]. The energy–based P system that simulates a transformed Fredkin circuit is the same as described in the previous section, but now it contains only the subsystem which simulates a full layer of Fredkin gates. If desired, also the membrane which encloses such subsystem can be removed, thus lowering the depth of the membrane hierarchy by 1. The new system has again $\lfloor n/3 \rfloor$ subsystems, each one simulating a Fredkin gate. Of course, the rules in the skin must be modified so that they dispatch the objects directly to the correct subsystem.

4 UREM P Systems

Let us now consider *UREM P systems* [4], that is, P systems with unit rules and energy assigned to membranes. A UREM P system of degree $d+1$ is a construct Π of the form $\Pi = (A, \mu, e_0, \ldots, e_d, w_0, \ldots, w_d, R_0, \ldots, R_d)$, where:

- A is an alphabet of *objects*;
- μ is a *membrane structure*, with the membranes labelled by numbers $0, \ldots, d$ in a one-to-one manner;
- $e_0, \ldots, e_d$ are the initial energy values assigned to the membranes $0, \ldots, d$. In what follows we assume that $e_0, \ldots, e_d$ are non–negative integers;
- $w_0, \ldots, w_d$ are multisets over A associated with the regions $0, \ldots, d$ of μ;
- $R_0, \ldots, R_d$ are finite sets of *unit rules* associated with the membranes $0, \ldots, d$. Each rule or R_i has the form $(\alpha_i : a, \Delta e, b)$, where $\alpha \in \{in, out\}$, $a, b \in A$, and $|\Delta e|$ is the amount of energy that — for $\Delta e \geq 0$ — is added to or — for $\Delta e < 0$ — is subtracted from e_i (the energy assigned to membrane i) by the application of the rule.

The *initial configuration* of Π consists of $e_0, \ldots, e_d$ and $w_0, \ldots, w_d$. The transition from a configuration to another one is performed by nondeterministically choosing one rule from some R_i and applying it (hence we consider the *sequential* mode of applying the rules). Applying $(in_i : a, \Delta e, b)$ means that an object a (being in the membrane immediately outside of i) is changed into b while entering membrane i, thereby changing the energy value e_i of membrane i by Δe. On the other hand, the application of a rule $(out_i : a, \Delta e, b)$ changes object a into b while leaving membrane i, and changes the energy value e_i by Δe. The rules can be applied only if the amount e_i of energy assigned to membrane i fulfills the requirement $e_i + \Delta e \geq 0$. Moreover, we use a sort of *local priorities*: if there are two or more applicable rules in membrane i, then one of the rules with $\max |\Delta e|$ has to be used.

A sequence of transitions that starts from the initial configuration is called a *computation*; it is *successful* if and only if it halts, that is, if and only if a configuration is reached in which no rule can be applied. The *result* of a successful computation is considered to be the distribution of energies among the membranes in the halting configuration. A non–halting computation does not produce a result. If we consider the energy distribution of the membrane structure as the input to be analysed, we obtain a model for accepting sets of (vectors of) non–negative integers.

4.1   Computational Power

The following result, proved in [4], establishes computational completeness for this model of P systems.

Theorem 3. *Every partial recursive function $f : \mathbb{N}^\alpha \to \mathbb{N}^\beta$ ($\alpha \geq 1$, $\beta \geq 1$) can be computed by a UREM P system with (at most) $\max\{\alpha, \beta\} + 3$ membranes.*

As in the case of energy–based P systems, the proof of this proposition is obtained by simulating register machines. In the simulation, a P system is defined which contains one subsystem for each register of the simulated machine. The contents of the register are expressed as the energy value e_i assigned to the i-th subsystem. A single object is present in the system at every computation step, which stores the label of the instruction of the program P currently simulated. Increment instructions are simulated in two steps by using the rules $(in_i : p_j, 1, \widetilde{p_j})$ and $(out_i : \widetilde{p_j}, 0, p_k)$. Decrement instructions are also simulated in two steps, by using the rules $(in_i : p_j, 0, \widetilde{p_j})$ and $(out_i : \widetilde{p_j}, -1, p_k)$ or $(out_i : \widetilde{p_j}, 0, p_l)$. The use of priorities associated to these last rules is crucial to correctly simulate a decrement instruction. For the details of the proof we refer the reader to [4].

When taking $\beta = 0$ in the proof of the above proposition, we get the *accepting* variant of P systems with unit rules and energy assigned to membranes:

Corollary 3. *Let $L \subseteq \mathbb{N}^\alpha$, $\alpha \geq 1$, be a recursively enumerable set of (vectors of) non–negative integers. Then L can be accepted by a UREM P system having (at most) $\alpha + 3$ membranes.*

The above results were obtained by simulating deterministic register machines by means of *deterministic* UREM P systems, where at each step only one rule is enabled and can be applied. As we did with energy–based P systems, for the *generative* case we have to pass to a *nondeterministic* choice of rules, and simulate nondeterministic register machines. Under this setting, the following corollary is also a simple consequence of Theorem 3, by taking $\alpha = 0$. As a technical detail we mention that the nondeterministic INC instruction $j : (INC(i), k, \ell)$ is simulated in two steps using the rules $(in_i : p_j, 1, \widetilde{p_j})$ and then $(out_i : \widetilde{p_j}, 0, p_k)$ or $(out_i : \widetilde{p_j}, 0, p_\ell)$.

Corollary 4. *Let $L \subseteq \mathbb{N}^\beta$, $\beta \geq 1$, be a recursively enumerable set of (vectors of) non–negative integers. Then L can be generated by a UREM P system having (at most) $\beta + 3$ membranes.*

Once again, when omitting the priority feature we do not get systems with universal computational power. This time, however, we obtain a characterization of the family $PsMAT^\lambda$ of Parikh sets generated by context–free matrix grammars, without occurrence checking and with λ-rules. The proof is quite involved, and hence we refer the reader to [4,11].

However, even without the priority feature UREM P systems can obtain universal computational power, provided that their rules are applied in the maximally parallel mode instead of the sequential mode:

Theorem 4. *Each partial recursive function* $f : \mathbb{N}^\alpha \to \mathbb{N}^\beta$ $(\alpha \geq 1, \beta \geq 1)$ *can be computed by a UREM P system with (at most)* $\max\{\alpha, \beta\} + 4$ *membranes when working in the maximally parallel mode without priorities on the rules.*

Once again, the proof is obtained by simulating register machines. This time, however, the simulation is more complicated, and requires the use of an auxiliary membrane which is used as a "pacemaker" to drive the correct simulation of INC and DEC instructions. We refer the reader to [11] for the details.

The following results are immediate consequences of Theorem 4 as Corollaries 3 and 4 were immediate consequences of Theorem 3:

Corollary 5. *Let* $L \subseteq \mathbb{N}^\alpha$, $\alpha \geq 1$, *be a recursively enumerable set of (vectors of) non–negative integers. Then* L *can be accepted by a UREM P system with (at most)* $\alpha + 4$ *membranes in the maximally parallel mode without priorities on the rules.*

Corollary 6. *Let* $L \subseteq \mathbb{N}^\beta$, $\beta \geq 1$, *be a recursively enumerable set of (vectors of) non–negative integers. Then* L *can be generated by a UREM P system with (at most)* $\beta + 4$ *membranes in the maximally parallel mode without priorities on the rules.*

5   Conclusions

In this paper we have reviewed some results obtained in the last few years, concerning the computational power of two models of computation defined in the framework of membrane computing: energy–based P systems and UREM P systems. Such models are inspired from the functioning of some physical laws, that consider the computation devices as physical objects that manipulate energy during their computations.

We believe that these P systems have the potential to generate further stimulating research. Two spin–offs of UREM P systems we have not mentioned in this paper are *tissue–like* UREM P systems, whose study has begun in [11], and *quantum–like* UREM P systems, introduced in [9]. A tissue–like version of energy–based P systems is missing, as well as a comparison with other models of P systems that use energy in their computation steps (such as [15,3,6]).

Acknowledgements

This work was partially supported by the italian FIAR 2007 project "Modelli di calcolo naturale e applicazioni alla Systems Biology".

References

1. Alhazov, A., Freund, R., Leporati, A., Oswald, M., Zandron, C.: (Tissue) P systems with unit rules and energy assigned to membranes. Fundamenta Informaticae 74, 391–408 (2006)
2. Fredkin, E., Toffoli, T.: Conservative logic. International Journal of Theoretical Physics 21(3-4), 219–253 (1982)
3. Freund, R.: Energy–controlled P systems. In: Păun, G., Rozenberg, G., Salomaa, A., Zandron, C. (eds.) WMC 2002. LNCS, vol. 2597, pp. 247–260. Springer, Heidelberg (2003)
4. Freund, R., Leporati, A., Oswald, M., Zandron, C.: Sequential P systems with unit rules and energy assigned to membranes. In: Margenstern, M. (ed.) MCU 2004. LNCS, vol. 3354, pp. 200–210. Springer, Heidelberg (2005)
5. Freund, R., Oswald, M.: GP systems with forbidding context. Fundamenta Informaticae 49(1-3), 81–102 (2002)
6. Frisco, P.: The conformon–P system: a molecular and cell biology–inspired computability model. Theoretical Computer Science 312, 295–319 (2004)
7. Karp, R., Miller, R.: Parallel program schemata. Journal of Computer and System Science 3(4), 167–195 (1968); Also RC2053, IBM T.J. Watson Research Center, New York (April 1968)
8. Leporati, A., Besozzi, D., Cazzaniga, P., Pescini, D., Ferretti, C.: Computing with energy and chemical reactions. Natural Computing (2009), doi:10.1007/s11047-009-9160-x
9. Leporati, A., Mauri, G., Zandron, C.: Quantum sequential P systems with unit rules and energy assigned to membranes. In: Freund, R., Păun, G., Rozenberg, G., Salomaa, A. (eds.) WMC 2005. LNCS, vol. 3850, pp. 310–325. Springer, Heidelberg (2006)
10. Leporati, A., Zandron, C., Mauri, G.: Simulating the Fredkin gate with energy-based P systems. Journal of Universal Computer Science 10(5), 600–619 (2004)
11. Leporati, A., Zandron, C., Mauri, G.: Reversible P systems to simulate Fredkin circuits. Fundamenta Informaticae 74, 529–548 (2006)
12. Minsky, M.L.: Finite and Infinite Machines. Prentice Hall, Englewood Cliffs (1967)
13. Păun, Gh.: Computing with membranes. Journal of Computer and System Sciences 1(61), 108–143 (2000); See also Turku Centre for Computer Science – TUCS Report No. 208 (1998)
14. Păun, Gh.: Membrane Computing. An Introduction. Springer, Berlin (2002)
15. Păun, Gh., Suzuki, Y., Tanaka, H.: P systems with energy accounting. International Journal Computer Math. 78(3), 343–364 (2001)
16. Peterson, J.L.: Petri Net Theory and the Modeling of Systems. Prentice Hall, Englewood Cliffs (1981)
17. The P systems, http://ppage.psystems.eu

A Computational Complexity Theory in Membrane Computing

Mario J. Pérez–Jiménez

Research Group on Natural Computing
Department of Computer Science and Artificial Intelligence
University of Sevilla
Avda. Reina Mercedes s/n, 41012 Sevilla, Spain
marper@us.es

Abstract. In this paper, a computational complexity theory within the framework of Membrane Computing is introduced. Polynomial complexity classes associated with different models of cell-like and tissue-like membrane systems are defined and the most relevant results obtained so far are presented. Many attractive characterizations of **P** $\neq$ **NP** conjecture within the framework of a bio-inspired and non-conventional computing model are deduced.

1 Introduction

The main objective of Computability Theory is to define the informal idea of mechanical/algorithmic problems resolution in a rigorous way. Each formal definition of the said concept provides a computing model. However, a basic question is to determine the class of all the problems that can be solved by a computing model when using the algorithms defined in it. In any computing model which captures the informal idea of algorithm, there are undecidable problems, that is, problems that cannot be solved by using the algorithms of the model.

Analyzing an algorithm which solves a problem consists of determining an upper bound for the minimal resource requirements with which the problem can be solved. The said upper bound will be a function of the size of the instance of the problem. One of the main goals of Computational Complexity Theory is to provide bounds on the amount of resources necessary for every mechanical procedure (algorithm) that solves a given problem.

Usually, complexity theory deals with *decision problems* which are problems that require a "*yes*" or "*no*" answer. A *decision problem*, X, is a pair (I_X, θ_X) such that I_X is a language over a finite alphabet (whose elements are called *instances*) and θ_X is a total boolean function (that is, a predicate) over I_X.

Many abstract problems are not decision problems. For example, in combinatorial optimization problems some value must be optimized (minimized or maximized). In order to deal with such problems, they can be transformed into roughly equivalent decision problems by supplying a target/threshold value for the quantity to be optimized, and then asking whether this value can be attained.

G. Păun et al. (Eds.): WMC 2009, LNCS 5957, pp. 125–148, 2010.
© Springer-Verlag Berlin Heidelberg 2010

A natural correspondence between decision problems and languages can be established as follows. Given a decision problem $X = (I_X, \theta_X)$, its associated language is $L_X = \{w \in I_X : \theta_X(w) = 1\}$. Conversely, given a language L, over an alphabet Σ, its associated decision problem is $X_L = (I_{X_L}, \theta_{X_L})$, where $I_{X_L} = \Sigma^*$, and $\theta_{X_L} = \{(x,1) : x \in L\} \cup \{(x,0) : x \notin L\}$.

The solvability of decision problems is defined through the recognition of the languages associated with them. Let M be a Turing machine with a working alphabet Γ and L a language over Γ. Assume that the result of any halting computation of M is *yes* or *no*. If M is a *deterministic* device, then we say that M *recognizes* or *decides* L whenever, for any string u over Γ, if $u \in L$, then the answer of M on input u is *yes* (that is, M accepts u), and the answer is *no* otherwise (that is, M rejects u). If M is a *non-deterministic* device, then we say that M *recognizes* or *decides* L if for any string u over Γ, $u \in L$ if and only if there exists a computation of M with input u such that the answer is *yes*.

Throughout this paper, it is assumed that each abstract problem has an associated fixed *reasonable encoding scheme* that describes the instances of the problem by means of strings over a finite alphabet. We do not define *reasonable* in a formal way, however, following [8], instances should be encoded in a concise way, without irrelevant information, and where relevant numbers are represented in binary form (or any fixed base other than 1). It is possible to use multiple reasonable encoding schemes to represent instances, but it is proved that the input sizes differ at most by a polynomial. The *size* $|u|$ of an instance u is the length of the string associated with it, in some reasonable encoding scheme.

Membrane computing is a branch of natural computing initiated by Gh. Păun at the end of 1998 [21]. P systems take multisets as input, usually in a unary fashion. Hence, it is important to be careful when asserting that a problem is polynomial-time solvable by membrane systems. In this context, polynomial-time solutions to **NP**–complete problems in the framework of membrane computing can be considered as *pseudo-polynomial* time solutions in the classical sense (see [8] and [26] for details).

The paper is organized as follows. In the next section, basic concepts are introduced related to cell-like membrane systems that are necessary to define the solution of decision problems in polynomial time. In Section 3, limitations to basic transition P systems are described from the point of view of computational efficiency. Section 4 presents the most relevant results on P systems with active membranes both with and without polarization. Section 5 is devoted to the study of polarizationless tissue P systems with active membranes, and results which provide borderlines between efficiency and non-efficiency are presented. The paper ends with the proposal of several open problems.

2 Cell–Like Recognizer Membrane Systems

Membrane Computing is a young branch of Natural Computing providing distributed parallel computational devices called *membrane systems*, which are inspired in some basic biological features of living cells, as well as in the cooperation of cells in tissues, organs and organisms.

In this area there are basically two ways to consider computational devices: cell–like membrane systems (*P systems*) and tissue–like membrane systems (*tissue P systems*). The first one uses membranes arranged hierarchically, inspired from the structure of the cell, and the second one uses membranes placed in the nodes of a graph, inspired from the cell inter–communication in tissues.

In the last years several computing models using powerful tools from Nature have been developed (because of this, they are known as *bio-inspired* models) and several solutions in polynomial time to **NP**–complete problems have been presented, making use of non-determinism and/or of an exponential amount of space. This is the reason why a practical implementation of such models (in biological, electronic, or other media) could provide a significant advance in the resolution of computationally hard problems.

Definition 1. *A P system (without input) of degree $q \geq 1$ is a tuple of the form $\Pi = (\Gamma, H, \mu, \mathcal{M}_1, \ldots, \mathcal{M}_q, R, i_{out})$, where:*

1. *Γ is a working alphabet of objects, and H is a finite set of labels;*
2. *μ is a membrane structure (a rooted tree) consisting of q membranes injectively labeled by elements of H;*
3. *$\mathcal{M}_1, \ldots, \mathcal{M}_q$ are strings over Γ describing the initial multisets of objects placed in the q initial regions of μ;*
4. *R is a finite set of developmental rules;*
5. *$i_{out} \in H$ or $i_{out} = env$ indicates the output region: in the case $i_{out} \in H$, for a computation to be successful there must be exactly one membrane with label i_{out} present in the halting configuration; in the case $i_{out} = env$, i_{out} is usually omitted from the tuple.*

Many variants of P systems can be obtained depending on the kind of *developmental rules* and the semantics which are considered. The *length of a rule* is the number of symbols necessary to write it, both its left and right sides.

If h is the label of a membrane, then $f(h)$ denotes the label of the father of the membrane labeled by h. We assume the convention that the father of the skin membrane is the environment (*env*).

Definition 2. *A P system with input membrane is a tuple (Π, Σ, i_{in}), where: (a) Π is a P system; (b) Σ is an (input) alphabet strictly contained in Γ such that the initial multisets are over the alphabet $\Gamma \setminus \Sigma$; and (c) i_{in} is the label of a distinguished (input) membrane.*

The difference between P systems with and without input membrane is not related to their computations, but only to their initial configurations. A P system Π without input has a single *initial configuration* $(\mu, \mathcal{M}_1, \ldots, \mathcal{M}_q)$. A P system (Π, Σ, h_i) with input has many *initial configurations*: for each multiset $m \in \Sigma^*$, the *initial configuration* associated with m is $(\mu, \mathcal{M}_1, \ldots, \mathcal{M}_{h_i} \cup m, \ldots, \mathcal{M}_q)$.

In order to solve decision problems, we define *recognizer P system*.

Definition 3. *A recognizer P system is a P system such that: (a) the working alphabet contains two distinguished elements* yes *and* no*; (b) all computations* *halt; and (c) if C is a computation of the system, then either object* yes *or object* no *(but not both) must have been sent to the output region of the system, and only at the last step of the computation.*

For recognizer P systems, a computation C is said to be an *accepting computation* (respectively, *rejecting computation*) if the object *yes* (respectively, *no*) appears (only) in the output region associated with the corresponding halting configuration of C.

For technical reasons all computations are required to halt, but this condition can often be removed without affecting computational efficiency.

Throughout this paper, R denotes an arbitrary class of recognizer P systems.

2.1   Uniform Families of P Systems

Many formal machine models (e.g. Turing machines or register machines) have an infinite number of memory locations. At the same time, P systems, or logic circuits, are computing devices of finite size and they have a finite description with a fixed amount of initial resources (number of membranes, objects, gates, etc.). For this reason, in order to solve a decision problem a (possibly infinite) family of P systems is considered.

The concept of solvability in the framework of P systems also takes into account the pre-computational process of (efficiently) constructing the family that provides the solution. In this paper, the terminology *uniform family* is used to denote that this construction is performed by a *single* computational machine.

In the case of P systems with input membrane, the term uniform family is consistent with the usual meaning for Boolean circuits: a family $\mathbf{\Pi} = \{\Pi(n) : n \in \mathbf{N}\}$ is uniform if there exists a deterministic Turing machine which constructs the system $\Pi(n)$ from $n \in \mathbf{N}$ (that is, which on input 1^n outputs $\Pi(n)$). In such a family, the P system $\Pi(n)$ will process all the instances of the problem with numerical parameters (reasonably) encoded by n – the common case is that $\Pi(n)$ processes all instances of size n. Note that this means that, for these families of P systems with input membrane, further pre–computational processes are needed in order to (efficiently) determine which P system (and from which input) deals with a given instance of the problem. The concept of *polynomial encoding* introduced below tries to capture this idea.

In the case of P systems without input membrane a new notion arises: a family $\mathbf{\Pi} = \{\Pi(w) : w \in I_X\}$ *associated with a decision problem* $X = (I_X, \theta_X)$ is *uniform* (some authors [15,35,38] use the term *semi-uniform* here) if there exists a deterministic Turing machine which constructs the system $\Pi(w)$ from the instance $w \in I_X$. In such a family, each P system usually processes only one instance, and the numerical parameters and syntactic specifications of the latter are part of the definition of the former.

It is important to point out that, in both cases, the family should be constructed in an efficient way. This requisite was first included within the term uniform family (introduced by Gh. Păun [22]), but nowadays it is preferred to use the term *polynomially uniform by Turing machines* to indicate a uniform (by a single Turing machine) and effective (in polynomial time) construction of the family.

Definition 4. *A family* $\Pi = \{\Pi(w) : w \in I_X\}$ *(respectively,* $\Pi = \{\Pi(n) : n \in \mathbf{N}\}$*) of recognizer membrane systems without input membrane (resp., with input membrane) is* polynomially uniform by Turing machines *if there exists a deterministic Turing machine working in polynomial time which constructs the system $\Pi(w)$ (resp., $\Pi(n)$) from the instance $w \in I_X$ (resp., from $n \in \mathbf{N}$).*

2.2 Confluent P Systems

In order for recognizer P systems to capture the true algorithmic concept, a condition of *confluence* is imposed, in the sense that all possible successful computations must give the same answer. This contrasts with the standard notion of accepting computations for non-deterministic (classic) models.

Definition 5. *Let $X = (I_X, \theta_X)$ be a decision problem, and $\Pi = \{\Pi(w) : w \in I_X\}$ be a family of recognizer P systems without input membrane.*

- Π *is said to be* sound with respect to X *if the following holds: for each instance of the problem, $w \in I_X$, if there exists an accepting computation of $\Pi(w)$, then $\theta_X(w) = 1$.*
- Π *is said to be* complete with respect to X *if the following holds: for each instance of the problem, $w \in I_X$, if $\theta_X(w) = 1$, then every computation of $\Pi(w)$ is an accepting computation.*

The concepts of soundness and completeness can be extended to families of recognizer P systems with input membrane in a natural way. However, an efficient process of selecting P systems from instances must be made precise.

Definition 6. *Let $X = (I_X, \theta_X)$ be a decision problem, and $\Pi = \{\Pi(n) : n \in \mathbf{N}\}$ a family of recognizer P systems with input membrane. A* polynomial encoding *of X in Π is a pair (cod, s) of polynomial–time computable functions over I_X such that for each instance $w \in I_X$, $s(w)$ is a natural number (obtained by means of a reasonable encoding scheme) and $cod(w)$ is an input multiset of the system $\Pi(s(w))$.*

Polynomial encodings are stable under polynomial–time reductions [29].

Proposition 1. *Let X_1, X_2 be decision problems, r a polynomial–time reduction from X_1 to X_2, and (cod, s) a polynomial encoding from X_2 to Π. Then, $(cod \circ r, s \circ r)$ is a polynomial encoding from X_1 to Π.*

Next, the concepts of soundness and completeness are defined for families of recognizer P systems with input membrane.

Definition 7. *Let $X = (I_X, \theta_X)$ be a decision problem, $\Pi = \{\Pi(n) : n \in \mathbf{N}\}$ a family of recognizer P systems with input membrane, and (cod, s) a polynomial encoding of X in Π.*

- Π *is said to be* sound with respect to (X, cod, s) *if the following holds: for each instance of the problem,* $w \in I_X$, *if there exists an accepting computation of* $\Pi(s(w))$ *with input* $cod(w)$, *then* $\theta_X(w) = 1$.
- Π *is said to be* complete with respect to (X, cod, s) *if the following holds: for each instance of the problem,* $w \in I_X$, *if* $\theta_X(w) = 1$, *then every computation of* $\Pi(s(w))$ *with input* $cod(w)$ *is an accepting computation.*

Notice that if a family of recognizer P systems is sound and complete, then every P system of the family is confluent, in the sense previously mentioned.

2.3 Semi-uniform Solutions versus Uniform Solutions

The first results showing that membrane systems could solve computationally hard problems in polynomial time were obtained using P systems without input membrane. In that context, a specific P system is associated with each instance of the problem. In other words, the syntax of the instance is part of the description of the associated P system. Thus this P system can be considered *special purpose*.

Definition 8. *A decision problem* X *is solvable in polynomial time by a family of recognizer P systems without input membrane* $\Pi = \{\Pi(w) : w \in I_X\}$, *denoted by* $X \in \mathbf{PMC}_{\mathcal{R}}^*$, *if the following holds:*

- *The family* Π *is polynomially uniform by Turing machines.*
- *The family* Π *is polynomially bounded; that is, there exists a natural number* $k \in \mathbf{N}$ *such that for each instance* $w \in I_X$, *every computation of* $\Pi(w)$ *performs at most* $|w|^k$ *steps.*
- *The family* Π *is sound and complete with respect to* X.

The family Π is said to provide a *semi–uniform solution* to the problem X.

Next, recognizer P systems with input membrane are defined to solve problems in a *uniform* way in the following sense: all instances of a decision problem of the same *size* (via a given reasonable encoding scheme) are processed by the same system, to which an appropriate input is supplied.

Definition 9. *A decision problem* $X = (I_X, \theta_X)$ *is solvable in polynomial time by a family of recognizer P systems with input membrane* $\Pi = \{\Pi(n) : n \in \mathbf{N}\}$, *denoted by* $X \in \mathbf{PMC}_{\mathcal{R}}$, *if the following holds:*

- *The family* Π *is polynomially uniform by Turing machines.*
- *There exists a polynomial encoding* (cod, s) *of* X *in* Π *such that:*
  - *The family* Π *is polynomially bounded with respect to* (X, cod, s); *that is, there exists a natural number* $k \in \mathbf{N}$ *such that for each instance* $w \in I_X$, *every computation of the system* $\Pi(s(w))$ *with input* $cod(w)$ *performs at most* $|w|^k$ *steps.*
  - *The family* Π *is sound and complete with respect to* (X, cod, s).

The family Π is said to provide a *uniform solution* to the problem X.

As a direct consequence of working with recognizer membrane systems, these complexity classes are closed under complement. Moreover, they are closed under polynomial–time reductions [29].

Obviously, every uniform solution of a decision problem provides a semi–uniform solution using the same amount of computational resources. That is, $\mathbf{PMC}_{\mathcal{R}} \subseteq \mathbf{PMC}^*_{\mathcal{R}}$, for any class $\mathcal{R}$ of recognizer P systems.

Remark: It is interesting to distinguish the concept of *polynomially uniform by Turing machines* from the concepts of *semi–uniform* and *uniform* solutions. The first concept is related with the resources required to construct the family of P systems solving a decision problem. The last two refer to the way in which the family processes the instances. In semi-uniform solutions, every instance is processed by a special purpose P system. While in uniform solutions, each P system processes all instances of a given size.

3   Efficiency of Basic Transition P Systems

In this section, the computational efficiency of P systems whose membrane structure does not increase is studied.

First of all, in order to formally define what means that a family of P systems simulates a Turing machine, we shall introduce for each Turing machine a decision problem associated with it.

Definition 10. *Let M be a Turing machine with input alphabet Σ_M. The decision problem associated with M is the problem $X_M = (I_M, \theta_M)$, where $I_M = \Sigma_M^*$, and for every $w \in \Sigma_M^*$, $\theta_M(w) = 1$ if and only if M accepts w.*

Obviously, the decision problem X_M is solvable by the Turing machine M.

Definition 11. *We say that a Turing machine M is simulated in polynomial time by a family of recognizer P systems from $\mathcal{R}$ if $X_M \in \mathbf{PMC}_{\mathcal{R}}$.*

A *basic transition* P system is a P system with only evolution, communication, and dissolution rules, which do not increase the size of the membrane structure. Let $\mathcal{T}$ denote the class of recognizer basic transition P systems.

M.A. Gutiérrez–Naranjo et al. [12] gave an efficient simulation of deterministic Turing machines by recognizer basic transition P systems.

Proposition 2. (Sevilla theorem) *Every deterministic Turing machine working in polynomial time can be simulated in polynomial time by a family of recognizer basic transition P systems with input membrane.*

They also proved that each confluent basic transition P system can be (efficiently) simulated by a deterministic Turing machine [12]. As a consequence, these P systems efficiently solve at most tractable problems.

Proposition 3. *If a decision problem is solvable in polynomial time by a family of recognizer basic transition P systems with input membrane, then there exists a deterministic Turing machine solving it in polynomial time.*

These results are also verified for recognizer basic transition P systems without input membrane. Therefore, the following holds.

Theorem 1. $\mathbf{P} = \mathbf{PMC}_{\mathcal{T}} = \mathbf{PMC}_{\mathcal{T}}^*$.

Thus, the ability of a P system in $\mathcal{T}$ to create exponential workspace (in terms of number of objects) in polynomial time (e.g. via evolution rules of the type $[a \rightarrow a^2]_h$) is not enough to efficiently solve **NP**–complete problems (unless $\mathbf{P} = \mathbf{NP}$). Theorem 1 provides a tool to attack conjecture $\mathbf{P} = \mathbf{NP}$ in the framework of membrane computing.

Corollary 1. $\mathbf{P} \neq \mathbf{NP}$ *if and only if every, or at least one, **NP***–*complete problem is not in* $\mathbf{PMC}_{\mathcal{T}} = \mathbf{PMC}_{\mathcal{T}}^*$.

4 P Systems with Active Membranes

P systems with active membranes having associated electrical charges with membranes were first introduced by Gh. Păun [23]. Replication is one of the most important functions of a cell and, in ideal circumstances, a cell produces two identical copies by division (mitosis). Bearing in mind that the reactions which take place in a cell are related to membranes, rules for membrane division are considered.

Definition 12. *A P system with active membranes of degree $q \geq 1$ is a tuple* $\Pi = (\Gamma, H, \mu, \mathcal{M}_1, \ldots, \mathcal{M}_q, R, i_{out})$, *where:*

1. *Γ is a working alphabet of objects, and H is a finite set of labels for membranes;*
2. *μ is a membrane structure (a rooted tree) consisting of q membranes injectively labeled by elements of H, and with electrical charges $(+, -, 0)$ associated with them;*
3. *$\mathcal{M}_1, \ldots, \mathcal{M}_q$ are strings over Γ describing the initial multisets of objects placed in the q initial regions of μ;*
4. *R is a finite set of rules, of the following forms:*
   (a) *$[a \rightarrow u]_h^\alpha$, for $h \in H, \alpha \in \{+, -, 0\}$, $a \in \Gamma$, $u \in \Gamma^*$ (object evolution rules).*
   (b) *$a [\ ]_h^{\alpha_1} \rightarrow [b]_h^{\alpha_2}$, for $h \in H$, $\alpha_1, \alpha_2 \in \{+, -, 0\}$, $a, b \in \Gamma$ (send–in communication rules).*
   (c) *$[a]_h^{\alpha_1} \rightarrow [\ ]_h^{\alpha_2} b$, for $h \in H$, $\alpha_1, \alpha_2 \in \{+, -, 0\}$, $a, b \in \Gamma$ (send–out communication rules).*
   (d) *$[a]_h^\alpha \rightarrow b$, for $h \in H$, $\alpha \in \{+, -, 0\}$, $a, b \in \Gamma$ (dissolution rules).*
   (e) *$[a]_h^{\alpha_1} \rightarrow [b]_h^{\alpha_2} [c]_h^{\alpha_3}$, for $h \in H$, $\alpha_1, \alpha_2, \alpha_3 \in \{+, -, 0\}$, $a, b, c \in \Gamma$ (division rules for elementary membranes).*
   (f) *$[[\ ]_{h_1}^{\alpha_1} \cdots [\ ]_{h_k}^{\alpha_1} [\ ]_{h_{k+1}}^{\alpha_2} \cdots [\ ]_{h_n}^{\alpha_2}]_h^\alpha \rightarrow [[\ ]_{h_1}^{\alpha_3} \cdots [\ ]_{h_k}^{\alpha_3}]_h^\beta [[\ ]_{h_{k+1}}^{\alpha_4} \cdots [\ ]_{h_n}^{\alpha_4}]_h^\gamma$, for $k \geq 1$, $n > k$, $h, h_1, \ldots, h_n \in H$, $\alpha, \beta, \gamma, \alpha_1, \ldots, \alpha_4 \in \{+, -, 0\}$ and $\{\alpha_1, \alpha_2\} = \{+, -\}$ (division rules for non–elementary membranes).*
5. *$i_{out} \in H$ or $i_{out} = env$ indicates the output region.*

These rules are applied as usual (see [22] for details).

Note that these P systems have some important features: (a) they use three electrical charges; (b) the polarization of a membrane, but not the label, can be modified by the application of a rule; and (c) they do not use cooperation neither priorities.

In the framework of P systems without input membrane, C. Zandron et al. [40] proved that confluent recognizer P systems with active membranes making use of no membrane division rule, can be efficiently simulated by a deterministic Turing machine.

Proposition 4. (Milano theorem) *A deterministic P system with active membranes but without membrane division can be simulated by a deterministic Turing machine with a polynomial slowdown.*

Let $\mathcal{NAM}$ be the class of recognizer P systems with active membranes which do not make use of division rules. As a consequence of the previous result, the following holds:

Corollary 2. $\mathbf{PMC}^*_{\mathcal{NAM}} \subseteq \mathbf{P}$.

A.E. Porreca [34] provides a simple proof of each tractable problem being able to be solved (in a semi–uniform way) by a family of recognizer P systems with active membranes (without polarizations) operating in exactly one step and using only send–out communication rules. That proof can be easily adapted to uniform solutions.

Proposition 5. $\mathbf{P} \subseteq \mathbf{PMC}_{\mathcal{NAM}}$.

Thus, we have a version of Theorem 1 for the class $\mathcal{NAM}$.

Theorem 2. $\mathbf{P} = \mathbf{PMC}_{\mathcal{NAM}} = \mathbf{PMC}^*_{\mathcal{NAM}}$.

The first efficient solutions to **NP**–complete problems by using P systems with active membranes were given in a *semi–uniform* way (where the P systems of the family depend on the syntactic structure of the instance) by S.N. Krishna et al. (Hamiltonian Path, Vertex Cover [13]), A. Obtulowicz (SAT [16]), A. Păun (Hamiltonian Path [20]), Gh. Păun (SAT [23,24]), and C. Zandron et al. (SAT, Undirected Hamiltonian Path [40]).

Let $\mathcal{AM}(+n)$ (respectively, $\mathcal{AM}(-n)$) be the class of recognizer P systems with active membranes using division rules for elementary and non–elementary membranes (respectively, only for elementary membranes).

In the framework of $\mathcal{AM}(-n)$, efficient *uniform* solutions to weakly **NP**–complete problems (Knapsack [28], Subset Sum [27], Partition [10]), and strongly **NP**–complete problems (SAT [33], Clique [4], Bin Packing [31], Common Algorithmic Problem [30]) have been obtained.

Proposition 6. SAT $\in \mathbf{PMC}_{\mathcal{AM}(-n)}$.

Since $\mathbf{PMC}_{\mathcal{R}}$ is closed under complement and polynomial–time reductions, for any class $\mathcal{R}$ of recognizer P systems, the following result is obtained.

Proposition 7. NP $\cup$ co-NP $\subseteq$ PMC$_{\mathcal{AM}(-n)}$.

In the framework of $\mathcal{AM}(+n)$, P. Sosík [38] gave an efficient *semi–uniform* solution to QBF-SAT (satisfiability of quantified propositional formulas), a well known **PSPACE**–complete problem [8]. Hence, the following is deduced.

Proposition 8. PSPACE $\subseteq$ PMC$^*_{\mathcal{AM}(+n)}$.

This result has been extended by A. Alhazov et al. [5] showing that QBF-SAT can be solved in a linear time and in a *uniform* way by a family of recognizer P systems with active membranes (without using dissolution rules) and using division rules for elementary and non–elementary membranes.

Proposition 9. PSPACE $\subseteq$ PMC$_{\mathcal{AM}(+n)}$.

A.E. Porreca et al. [35] described a (deterministic and efficient) algorithm simulating a single computation of any confluent recognizer P system with active membranes and without input. Such P systems can be simulated by a deterministic Turing machine working with exponential space, and spending a time of the order $O(2^{p(n)})$, for some polynomial $p(n)$. Thus,

Proposition 10. PMC$^*_{\mathcal{AM}(+n)}$ $\subseteq$ EXP.

Therefore, **PMC$_{\mathcal{AM}(+n)}$** and **PMC$^*_{\mathcal{AM}(+n)}$** are two membrane computing complexity classes between **PSPACE** and **EXP**.

Corollary 3. PSPACE $\subseteq$ PMC$_{\mathcal{AM}(+n)}$ $\subseteq$ PMC$^*_{\mathcal{AM}(+n)}$ $\subseteq$ EXP.

P. Sosík et al. [37] have proven that the reverse inclusion of Proposition 8 holds as well. Nevertheless, the concept of *uniform family* of P systems considered in that paper is different from that of Definition 4, although maybe the proof can be adapted to fit into the framework presented in this paper. In this case the following would hold: **PSPACE = PMC$^*_{\mathcal{AM}(+n)}$.**

Previous results show that the usual framework of P systems with active membranes for solving decision problems is too powerful from the computational complexity point of view. Therefore, it would be interesting to investigate weaker models of P systems with active membranes able to characterize classical complexity classes below **NP** and providing borderlines between efficiency and non–efficiency.

Efficient (semi–uniform and/or uniform) solutions to computationally hard problems have been obtained within different apparently weaker variants of P systems with active membranes:

- P systems with separation rules instead of division rules, in two different cases: first one, using polarizations without changing membrane labels; and second one, without polarizations but allowing change of membrane labels (SAT, uniform solution [18]).
- P systems using division for elementary membranes, without changing membrane labels, without polarizations, but using bi–stable catalysts (SAT, uniform solution [32]).

- P systems using division for elementary membranes, without label changing, but using only two electrical charges (SAT, uniform solution [2], Subset Sum, uniform solution [36]).
- P systems without polarizations, without label changing, without division, but using three types of membrane rules: separation, merging, and release (SAT, semi–uniform solution [17]).
- P systems without dissolution nor polarizations, but allowing to change the labels of membranes in division rules (SAT, uniform solution [3]).
- P systems without dissolution nor polarizations, but allowing to change the labels of membranes in send–out rules (SAT, uniform solution [3]).
- P systems without polarizations, but using division for elementary and non–elementary membranes (SAT, semi–uniform solution [3]).

4.1 Polarizationless P Systems with Active Membranes

Next, several classes of recognizer P systems with active membranes without electrical charges and with different kinds of membrane division rules are studied from a computational complexity point of view.

Definition 13. *A polarizationless P system with active membranes of degree $q \geq 1$ is a tuple $\Pi = (\Gamma, H, \mu, \mathcal{M}_1, \ldots, \mathcal{M}_q, R, i_{out})$, where:*

1. *Γ is a working alphabet of objects, and H is a finite set of labels for membranes;*
2. *μ is a membrane structure (a rooted tree) consisting of q membranes injectively labeled by elements of H;*
3. *$\mathcal{M}_1, \ldots, \mathcal{M}_q$ are strings over Γ describing the multisets of objects placed in the q initial regions of μ;*
4. *R is a finite set of developmental rules, of the following forms:*
   (a) $[a \rightarrow u]_h$, for $h \in H$, $a \in \Gamma$, $u \in \Gamma^*$ (object evolution rules).
   (b) $a[\ ]_h \rightarrow [b]_h$, for $h \in H$, $a, b \in \Gamma$ (send–in communication rules).
   (c) $[a]_h \rightarrow [\ ]_h\, b$, for $h \in H$, $a, b \in \Gamma$ (send–out communication rules).
   (d) $[a]_h \rightarrow b$, for $h \in H$, $a, b \in \Gamma$ (dissolution rules).
   (e) $[a]_h \rightarrow [b]_h\,[c]_h$, for $h \in H$, $a, b, c \in \Gamma$ (division rules for elementary or weak division rules for non-elementary membranes).
   (f) $[[\ ]_{h_1} \cdots [\ ]_{h_k}\,[\ ]_{h_{k+1}} \cdots [\ ]_{h_n}]_h \;\rightarrow\; [[\ ]_{h_1} \cdots [\ ]_{h_k}]_h\,[[\ ]_{h_{k+1}} \cdots [\ ]_{h_n}]_h$, where $k \geq 1$, $n > k$, $h, h_1, \ldots, h_n \in H$ (strong division rules for non-elementary membranes).
5. *$i_{out} \in H$ or $i_{out} = env$ indicates the output region.*

These rules are applied according to usual principles of polarizationless P systems (see [11] for details).

Notice that in this polarizationless framework there is no cooperation, priority, nor changes of the labels of membranes. Besides, throughout this paper, rules of type (f) are used only for $k = 1, n = 2$, that is, rules of the form (f) $[[\ ]_{h_1}[\ ]_{h_2}]_h \rightarrow [[\ ]_{h_1}]_h\,[[\ ]_{h_2}]_h$. They can also be restricted to the case where they are controlled by the presence of a specific membrane, that is, rules of the form (g) $[[\,]_{h_1}[\,]_{h_2}[\,]_p]_h \rightarrow [[\,]_{h_1}[\,]_p]_h [[\,]_{h_2}[\,]_p]_h$.

The class of recognizer polarizationless P systems with active membranes (resp., which do not make use of division rules) is denoted by $\mathcal{AM}^0$ (resp., $\mathcal{NAM}^0$), and $\mathcal{AM}^0(\alpha, \beta, \gamma, \delta)$, where $\alpha \in \{-d, +d\}$, $\beta \in D = \{-n, +nw, +ns, +nsw, +nsr\}, \gamma \in \{-e, +e\}$, and $\delta \in \{-c, +c\}$, denotes the class of all recognizer P systems with polarizationless active membranes such that:

(a) if $\alpha = +d$ (resp., $\alpha = -d$) then dissolution rules are permitted (resp., forbidden);
(b) if $\beta = +nw$ or $+ns$ (resp., $\beta = +nsw$) then division rules for elementary and non–elementary membranes, weak or strong (resp., weak and strong) are permitted; if $\beta = +nsr$ then division rules of the types (e), (f) and (g) are permitted; if $\beta = -n$ then only division rules for elementary membranes are permitted.
(c) if $\gamma = +e$ (resp., $\gamma = -e$) then evolution rules are permitted (resp., forbidden);
(d) if $\delta = +c$ (resp., $\delta = -c$) then communication rules are permitted (resp., forbidden).

Proposition 5 can be adapted to polarizationless P systems with active membranes which do not make use of division nor evolution rules, providing a lower bound about their efficiency.

Proposition 11. P $\subseteq$ PMC$_{\mathcal{NAM}^0(-d,-e,+c)}$.

4.2 A Conjecture of Păun

At the beginning of 2005, Gh. Păun (problem **F** from [25]) wrote:

My favorite question (related to complexity aspects in P systems with active membranes and with electrical charges) is that about the number of polarizations. Can the polarizations be completely avoided? The feeling is that this is not possible – and such a result would be rather sound: passing from no polarization to two polarizations amounts to passing from non–efficiency to efficiency.

This so–called Păun's conjecture can be formally formulated in terms of membrane computing complexity classes as follows:

$$\mathbf{P} = \mathbf{PMC}^{[*]}_{\mathcal{AM}^0(+d,-n,+e,+c)}$$

where the notation $\mathbf{PMC}^{[*]}_{\mathcal{R}}$ indicates that the result holds for both $\mathbf{PMC}_{\mathcal{R}}$ and $\mathbf{PMC}^*_{\mathcal{R}}$.

Let Π be a recognizer polarizationless P system with active membranes which do not make use of dissolution rules. A directed graph can be associated with Π verifying the following property: every accepting computation of Π is characterized by the existence of a path in the graph between two specific nodes.

Each rule of Π can be considered as a *dependency relation* between the object triggering the rule and the object(s) produced by its application. We can consider a general pattern for rules of types $(a), (b), (c), (e)$ in the form $(a, h) \rightarrow (a_1, h')(a_2, h') \ldots (a_s, h')$, where the rules of type (a) correspond to the case $h = h'$, the rules of type (b) correspond to the case $h = f(h')$ and $s = 1$, the rules of type (c) correspond to the case $h' = f(h)$ and $s = 1$, and the rules of type (e) correspond to the case $h = h'$ and $s = 2$. A formal definition of the *dependency graph* associated with a P system can be found in [11].

Note that a P system can dynamically evolve according to its rules, but the dependency graph associated with it is static. Furthermore, rules of the kind (f) and (g) do not provide any node nor arc to the dependency graph.

Let Δ_Π be the set of all pairs $(a, h) \in \Gamma \times H$ such that there exists a path (within the dependency graph) from (a, h) to $(\mathbf{yes}, env)$ – the environment is considered to be the output region, although the results obtained are also valid for any output membrane.

In [11] the following results are shown.

Proposition 12. *Let Π be a recognizer polarizationless P systems with active membranes not using dissolution rules, and where every kind of division rules is permitted. Then,*

- *There exists a Turing machine that constructs the dependency graph associated with Π in a time bounded by a polynomial function depending on the total number of rules and the maximum length of the rules.*
- *There exists a Turing machine that constructs the set Δ_Π in a time bounded by a polynomial function depending on the total number of rules and the maximum length of the rules.*

Given a family $\mathbf{\Pi} = \{\Pi(n) : n \in \mathbf{N}\}$ of recognizer P systems solving a decision problem in a uniform way (with (cod, s) being the associated polynomial encoding), the acceptance of a given instance of the problem, w, can be characterized by using the set $\Delta_{\Pi(s(w))}$ associated with $\Pi(s(w))$.

Let $\overline{\mathcal{M}_j} = \{(a, j) : a \in \mathcal{M}_j\}$, for $1 \leq j \leq q$ and $\overline{m} = \{(a, h_i) : a \in m\}$, for each input multiset m over Σ (recall that h_i is the label of the input membrane). Then, the following holds [11]:

Proposition 13. *Let $X = (I_X, \theta_X)$ be a decision problem, and $\mathbf{\Pi} = \{\Pi(n) : n \in \mathbf{N}\}$ a family of recognizer polarizationless P systems and not using dissolution rules solving X in a uniform way. Let (cod, s) be a polynomial encoding associated with that solution. Then, for each instance w of the problem X the following statements are equivalent:*

*(a) $\theta_X(w) = 1$ (that is, the answer to the problem is **yes** for w).*

(b) $\Delta_{\Pi(s(w))} \cap \left(\overline{cod(w)} \cup \bigcup_{j=1}^{q} \overline{\mathcal{M}_j} \right) \neq \emptyset$, where $\mathcal{M}_1, \ldots, \mathcal{M}_q$ are the initial multisets of $\Pi(s(w))$.

A similar result holds for semi–uniform solutions [11] and the following theorem can be deduced.

Theorem 3. $\mathbf{P} = \mathbf{PMC}^{[*]}_{\mathcal{AM}^0\,(-d,\beta,+e,+c)}$, *where* $\beta \in D$.

Thus, polarizationless P systems with active membranes which do not make use of dissolution rules are non–efficient in the sense that their cannot solve **NP**–complete problems in polynomial time (unless **P=NP**).

Let us now consider polarizationless P systems with active membranes making use of dissolution rules. Will it be possible to solve **NP**–complete problems in that framework?

N. Murphy et al. [15] gave a negative answer in the case that division rules are used only for elementary membranes and being *symmetric*, in the following sense $[\,a\,]_h \to [\,b\,]_h[\,b\,]_h$.

Theorem 4. $\mathbf{P} = \mathbf{PMC}^{[*]}_{\mathcal{AM}^0\,(+d,-n(sym),+e,+c)}$.

D. Woods et al. [39] have recently provide a **P** upper bound on polarizationless P systems with dissolution and division only for elementary membranes, without evolution and communication rules, where at the initial timestep, the depth of membrane nesting is equal to the total number of membranes.

Theorem 5. *If* $\mathcal{D}$ *is the class of systems in* $\mathcal{AM}^0\,(+d,-n,-e,-c)$, *having an initial membrane structure that is a single (linear) path, then* $\mathbf{P} = \mathbf{PMC}^{[*]}_{\mathcal{D}}$.

Several authors [3,11] gave a positive answer when division for non–elementary membranes, in the strong sense, is permitted. The mentioned papers provide semi–uniform solutions in a linear time to SAT and Subset Sum, respectively. Thus, we have the following result:

Proposition 14. $\mathbf{NP} \cup \mathbf{co\text{-}NP} \subseteq \mathbf{PMC}^{*}_{\mathcal{AM}^0\,(+d,+ns,+e,+c)}$.

As a consequence of Theorems 3 and 14, a *partial negative* answer to Păun's conjecture is given: assuming that $\mathbf{P} \neq \mathbf{NP}$ and making use of dissolution rules and division rules for elementary and non–elementary membranes, computationally hard problems can be efficiently solved avoiding polarizations. The answer is partial because efficient solvability of **NP**–complete problems by polarizationless P systems with active membranes making use of dissolution rules and division *only* for elementary membranes is unknown.

The result of Theorem 14 was improved by A. Alhazov et al. [1] giving a family of recognizer polarizationless P systems with active membranes using dissolution rules and division for elementary and (strong) non–elementary membranes solving QBF-SAT in a *uniform* way and in a linear time. Then,

Proposition 15. $\mathbf{PSPACE} \subseteq \mathbf{PMC}_{\mathcal{AM}^0\,(+d,+ns,+e,+c)}$.

Next, we present some results about the efficiency of polarizationless P systems with active membranes when evolution rules and/or communication rules are forbidden.

First, one can adapt a solution given in [3] to provide a semi-uniform solution to SAT in a linear time by a family of recognizer polarizationless P systems with active membranes by using evolution, dissolution and division rules for elementary and non–elementary membranes (both in the strong and weak versions), and avoiding communication rules. That is, we have the following:

Proposition 16. $\mathbf{NP} \cup \mathbf{co\text{-}NP} \subseteq \mathbf{PMC}^*_{\mathcal{AM}^0\,(+d,\beta,+e,-c)}$, *where* $\beta \in \{+nw, +ns\}$.

Evolution and communication rules can be avoided without loss of efficiency. Indeed, in [41] a semi–uniform solution to 3-SAT in a linear time by a family of polarizationless recognizer P systems with active membranes by using only dissolution rules and division rules for elementary and non–elementary membranes of the types (e) and (f), is presented. Thus, the following holds:

Proposition 17. $\mathbf{NP} \cup \mathbf{co\text{-}NP} \subseteq \mathbf{PMC}^*_{\mathcal{AM}^0\,(+d,+nsw,-e,-c)}$.

Moreover, Proposition 17 can be extended when non–elementary membrane division controlled by the presence of a membrane is allowed. In [14] it was presented a semi–uniform solution to QBF-3-SAT in a linear time by a family of polarizationless recognizer P systems with active membranes by using only dissolution rules and division rules of the types (e), (f) and (g). Thus, the following holds:

Proposition 18. $\mathbf{PSPACE} \subseteq \mathbf{PMC}^*_{\mathcal{AM}^0\,(+d,+nsr,-e,-c)}$.

Figure 1 graphically summarize the results known related with complexity classes associated with polarizationless P systems with active membranes making use

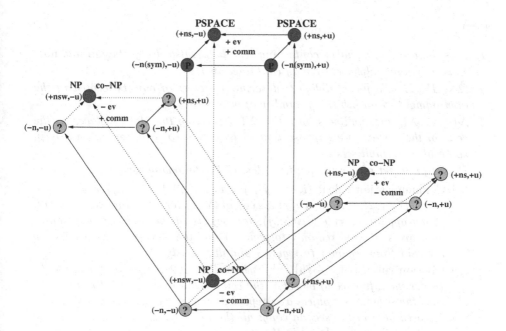

Fig. 1. Polarizationless active membranes *using dissolution rules*

of dissolution rules. In the picture, $-u$ (resp. $+u$) means semi–uniform (resp. uniform) solutions, $-n$ (resp. $+ns$ or $+nsw$)) means using division only for elementary membranes (resp. division for elementary and non–elementary membranes in the *strong* version or *strong* and *weak* version), $-n(sym)$ means using division only for elementary membranes and being *symmetric*, $-ev$ (resp. $+ev$) means that evolution rules are forbidden (resp. permitted), and $-comm$ (resp. $+comm$) means that communication rules are forbidden (resp. permitted). A standard class inside (respectively, over) a dark node means that the corresponding membrane computing class is equal (resp., is a lower bound) to the standard class.

5   Tissue–Like Recognizer P Systems with Cell Division

In this section, we consider computational devices inspired in cell inter–communication in tissues and we add the ingredient of cell division rules as we did to polarizationless P systems with active membranes (and with input membrane).

Definition 14. *A polarizationless tissue–like membrane system with cell division of degree $q \geq 1$ is a tuple*

$$\Pi = (\Gamma, \Sigma, \Omega, \mathcal{M}_1, \ldots, \mathcal{M}_q, R, i_{in}, i_{out})$$

where:

1. *Γ is the working alphabet containing two distinguished objects* yes *and* no*;*
2. *Σ is an (input) alphabet strictly contained in Γ.*
3. *$\Omega \subseteq \Gamma \setminus \Sigma$ is a finite alphabet, describing the set of objects located in the environment in an arbitrary number of copies each;*
4. *$\mathcal{M}_1, \ldots, \mathcal{M}_q$ are multisets over $\Gamma - \Sigma$, describing the objects placed in the cells of the system (we suppose that at least one copy of* yes *and* no *is in some of these multisets);*
5. *R is a finite set of developmental rules, of the following forms:*
   (a) *Communication rules: $(i, u/v, j)$, for $i, j \in \{0, 1, 2, \ldots, q\}, i \neq j$, and $u, v \in \Gamma^*; 1, 2, \ldots, q$ identify the cells of the system, 0 is the environment: When applying a rule $(i, u/v, j)$, the objects of the multiset represented by u are sent from region i to region j and the objects of the multiset v are sent from region j to region i simultaneously;*
   (b) *Division rules: $[a]_i \to [b]_i[c]_i$, where $i \in \{1, 2, \ldots, q\}$ and $a, b, c \in \Gamma$: Under the influence of object a, the cell labeled by i is divided in two cells with the same label; object a is replaced by b in the first copy, object a is replaced by c in the second copy; all the other objects are replicated and copies of them are placed in the two new cells;*
6. *$i_{in} \in \{1, \ldots, q\}$ is the input cell, and $i_{out} \in \{0, 1, \ldots, q\}$ is the output cell.*

Definition 15. *A polarizationless tissue–like membrane system with cell separation of degree $q \geq 1$ is a tuple*

$$\Pi = (\Gamma, \Sigma, \Omega, O_1, O_2, \mathcal{M}_1, \dots, \mathcal{M}_q, R, i_{in}, i_{out})$$

where:

1. Γ *is the working alphabet containing two distinguished objects* yes *and* no;
2. Σ *is an (input) alphabet strictly contained in Γ.*
3. $\Omega \subseteq \Gamma \setminus \Sigma$ *is a finite alphabet, describing the set of objects located in the environment in an arbitrary number of copies each;*
4. $\{O_1, O_2\}$ *is a partition of Γ, that is, $\Gamma = O_1 \cup O_2$, $O_1, O_2 \neq \emptyset$, $O_1 \cap O_2 = \emptyset$*
5. $\mathcal{M}_1, \dots, \mathcal{M}_q$ *are multisets over $\Gamma - \Sigma$, describing the objects placed in the cells of the system (we suppose that at least one copy of* yes *and* no *is in some of these multisets);*
6. R *is a finite set of developmental rules, of the following forms:*
   (a) Communication rules: $(i, u/v, j)$, *for $i, j \in \{0, 1, 2, \dots, q\}, i \neq j$, and $u, v \in \Gamma^*; 1, 2, \dots, q$ identify the cells of the system, 0 is the environment.*
   (b) Separation rules: $[a]_i \to [O_1]_i[O_2]_i$, *where $i \in \{1, 2, \dots, q\}$ and $a \in \Gamma$, and $i \neq i_0$: In reaction with an object a, the cell is separated into two cells with the same label; at the same time, object a is consumed; the objects from O_1 are placed in the first cell, those from O_2 are placed in the second cell; the output cell i_0 cannot be separated.*
7. $i_{in} \in \{1, \dots, q\}$ *is the input cell, and $i_{out} \in \{0, 1, \dots, q\}$ is the output cell.*

Let m be a multiset over Σ. The *initial configuration of Π with input m* is tuple $(\mathcal{M}_1, \dots, \mathcal{M}_{i_{in}} \cup m, \dots, \mathcal{M}_q)$.

The rules of a tissue–like membrane system as the one above are used in a non-deterministic maximally parallel way as customary in membrane computing. At each step, we apply a set of rules which is maximal (no further rule can be added), with the following important restriction: if a cell is divided (respectively, separated), then the division rule (respectively, separation rule) is the only one which is applied for that cell at that step, and so its objects do not participate in any communication rule.

All computations start from an initial configuration and proceed as stated above; only halting computations give a result, which is encoded by the number of objects in the output cell i_{out} in the last configuration. From now on, we will consider that the output is collected in the environment (that is, $i_{out} = 0$, and thus, we will omit i_{out} in the definition of tissue P systems). In this way, if Π is a tissue P system and $\mathcal{C} = \{C_i\}_{i<r}$ is a halting computation of Π, then the answer of the computation $\mathcal{C}$ is

$$Output(\mathcal{C}) = \Psi_{\Gamma \setminus \Omega}(M_{r-1,0})$$

where Ψ is the Parikh function, and $M_{r-1,0}$ is the multiset over $\Gamma \setminus \Omega$ associated with the environment at the halting configuration C_{r-1}.

Definition 16. *A polarizationless tissue–like membrane system is said to be a recognizer system (recognizer tissue P system, in short) if: (a) the working alphabet contains two distinguished elements* yes *and* no; *(b) all computations halt; and (c) if* $\mathcal{C}$ *is a computation of the system, then either object* yes *or object* no *(but not both) must have been sent to the output region of the system, and only at the last step of the computation.*

Given a recognizer tissue P system, and a computation $\mathcal{C} = \{C_i\}_{i<r}$ of Π ($r \in \mathbb{N}$), we define the result of $\mathcal{C}$ as follows:

$$Output(\mathcal{C}) = \begin{cases} \text{yes, if } \Psi_{\{\text{yes,no}\}}(M_{r-1,0}) = (1,0) \\ \qquad \wedge \ \Psi_{\{\text{yes,no}\}}(M_{k,0}) \ = (0,0) \text{ for } k = 0,\ldots,r-2 \\ \text{no, if } \Psi_{\{\text{yes,no}\}}(M_{r-1,0}) = (0,1) \\ \qquad \wedge \ \Psi_{\{\text{yes,no}\}}(M_{k,0}) \ = (0,0) \text{ for } k = 0,\ldots,r-2 \end{cases}$$

That is, $\mathcal{C}$ is an accepting computation (respectively, rejecting computation) if the object yes (respectively, no) appears (only) in the environment associated with the halting configuration.

We denote by $\mathcal{TDC}$ and $\mathcal{TDC}(k)$ the class of recognizer tissue P systems with cell division, the latter using communication rules with length at most k. We denote by $\mathcal{TSC}$ and $\mathcal{TSC}(k)$ the class of recognizer tissue P systems with cell separation, the latter using communication rules with length at most k. We also denote by $\mathcal{TC}$ the class of recognizer tissue P systems without cell division nor cell separation.

The concepts of polynomially uniform by Turing machines, polynomial encoding, polynomially bounded, soundness and completeness introduced at definitions 4, 6, 7 and 9 can be naturally generalized to the framework of recognizer tissue P systems.

Definition 17. *We say that a decision problem* $X = (I_X, \theta_X)$ *is solvable in polynomial time by a family* $\mathbf{\Pi} = \{\Pi(n) : n \in \mathbb{N}\}$ *of recognizer tissue P systems if the following holds:*

- *The family* $\mathbf{\Pi}$ *is polynomially uniform by Turing machines, that is, there exists a deterministic Turing machine which constructs the system* $\Pi(n)$ *from* $n \in \mathbb{N}$ *in polynomial time with respect to* n.
- *There exists a pair* (cod, s) *of polynomial-time computable functions over* I_X *(called a polynomial encoding of* I_X *in* $\mathbf{\Pi}$*) such that:*
  - *For each instance* $w \in I_X$, $s(w)$ *is a natural number (obtained by means of a reasonable encoding scheme), and* $cod(w)$ *is an input multiset of the system* $\Pi(s(w))$.
  - *The family* $\mathbf{\Pi}$ *is polynomially bounded with regard to* (X, cod, s); *that is, there exists a polynomial function* p, *such that for each* $w \in I_X$ *every computation of* $\Pi(s(w))$ *with input* $cod(w)$ *is halting and, moreover, it performs at most* $p(|w|)$ *steps.*
  - *The family* $\mathbf{\Pi}$ *is sound with regard to* (X, cod, s); *that is, for each* $w \in I_X$, *if there exists an accepting computation of* $\Pi(s(w))$ *with input* $cod(w)$, *then* $\theta_X(w) = 1$.

- *The family* Π *is* complete *with regard to* (X, cod, s)*; that is, for each* $w \in I_X$*, if* $\theta_X(w) = 1$*, then every computation of* $\Pi(s(w))$ *with input* $cod(w)$ *is an accepting one.*

From the soundness and completeness conditions above we deduce that every P system $\Pi(n)$ is *confluent*, in the following sense: every computation of a system with the *same* input multiset must always give the *same* answer.

We denote by $\mathbf{PMC}_{\mathcal{R}}$ the set of all decision problems which can be solved by means of recognizer tissue P systems from $\mathcal{R}$ in polynomial time. This class is closed under complement and polynomial–time reductions (see [29] for a similar result for cell-like P systems).

In [19], a polynomial time solution for the SAT problem was given by using a family of recognizer tissue P systems with cell separation and communication rules of length at most 6. Then

Proposition 19. $\mathbf{NP} \cup \mathbf{co\text{-}NP} \subseteq \mathbf{PMC}_{\mathcal{TSC}(6)}$.

In [7], a polynomial time solution for the Vertex Cover problem was given by using a family of recognizer tissue P systems with cell division and communication rules of length at most 3. Then

Proposition 20. $\mathbf{NP} \cup \mathbf{co\text{-}NP} \subseteq \mathbf{PMC}_{\mathcal{TDC}(3)}$.

5.1 Allowing Communication Rules of Length at Most 1

For recognizer tissue P systems with cell division and communication rules with length at most 1, it can be generalized the concept of dependency graph in a natural way.

We can consider a general pattern $(a, i) \rightarrow (b_1, j) \ldots (b_s, j)$ where $i, j \in \{0, 1, 2, \ldots, q\}, i \neq j$, and $a, b \in \Gamma$. Communication rules correspond to the case $s = 1$ and $b_1 = a$, and division rules correspond to the case $s = 2$ and $j = i \neq 0$. The above pattern can be interpreted as follows: from the object a in the cell (or in the environment) labeled by i we can *reach* objects $b_1, \ldots, b_s$ in the cell (or in the environment) labeled by j.

By using the concept of dependency graph associated with tissue P systems with cell division and communication rules with length at most 1, it is proved that this kind of tissue P systems can only efficiently solve tractable problems (see [9], for details).

Theorem 6. $\mathbf{P} = PMC_{\mathcal{TDC}(1)}$.

From Proposition 20 and Theorem 6, we deduce that in the framework of recognizer tissue P systems with cell division the length of the communication rules provides a borderline between efficiency and non-efficiency. Specifically, a frontier is obtained when passing from length 1 to length 3.

Similarly, we can consider a dependency graph associated with a tissue P system with cell separation and communication rules with length at most 1.

In this case, separation rules do not provide any arc to the graph. In [19], it is proved that this kind of tissue P systems can only efficiently solve tractable problems. That is,

Theorem 7. $\mathbf{P} = PMC_{\mathcal{TSC}(1)}$.

Hence, in the framework of recognizer tissue P systems with cell separation, the lengths of communication rules provide a borderline between efficiency and non-efficiency. Specifically, there is a frontier when passing from length 1 to length 6.

6   Efficiency of Tissue P Systems without Cell Division

A family of recognizer tissue P systems with symport/antiport rules which solves a decision problem can be efficiently simulated by a family of basic recognizer P systems solving the same problem. This simulation allows us to transfer the result about the limitations in computational power, from the model of basic cell–like P systems to this kind of tissue–like P systems.

Definition 18. *Let Π and Π' be recognizer cellular systems (cell–like and/or tissue–like). We say that Π' efficiently simulates Π if the following holds:*

- *Π' can be constructed from Π by a deterministic Turing machine working in polynomial time.*
- *There exists a bijective function, f, from the set $\mathbf{Comp}(\Pi)$ of computations of Π onto the set $\mathbf{Comp}(\Pi')$ of computations of Π' such that:*
  - *A computation $\mathcal{C} \in \mathbf{Comp}(\Pi)$ is an accepting computation if and only if $f(\mathcal{C}) \in \mathbf{Comp}(\Pi')$ is an accepting one.*
  - *There exists a polynomial $p(n)$ such that for each $\mathcal{C} \in \mathbf{Comp}(\Pi)$ we have $|f(\mathcal{C})| \leq p(|\mathcal{C}|)$.*

Next, for every recognizer tissue P system with symport/antiport rules we design a basic recognizer P systems efficiently simulating it, according to Definition 18.

Definition 19. *Let $\Pi = (\Gamma, \Sigma, \Omega, \mathcal{M}_1, \ldots, \mathcal{M}_q, \mathcal{R}, i_{in})$ be a recognizer tissue P system of degree $q \geq 1$ with communication rules and without cell division. Let us consider the basic recognizer P system $S(\Pi) = (\Gamma', \Sigma', \mu, \mathcal{M}'_1, \mathcal{R}', i'_{in})$ defined as follows:*

- *$\Gamma' = \{(a,i) : a \in \Gamma \wedge i \in \{1, \ldots, q\}\} \cup \{(a,0) : a \in \Gamma \setminus \Omega\} \cup \{\mathsf{yes}, \mathsf{no}\}$.*
  The objects of $S(\Pi)$ are ordered pairs encoding objects of Π and cells where the objects are placed. From the environment, we only consider objects with finite multiplicity, that is, belonging to $\Gamma \setminus \Omega$.
- *$\Sigma' = \{(a, i_{in}) : a \in \Sigma\}$.*
- *$\mu = [\,]_1$.*
- *$\mathcal{M}'_1 = \displaystyle\sum_{i=1}^{q} \sum_{a \in \Gamma \setminus \Sigma} (a, i)^{\mathcal{M}_i(a)}$.*
  For each cell i of Π and for each object $a \in \Gamma \setminus \Sigma$ belonging to that cell, we consider in the membrane of $S(\Pi)$ the pair (a, i) with the same multiplicity.

– *In the set $\mathcal{R}'$ the following rules associated with $S(\Pi)$ are included:*
- *For each rule $r_{_\Pi} \equiv (i, a_1 \ldots a_m \,/\, b_1 \ldots b_n, j) \in \mathcal{R}$ with $i, j \neq 0$, associated with Π, we consider the following rule (denoted by $r_{_{S(\Pi)}}$)*
$$(a_1, i) \ldots (a_m, i)(b_1, j) \ldots (b_n, j) \to (b_1, i) \ldots (b_n, i)(a_1, j) \ldots (a_m, j)$$
- *For each rule $r_{_\Pi} \equiv (i, a_1 \ldots a_m \,/\, b_1 \ldots b_n, 0) \in \mathcal{R}$ with $i \neq 0$, associated with Π, we consider the following rule (denoted by $r_{_{S(\Pi)}}$)*
$$(a_1, i) \ldots (a_m, i)(b_1, 0) \ldots (b_s, 0) \to (b_1, i) \ldots (b_n, i)(a_1, 0) \ldots (a_r, 0)$$
where $a_1, \ldots, a_r, b_1, \ldots, b_s \notin \Omega$ and $a_{r+1}, \ldots, a_m, b_{s+1}, \ldots, b_n \in \Omega$.
- *For each rule $r_{_\Pi} \equiv (0, a_1 \ldots a_m \,/\, b_1 \ldots b_n, i) \in \mathcal{R}$ with $i \neq 0$, associated with Π, we consider the following rule (denoted by $r_{_{S(\Pi)}}$)*
$$(a_1, 0) \ldots (a_r, 0)(b_1, i) \ldots (b_n, i) \to (b_1, 0) \ldots (b_s, 0)(a_1, i) \ldots (a_m, i)$$
where $a_1, \ldots, a_r, b_1, \ldots, b_s \notin \Omega$ and $a_{r+1}, \ldots, a_m, b_{s+1}, \ldots, b_n \in \Omega$.
- $(\text{yes}, 0) \to (\text{yes}, out)$; $(\text{no}, 0) \to (\text{no}, out)$.
  These rules translate the answer provided by the system Π to an answer for the system $S(\Pi)$.
– $i'_{in} = 1$, *that is, the membrane of the system is the input membrane.*

Proposition 21. *Let Π be a recognizer tissue P system with communication rules and without cell division. The system $S(\Pi)$ is a basic recognizer P system that efficiently simulates Π.*

This result provides us a limitation concerning the efficiency of tissue P systems with communication rules and without cell division. Within this framework, it is only possible to efficiently solve tractable problems, that is, problems belonging to the complexity class **P** [6].

Theorem 8. $\mathbf{P} = \mathbf{PMC}_{\mathcal{TC}}$.

7 Conclusions

In this paper, we have described the basic concepts and the main results that pertain to pioneering computational complexity in the membrane computing field.

We conclude by presenting new research directions within membrane computing complexity theory by listing some of the current open questions.

(A) Are there significant differences between uniform and semi–uniform solutions? Namely, is there some class $\mathcal{R}$ of recognizer P systems such that the inclusion $\mathbf{PMC}_{\mathcal{R}} \subseteq \mathbf{PMC}^*_{\mathcal{R}}$ is strict?

(B) Efficient uniform solutions to **NP**–complete problems have been given by models of $\mathcal{AM}(-n)$. Is it possible to efficiently solve **PSPACE**–complete problems by using families of P systems from $\mathcal{AM}(-n)$?

(C) What is the efficiency of P systems with active membranes and electrical charges where evolution and communication rules are forbidden? Are there any relations with the results obtained for polarizationless P systems?

(D) Dissolution rules provide a borderline between tractability and intractability in the framework of polarizationless P systems with active membranes making use of division rules for elementary and non–elementary membranes. What happens if division for only elementary membranes is allowed? Is $\mathbf{P} = \mathbf{PMC}^{[*]}_{\mathcal{AM}^0(+d,-n,+e,+c)}$ true?

(E) It is well known that $\mathbf{PSPACE} \subseteq \mathbf{PMC}^*_{\mathcal{AM}^0(+d,+nsr,-e,-c)}$. Determine an upper bound for that membrane computing complexity class.

(F) It is known that $\mathbf{P} = \mathbf{PMC}_{\mathcal{TDC}(1)}$ and $\mathbf{NP} \cup \mathbf{co\text{-}NP} \subseteq \mathbf{PMC}_{\mathcal{TDC}(3)}$. What about the complexity class $\mathbf{PMC}_{\mathcal{TDC}(2)}$? In the solution provided in [7], antiport rules of length at most 3 were used. Would it be possible to provide another efficient solution in which all rules of length at most 3 were symport rules?

(G) It is known that $\mathbf{P} = \mathbf{PMC}_{\mathcal{TSC}(1)}$ and $\mathbf{NP} \cup \mathbf{co\text{-}NP} \subseteq \mathbf{PMC}_{\mathcal{TSC}(6)}$. What about the complexity class $\mathbf{PMC}_{\mathcal{TSC}(k)}$, for $2 \leq k \leq 5$? In the solution provided in [19], antiport rules of length at most 6 were used. Would it be possible to provide another efficient solution in which all rules of length at most 6 were symport rules?

Acknowledgements

The authors acknowledge the support of the project TIN2006–13425 of the Ministerio de Educación y Ciencia of Spain, cofinanced by FEDER funds, and the support of the Project of Excellence with *Investigador de Reconocida Valía* of the Junta de Andalucía, grant P08-TIC-04200.

References

1. Alhazov, A., Pérez–Jiménez, M.J.: Uniform solution of QSAT using polarizationless active membranes. In: Durand-Lose, J., Margenstern, M. (eds.) MCU 2007. LNCS, vol. 4664, pp. 122–133. Springer, Heidelberg (2007)

2. Alhazov, A., Freund, R., Păun, Gh.: P systems with active membranes and two polarizations. In: Păun, Gh., et al. (eds.) Proc. Second Brainstorming Week on Membrane Computing, Report RGNC 01/04, Sevilla, pp. 20–35 (2004)

3. Alhazov, A., Pan, L., Păun, G.: Trading polarizations for labels in P systems with active membranes. Acta Informaticae 41(2-3), 111–144 (2004)

4. Alhazov, A., Martín–Vide, C., Pan, L.: Solving graph problems by P systems with restricted elementary active membranes. In: Jonoska, N., Păun, Gh., Rozenberg, G. (eds.) Aspects of Molecular Computing. LNCS, vol. 2950, pp. 1–22. Springer, Heidelberg (2003)

5. Alhazov, A., Martín–Vide, C., Pan, L.: Solving a PSPACE–complete problem by recognizing P systems with restricted active membranes. Fundamenta Informaticae 58, 67–77 (2003)

6. Díaz–Pernil, D., Gutiérrez–Naranjo, M.A., Pérez-Jiménez, M.J., Romero–Jiménez, A.: Efficient simulation of tissue–like P systems by transition cell–like P systems. Natural Computing, http://dx.doi.org/10.1007/s11047-008-9102-z

7. Díaz–Pernil, D., Pérez–Jiménez, M.J., Riscos–Núñez, A., Romero–Jiménez, A.: Computational efficiency of cellular division in tissue-like membrane systems. Romanian Journal of Information Science and Technology 11(3), 229–241 (2008)

8. Garey, M.R., Johnson, D.S.: Computers and Intractability. A Guide to the Theory of NP-completeness. W.H. Freeman and Company, New York (1979)

9. Gutiérrez–Naranjo, R., Pérez–Jiménez, M.J., Rius–Font, M.: Characterizing tractability by tissue–like P systems. In: Gutiérrez–Escudero, R., et al. (eds.) Proc. Seventh Brainstorming Week on Membrane Computing, Fénix Editora, Seville, pp. 169–180 (2009)

10. Gutiérrez-Naranjo, M.A., Pérez-Jiménez, M.J., Riscos-Núñez, A.: A fast P system for finding a balanced 2-partition. Soft Computing 9(9), 673–678 (2005)

11. Gutiérrez–Naranjo, M.A., Pérez–Jiménez, M.J., Riscos–Núñez, A., Romero–Campero, F.J.: On the power of dissolution in P systems with active membranes. In: Freund, R., Păun, G., Rozenberg, G., Salomaa, A. (eds.) WMC 2005. LNCS, vol. 3850, pp. 224–240. Springer, Heidelberg (2006)

12. Gutiérrez–Naranjo, M.A., Pérez–Jiménez, M.J., Riscos–Núñez, A., Romero–Campero, F.J., Romero–Jiménez, A.: Characterizing tractability by cell–like membrane systems. In: Subramanian, K.G., et al. (eds.) Formal Models, Languages and Applications, pp. 137–154. World Scientific, Singapore (2006)

13. Krishna, S.N., Rama, R.: A variant of P systems with active membranes: Solving NP–complete problems. Romanian Journal of Information Science and Technology 2(4), 357–367 (1999)

14. Leporati, A., Ferretti, C., Mauri, G., Pérez–Jiménez, M.J., Zandron, C.: Complexity aspects of polarizationless membrane systems. Natural Computing, http://dx.doi.org/10.1007/s11047-008-9100-1

15. Murphy, N., Woods, D.: Active membrane systems without charges and using only symetric elementary division characterise **P**. In: Eleftherakis, G., Kefalas, P., Păun, G., Rozenberg, G., Salomaa, A. (eds.) WMC 2007. LNCS, vol. 4860, pp. 367–384. Springer, Heidelberg (2007)

16. Obtulowicz, A.: Deterministic P systems for solving SAT problem. Romanian Journal of Information Science and Technology 4(1-2), 551–558 (2001)

17. Pan, L., Alhazov, A., Ishdorj, T.-O.: Further remarks on P systems with active membranes, separation, merging, and release rules. In: Păun, Gh., et al. (eds.) Proc. Second Brainstorming Week on Membrane Computing, Report RGNC 01/04, Sevilla, pp. 316–324 (2004)

18. Pan, L., Ishdorj, T.-O.: P systems with active membranes and separation rules. Journal of Universal Computer Science 10(5), 630–649 (2004)

19. Pan, L., Pérez-Jiménez, M.J.: Computational complexity of tissue–like P systems with cell separation (submitted, 2009)

20. Păun, A.: On P systems with membrane division. In: Antoniou, I., et al. (eds.) Unconventional Models of Computation, pp. 187–201. Springer, London (2000)

21. Păun, Gh.: Computing with membranes. Journal of Computer and System Sciences 61(1), 108–143 (2000); Turku Center for CS-TUCS Report No. 208 (1998)

22. Păun, Gh.: Membrane Computing. An Introduction. Springer, Berlin (2002)

23. Păun, Gh.: P systems with active membranes: Attacking **NP**–complete problems. Journal of Automata, Languages and Combinatorics 6(1), 75–90 (2001)

24. Păun, Gh.: Computing with membranes. Attacking **NP**–complete problems. In: Antoniou, I., et al. (eds.) Unconventional Models of Computation, pp. 94–115. Springer, London (2000)

25. Păun, Gh.: Further twenty six open problems in membrane computing. In: Gutiérrez-Naranjo, M.A., et al. (eds.) Proc. Third Brainstorming Week on Membrane Computing, Report RGNC 01/04, Fénix Editora, Sevilla, pp. 249–262 (2005)

26. Pérez–Jiménez, M.J.: An approach to computational complexity in membrane computing. In: Mauri, G., Păun, G., Jesús Pérez-Jímenez, M., Rozenberg, G., Salomaa, A. (eds.) WMC 2004. LNCS, vol. 3365, pp. 85–109. Springer, Heidelberg (2005)

27. Pérez-Jiménez, M.J., Riscos-Núñez, A.: Solving the Subset-Sum problem by active membranes. New Generation Computing 23(4), 367–384 (2005)

28. Pérez-Jiménez, M.J., Riscos-Núñez, A.: A linear–time solution to the Knapsack problem using P systems with active membranes. In: Martín-Vide, C., Mauri, G., Păun, G., Rozenberg, G., Salomaa, A. (eds.) WMC 2003. LNCS, vol. 2933, pp. 250–268. Springer, Heidelberg (2004)

29. Pérez-Jiménez, M.J., Romero-Jiménez, A., Sancho-Caparrini, F.: A polynomial complexity class in P systems using membrane division. Journal of Automata, Languages and Combinatorics 11(4), 423–434 (2006); A preliminary version in Csuhaj-Varjú, E., et al. (eds.) Proc. Fifth International Workshop on Descriptional Complexity of Formal Systems, DCFS 2003, Budapest, Hungary, July 12-14, pp. 284–294 (2003)

30. Pérez-Jiménez, M.J., Romero–Campero, F.J.: Attacking the common algorithmic problem by recognizer P systems. In: Margenstern, M. (ed.) MCU 2004. LNCS, vol. 3354, pp. 304–315. Springer, Heidelberg (2005)

31. Pérez–Jiménez, M.J., Romero–Campero, F.J.: An efficient family of P systems for packing items into bins. Journal of Universal Computer Science 10(5), 650–670 (2004)

32. Pérez–Jiménez, M.J., Romero-Campero, F.J.: Trading polarizations for bi-stable catalysts in P systems with active membranes. In: Mauri, G., Păun, G., Jesús Pérez-Jímenez, M., Rozenberg, G., Salomaa, A. (eds.) WMC 2004. LNCS, vol. 3365, pp. 373–388. Springer, Heidelberg (2005)

33. Pérez–Jiménez, M.J., Romero–Jiménez, A., Sancho–Caparrini, F.: Complexity classes in cellular computing with membranes. Natural Computing 2(3), 265–285 (2003)

34. Porreca, A.E.: Computational Complexity Classes for Membrane Systems, Master Degree Thesis, Universita' di Milano-Bicocca, Italy (2008)

35. Porreca, A.E., Mauri, G., Zandron, C.: Complexity classes for membrane systems. Informatique théorique et applications 40(2), 141–162 (2006)

36. Riscos–Núñez, A.: Cellular Programming: efficient resolution of NP–complete numerical problems. PhD. Thesis, University of Sevilla, Spain (2004)

37. Sosík, P., Rodríguez–Patón, A.: Membrane computing and complexity theory: A characterization of PSPACE. Journal of Computer and System Sciences 73, 137–152 (2007)

38. Sosík, P.: The computational power of cell division. Natural Computing 2(3), 287–298 (2003)

39. Woods, D., Murphy, N., Pérez-Jiménez, M.J., Riscos-Núñez, A.: Membrane dissolution and division in P. LNCS, vol. 5715, pp. 263–277. Springer, Heidelberg (2009)

40. Zandron, C., Ferretti, C., Mauri, G.: Solving NP–complete problems using P systems with active membranes. In: Antoniou, I., Calude, C.S., Dinneen, M.J. (eds.) Unconventional Models of Computation, pp. 289–301. Springer, Heidelberg (2000)

41. Zandron, C., Leporati, A., Ferretti, C., Mauri, G., Pérez–Jiménez, M.J.: On the computational efficiency of polarizationless recognizer P systems with strong division and dissolution. Fundamenta Informaticae 87(1), 79–91 (2008)

Evolving by Maximizing the Number of Rules: Complexity Study

Oana Agrigoroaiei, Gabriel Ciobanu, and Andreas Resios

Romanian Academy, Institute of Computer Science
Blvd. Carol I no.8, 700505 Iaşi, Romania, and
"A.I.Cuza" University, Blvd. Carol I no.11, 700506 Iaşi, Romania
oanaag@iit.tuiasi.ro, gabriel@info.uaic.ro, andreas.resios@iit.tuiasi.ro

Abstract. This paper presents the complexity of finding a multiset of rules in a P system in such a way to have a maximal number of rules applied. It is proved that the decision version of this problem is **NP**-complete. We study a number of subproblems obtained by considering that a rule can be applied at most once, and by considering the number of objects in the alphabet of the membrane as being fixed. When considering P systems with simple rules, the corresponding decision problem is in **P**. When considering P systems having only two types of objects, and P systems in which a rule is applied at most once, their corresponding decision problems are **NP**-complete. We compare these results with those obtained for $maxO$ evolution.

1 Introduction

The reader is assumed to have basic knowledge of elements of membrane computing, which can be found in [6]. Here we just mention the main biological inspiration of P systems, and some terminology concerning the variants of maximal parallelism we consider in this paper.

P systems are inspired by the structure and the functioning of the living cells. Inside the cell, several membranes define compartments where specific biochemical processes take place. Each compartment contains substances (ions, small molecules, macromolecules) and specific reactions. The substances are represented by multisets of objects, and the reactions by rules of form $u \rightarrow v$, where u and v are multisets of objects. The multisets are represented by strings, with the understanding that all permutations of a string represent the same multiset. We denote by O the alphabet of objects, and by R_i the set of rules associated with a compartment i. When such a system is evolving, the objects and the rules are chosen in a nondeterministic manner, and the rules are applied in parallel.

The most investigated way of using the rules in a P system is the maximal parallelism: in each membrane a multiset of rules is chosen to be applied to the objects from that membrane; the multiset is maximal in the sense of inclusion, i.e., no further rule can be added such that the enlarged multiset is still applicable. We use "$maxP$" to refer to this evolution strategy.

G. Păun et al. (Eds.): WMC 2009, LNCS 5957, pp. 149–157, 2010.

Another natural idea is to apply the rules in such a way to have a maximal number of objects consumed in each membrane; this manner of evolution is denoted by "*maxO*". This strategy was explicitly considered in [1,2], where it is proved that the problem of finding a multiset of rules which consumes a maximal number of objects is **NP**-complete.

Yet a third idea is to apply the rules in such a way to have a maximal number of rules applied. We denote this type of evolution by "*maxR*". Note that any evolution of either type *maxR* or type *maxO* is also of type *maxP*.

The computing power of these strategies of applying a multiset of rules in membranes is studied in [3]. Specifically, P systems having multiset rewriting rules (with cooperative rules), symport/antiport rules, and active membranes are considered. The universality of the system is proved for any combination of type of system and type of evolution.

Two variants of membrane systems called *simple P systems* and *maximum cooperative P systems* are considered in two previous papers [1,2]. These systems evolve at each step by consuming the maximum number of objects. The problem of distributing objects to rules in order to achieve a maximum consuming and non-deterministic evolution of simple P systems is studied in [1]; using the knapsack problem, the decision version of the resource mapping problem for simple P systems is proved to be **NP**-complete. The integer linear programming problem is used in [2] to prove that the resource mapping problem for maximum cooperative P systems is also **NP**-complete.

In this paper we study the complexity of finding a multiset of rules which evolves using the *maxR* strategy. We study a number of subproblems obtained by considering the number of objects in the alphabet of the membrane as being fixed, and by considering that a rule can be applied at most once. We compare the results with those obtained for *maxO* strategy.

2   *maxR* Complexity

We recall a number of notations for multisets and P systems. We represent multisets as strings of elements over their support alphabet together with their multiplicities (for example $w = a^2 b^5 c$ is a multiset over $\{a, b, c, d\}$). The union $v + w$ of two multisets over a set O is given by the sum of multiplicities for each element of O. We define $w(a) \in \mathbb{N}$ to be the multiplicity of a in w. We say that $w \leq w'$ if $w(a) \leq w'(a)$ for each element a of the multiset w. In this case we define $w' - w$ to be the multiset obtained by subtracting the multiplicity in w of an element from its multiplicity in w'. We use the notation $i = \overline{1, n}$ to denote $i \in \{1, \ldots, n\}$.

Definition 1. *A transition P system of degree n (n ≥ 1) is a construct*
$$\Pi = (O, \mu, w_1, \ldots, w_n, R_1, \ldots, R_n), \text{ where}$$

- *O is an alphabet of objects;*
- *μ is a membrane structure, with the membranes labelled by natural numbers $1, \ldots, m$ in a one-to-one manner;*

- w_i are multisets over O associated with the regions $1, \ldots, m$ defined by μ;
- $R_1, \ldots, R_m$ are finite sets of rules associated with the membranes with labels $1, \ldots, m$; the rules have the form $u \rightarrow v$, where u is a non-empty multiset of objects, and v a multiset over messages of the form $(a, here), (a, out), (a, in_j)$.

For a rule $r = u \rightarrow v$ we use the notations $lhs(r) = u$ and $rhs(r) = v$. These notations are extended naturally to multisets of rules: given a multiset of rules $\mathcal{R}$, the left hand side of the multiset $lhs(\mathcal{R})$ is obtained by adding the left hand sides of the rules in the multiset, considered with their multiplicities. A configuration of the system is given by the membrane structure and the multisets contained in each membrane.

We define the three evolution strategies as follows:

Definition 2. For $i = \overline{1,n}$, a multiset $\mathcal{R}$ of rules over R_i is applicable (in membrane i) with respect to the multiset w_i if $lhs(\mathcal{R}) \leq w_i$ and for each message (a, in_j) present in $rhs(\mathcal{R})$ we have that j is one of the children of membrane i.

A multiset $\mathcal{R}$ of rules over R_i which is applicable with respect to the multiset w_i is called:

- *maxP-applicable with respect to w_i if there is no rule r in R_i such that $\mathcal{R}+r$ is applicable with respect to w_i;*
- *maxO-applicable with respect to w_i if for any other multiset $\mathcal{R}'$ of rules which is applicable with respect to w_i we have that*

$$\sum_{a \in O} lhs(\mathcal{R})(a) \geq \sum_{a \in O} lhs(\mathcal{R}')(a);$$

- *maxR-applicable with respect to w_i if for any other multiset $\mathcal{R}'$ of rules which is applicable with respect to w_i we have that*

$$\sum_{r \in R_i} \mathcal{R}(r) \geq \sum_{r \in R_i} \mathcal{R}'(r).$$

In other words, when choosing the *maxP* evolution strategy we only apply multisets of rules which are maximal with respect to inclusion; when choosing *maxO* we only apply multisets of rules which are maximal with respect to the number of objects (considered with their multiplicities) in the left hand side of the multiset; when choosing *maxR* we only apply multisets of rules which are maximal with respect to the number of rules in the multiset (considered with their multiplicities). Note that any multiset of rules which is either *maxR* or *maxO*-applicable is also *maxP*-applicable. P systems generally employ the *maxP* evolution strategy; however, *maxO* and *maxR* represent convincing alternatives.

As it is mentioned in [3], maximizing the number of objects or the number of rules can be related to the idea of energy for controlling the evolutions of P systems. In the same paper, the complexity of finding the multiset of rules in a P system in the case of *maxR* was presented as an open problem.

We denote by P_O and P_R the problems of finding a $maxO$ or $maxR$-applicable multiset of rules, with respect to a given multiset of objects w. We can consider similar problems for the entire system, but they are solved by splitting the problems into smaller ones, one for each membrane. Thus for our purposes we can just consider the systems containing only one membrane, i.e., the degree of the P systems is $n = 1$. In other words, all multisets of rules we consider from now on are over a set of rules R. We use the following notations:

- m is the cardinal of the alphabet O, and we consider the objects to be denoted by $o_1, \ldots, o_m$;
- d is the number of rules associated to the membrane, and the rules are denoted by $r_1, \ldots, r_d$;
- C_a is the multiplicity of o_a in the multiset w which is in the membrane;
- $k_{i,a}$ is the multiplicity of o_a in the left hand side of the rule r_i.

The problem P_O can be described as an integer linear programming problem. Given the positive integers $m, d, k_{i,a}, C_a$ for $i = \overline{1,d}$ and $a = \overline{1,m}$, find positive integers x_i such that

- $\sum_{i=\overline{1,d}} (\sum_{a=\overline{1,m}} k_{i,a}) x_i$ is maximal;
- $\sum_{i=\overline{1,d}} x_i \cdot k_{i,a} \leq C_a$ for all $a = \overline{1,m}$.

The decision version of this problem was shown to be **NP**-complete in [1,2]. The proofs were based on the knapsack problem and integer linear programming [4,5].

The problem P_R can be described as follows. Given the positive integers $m, d, k_{i,a}, C_a$ for $i = \overline{1,d}$ and $a = \overline{1,m}$, find positive integers x_i such that

- $\sum_{i=\overline{1,d}} x_i$ is maximal;
- $\sum_{i=\overline{1,d}} x_i \cdot k_{i,a} \leq C_a$ for all $a = \overline{1,m}$.

The decision version of P_R is denoted by DP_R: being given positive integers $m, d, t, k_{i,a}$ and C_a, find whether there exist positive integers x_i such that

- $\sum_{i=\overline{1,d}} x_i \geq t$;
- $\sum_{i=\overline{1,d}} x_i \cdot k_{i,a} \leq C_a$ for all $a = \overline{1,m}$.

The length of this instance of the problem can be considered to be $m + d + \max_{a,i} \{\log C_a, \log k_{i,a}\}$.

Proposition 1. DP_R is **NP**-complete.

Proof. First, we prove that DP_R is in **NP**. To show this we construct a Turing machine that computes the result in nondeterministic polynomial time by either accepting (output YES) or rejecting (output NO) the input string. The machine operates as follows:

1. nondeterministically assign values for x_i, $i = \overline{1,d}$;
2. if the assigned values verify the constraints,
3. and $\sum_{i=\overline{1,d}} x_i \geq t$, then output YES;
4. in any other case output NO.

It is easy to see that the number of steps performed by the machine is polynomial with respect to the input size. Thus DP_R is in **NP**.

Secondly, we construct a polynomial-time reduction from $3CNFSAT$ to DP_R. The $3CNFSAT$ problem asks whether a formula ϕ given in conjunctive normal form with 3 variables per clause is satisfiable, i.e., if there exists a variable assignment which makes the formula true [4]. Consider a formula ϕ with variables $x_1, \ldots, x_r$ and clauses $c_1, \ldots, c_s$. We describe a corresponding instance of DP_R:

- $d = 2r, m = r + s, t = r$;
- for each variable x_i of ϕ we consider two variables y_i and z_i together with an inequality $y_i + z_i \leq 1$ in the instance of DP_R;
- for each clause c_a we consider the inequality

$$\sum_{i=\overline{1,r}} q_{i,a} y_i + \sum_{i=\overline{1,r}} l_{i,a} z_i \leq 2$$

such that:
- $q_{i,a} = 0$, $l_{i,a} = 1$ if the literal x_i appears in c_a;
- $q_{i,a} = 1$, $l_{i,a} = 0$ if the literal $\neg x_i$ appears in c_a;
- $q_{i,a} = l_{i,a} = 0$ if neither x_i nor $\neg x_i$ appear in c_a.

Since we consider $t = r$, the first inequality in this instance of DP_R becomes $\sum_{i=\overline{1,r}} y_i + z_i \geq r$. This can be computed in polynomial time with respect to the size of the input. The idea behind the reduction is to set $x_i = 1$ if and only if $y_i = 1$, $z_i = 0$, and $x_i = 0$ if and only if $y_i = 0$, $z_i = 1$.

For example, consider the formula $\phi = c_1 \wedge c_2 \wedge c_3 \wedge c_4$ with $c_1 = x_1 \vee \neg x_2 \vee x_3$, $c_2 = \neg x_1 \vee \neg x_2 \vee \neg x_3$, $c_3 = x_1 \vee \neg x_2 \vee \neg x_3$ and $c_4 = \neg x_1 \vee x_2 \vee x_3$. The corresponding instance of DP_R is:

Find y_i, z_i ($i = \overline{1,3}$) positive integers such that $\sum_{i=\overline{1,3}} y_i + z_i \geq 3$, $y_i + z_i \leq 1$ and:

$$\begin{cases} z_1 + y_2 + z_3 \leq 2 \\ y_1 + y_2 + y_3 \leq 2 \\ z_1 + y_2 + y_3 \leq 2 \\ y_1 + z_2 + z_3 \leq 2 \end{cases}$$

We notice that $y_i + z_i = 1$ and that a solution is $y_1 = 0, y_2 = 0, z_3 = 0$, together with the corresponding values for z_1, z_2, y_3. This means that we consider the assignment $x_1 = 0, x_2 = 0, x_3 = 1$ for which the formula ϕ is satisfiable.

We now prove that a formula ϕ is satisfiable if and only if there is a vector $(y_1, \ldots, y_r, z_1, \ldots, z_r)$ of positive integers which is a solution for the above instance of DP_R. First, suppose there is a satisfying assignment for ϕ. If $x_i = 1$ we set $y_i = 1, z_i = 0$, and if $x_i = 0$ we set $y_i = 0, z_i = 1$. Thus we have $y_i + z_i \leq 1$, for all $i = \overline{1,r}$, and also $\sum_{i=\overline{1,r}} y_i + z_i \geq r$. Consider now one of the inequalities

$$\sum_{i=\overline{1,r}} q_{i,a} y_i + \sum_{i=\overline{1,r}} l_{i,a} z_i \leq 2.$$

We notice that it contains in its left hand side exactly three variables with coefficient 1, one for each literal appearing in C_a. If the literal with value 1 in

C_a is x_j, then its corresponding variable is z_j which is 0. If the literal with value 1 in C_a is $\neg x_j$, then its corresponding variable is y_j which is 0. Thus there are at most two terms equal to 1, meaning that the inequality is satisfied.

Now suppose there is a solution $(y_1, \ldots, y_r, z_1, \ldots, z_r)$ for the DP_R instance. Since $y_i + z_i \leq 1$ for all $i = \overline{1,r}$ and $\sum_{i=\overline{1,r}} y_i + z_i \geq r$, it follows that $y_i + z_i = 1$ for all i. We consider the assignment $x_i = 1$ if $y_i = 1, z_i = 0$, and $x_i = 0$ if $y_i = 0, z_i = 1$. As previously noted, the inequality corresponding to a clause c_a has exactly three variables, each with coefficient 1, in its left hand side. Thus at least one of them must be equal to 0. If that variable is z_j, it means that the literal x_j, with assignment $x_j = 1$, appears in C_a. If that variable is y_j, it means that the literal $\neg x_j$, with assignment $x_j = 0$, appears in C_a. Thus ϕ is satisfied.

We can also consider the problem $1DP_R$ obtained from DP_R by restricting the possible values of the variables to 0 or 1. This corresponds to requesting that in a membrane a rule can be applied at most once. Then exactly the same reduction can be made from $3CNFSAT$ to $1DP_R$, and so placing $1DP_R$ in the category of **NP**-complete problems.

3   Certain Subproblems

We denote by DP_R^k the problem obtained from DP_R by considering $m = k$ fixed. A similar notation is used for DP_O^k. We start by looking at the case of a P system which has only simple rules, i.e., rules which have only one type of object in their right hand side. Then DP_R^1 describes the decision version of the problem of finding a multiset of simple rules which is $maxR$-applicable: given $d, t, k_{i,1}$ and C_1, find x_i such that $\sum_{i=\overline{1,d}} x_i \geq t$ and $\sum_{i=\overline{1,d}} x_i \cdot k_{i,1} \leq C_1$.

Proposition 2. DP_R^1 *is in* **P**.

Proof. Note that all $k_{i,1} \neq 0$ by definition, because rules always have a non-empty left hand side. Let j be chosen such that $k_{j,1} = min_{i=\overline{1,d}}\{k_{i,1}\}$. A solution is given by setting $x_j = \left\lfloor \frac{C}{k_{j,1}} \right\rfloor$ (the integer part of $\frac{C}{k_{j,1}}$), and $x_i = 0$ for $i \neq j$.

We can also consider the problem $1DP_R^1$ as obtained by restricting the possible values of x_i to 0 or 1. This problem is in **P**, and this can be seen by the following algorithm. First we renumber the coefficients $k_{i,1}$ (together with the variables x_i) such that $k_{1,1} \leq k_{2,1} \leq \ldots \leq k_{d,1}$; then we set $s_1 = k_{1,1}$ and $s_{i+1} = s_i + k_{i+1,1}$. If $s_d \leq C_1$, then the maximum value for $\sum_i x_i$ is d. Otherwise, there exists an unique j such that $s_j \leq C_1 < s_{j+1}$. Therefore the maximum value for $\sum_i x_i$ is j, because whenever we choose $j + 1$ different coefficients $k_{r_1,1}, k_{r_2,1}, \ldots k_{r_{j+1},1}$ randomly, their sum is greater than s_{j+1}.

We now consider a membrane whose $maxR$ evolution has only two types of objects, i.e., $\#O = 2$. The corresponding decision problem is DP_R^2.

Proposition 3. DP_R^2 *is* **NP**-*complete.*

To prove this result we consider the following auxiliary problem AP:
For s, r, k positive integers, there are positive integers $x_1, \ldots, x_s$ such that

$$\sum_{i=\overline{1,s}} x_i = r, \quad \sum_{i=\overline{1,s}} k_i x_i = k.$$

Note that if we restrict this problem by imposing the condition that all $x_i \in \{0, 1\}$, then we obtain a subproblem of the subset sum problem, namely: given a set S of positive integers $S = \{k_i \mid i = \overline{1,s}\}$, is there a subset of S with r elements such that the sum of its elements equals k? This provides a strong hint that AP is **NP**-complete. The proof of Proposition 3 is based on constructing a polynomial-time reduction from $X3C$ to AP, and another one from AP to DP_R^2.

Proof. First, let us note that both DP_R^2 and AP are in **NP**. This can be easily proved by constructing a Turing machine similar to the one used in the proof of Proposition 1. Secondly, we provide a a polynomial-time reduction from $X3C$ to AP. The *exact cover by 3-sets* $(X3C)$ problem asks if, given a set X with $3q$ elements and a collection C of 3-element subsets of X, there is a subcollection C' of C which is an *exact cover* for X, i.e., any element of X belongs to exactly one element of C' [4].

To reduce $X3C$ to AP we do the following. Let l be the number of elements of C and consider an indexing $c_1, \ldots, c_l$ of the elements of C. For each c_i we consider a variable x_i in the AP problem, thus setting $s = l$. To construct the coefficients k_i, we employ the notations $e_{ij} = \#c_i \cap c_j$ $(i, j = \overline{1, l})$ and $M = 3q+1$. We set $s = l, r = q, k_i = \sum_{j=\overline{1,l}} e_{ij} \cdot M^{l-j}$ and $k = \sum_{j=\overline{1,l}} 3 \cdot M^{l-j}$. For a solution C' of $X3C$ we set $x_i = 1$ whenever $c_i \in C'$, and $x_i = 0$ otherwise. We prove that this yields a solution of the constructed instance of AP; moreover, that any solution of the instance has $x_i \in \{0, 1\}$ and produces a solution of $X3C$.

Example. Consider the problem $X3C$ for $X = \{1, \ldots, 9\}$ and $c_1 = \{1, 2, 3\}$, $c_2 = \{1, 3, 4\}$, $c_3 = \{4, 5, 6\}$, $c_4 = \{1, 6, 8\}$, $c_5 = \{4, 7, 9\}$ and $c_6 = \{7, 8, 9\}$. Then $M = 10$, and the coefficients k_i are written in base 10 such that they have a digit for each variable x_j:

	x_1	x_2	x_3	x_4	x_5	x_6
k_1	3	2	0	1	0	0
k_2	2	3	1	1	1	0
k_3	0	1	3	1	1	0
k_4	1	1	1	3	0	1
k_5	0	1	1	0	3	2
k_6	0	0	0	1	2	3
k	3	3	3	3	3	3

We can see that an exact cover of X is given by c_1, c_3, c_6. Looking at this example, we see why any solution to AP has all $x_i \in \{0, 1\}$: all coefficients have at least a digit equal to 3, and the basis M is chosen such that when adding coefficients no carries can occur from lower digits to higher digits.

We first prove that a solution C' for $X3C$ gives a solution for AP. Let $I = \{i \mid c_i \in C'\}$. Since C' is an exact cover for X, it follows that I has q elements, and that $e_{ij} = 0$ for $i, j \in I, i \neq j$. Moreover, if $j \notin I$ then we have that $c_j = c_j \cap (\cup_{i \in I} c_i) = \cup(c_j \cap c_i)$, and so $\sum_{i \in I} e_{ij} = 3$. Since $x_i = 1$ for $i \in I$ and $x_i = 0$ for $i \notin I$, it follows that indeed $\sum_{i=\overline{1,m}} x_i = q$. We also have

$$\sum_{i=\overline{1,l}} k_i x_i = \sum_{i \in I} (\sum_{j=\overline{1,l}} e_{ij} M^{l-j}) =$$

$$= \sum_{i \in I} (e_{ii} M^{l-i} + \sum_{j \notin I} e_{ij} M^{l-j}) = \sum_{i \in I} 3 \cdot M^{l-i} + \sum_{j \notin I} (\sum_{i \in I} e_{ij}) M^{l-j}.$$

Using the previous remarks, we obtain that the term of the second sum is $3 \cdot M^{l-j}$, and so $\sum_{i=\overline{1,m}} k_i x_i = k$.

Secondly, let us consider a solution $(x_i)_{i=\overline{1,s}}$ for the instance of AP with s, r, k_i, k as above. Let $I = \{i \mid x_i = 1\}$. We prove that if $j \notin I$ then $x_j = 0$, and that $e_{ij} = 0$ for $i, j \in I, i \neq j$. This is sufficient to prove that $C' = \{c_i \mid i \in I\}$ is an exact cover, because this follows from the above statement that C' has exactly q elements and $c \cap c' = \emptyset$ for all $c, c' \in C', c \neq c'$. We have

$$\sum_{i=\overline{1,l}} 3 \cdot M^{l-j} = k = \sum_{i=\overline{1,l}} k_i x_i = \sum_{j=\overline{1,l}} (\sum_{i=\overline{1,l}} e_{ij} x_i) M^{l-j} \qquad (1)$$

Since $\sum_{i=\overline{1,l}} e_{ij} x_i \leq \sum_{i=\overline{1,l}} 3x_i = 3q < M$, the two sides of equation (1) represent two decompositions in base M of the same number k. Therefore we have $\sum_{i=\overline{1,l}} e_{ij} x_i = 3$ for any $j = \overline{1,l}$. For $i = j$ we get $e_{ii} x_i = 3x_i \leq 3$, i.e., all $x_i \in \{0,1\}$. Thus $3 = \sum_{i \in I} e_{ij}$. Considering $j \in I$ we obtain that $3 = 3 + \sum_{i \in I, i \neq j} e_{ij}$, namely that $e_{ij} = 0$ for $i, j \in I, i \neq j$, and so concluding the second part of the reduction.

We still should show that AP reduces to DP_R^2. We recall the data of DP_R^2: Given $d, t, C_1, C_2, k_{i,1}, k_{i,2}$ $(i = \overline{1,d})$ are there positive integers $x_1, \ldots, x_d$ such that

$$\begin{cases} \sum_{i=\overline{1,d}} x_i \geq t \\ \sum_{i=\overline{1,d}} k_{i,1} x_i \leq C_1 \\ \sum_{i=\overline{1,d}} k_{i,2} x_i \leq C_2 \ ? \end{cases} \qquad (2)$$

The reduction is as follows: let $K = max_{i=\overline{1,d}} k_i$ and set $d = s, t = r, k_{i,1} = k_i, k_{i,2} = K - k_i, C_1 = k, C_2 = Kr - k$. If $x_1, \ldots, x_s$ is a solution for the instance of AP, it is also a solution for this instance of DP_R^2. Reversely, if $x_1, \ldots, x_s$ is a solution for this instance of DP_R^2, we sum the last two inequalities of (2), obtaining $\sum_{i=\overline{1,s}} K \cdot x_i \leq Kr$. Since $\sum_{i=\overline{1,d}} x_i \geq t$, we obtain that $\sum_{i=\overline{1,s}} x_i = r$ and also that $\sum_{i=\overline{1,s}} k_i x_i = k$.

We contrast these results with those for DP_O and its analogous subproblems. Both DP_O and DP_R are **NP**-complete. However we obtain significant differences when restricting to the case of P systems with simple rules. Namely, while DP_O^1

is **NP**-complete, DP_R^1 is in **P**. When we employ cooperative rules with a fixed maximum number $k > 1$ of objects in the left hand side, the decision problems thus obtained, DP_O^k and DP_R^k, are all **NP**-complete.

4 Conclusion

The most investigated way of applying the rules in a P system is the maximal parallelism ($maxP$ strategy). Two other strategies of applying the rules are also possible. One strategy is to maximize the number of objects consumed in each membrane ($maxO$ strategy), and the other is to maximize the number of rules applied in each membrane ($maxR$ strategy).

The $maxO$ strategy was explicitly considered in [1] and [2] where it is proved that the problem of finding a multiset of rules which consume a maximal number of objects is **NP**-complete for both simple P systems and cooperative P systems.

In this paper we consider the $maxR$ strategy, and study the complexity of finding the multiset of rules in a P system in such a way to have a maximal number of rules applied. We prove that the decision version of this problem is **NP**-complete. However, in contrast to the results for $maxO$ strategy, the problem for P systems with simple rules is in **P**.

Together with the results presented in [1,2,3], this paper provides the possibility of studying complexity and computability for new classes of P systems. It also facilitates a complexity comparison between various classes of P systems.

Acknowledgements

This work has been partially supported by research grants CNCSIS IDEI 402/2007 and CNMP D1/1052/2007.

References

1. Ciobanu, G., Resios, A.: Computational complexity of simple P systems. Fundamenta Informaticae 87, 49–59 (2008)
2. Ciobanu, G., Resios, A.: Complexity of evolution in maximum cooperative P systems. Natural Computing (2009) (accepted, available online)
3. Ciobanu, G., Marcus, S., Păun, Gh.: New strategies of using the rules of a P system in a maximal way. Romanian Journal of Information Science and Technology 12, 157–173 (2009)
4. Garey, M.R., Johnson, D.S.: Computers and Intractability: A Guide to the Theory of **NP**-Completeness. Freeman & Co., New York (1979)
5. Papadimitriou, C.H., Steiglitz, K.: Combinatorial Optimization: Algorithms and Complexity. Dover Publications, New York (1998)
6. Păun, Gh.: Computing with Membranes: An Introduction. Springer, Heidelberg (2002)

On Reversibility and Determinism in P Systems

Artiom Alhazov[1,2] and Kenichi Morita[1]

[1] IEC, Department of Information Engineering
Graduate School of Engineering, Hiroshima University
Higashi-Hiroshima 739-8527 Japan
morita@iec.hiroshima-u.ac.jp
[2] Institute of Mathematics and Computer Science
Academy of Sciences of Moldova
Academiei 5, Chişinău MD-2028 Moldova
artiom@math.md

Abstract. Membrane computing is a formal framework of distributed parallel computing. In this paper we study the reversibility and maximal parallelism of P systems from the computability point of view. The notions of reversible and strongly reversible systems are considered. The universality is shown for reversible P systems with either priorities or inhibitors, and a negative conjecture is stated for reversible P systems without such control. Strongly reversible P systems without control have shown to only generate sub-finite sets of numbers; this limitation does not hold if inhibitors are used.

Another concept considered is strong determinism, which is a syntactic property, as opposed to the determinism typically considered in membrane computing. Strongly deterministic P systems without control only accept sub-regular sets of numbers, while systems with promoters and inhibitors are universal.

1 Introduction

Reversibility is an important property of computational systems. It has been well studied for circuits of logical elements ([4]), circuits of memory elements ([9]), cellular automata ([10]), Turing machines ([2], [12]), register machines ([8]). Reversibility as a syntactical property is closely related to the microscopic physical reversibility, and hence it assumes better miniaturization possibilities for potential implementation.

A slightly different view on reversible systems is given for type-0 grammars ([11]). In this case, the so-called uniquely parsable grammars are studied. In very simple words, this property (still being syntactical) implies that the generation of any word in the language is unique (modulo the order of applying the rules in case when the composition of applying them is commutative). The advantage of having such a property is that it is easier to analyze their behavior.

Clearly, this reason is valid even if the property of reversibility becomes undecidable (just like the property of determinism in certain membrane systems). Moreover, reversibility essentially is backward determinism. Reversible P systems

G. Păun et al. (Eds.): WMC 2009, LNCS 5957, pp. 158–168, 2010.

already were considered ([6]), but the model is energy-based (so the parallelism is invariant-driven rather than maximal) and the main result is the simulation of the Fredkin gate and thus of reversible circuits (so construction of a universal system in this way would use an infinite structure). In this paper we focus on the interplay between maximal parallelism and such fundamental notions as reversibility and determinism, from the viewpoint of computability.

It is interesting that the description of some computational systems includes the initial configuration (grammars, membrane systems), while it is not the case for many others (cellular automata, Turing machines). We generalize reversibility and determinism in such a way that these properties do not depend on the initial configurations, and call them strong. Finally, we present a number of results. In particular, we show that the power of strongly deterministic systems is weaker than that of deterministic systems, and we conjecture that also the power of strongly reversible systems is weaker than that of reversible systems.

2   Definitions

In this paper we illustrate the reversibility and determinism concepts on P systems with symport/antiport rules and one membrane, sometimes with promoters, inhibitors or priorities. For simplicity, we also assume that the environment contains an unbounded supply of all objects[1]. The system thus can be defined by the alphabet, the initial multiset, the set of rules associated to the membrane and the set of terminal objects. Throughout this paper we represent multisets by strings. The union of multisets is defined by adding multiplicities of the symbols. A comprehensive bibliography of membrane computing can be found at [14].

We write an antiport rule sending a multiset x out and bringing a multiset y in as x/y, and the symport case corresponds to $y = \lambda$. If a rule has a promoter a, we write it as $x/y|_a$. If a rule has an inhibitor a, we write it as $x/y|_{\neg a}$. The priority relationship is denoted by $>$. It is not difficult to generalize the definitions for the models with multiple membranes and changing membrane structure, but it is not important here.

We can define a P system in the above-mentioned normal form as $\Pi = (O, T, w, R)$, where O is the object alphabet, T is the terminal subalphabet, w is the initial multiset, and R is the set of rules. In the accepting case, T is replaced by Σ, which is the input subalphabet, the computation starts when an arbitrary multiset over Σ is added to w.

Consider a P system Π with alphabet O. In our setting, a configuration is defined by the multiset of objects inside the membrane, represented by some string

---

[1] It is well-known that symport/antiport systems can be represented as cooperative rewriting on objects of the form (object,region). It is also known that, in case the environment contains an unbounded supply of all objects, a rewriting rule $u \rightarrow v$ is equivalent to a symport/antiport rule u/v. Therefore, one-membrane full-environment is a normal form for symport/antiport P systems. Clearly, symport-in rules are not allowed. Moreover, transition into this normal form preserves properties we consider in this paper, so in the following we only consider this case.

$u \in O^*$. The space $\mathcal{C}$ of configurations (i.e., of multisets over O) is essentially $|O|$-dimensional space with non-negative integer coordinates. We use the usual definitions of maximally parallel transition ([13]): no rule is applicable together with a chosen multiset of rules. It induces an infinite graph of $\mathcal{C}$. Notice that the halting configurations (and only them) have out-degree zero.

Throughout this paper by reachable we mean reachable from the initial configuration. We now define two properties; extending the requirement from reachable configurations to all configuration, we obtain their strong variants (in case of accepting systems the initial configurations are obtained by adding to a fixed multiset arbitrary multisets over a fixed subalphabet; the extension is natural).

Definition 1. *We call Π **strongly reversible** if every configuration has in-degree at most one. We call Π **reversible** if every reachable configuration has in-degree at most one. We call Π **strongly deterministic** if every configuration has out-degree at most one. It is common in membrane computing to call Π **deterministic** if every reachable configuration has out-degree at most one.*

Note: it is crucial that in-degree is the number of all preimages, not just those that are reachable; otherwise the concept of reversibility becomes trivial.

A property equivalent to reversibility is determinism of a dual P system ([1]). We underline that the not-strong properties refer to the actual computation of the system, where the strong ones do not depend on the initial configuration.

By a computation we mean a sequence of (maximally parallel) transitions, starting in the initial configuration, and ending in some halting configuration if it is finite. The result of a halting computation is the number of terminal objects inside the membrane when the system halts (or the number of input objects when the system starts, in the accepting case). The set $N(\Pi)$ of numbers generated by a P system Π is the set of results of all its computations. The family of number sets generated by reversible P systems with features α is denoted by $NROP_1(\alpha)_T$, where $\alpha \subseteq \{sym_*, anti_*, pro, inh, Pri\}$ and the braces of the set notation are omitted. Subscript T means that only terminal objects contribute to the result of computations; if $T = O$, we omit specifying it in the description and we then also omit the subscript T in the notation. To bound the weight (i.e., maximal number of objects sent in a direction) of symport or antiport rules, the associated $*$ is replaced by the actual number. In the case of accepting systems, we write N_a instead of N, and subscript T has no meaning. For strongly reversible systems, we replace in the notation R by R_s. For deterministic (strongly deterministic) systems, we replace R by D (D_s, respectively).

2.1   Register Machines

In this paper we consider register machines with increment, unconditional decrement and test instructions, [8], see also [7].

A register machine is defined by a tuple $M = (n, Q, q_0, q_f, I)$ where

- n is the number of registers;
- I is a set of instructions bijectively labeled by elements of Q;

- $q_0 \in Q$ is the initial label;
- $q_f \in Q$ is the final label.

The allowed instructions are:

- $(q : i?, q', q'')$ - jump to instruction q'' if the contents of register i is zero, otherwise proceed to instruction q';
- $(q : i+, q', q'')$ - add one to the contents of register i and proceed to either instruction q' or q'', non-deterministically;
- $(q : i-, q', q'')$ - subtract one from the contents of register i and proceed to either instruction q' or q'', non-deterministically;
- $(q_f : halt)$ - halt the computation; it is a unique instruction with label q_f.

As for subtract instructions, the computation is blocked if the contents of the corresponding register is zero. Without restricting generality, we can assume that a test of a register always precedes its subtraction. (A popular model where test and subtraction are combined in a conditional subtraction instruction is not suitable for defining reversibility.) A configuration of a register machine is defined by the current instruction and the contents of all registers, which are non-negative integers.

If $q' = q''$ for every instruction $(q : i+, q', q'')$ and for every instruction $(q : i-, q', q'')$, then the machine is called deterministic. Clearly, this is necessary and sufficient for the global transition (partial) mapping not to be multi-valued.

A register machine is called reversible if there is more than one instruction leading to some instruction q, then exactly two exist, they test the same register, one leads to q if the register is zero and the other one leads to q if the register is positive. It is not difficult to check that this requirement is a necessary and sufficient condition for the global transition mapping to be injective. Let us formally state the reversibility of a register machine: for any two different instructions $(q_1 : i_1 \alpha_1, q_1', q_1'')$ and $(q_2 : i_2 \alpha_2, q_2', q_2'')$, it holds that $q_1' \neq q_2'$ and $q_1'' \neq q_2''$. Moreover,

$$\text{if } q_1' = q_2'' \text{ or } q_1'' = q_2', \text{ then } \alpha_1 = \alpha_2 =? \text{ and } i_1 = i_2.$$

It has been shown ([8]) that reversible register machines are universal (a straightforward simulation of, e.g., reversible Turing Machines [2], would not be reversible). It follows that non-deterministic reversible register machines can generate any recursively enumerable set of non-negative integers as a value of the first register by all its possible computations starting from all registers having zero value.

3 Examples and Universality

We now present a few examples to illustrate the definitions.

Example 0. Consider a P system $\Pi_0 = (\{a, b\}, a, \{a/ab\})$. It is strongly reversible (for a preimage, remove as many copies of b as there are copies of a, in case it is possible and there is at least one copy of a), but no halting configuration is reachable. Therefore, $\emptyset \in NR_s OP_1(anti_2)$.

Example 1. Consider a P system $\Pi_1 = (\{a, b, c\}, a, \{a/ab, \ a/c\})$. It generates the set of positive integers since the reachable halting configurations are cb^*, and it is reversible (for the preimage, replace c with a or ab with a), but not strongly reversible (e.g., $aa \Rightarrow cc$ and $ac \Rightarrow cc$). Hence, $\mathbb{N}_+ \in NROP(anti_2)$.

Example 2. Consider a P system $\Pi_2 = (\{a, b\}, aa, \{aa/ab, \ ab/bbb\})$. It is reversible ($aa$ has in-degree 0, while ab and bbb have in-degree 1, and no other configuration is reachable), but not strongly reversible (e.g., $aab \Rightarrow abbb$ and $aabb \Rightarrow abbb$).

Example 3. Any P system containing a rule x/λ, $x \in O^+$ is not reversible. Therefore, symport rules cannot be actually used in reversible P systems with one membrane.

Example 4. Any P system containing rules x_1/y, x_2/y that applied at least one of them in some computation is not reversible.

We now show that reversible P systems with either inhibitors or priorities are universal.

Theorem 1. $NROP_1(anti_2, Pri)_T = NROP_1(anti_2, inh)_T = NRE$.

Proof. We reduce the theorem statement to the claim that such P systems simulate the work of any reversible register machine $M = (n, Q, q_0, q_f, I)$. Consider a P system

$$\Pi = (O, \{r_1\}, q_0, R), \text{ where}$$
$$O = \{r_i \mid 1 \leq i \leq n\} \cup Q,$$
$$R = \{q/q'r_i, q/q''r_i \mid (q : i+, q', q'') \in I\}$$
$$\cup \{qr_i/q', qr_i/q'' \mid (q : i-, q', q'') \in I\} \cup R_t,$$
$$R_t = \{q/q''|_{\neg r_i}, qr_i/q'r_i \mid (q : i?, q', q'') \in I\}.$$

Inhibitors can be replaced by priorities by redefining R_t as follows.

$$R_t = \{qr_i/q'r_i > q/q'' \mid (q : i?, q', q'') \in I\}.$$

Since there is a bijection between the configurations of Π containing one symbol from Q and the configurations of M, the reversibility of Π follows from the correctness of the simulation, the reversibility of M and from the fact that the number of symbols from Q is preserved by transitions of Π.   □

The universality leads to the following undecidability.

Corollary 1. *It is undecidable whether a system from the class of P systems with either inhibitors or priorities is reversible.*

Proof. We recall that the halting problem for register machines is undecidable. Add instructions q_f/F_1, q_f/F_2, F_1/F, F_2/F to the construction presented above, where F_1, F_2, F are new objects; the system is now reversible if and only if some configuration containing F is reachable, i.e., when the underlying register machine does not halt, which is undecidable.   □

A more restricted property of strong reversibility is much easier to check, see [5], since checking that at most one preimage exists for any configuration is no longer related to the reachability.

4 Limitations

The construction in Theorem 1 uses both cooperation and additional control. It is natural to ask whether both inhibitors and priorities can be avoided. Yet, consider the following situation. Let $(p : i?, s, q''), (q : i?, q', s) \in I$. It is usual for reversible register machines to have this, since the preimage of configuration containing a representation of instruction s depends on register i. Nevertheless, P systems with maximal parallelism without additional control can only implement a zero-test by try-and-wait-then-check strategy. In this case, the object containing the information about the register p finds out the result of checking after a possible action of the object related to the register. Therefore, when the instruction represented in the configuration of the system changes to s, it obtains an erroneous preimage representing instruction q. This leads to the following

Conjecture 1. Reversible P systems without priorities and without inhibitors are not universal.

Now consider strongly reversible P systems. The following theorem establishes a very serious limitation on such systems if no additional control is used.

Theorem 2. *In strongly reversible P systems without priorities and without inhibitors, every configuration is either halting or induces only infinite computation(s).*

Proof. If the right-hand side of every rule contains a left-hand side of some rule, then the claim holds. Otherwise, let x/y be a rule of the system such that y does not contain the left-hand side of any rule. Then $x \Rightarrow y$ and y is a halting configuration. It is not difficult to see that $xy \Rightarrow yy$ (objects y are idle) and $xx \Rightarrow yy$ (the rule can be applied twice). Therefore, such a system is not strongly reversible, which proves the theorem.   □

Therefore, the strongly reversible systems without additional control can generate at most singletons, i.e., $NR_sOP_1(anti_*)_T = \{\emptyset\} \cup \{\{n\} \mid n \in \mathbb{N}\}$, and only in a degenerate way, i.e., without actual computing.

It turns out that the theorem above does not hold if inhibitors are used. Consider a system $\Pi_3 = (\{q, f, a\}, q, \{q/qaa|_{\neg f}\}, \{q/f|_{\neg f}\})$. If at least one object f is present or no objects q are present, such a configuration is a halting one. Otherwise, all objects q are used by the rules of the system. Therefore, the only possible transitions in the space of all configurations are of the form $q^{m+n}a^{p-2m} \Rightarrow q^m f^n a^p$, $m+n > 0$, $p \geq 2m$ and the system is strongly reversible. Notice that $N(\Pi) = \{2k + 1 \mid k \geq 0\}$, since starting from q we apply the first rule for $k \geq 0$ steps and eventually the second rule.

5   Strong Determinism

The concept of determinism common to membrane computing essentially means that such a system, starting from the fixed configuration, has a unique computation. As it will be obvious later, this property is often not decidable. Of course, this section only deals with accepting systems.

First, we recall from [3] that deterministic symport/antiport P systems with restrictions mentioned in the preliminaries (one membrane, infinite supply of all objects in the environment) are still universal, by simulation of register machines.

In general, if a certain class of non-deterministic P systems is universal even in a deterministic way, then the determinism is undecidable for that class. This applies to our model of one-membrane all-objects-in-environment P systems with symport/antiport, similarly to Corollary 1.

Corollary 2. *It is undecidable whether a given symport/antiport P system is deterministic.*

Proof. Consider an arbitrary register machine M. There is a deterministic P system Π simulating M. Without restricting generality we assume that an object q_f appears in the configuration of Π if and only if it halts. Add instructions q_f/F_1 and q_f/F_2 to the set of rules, where F_1, F_2 are new objects; the system is now deterministic if and only if some configuration with q_f is reachable, i.e., when the underlying register machine does not halt, which is undecidable.   □

On the contrary, the strong determinism we now consider means that a system has no choice of transitions from any configuration. We now claim that it is a syntactic property. To formulate the claim, we need the following notions. We call the *domain* of a rule x/y, $x/y|_a$ or $x/y|_{\neg a}$ the set of objects in x (the multiplicities of objects in x are not relevant for the results in this paper). We say that two rules are mutually excluded by promoter/inhibitor conditions if the inhibitor of one is either the promoter of the other rule, or is in the domain of the other rule.

Theorem 3. *A P system is strongly deterministic if and only if any two rules with intersecting domains are either mutually excluded by promoter/ inhibitor conditions, or are in a priority relation.*

Proof. Clearly, any P system with only one rule is strongly deterministic, because the degree of parallelism is defined by exhausting the objects from the domain of this rule.

The forward implication of the theorem holds because the rules with non-intersecting domains do not compete for the objects, while mutually excluding promoter/inhibitor conditions eliminate all competing rules except one, and so does the priority relation. In the result, for any configuration the set of objects is partitioned in disjoint domains of applicable rules, and the number of applications of different rules can be computed independently.

We now proceed with the converse implication. Assume that rules p, p' of the system intersect in the domain, are not in a priority relation, and are not mutually excluded by the promoter/inhibitor conditions. Let x, x' be the multisets of objects to be sent out by rules p, p', respectively. Then consider the multiset C, which is the minimal multiset including x, x', and the configuration C', defined as the minimal multiset including C' and promoters of p, p', if any.

Starting from C', there are enough objects for applying either p or p'. Since the rules neither are mutually excluded nor are in a priority relation, both rules are applicable. However, both cannot be applied together because the rules intersect in the domain and thus the multiset C is strictly included in the union of x, x' (and C' is only different from C if either promoter of p, p' does not belong to C). The sufficiency of the condition of this theorem follows from contradicting the strong determinism.   □

Corollary 3. *A P system without promoters, inhibitors, and without priority is strongly deterministic if and only if the domains of all rules are disjoint.*

We show an interesting property of strongly deterministic P systems without additional control. To define it, we use the following notion for deterministic P systems. Let $C \Rightarrow^{\rho_1} C_1 \Rightarrow^{\rho_2} C_2 \cdots \Rightarrow^{\rho_n} C_n$, where ρ_i are multisets of applied rules, $1 \le i \le n$. We define the multiset of rules applied starting from configuration C in n steps as

$$m(C, n) = \bigcup_{i=1}^{n} \rho_i.$$

We write $lhs(x/y) = x$ and $rhs(x/y) = y$, and extend this notation to the multiset of rules by taking the union of the corresponding multisets. For instance, if $C \Rightarrow^{\rho} C_1$, then $C_1 = C \cup rhs(\rho) \setminus lhs(\rho)$.

Lemma 1. *Consider a strongly deterministic P system Π without promoters, inhibitors and without priorities. Consider also two configurations C, C' with $C \subsetneq C'$ and a number n. Then, $m(C, n) \subseteq m(C', n)$.*

Proof. We prove the statement by induction. It holds for $n = 1$ step because strongly deterministic systems are deterministic, and if the statement did not hold, then neither would the determinism.

Assume the statement holds for $n - 1$ steps, and

$$C \Rightarrow^{\rho_1} C_1 \Rightarrow^{\rho_2} C_2 \cdots \Rightarrow^{\rho_n} C_n,$$
$$C' \Rightarrow^{\rho'_1} C'_1 \Rightarrow^{\rho'_2} C'_2 \cdots \Rightarrow^{\rho'_n} C'_n.$$

Then, after $n-1$ steps the difference between the configurations can be described by $C'_{n-1} = C_{n-1} \cup D_1 \cup D_2 \setminus D_3$, where

- $D_1 = C' \setminus C$,
- $D_2 = rhs(m(C', n - 1) \setminus m(C, n - 1))$,
- $D_3 = lhs(m(C', n - 1) \setminus m(C, n - 1))$.

Therefore, $C_{n-1} \setminus C'_{n-1} \subsetneq D_3$. Because of the strong determinism property, these objects will either be consumed by some rules from $m(C', n-1) \setminus m(C, n-1)$, or remain idle. Therefore, $m(C_{n-1}, 1) \subseteq m(C'_{n-1}, 1) \cup (m(C', n-1) \setminus m(C, n-1))$. It follows that $m(C, n) \subseteq m(C', n)$, concluding the proof. $\qquad\square$

Example 5: For a P system $\Pi = (\{a\}, a, \{p : a^3/a\})$,

$$a^{15} \Rightarrow^{p^5} a^5 \Rightarrow^p a^4 \Rightarrow^p a.$$
$$a^{14} \Rightarrow^{p^4} a^6 \Rightarrow^{p^2} a^2.$$

We now establish an upper bound for the power of strongly deterministic P systems without additional control: any P system without promoters, inhibitors or priorities accepts either the set of all non-negative integers, or a finite set of all numbers bounded by some number.

Theorem 4. $N_a D_s OP_1(sym_*, anti_*) = \{\emptyset, \mathbb{N}\} \cup \{\{k \mid 0 \le k \le n\} \mid n \in \mathbb{N}\}$.

Proof. A computation starting from a configuration C is not accepting if it does not halt, i.e., if $\lim_{n \to \infty} m(C, n) = \infty$. Due to Lemma 1, if the computation starting from C is accepting, then any computation starting from a submultiset $C' \subseteq C$ would also be accepting. This also implies that if the computation starting from C is not accepting, then neither is any computation starting from a multiset containing C. Therefore, the set of numbers accepted by a strongly deterministic P system without additional control can be identified by the largest number of input objects leading to acceptance, unless the system accepts all numbers or none.

The converse can be shown by the following P systems.

- System $(\{a\}, \{a\}, a, \{a/a\})$ accepts $\emptyset$ because of the infinite loop in its computation;
- system $(\{a\}, \{a\}, a, \{a/\lambda\})$ accepts $\mathbb{N}$, i.e., anything, because it halts after erasing everything in one step; and
- for any $n \in \mathbb{N}$ there is a system $(\{a\}, \{a\}, \lambda, \{a^{n+1}/a^{n+1}\})$ accepting $\{k \mid 0 \le k \le n\}$, because the system starts in a final configuration if and only if the input does not exceed n, and enters an infinite loop otherwise. $\qquad\square$

Theorem 4 shows that the computational power of strongly deterministic P systems without additional control is, in a certain sense, degenerate (it is subregular). We now show that the use of promoters and inhibitors lead to universality of even the strongly deterministic P systems.

Theorem 5. $N_a D_s OP_1(sym_2, anti_2, pro, inh) = NRE$.

Proof. We reduce the theorem statement to the claim that such P systems simulate any deterministic register machine $M = (n, Q, q_0, q_f, I)$. Without restricting generality, we assume that every subtracting instruction is preceded by the testing instruction. Consider a P system

$$\Pi = (O, \{r_1\}, q_0, R), \text{ where}$$
$$O = \{r_i, d_i \mid 1 \le i \le n\} \cup \{q, q_1 \mid q \in Q\},$$
$$R = \{q/q'r_i \mid (q : i+, q', q') \in I\}$$
$$\cup \ \{q/q_1 d_i, q_1/q', \ d_i r_i/\lambda \mid (q : i-, q', q') \in I\}$$
$$\cup \ \{q/q'|_{r_i}, q/q''|_{\neg r_i} \mid (q : i?, q', q'') \in I\}.$$

All rules using objects q, q' have disjoint domains, except the ones in the last line, simulating the zero/non-zero test. However, they exclude each other by the same object which serves as promoter and inhibitor. Subtraction of register i is handled by producing object d_i, which will "annihilate" (i.e., be deleted together with) with r_i. Therefore, different instructions subtracting the same r_i are implemented by the same rule $d_i r_i/\lambda$, hence all rules using objects d_i, r_i have different domains. It follows from Theorem 3 that the system is strongly deterministic, concluding the proof.   □

6 Conclusions

We outlined the concepts of reversibility, strong reversibility and strong determinism for P systems, concentrating on the case of symport/antiport rules (possibly with control such as priorities or inhibitors) with one membrane, assuming that the environment contains an unbounded supply of all objects, see Table 1. We added the universality of the usual deterministic systems without control from [3] for comparison.

Table 1. The power of P systems with different properties, depending on the features. U - universal, E - degenerate, ? - open, C - conjectured to be non-universal.

Property	$npro, ninh, nPri$	Pri	inh	pro, inh
$D(\text{acc})$	U	U	U	U
$D_s(\text{acc})$	E (Th. 4)	?	?	U (Th. 5)
$R(\text{gen})$	C (Conj. 1)	U (Th. 1)	U (Th. 1)	U (Th. 1)
$R_s(\text{gen})$	E (Th. 2)	C	C	C

We showed that reversible P systems with control are universal, and we conjectured that this result does not hold without control. Moreover, the strongly reversible P systems without control do not halt unless the starting configuration is halting, but this is no longer true if inhibitors are used.

We also gave a syntactic characterization for the strong determinism property. Moreover, we showed that a corresponding system without control either accepts all natural numbers, or a finite set of numbers. With the help of promoters and inhibitors the corresponding systems become universal.

Showing related characterizations might be quite interesting. Some other problems are still open, e.g., cells with "C" and "?" in Table 1. Another interesting direction is to formulate the properties of reversibility, strong reversibility

and strong determinism for P systems with dynamic membrane structure, i.e.,
P systems with active membranes, and to characterize their power.

Acknowledgments. Artiom Alhazov gratefully acknowledges the support of the
Japan Society for the Promotion of Science and the Grant-in-Aid for Scientific
Research, project 20·08364. He also acknowledges the support by the Science
and Technology Center in Ukraine, project 4032.

References

1. Agrigoroaiei, O., Ciobanu, G.: Dual P Systems. In: Corne, D.W., Frisco, P., Paun,
   G., Rozenberg, G., Salomaa, A. (eds.) WMC 2008. LNCS, vol. 5391, pp. 95–107.
   Springer, Heidelberg (2009)
2. Bennett, C.H.: Logical reversibility of computation. IBM Journal of Research and
   Development 17, 525–532 (1973)
3. Calude, C., Păun, Gh.: Bio-steps beyond Turing. BioSystems 77, 175–194 (2004)
4. Fredkin, E., Toffoli, T.: Conservative logic. Int. J. Theoret. Phys. 21, 219–253
   (1982)
5. Ibarra, O.H.: On strong reversibility in P systems and related problems
   (manuscript)
6. Leporati, A., Zandron, C., Mauri, G.: Reversible P systems to simulate Fredkin
   circuits. Fundam. Inform. 74(4), 529–548 (2006)
7. Minsky, M.L.: Computation: Finite and Infinite Machines. Prentice-Hall, Englewood Cliffs (1967)
8. Morita, K.: Universality of a reversible two-counter machine. Theoret. Comput.
   Sci. 168, 303–320 (1996)
9. Morita, K.: A simple reversible logic element and cellular automata for reversible
   computing. In: Margenstern, M., Rogozhin, Y. (eds.) MCU 2001. LNCS, vol. 2055,
   pp. 102–113. Springer, Heidelberg (2001)
10. Morita, K.: Simple universal one-dimensional reversible cellular automata. J. Cel-
    lular Automata 2, 159–165 (2007)
11. Morita, K., Nishihara, N., Yamamoto, Y., Zhang, Zh.: A hierarchy of uniquely
    parsable grammar classes and deterministic acceptors. Acta Inf. 34(5), 389–410
    (1997)
12. Morita, K., Yamaguchi, Y.: A universal reversible Turing machine. In: Durand-
    Lose, J., Margenstern, M. (eds.) MCU 2007. LNCS, vol. 4664, pp. 90–98. Springer,
    Heidelberg (2007)
13. Păun, G.: Membrane Computing. An Introduction. Springer, Berlin (2002)
14. P systems webpage, http://ppage.psystems.eu/

Typed Membrane Systems

Bogdan Aman and Gabriel Ciobanu

Romanian Academy, Institute of Computer Science, and
A.I.Cuza University of Iaşi, Romania
baman@iit.tuiasi.ro, gabriel@info.uaic.ro

Abstract. We introduce and study typing rules and a type inference
algorithm for membrane systems with symport/antiport evolution rules.
The main results are given by a subject reduction theorem and the completeness of type inference. We exemplify how the type system is working
by presenting a typed description of the sodium-potassium pump.

1 Introduction

Membrane systems (also called P systems) were introduced by Gh. Păun, and
several variants are presented in his monograph [7]. Membrane systems are parallel and nondeterministic computing models inspired by the compartments of
eukaryotic cells and by their biochemical reactions. The structure of the cell is
represented by a set of hierarchically embedded regions, each one delimited by
a surrounding boundary (called membrane), and all of them contained inside
an external special membrane called *skin*. The molecular species (ions, proteins,
etc.) floating inside cellular compartments are represented by multisets of objects
described by means of symbols or strings over a given alphabet, objects which
can be modified or communicated between adjacent compartments. Chemical
reactions are represented by evolution rules which operate on the objects, as
well as on the compartmentalized structure (by dissolving, dividing, creating,
or moving membranes). A membrane system can perform computations in the
following way: starting from an initial configuration which is defined by the multisets of objects initially placed inside the membranes, the system evolves by
applying the evolution rules of each membrane in a nondeterministic and maximally parallel manner. A rule is applicable when all the objects that appear
in its left hand side are available in the region where the rule is placed. The
maximally parallel way of using the rules means that in each step, in each region
of the system, we apply a maximal multiset of rules, namely a multiset of rules
such that no further rule can be added to this multiset. A halting configuration
is reached when no rule is applicable. The result is represented by the number
of objects from a specified membrane.

Several variants of membrane systems are inspired by different aspects of
living cells (symport and antiport-based communication through membranes,
catalytic objects, membrane charge, etc.). Their computing power and efficiency
have been investigated using the approaches of formal languages and grammars,

G. Păun et al. (Eds.): WMC 2009, LNCS 5957, pp. 169–181, 2010.

register machines and complexity theory. An updated bibliography can be found on the P systems web page http://ppage.psystems.eu.

Membrane systems are known to be Turing complete [7]. They are also used to model biological systems and their evolutions [5]. In this paper we define a typing system and a type inference algorithm for membrane systems with symport/antiport evolution rules. To exemplify how the newly introduced type system works, we add types in the description of the sodium-potassium pump.

The cells of the human body have different *types* depending on the morphological or functional form [1]. A complete list of distinct cell types in the adult human body may include about 210 distinct types. The chemical reactions inside cells are usually expressed by using types of the components; for instance, a reaction between an *acid* and a *carbonate* forms *salt, carbon dioxide* and *water* as the only products. In this paper we enrich the symport/antiport membrane systems with a *type discipline*. The key technical tools are type inference and principal typing [10]; we associate to each reduction rule a minimal set of conditions that must be satisfied in order to assure that applying this rule to a correct membrane system, we get a correct membrane system as well. The type system for membrane systems with symport/antiport rules is (up to our knowledge) the first attempt to control the evolution of membrane systems using typing rules. The presentation of the typed sodium-potassium pump is an example how to introduce and use types in membrane systems.

The structure of the paper is as follows. A type system for membranes with symport/antiport rules is introduced in Section 2. In Section 3 we extend the description of the sodium-potassium pump using P systems [3] with the newly introduced type system; this section ends with an example of a rule that would be considered ill-typed for the pump. Conclusion and references end the paper.

2   Type Theory

Type theory is fundamental both in logic and computer science. Theory of types was introduced by B.Russell in [9] in order to solve some contradictions of set theory. In computer science, type theory refers to the design, analysis and study of type systems. A type system controls the ways typed programs may behave, and makes behaviour outside these rules illegal. For instance, type systems consider sets of program values called types, and makes certain program behaviours illegal on the basis of these types. Practical applications of type theory are related to the type checking algorithms in the semantic analysis phase of compilers for programming languages. More information on type systems in programming can be found in [8].

Generally, a type system is used to prevent the occurrences of errors during the evolution of a system. A type inference procedure determines the minimal requirements to accept a system or a component as well-typed. In this paper, we investigate the application of these concepts to a biologically inspired formalism, namely to membrane systems.

We use membrane systems with symport/antiport rules. From biological observations we know that there are many cases where two chemicals pass through a membrane at the same time, with the help of each other, either in the same direction, or in opposite directions; in the former case we say that we have a *symport*, in the latter case we have an *antiport*. Symport is standardly described by rules of the form (ab, in) and (ab, out) associated with a membrane, that state that the objects a and b can enter, respectively, exit the membrane together; antiport is described by rules of form $(a, out; b, in)$ associated with a membrane, that state that a exits at the same time when b enters the membrane. Inspired by the rules for active membranes [7], and the notation used in [3], we denote the symport rules by $ab[_l \to [_l ab$ or $[_l ab \to ab[_l$, and the antiport rules by $b[_l a \to a[_l b$. Generalizing such kinds of rules, we can consider rules of the unrestricted forms $u[_l \to [_l u$ or $[_l v \to v[_l$ (generalized symport rules), and $u[_l v \to v[_l u$ (generalized antiport rules), where u, v are strings representing multisets of objects, without any restriction on the length, and l is the label of the membrane in which the rules are placed. It is worth to note that an antiport rule with one of u, v empty is nothing more than a symport rule.

Definition 1. *A P system with symport/antiport rules is a construct*

$$\Pi = (V, H, \mu, w_1, \ldots, w_n, R_1, \ldots, R_n, i_O), \ where$$

- $n \geq 1$ *(the initial degree of the system);*
- V *is an alphabet (its elements are called* objects*);*
- H *is a finite set of* labels *for membranes;*
- $\mu \subseteq H \times H$ *describes the* membrane structure, *such that $(i, j) \in \mu$ denotes that the membrane labelled by j is contained in the membrane labelled by i; we distinguish the external membrane (usually called the "skin" membrane) and several internal membranes;*
- w_i, $1 \leq i \leq n$, *is a string over V, describing the* multiset of objects *placed initially in membrane i;*
- R_i, $1 \leq i \leq n$, *is a finite set of symport and antiport rules over V associated to membrane i;*
- i_O, $1 \leq i_O \leq n$ *is the* output membrane.

We denote by $\mathcal{M}(\Pi)$ the set of configurations obtained along all the possible evolutions of a system Π.

Definition 2. *For a P system Π, if M and N are two configurations from $\mathcal{M}(\Pi)$, we say that M reduces to N (denoted by $M \to N$) if there exists a rule in a R_i ($1 \leq i \leq n$) applicable to configuration M such that we obtain configuration N.*

2.1 Typed Membrane Systems

We introduce typing rules for the class of membrane systems with symport/antiport rules in Table 1 and Table 2. We use *obj* to denote objects, u and v to denote multisets of objects, and *mem* to denote membranes.

The main judgements are of the form $\Gamma \vdash E : T$ indicating that an element E (object or membrane) is well-typed having the type T relative to a typing environment Γ.

The steps for defining a type system are as follows:

1. For each object *obj* we establish a certain type T.
2. A membrane *mem* has a type $\{S, D^\uparrow, D^\downarrow, L\}$, where:
   - S is a set of object types representing the objects that are allowed to stay in membrane *mem* during all the possible evolutions of the system;
   - $D^\uparrow$ is a set of sets of object types representing the objects that are allowed to be communicated up through membrane *mem* during all the possible evolutions of the system;
   - $D^\downarrow$ is a set of sets of object types representing the objects that are allowed to be communicated down through membrane *mem* during all the possible evolutions of the system;
   - L is a set of labels denoting certain states of the membrane *mem* during all the possible evolutions of the systems.

These steps are applied in the example of Section 3.

We denote by T_u the set of types of a multiset of objects u.

Table 1. Typing Objects and Membranes

$$\frac{obj : T \in \Gamma}{\Gamma \vdash obj : T}(\mathbf{R1})$$

$$\frac{mem : \{S_{mem}, D^\uparrow_{mem}, D^\downarrow_{mem}, L_{mem}\} \in \Gamma \quad [u_1 \dots u_i \ mem_1 \dots mem_j]^l_{mem}}{\Gamma \vdash u_1 : T_{u_1} \dots \Gamma \vdash u_i : T_{u_i} \quad \{T_{u_1}, \dots, T_{u_i}\} \subseteq S_{mem} \quad l \in L_{mem}}{\Gamma \vdash mem_1 : \{S_1, D^\uparrow_1, D^\downarrow_1, L_1\} \dots \Gamma \vdash mem_j : \{S_j, D^\uparrow_j, D^\downarrow_j, L_j\}}{\Gamma \vdash mem : \{S_{mem}, D^\uparrow_{mem}, D^\downarrow_{mem}, L_{mem}\}}(\mathbf{R2})$$

Rule (**R1**) tells that an object *obj* is well-typed if it appears in Γ with type T. Rule (**R2**) tells that membrane *mem* is well-typed if:

- it appears in Γ with type $\{S_{mem}, D^\uparrow_{mem}, D^\downarrow_{mem}, L_{mem}\}$;
- it contains some membranes $mem_1, \dots, mem_j$ (possibly none) which are well-typed in the environment Γ;
- it contains some multisets of objects $u_1, \dots, u_i$ which are well-typed in Γ;
- $u_1 \dots u_i$ can stay in the membrane *mem*;
- a label l can be associated to the membrane *mem*.

Lemma 1 (Generation Lemma).

1. If $\Gamma \vdash obj : T$, then $obj : T \in \Gamma$.
2. If $\Gamma \vdash mem : \{S_{mem}, D^\uparrow_{mem}, D^\downarrow_{mem}, L_{mem}\}$, then we have that mem has the form $[u_1 \dots u_i \ mem_1 \dots mem_j]^l_{mem}$ with $\Gamma \vdash mem_1 : \{S_1, D^\uparrow_1, D^\downarrow_1, L_1\}$ ... $\Gamma \vdash mem_j : \{S_j, D^\uparrow_j, D^\downarrow_j, L_j\}$, $\Gamma \vdash u_1 : T_{u_1} \dots \Gamma \vdash u_i : T_{u_i}$, $l \in L_{mem}$, $\{T_{u_1}, \dots, T_{u_i}\} \subseteq S_{mem}$, $mem : \{S_{mem}, D^\uparrow_{mem}, D^\downarrow_{mem}, L_{mem}\} \in \Gamma$.

Proof. By induction on the depth of the membrane structure.   □

In Table 2 we describe the type conditions the rules from the class of membrane systems with symport/antiport rules must fulfill such that the evolution takes place as expected. We consider in what follows that $(mem1, mem2) \in \mu$, namely $mem1$ is the parent membrane of $mem2$. In these rules, l' can be the same l (meaning that the state of the membrane does not change).

Table 2. Typing Evolution Rules in P Systems

$$\frac{\begin{array}{c} u[^l_{mem2} \to [^{l'}_{mem2}u \quad \Gamma \vdash u : T_u \quad T_u \subseteq D^{\downarrow}_{mem2} \\ T_u \subseteq S_{mem1} \quad T_u \subseteq S_{mem2} \quad l' \in L_{mem2} \\ mem1 : \{S_{mem1}, D^{\uparrow}_{mem1}, D^{\downarrow}_{mem1}, L_{mem1}\} \in \Gamma \\ mem2 : \{S_{mem2}, D^{\uparrow}_{mem2}, D^{\downarrow}_{mem2}, L_{mem2}\} \in \Gamma \end{array}}{\begin{array}{c} \Gamma \vdash mem1 : \{S_{mem1}, D^{\uparrow}_{mem1}, D^{\downarrow}_{mem1}, L_{mem1}\} \\ \Gamma \vdash mem2 : \{S_{mem2}, D^{\uparrow}_{mem2}, D^{\downarrow}_{mem2}, L_{mem2}\} \end{array}} \text{(R3)}$$

$$\frac{\begin{array}{c} [^l_{mem2}u \to u[^{l'}_{mem2} \quad \Gamma \vdash u : T_u \quad T_u \subseteq D^{\uparrow}_{mem2} \\ T_u \subseteq S_{mem1} \quad T_u \subseteq S_{mem2} \quad l' \in L_{mem2} \\ mem1 : \{S_{mem1}, D^{\uparrow}_{mem1}, D^{\downarrow}_{mem1}, L_{mem1}\} \in \Gamma \\ mem2 : \{S_{mem2}, D^{\uparrow}_{mem2}, D^{\downarrow}_{mem2}, L_{mem2}\} \in \Gamma \end{array}}{\begin{array}{c} \Gamma \vdash mem1 : \{S_{mem1}, D^{\uparrow}_{mem1}, D^{\downarrow}_{mem1}, L_{mem1}\} \\ \Gamma \vdash mem2 : \{S_{mem2}, D^{\uparrow}_{mem2}, D^{\downarrow}_{mem2}, L_{mem2}\} \end{array}} \text{(R4)}$$

$$\frac{\begin{array}{c} v[^l_{mem2}u \to u[^{l'}_{mem2}v \quad \Gamma \vdash u : T_u \quad \Gamma \vdash v : T_v \quad T_u \subseteq D^{\uparrow}_{mem2} \quad T_v \subseteq D^{\downarrow}_{mem2} \\ T_u \subseteq S_{mem1} \quad T_u \subseteq S_{mem2} \quad T_v \subseteq S_{mem1} \quad T_v \subseteq S_{mem2} \quad l' \in L_{mem2} \\ mem1 : \{S_{mem1}, D^{\uparrow}_{mem1}, D^{\downarrow}_{mem1}, L_{mem1}\} \in \Gamma \\ mem2 : \{S_{mem2}, D^{\uparrow}_{mem2}, D^{\downarrow}_{mem2}, L_{mem2}\} \in \Gamma \end{array}}{\begin{array}{c} \Gamma \vdash mem1 : \{S_{mem1}, D^{\uparrow}_{mem1}, D^{\downarrow}_{mem1}, L_{mem1}\} \\ \Gamma \vdash mem2 : \{S_{mem2}, D^{\uparrow}_{mem2}, D^{\downarrow}_{mem2}, L_{mem2}\} \end{array}} \text{(R5)}$$

Let us consider a P system Π, and M and N two configurations from $\mathcal{M}(\Pi)$. Then we have the following result:

Theorem 1 (Subject Reduction).
If M is such that all its objects and membranes are well-typed in an environment Γ, and $M \to N$ by a rule of Table 2, then N is such that all its objects and membranes are well-typed in the same environment Γ.

Proof (Sketch). Case $[^l_{mem2}u \to u[^{l'}_{mem2}$. If we apply this rule, the only structure that changes is $[[u \ldots]^l_{mem2} \cdots]_{mem1}$ which is transformed into $[[\ldots]^{l'}_{mem2}u \ldots]_{mem1}$. If $\Gamma \vdash mem1 : \{S_{mem1}, D^{\uparrow}_{mem1}, D^{\downarrow}_{mem1}, L_{mem1}\}$, then by Lemma 1 applied twice we have $\Gamma \vdash u : T_u$, $mem1 : \{S_{mem1}, D^{\uparrow}_{mem1}, D^{\downarrow}_{mem1}, L_{mem1}\} \in \Gamma$, $mem2 :$

$\{S_{mem2}, D^{\uparrow}_{mem2}, D^{\downarrow}_{mem2}, L_{mem2}\} \in \Gamma$. Since the rule can be applied, then we have that $T_u \subseteq S_{mem2}$, $l' \in L_{mem2}$ and $T_u \subseteq S_{mem1}$. By applying (**R4**) we get that $\Gamma \vdash mem1 : \{S_{mem1}, D^{\uparrow}_{mem1}, D^{\downarrow}_{mem1}, L_{mem1}\}$ and $\Gamma \vdash mem2 : \{S_{mem2}, D^{\uparrow}_{mem2}, D^{\downarrow}_{mem2}, L_{mem2}\}$ which means that all objects and membranes from N are well-typed in the environment Γ.

The other cases are treated similarly.   □

2.2   Type Inference Algorithm

For a P system Π, given a configuration $M \in \mathcal{M}(\Pi)$ in which all membranes and all objects are well-typed, we present a type inference algorithm computing the minimal types and environment for the objects and membranes occurring in M. The typing is principal in the sense of [10]: all the other possible typings which can be given to the membrane configuration M are equivalent by using appropriate operations of renaming. The inference algorithm is then proved to be sound and complete with respect to the rules of Subsection 2.1.

Types and type environments of the algorithm are related to the structure of the P system; it has therefore to put together distinct environments whenever the system contains parallel membranes.

The type reconstruction procedure is represented by a judgement

$$\vdash_I E : \langle W, \Gamma \rangle,$$

where E is an element (object or membrane), W is the type inferred for E from the environment Γ, and I represents the fact that this judgement results from the inference algorithm. As before, we consider that $mem1$ is the parent membrane of $mem2$. We define the domain of a set of typed names from Γ as

$$dom(\Gamma) = \{n \mid n : T \in \Gamma\}.$$

where T is the type of an object or membrane.

We say that two typed sets of names Γ and Γ' are *compatible* (and denote this by $\Gamma \bowtie \Gamma'$) if and only if $n : T \in \Gamma$ and $n : T' \in \Gamma'$, then $T = T'$. The disjoint union of Γ and Γ' is defined as

$$\Gamma \uplus \Gamma' = \{n : T \in \Gamma \wedge n \notin dom(\Gamma')\} \cup \{n : T' \in \Gamma' \wedge n \notin dom(\Gamma)\}.$$

We also define a function that returns the type of an object or a membrane with respect to a type environment Γ:

$$type(n, \Gamma) = \{n : T \mid n : T \in \Gamma\}$$

The inference procedure is defined in a natural semantic style. In all the type inference rules, the objects and membrane types which appear in conclusions are derived from those appearing in premises.

Using rules of the form (**I1**) to each object obj is attached a fresh type Obj. If to the same object obj we add two different types Obj_1 and Obj_2 when constructing the type of the whole membrane using rules from Table 3, by using the relation $\bowtie$, we have $Obj_1 = Obj_2$. Rule (**I2**) is used to update the types of membranes. Rules (**I3**), (**I4**) and (**I5**) are used to construct the types of the

Table 3. Type Inference Rules

$$\vdash_I obj : \langle Obj, obj : Obj \rangle \quad \textbf{(I1)}$$

$$\frac{\begin{array}{c}[u_1 \ldots u_i\, mem_1 \ldots mem_j]_{mem}^l \quad \Gamma_s \bowtie \Gamma_t, s \neq t, 1 \leq s,t \leq i+j \\ \vdash_I u_1 : \langle T_{u_1}, \Gamma_1 \rangle \ldots \vdash_I u_i : \langle T_{u_i}, \Gamma_i \rangle \\ \vdash_I mem : \langle T, \Gamma \rangle \quad \vdash_I mem_1 : \langle T_1, \Gamma_{i+1} \rangle \ldots \vdash_I mem_j : \langle T_j, \Gamma_{i+j} \rangle\end{array}}{\vdash_I mem : \langle T', \Gamma' \rangle} \quad \textbf{(I2)}$$

where $T' = \{S_{mem} \cup \{T_{u_1}, \ldots, T_{u_i}\}, D_{mem}^\uparrow, D_{mem}^\downarrow, L_{mem} \cup \{l\}\}$
if $T = \{S_{mem}, D_{mem}^\uparrow, D_{mem}^\downarrow, L_{mem}\}$
and $\Gamma' = \Gamma \cup (\biguplus_{k=1}^{i+j} \Gamma_k \backslash type(mem, \biguplus_{k=1}^{i+j} \Gamma_k)) \cup \{mem : T'\}$

$$\frac{\begin{array}{c}u[_{mem2}^l \rightarrow [_{mem2}^{l'} u \quad \Gamma \bowtie \Gamma_2 \quad \Gamma \bowtie \Gamma_1 \\ \vdash_I u : \langle T_u, \Gamma \rangle \quad \vdash_I mem1 : \langle \{S_{mem1}, D_{mem1}^\uparrow, D_{mem1}^\downarrow, L_{mem1}\}, \Gamma_1 \rangle \\ \vdash_I mem2 : \langle \{S_{mem2}, D_{mem2}^\uparrow, D_{mem2}^\downarrow, L_{mem2}\}, \Gamma_2 \rangle\end{array}}{\begin{array}{c}\vdash_I mem1 : \langle \{S_{mem1} \cup T_u, D_{mem1}^\uparrow, D_{mem1}^\downarrow, L_{mem1}\}, \Gamma_1' \rangle\end{array}} \quad \textbf{(I3)}$$

where $\Gamma_1' = ((\Gamma_1 \uplus \Gamma) \backslash type(mem1, \Gamma \uplus \Gamma_1)) \cup$
$\{mem1 : \{S_{mem1} \cup T_u, D_{mem1}^\uparrow, D_{mem1}^\downarrow, L_{mem1}\}\}$
$\vdash_I mem2 : \langle \{S_{mem2} \cup T_u, D_{mem2}^\uparrow, D_{mem2}^\downarrow \cup T_u, L_{mem2} \cup \{l'\}\}, \Gamma_2' \rangle$
where $\Gamma_2' = ((\Gamma_2 \uplus \Gamma) \backslash type(mem2, \Gamma_2 \uplus \Gamma)) \cup$
$\{mem2 : \{S_{mem2} \cup T_u, D_{mem2}^\uparrow, D_{mem2}^\downarrow \cup T_u, L_{mem2} \cup \{l'\}\}\}$

$$\frac{\begin{array}{c}[_{mem2}^l u \rightarrow u[_{mem2}^{l'} \quad \Gamma_1 \bowtie \Gamma_2 \quad \Gamma \bowtie \Gamma_1 \quad \Gamma \bowtie \Gamma_2\} \\ \vdash_I u : \langle T_u, \Gamma \rangle; \emptyset \quad \vdash_I mem1 : \langle \{S_{mem1}, D_{mem1}^\uparrow, D_{mem1}^\downarrow, L_{mem1}\}, \Gamma_1 \rangle \\ \vdash_I mem2 : \langle \{S_{mem2}, D_{mem2}^\uparrow, D_{mem2}^\downarrow, L_{mem2}\}, \Gamma_2 \rangle\end{array}}{\vdash_I mem1 : \langle \{S_{mem1} \cup T_u, D_{mem1}^\uparrow, D_{mem1}^\downarrow, L_{mem1}\}, \Gamma_1' \rangle} \quad \textbf{(I4)}$$

where $\Gamma_1' = ((\Gamma \uplus \Gamma_1 \uplus \Gamma_2) \backslash type(mem1, \Gamma \uplus \Gamma_1 \uplus \Gamma_2)) \cup$
$\{mem1 : \{S_{mem1} \cup T_u, D_{mem1}^\uparrow, D_{mem1}^\downarrow, L_{mem1}\}\}$
$\vdash_I mem2 : \langle \{S_{mem2} \cup T_u, D_{mem2}^\uparrow \cup T_u, D_{mem2}^\downarrow, L_{mem2} \cup \{l'\}\}, \Gamma_2' \rangle$
where $\Gamma_2' = ((\Gamma \uplus \Gamma_1 \uplus \Gamma_2) \backslash type(mem2, \Gamma \uplus \Gamma_1 \uplus \Gamma_2)) \cup$
$\{mem2 : \{S_{mem2} \cup T_u, D_{mem2}^\uparrow \cup T_u, D_{mem2}^\downarrow, L_{mem2} \cup \{l'\}\}\}$

$$\frac{\begin{array}{c}v[_{mem2}^l u \rightarrow u[_{mem2}^{l'} v \quad \Gamma_i \bowtie \Gamma_j, i \neq j \quad \vdash_I u : \langle T_u, \Gamma_3 \rangle \quad \vdash_I v : \langle T_v, \Gamma_4 \rangle \\ \vdash_I mem1 : \langle \{S_{mem1}, D_{mem1}^\uparrow, D_{mem1}^\downarrow, L_{mem1}\}, \Gamma_1 \rangle \\ \vdash_I mem2 : \langle \{S_{mem2}, D_{mem2}^\uparrow, D_{mem2}^\downarrow, L_{mem2}\}, \Gamma_2 \rangle\end{array}}{\vdash_I mem1 : \langle \{S_{mem1} \cup T_u, D_{mem1}^\uparrow, D_{mem1}^\downarrow, L_{mem1}\}, \Gamma_1' \rangle} \quad \textbf{(I5)}$$

where $\Gamma_1' = ((\biguplus_{i=1; i \neq 2}^{4} \Gamma_i) \backslash type(mem1, \biguplus_{i=1; i \neq 2}^{4} \Gamma_i)) \cup$
$\{mem1 : \{S_{mem1} \cup T_u, D_{mem1}^\uparrow, D_{mem1}^\downarrow, L_{mem1}\}\}$
$\vdash_I mem2 : \langle \{S_{mem2} \cup T_v, D_{mem2}^\uparrow \cup T_u, D_{mem2}^\downarrow \cup T_v, L_{mem2} \cup \{l'\}\}, \Gamma_2'$
where $\Gamma_2' = ((\biguplus_{i=2}^{4} \Gamma_i) \backslash type(mem2, \biguplus_{i=2}^{4} \Gamma_i)) \cup$
$\{mem2 : \{S_{mem2} \cup T_v, D_{mem2}^\uparrow \cup T_u, D_{mem2}^\downarrow \cup T_v, L_{mem2} \cup \{l'\}\}\}$

membranes with conditions given by symport and antiport rules considered as safe rules to be applied.

A subtyping relation $\leq$ is introduced to compare the environments. If we take two type environments $\Gamma = \{a : K, b : Na\}$ and $\Delta = \{a : K\}$, then $\Gamma \leq \Delta$.

Theorem 2 (Soundness of the Type Inference).
$$\text{If } \vdash_I E : \langle W, \Gamma \rangle, \text{ then } \Gamma \vdash E : W.$$

Proof. By induction on deductions in $\vdash_I$.

Case (**I1**): We have $\vdash_I obj : \langle Obj, obj : Obj \rangle$, from where it results that $obj : Obj \in \Gamma$. Applying rule (**R1**), it results that $\Gamma \vdash obj : Obj$.

Case (**I2**): We have

(i) a membrane $[u_1 \ldots u_i mem_1 \ldots mem_j]^l_{mem}$;
(ii) from $\vdash_I u_1 : \langle T_{u_1}, \Gamma_1 \rangle \ldots \vdash_I u_i : \langle T_{u_i}, \Gamma_i \rangle$; $\Gamma_k \bowtie \Gamma_t, 1 \leq k, t \leq i$; $\Gamma \leq \Gamma_k$, $1 \leq k \leq i$, applying the induction we have that $\Gamma \vdash u_1 : T_{u_1} \ldots \Gamma \vdash u_i : T_{u_i}$;
(iii) from $\vdash_I mem : \langle T, \Gamma_i \rangle$ $\vdash_I mem_1 : \langle T_1, \Gamma_{i+1} \rangle \ldots \vdash_I mem_j : \langle T_j, \Gamma_{i+j} \rangle$; $\Gamma_k \bowtie \Gamma_t, i + 1 \leq k, t \leq i + j$; $\Gamma \leq \Gamma_k$, $i + 1 \leq k \leq i + j$, applying the induction we have that $\Gamma \vdash mem_1 : \{S_1, D_1^\uparrow, D_1^\downarrow, L_1\} \ldots \Gamma \vdash mem_j : \{S_j, D_j^\uparrow, D_j^\downarrow, L_j\}$;
(iv) $mem : \{S_{mem}, D_{mem}^\uparrow, D_{mem}^\downarrow, L_{mem}\} \in \Gamma$;
(v) $\{T_{u_1}, \ldots, T_{u_i}\} \subseteq S_{mem}$, $l \in L_{mem}$.

Using (*i*), (*ii*), (*iii*), (*iv*) and (*v*), we can apply rule (**R2**), and so obtaining that $\Gamma \vdash mem : \{S_{mem}, D_{mem}^\uparrow, D_{mem}^\downarrow, L_{mem}\}$.

Case (**I3**): For membrane $mem2$ we have

(i) rule $u[^l_{mem2} \rightarrow [^{l'}_{mem2} u$;
(ii) from $\vdash_I u : \langle T_u, \Gamma \rangle$, $\Gamma \bowtie \Gamma_2'$, $\Gamma_2' \leq \Gamma$ we get $\Gamma_2' \vdash u : T_u$ by applying the induction;
(iii) $mem2 : \{S_{mem2}, D_{mem2}^\uparrow, D_{mem2}^\downarrow, L_{mem2}\} \in \Gamma_2'$;
(iv) $T_u \subseteq D_{mem2}^\downarrow$, $T_u \subseteq S_{mem2}$, $l' \in L_{mem2}$.

Using (*i*), (*ii*), (*iii*) and (*iv*), we can apply rule (**R3**), and so obtaining that $\Gamma \vdash mem2 : \{S_{mem2}, D_{mem2}^\uparrow, D_{mem2}^\downarrow, L_{mem2}\}$.

For membrane $mem1$ we have

(i) rule $u[^l_{mem2} \rightarrow [^{l'}_{mem2} u$;
(ii) from $\vdash_I u : \langle T_u, \Gamma \rangle$, $\Gamma \bowtie \Gamma_1'$, $\Gamma_1' \leq \Gamma$ we get $\Gamma_1' \vdash u : T_u$ by applying the induction;
(iii) $mem1 : \{S_{mem1}, D_{mem1}^\uparrow, D_{mem1}^\downarrow, L_{mem1}\} \in \Gamma_1'$;
(iv) $T_u \subseteq S_{mem1}$.

Using (*i*), (*ii*), (*iii*) and (*iv*), we can apply rule (**R3**), and so obtaining that $\Gamma \vdash mem1 : \{S_{mem1}, D_{mem1}^\uparrow, D_{mem1}^\downarrow, L_{mem1}\}$.

The other cases are treated in a similar manner.   □

Theorem 3 (Completeness of the Type Inference).
If $\Gamma \vdash E : W$, then $\vdash_I E : \langle W', \Gamma' \rangle$. Moreover, there is a renaming function σ such that:

1. $\sigma(W') = W$;
2. $\sigma(\Gamma') \leq \Gamma$.

Proof. By induction on the depth of the inference tree.

Case (**R1**): From (**R1**) we have that $\Gamma \vdash obj : Obj$, while from (**I1**) we have that $\vdash_I obj : \langle Obj', obj : Obj' \rangle$. If we consider the renaming function σ with $\sigma(Obj') = Obj$, $\sigma(obj : Obj') = obj : Obj$ we get that $\vdash_I obj : \langle Obj, obj : Obj \rangle$.

Case (**R2**): We have

(i) a membrane $[u_1 \ldots u_i mem_1 \ldots mem_j]^l_{mem}$;

(ii) from $\Gamma \vdash u_1 : T_{u_1} \ldots \Gamma \vdash u_i : T_{u_i}$, by applying the induction, we have that $\vdash_I u_1 : \langle T_{u_1}, \Gamma_1 \rangle \ldots \vdash_I u_i : \langle T_{u_i}, \Gamma_i \rangle$; $\sigma(\Gamma_k) \leq \Gamma$, $1 \leq k \leq i$;

(iii) from $\Gamma \vdash mem_1 : \{S_1, D_1^{\uparrow}, D_1^{\downarrow}, L_1\} \ldots \Gamma \vdash mem_j : \{S_j, D_j^{\uparrow}, D_j^{\downarrow}, L_j\}$, by applying the induction, we have that $\vdash_I mem : \langle T, \Gamma_i \rangle$, $\vdash_I mem_1 : \langle T_1, \Gamma_{i+1} \rangle \ldots \vdash_I mem_j : \langle T_j, \Gamma_{i+j} \rangle$; $\sigma(\Gamma_k) \leq \Gamma$, $i + 1 \leq k \leq i + j$;

(iv) $mem : T \in \Gamma$, where $T = \{S_{mem}, D_{mem}^{\uparrow}, D_{mem}^{\downarrow}, L_{mem}\}$;

(v) $\{T_{u_1}, \ldots, T_{u_i}\} \subseteq S_{mem}$, $l \in L_{mem}$.

Using (i), (ii), (iii), (iv) and (v), we can apply rule (**I2**) and so obtaining that

$$\vdash_I mem : \langle T', \Gamma' \rangle, \text{ with } T' = \{S_{mem}, D_{mem}^{\uparrow}, D_{mem}^{\downarrow}, L_{mem}\} \text{ and } \Gamma' = \Gamma \cup \biguplus_{k=1}^{i+j} \Gamma_k.$$

We have that $T' = T$ and $\sigma(\Gamma') \leq \Gamma$.

The other cases are treated in a similar manner.   □

3   Na-K Pump Modelled by Typed Membranes

The sodium-potassium pump is a primary active transport system driven by a cell membrane ATPase carrying sodium ions out and potassium ions in. The description given in Table 4 it is known as the Albers-Post model. According to this mechanism:

1. Na^+ and K^+ transport is similar to a ping-pong mechanism, meaning that the two ions species are transported sequentially;
2. Na-K pump essentially exists in two conformations, E1 and E2, which may be phosphorylated or dephosphorylated.

These conformations correspond to two mutually exclusive states in which the pump exposes ion binding sites alternatively on the cytoplasmic (E1) and extracellular (E2) sides of the membrane. Ion transport is mediated by transitions between these conformations. In Table 4 we use the following notations:

- $A + B$ means that A and B are present together and could react;
- $A \cdot B$ means that A and B are bound to each other non-covalently;
- $E_2 \sim P$ indicates that the phosphoryl group P is covalently bound to E_2;
- P_i is the inorganic phosphate group;
- $\rightleftharpoons$ indicates that the process can be reversible.

Table 4. The Albers-Post Model

$$
\begin{align}
E_1 + Na_{in}^+ &\rightleftharpoons Na^+ \cdot E_1 \tag{1}\\
Na^+ \cdot E_1 + ATP &\rightleftharpoons Na^+ \cdot E_1 \sim P + ADP \tag{2}\\
Na^+ \cdot E_1 \sim P &\rightleftharpoons Na^+ \cdot E_2 \sim P \tag{3}\\
Na^+ \cdot E_2 \sim P &\rightleftharpoons E_2 \sim P + Na_{out}^+ \tag{4}\\
E_2 \sim P + K_{in}^+ &\rightleftharpoons K^+ \cdot E_2 \sim P \tag{5}\\
K^+ \cdot E_2 \sim P &\rightleftharpoons K^+ \cdot E_2 + P_i \tag{6}\\
K^+ \cdot E_2 &\rightleftharpoons K^+ \cdot E_1 \tag{7}\\
K^+ \cdot E_1 &\rightleftharpoons K_{in}^+ + E_1 \tag{8}
\end{align}
$$

3.1   Modelling the Pump with Untyped P Systems

The environment and the inner region are characterized by multisets of symbols over the alphabet $V = \{Na, K, ATP, ADP, P\}$, representing the substances floating inside membranes. The conformations of the pump are described by means of labels attached to a membrane, that is $[\,|_l$ with $l \in L$, $L = \{E_1, E_2, E_1^P, E_2^P\}$. The labels E_1 and E_2 correspond to the dephosphorylated conformations of the pump, while E_1^P and E_2^P correspond to the phosphorylated conformations. Note an important aspect of this system: object P becomes part of the membrane label, hence it undergoes a structural modification by passing from being an element of the alphabet V to being a component of the membrane labels in the set L.

Initially, the multiset inside the region consists of n symbols Na of sodium, m symbols K of potassium, and s symbols ATP. The multiset from the environment consists of n' symbols Na and m' symbols K, while the bilayer does not contain any symbols.

Denoting by $R_{Na} = \dfrac{n'}{n}$, $R_K = \dfrac{m'}{m}$ the ratios of occurrences of sodium and potassium ions outside and inside the membrane at any given step, we use this values to describe the starting time for the functioning of the pump. We assume that the activation of the pump is triggered by a change in the values of the ratios evaluated at the current step. Once the following two conditions $R_{Na} > k_1$ and $R_K > k_2$ (for some fixed $k_1, k_2 \in \mathbf{R}$) are satisfied, the pump is activated. A description of the pump using P systems is presented in [3] as follows:

3.2   Modelling the Pump with Typed P Systems

The motivation for introducing a type system for P systems with symport/antiport rules, namely the class used to model the sodium-potassium pump, comes from the fact that we would like to increase the control in the evolution of the pump. This would mean that if we had a larger set of rules used in the description of the pump, only the ones assuring a correct evolution with respect to the restrictions imposed by the environment would be applied. In this way we increase the control over the evolution of the P system.

Table 5. The Membrane Systems Model

$$Env[Bilayer \mid Reg \mid Bilayer] \, Env$$

$r_1 \; : \; [|_{E_1} Na^3 \stackrel{(R_{Na} > k_1) \wedge (R_K > k_2)}{\longrightarrow} [Na^3|_{E_1}$

$r_2 \; : \; [Na^3|_{E_1} ATP \rightarrow [Na^3|_{E_1^P} ADP$

$r_3 \; : \; [Na^3|_{E_1^P} \rightarrow Na^3[|_{E_1^P}$

$r_4 \; : \; K^2[|_{E_2^P} \rightarrow [K^2|_{E_2^P}$

$r_5 \; : \; [K^2|_{E_2^P} \rightarrow [K^2|_{E_1} P$

$r_6 \; : \; [K^2|_{E_1} \rightarrow [|_{E_1} K^2$

For the case of the pump we consider the following type environment:

$$\Gamma = \{Na : \mathbf{Na}, K : \mathbf{K}, P : \mathbf{P}, ATP : \mathbf{ATP}, ADP : \mathbf{ADP},$$
$$skin : \{\{\mathbf{Na}, \mathbf{K}\}, \emptyset, \emptyset, \emptyset\}, mem1 : \{\{\mathbf{Na}, \mathbf{K}\}, \{\{\mathbf{Na}, \mathbf{Na}, \mathbf{Na}\}\}, \{\{\mathbf{K}, \mathbf{K}\}\}, \emptyset\},$$
$$mem2 : \{\{\mathbf{Na}, \mathbf{K}, \mathbf{P}, \mathbf{ATP}, \mathbf{ADP}\}, \{\{\mathbf{Na}, \mathbf{Na}, \mathbf{Na}\}\}, \{\{\mathbf{K}, \mathbf{K}\}\}, \{E_1, E_2, E_1^P, E_2^P\}\}\}$$

For the membrane configuration:

$$[K \ldots Na \ldots [[K \ldots Na \ldots ATP]^{E_1}_{mem2}]_{mem1}]_{skin}$$

and the environment Γ defined above, we have that

Lemma 2. $\Gamma \vdash skin : \{S_{skin}, D^{\uparrow}_{skin}, D^{\downarrow}_{skin}, L_{skin}\}$.

Proof

$$\dfrac{\dfrac{K : \mathbf{K} \in \Gamma \quad Na : \mathbf{Na} \in \Gamma}{\Gamma \vdash K : \mathbf{K} \quad \Gamma \vdash Na : \mathbf{Na}} \quad mem2 : \{S_{mem2}, D^{\uparrow}_{mem2}, D^{\downarrow}_{mem2}, L_{mem2}\} \in \Gamma}{\mathbf{K} \in S_{mem2} \quad \mathbf{Na} \in S_{mem2} \quad E_1 \in L_{mem2} \quad [K \ldots Na \ldots]^{E_1}_{mem2}}$$
$$\Gamma \vdash mem2 : \{S_{mem2}, D^{\uparrow}_{mem2}, D^{\downarrow}_{mem2}, L_{mem2}\}$$

$$\dfrac{\Gamma \vdash mem2 : \{S_{mem2}, D^{\uparrow}_{mem2}, D^{\downarrow}_{mem2}, L_{mem2}\}}{mem1 : \{S_{mem1}, D^{\uparrow}_{mem1}, D^{\downarrow}_{mem1}, L_{mem1}\} \in \Gamma \quad [[K \ldots Na \ldots]^{E_1}_{mem2}]_{mem1}}$$
$$\Gamma \vdash mem1 : \{S_{mem1}, D^{\uparrow}_{mem1}, D^{\downarrow}_{mem1}, L_{mem1}\}$$

$$\dfrac{K : \mathbf{K} \in \Gamma \quad Na : \mathbf{Na} \in \Gamma}{\Gamma \vdash K : \mathbf{K} \quad \Gamma \vdash Na : \mathbf{Na}} \quad skin : \{S_{skin}, D^{\uparrow}_{skin}, D^{\downarrow}_{skin}, L_{skin}\} \in \Gamma$$
$$\mathbf{K} \in S_{mem2} \quad \mathbf{Na} \in S_{mem2} \quad \Gamma \vdash mem1 : \{S_{mem1}, D^{\uparrow}_{mem1}, D^{\downarrow}_{mem1}, L_{mem1}\}$$
$$[K \ldots Na \ldots [[K \ldots Na \ldots ATP]^{E_1}_{mem2}]_{mem1}]_{skin}$$
$$\Gamma \vdash skin : \{S_{skin}, D^{\uparrow}_{skin}, D^{\downarrow}_{skin}, L_{skin}\}$$

□

The evolution rules of Table 6 provide the conditions which must be satisfied for a rule describing the evolution of the pump in order to be applied correctly.

In (**T1**), by using types in $[^{E_1}_{mem2} \mathbf{Na}^3 \rightarrow \mathbf{Na}^3 [^{E_1}_{mem2}$ we indicate that only three objects of type $\mathbf{Na}$ can pass through membrane $mem2$.

Table 6. Typing Evolution Rules for Na-K Pump

$$\frac{\mathbf{Na} \in S_{mem1} \quad \{\mathbf{Na}, \mathbf{Na}, \mathbf{Na}\} \in D_{mem2}^{\uparrow}}{[_{mem2}^{E_1}\mathbf{Na}^3 \rightarrow \mathbf{Na}^3[_{mem2}^{E_1}} \textbf{(T1)}$$

$$\frac{\mathbf{ADP} \in S_{mem2} \quad E_1^P \in L_{mem2}}{\mathbf{Na}^3[_{mem2}^{E_1}\mathbf{ATP} \rightarrow \mathbf{Na}^3[_{mem2}^{E_1^P}\mathbf{ADP}} \textbf{(T2)}$$

$$\frac{\mathbf{Na} \in S_{skin} \quad E_2^P \in L_{mem2} \quad \{\mathbf{Na}, \mathbf{Na}, \mathbf{Na}\} \in D_{mem1}^{\uparrow}}{[_{mem1}\mathbf{Na}^3[_{mem2}^{E_1^P} \rightarrow \mathbf{Na}^3[_{mem1}[_{mem2}^{E_2^P}} \textbf{(T3)}$$

$$\frac{\mathbf{K} \in S_{mem1} \quad \{\mathbf{K}, \mathbf{K}\} \in D_{mem1}^{\downarrow}}{\mathbf{K}^2[_{mem1}[_{mem2}^{E_2^P} \rightarrow [_{mem1}\mathbf{K}^2[_{mem2}^{E_2^P}} \textbf{(T4)}$$

$$\frac{\mathbf{P} \in S_{mem2} \quad E_1 \in L_{mem2}}{\mathbf{K}^2[_{mem2}^{E_2^P} \rightarrow \mathbf{K}^2[_{mem2}^{E_1}\mathbf{P}} \textbf{(T5)}$$

$$\frac{\mathbf{K} \in S_{mem2} \quad \{\mathbf{K}, \mathbf{K}\} \in D_{mem2}^{\downarrow}}{\mathbf{K}^2[_{mem2}^{E_1} \rightarrow [_{mem2}^{E_1}\mathbf{K}^2} \textbf{(T6)}$$

Remark 1. In programming, types are used to eliminate (statically) programs in which problems could appear during their execution. In the framework of membrane systems, types are used to increase the control and in this way assuring that no typing problem appears during the evolution of the system. As a consequence, all the ill-typed rules could be eliminated, and the description of the system is simplified. For example, let us consider the membrane $mem2$ which appears in the typed description of the pump with the type

$$mem2 : \{S_{mem2}, D_{mem2}^{\uparrow}, D_{mem2}^{\downarrow}, L_{mem2}\}$$

where we have $S_{mem2} = \{\mathbf{Na}, \mathbf{K}, \mathbf{P}, \mathbf{ATP}, \mathbf{ADP}\}$, $D_{mem2}^{\uparrow} = \{\{\mathbf{Na}, \mathbf{Na}, \mathbf{Na}\}\}$, $D_{mem2}^{\downarrow} = \{\{\mathbf{K}, \mathbf{K}\}\}$ and $L_{mem2} = \{E_1, E_2, E_1^P, E_2^P\}$.

Using this typing for $mem2$ membrane, a rule of the form:

$$\mathbf{K}[_{mem2}^{E_1} \rightarrow [_{mem2}^{E_1}\mathbf{K}$$

is not allowed because membrane $mem2$ contains in $D_{mem2}^{\downarrow}$ only a tuple of two elements of type $\mathbf{K}$, and so it does not allow single elements of type $\mathbf{K}$ to be sent inside it. In a similar manner, all the rules which do not satisfy the requirements of the environment are rejected.

4   Conclusion

The novelty of this paper is that it introduces types over P systems. In fact we enrich the symport/antiport P systems with typing rules which help to control the evolution of the systems. According to these typing rules, for the typed symport/antiport P systems we prove that if a system is well-typed and an evolution rule is applied, then the obtained system is also well-typed. Another contribution of the paper is the introduction of a type inference algorithm for symport/antiport P systems for which soundness and completeness are proved. We use types in the description of the sodium-potassium pump. This pump was modelled previously using untyped π-calculus [4] and untyped P systems [3].

The type systems can be used in defining more general and simpler rules for P systems. For example, if N_1 and N_2 are some basic types, by considering a set of typed objects $V = \{X_1 : N_1, X_2 : N_1, X_3 : N_1, A : N_2\}$, the evolution rules of the form $X_i \rightarrow X_j$, $X_j \rightarrow A$, $1 \leq i \leq 3$, $1 \leq j \leq 3$, can be replaced by rules of a more general form:

1. $N_1 \rightarrow N_1$ (any object of type N_1 can evolve in any object of type N_1);
2. $N_1 \rightarrow N_2$ (any object of type N_1 can evolve in any object of type N_2).

As related work, we can mention a type description of calculus of looping sequences together with a type inference algorithm presented in [2]. Types and abstract interpretations for systems biology are presented in [6].

Acknowledgments. This work was partially supported by CNCSIS research grants IDEI 402/2007 and TD 345/2008.

References

1. Alberts, B., Johnson, A., Lewis, J., Raff, M., Roberts, K., Walter, P.: Molecular Biology of the Cell, 5th edn. Garland Science, Taylor & Francis (2008)
2. Aman, B., Dezani-Ciancaglini, M., Troina, A.: Type disciplines for analysing biologically relevant properties. Electronic Notes in Theoretical Computer Science 227, 97–111 (2009)
3. Besozzi, D., Ciobanu, G.: A P system description of the sodium-potassium pump. In: Mauri, G., Păun, G., Jesús Pérez-Jímenez, M., Rozenberg, G., Salomaa, A. (eds.) WMC 2004. LNCS, vol. 3365, pp. 210–223. Springer, Heidelberg (2005)
4. Ciobanu, G., Ciubotariu, V., Tanasa, B.: A Pi-calculus model of the Na-K pump. Genome Informatics 13, 469–472 (2002)
5. Ciobanu, G., Păun, Gh., Pérez-Jiménez, M.J. (eds.): Applications of Membrane Computing. Springer, Heidelberg (2006)
6. Fages, F., Soliman, S.: Abstract interpretation and types for systems biology. Theoretical Computer Science 403, 52–70 (2008)
7. Păun, Gh.: Membrane Computing. An Introduction. Springer, Heidelberg (2002)
8. Pierce, B.: Types and Programming Languages. MIT Press, Cambridge (2002)
9. Russell, B.: The Principles of Mathematics, vol. I. Cambridge University Press, Cambridge (1903)
10. Wells, J.: The essence of principal typings. In: Widmayer, P., Triguero, F., Morales, R., Hennessy, M., Eidenbenz, S., Conejo, R. (eds.) ICALP 2002. LNCS, vol. 2380, pp. 913–925. Springer, Heidelberg (2002)

A P System Based Model of an Ecosystem of Some Scavenger Birds

Mónica Cardona[1], M. Angels Colomer[1],
Antoni Margalida[4], Ignacio Pérez-Hurtado[2],
Mario J. Pérez-Jiménez[2], and Delfí Sanuy[3]

[1] Dpt. of Mathematics, University of Lleida
Av. Alcalde Rovira Roure, 191. 25198 Lleida, Spain
{mcardona,colomer}@matematica.udl.es

[2] Research Group on Natural Computing
Dpt. of Computer Science and Artificial Intelligence, University of Sevilla
Avda. Reina Mercedes s/n, 41012 Sevilla, Spain
{perezh,marper}@us.es

[3] Dpt. of Animal Production, University of Lleida
Av. Alcalde Rovira Roure, 191. 25198 Lleida, Spain
dsanuy@prodan.udl.cat

[4] Bearded Vulture Study & Protection Group
Adpo. 43 E-25520 El Pont de Suert (Lleida), Spain
margalida@inf.entorno.es

Abstract. In [1], we presented a P system in order to study the evolution of the bearded vulture in the Pyrenees (NE Spain). Here, we present a new model that overcomes some limitations of the previous work incorporating other scavenger species and additional prey species that provide food for the scavenger intraguild and interact with the Bearded Vulture in the ecosystem. After the validation, the new model can be a useful tool for the study of the evolution and management of the ecosystem. P systems provide a high level computational modelling framework which integrates the structural and dynamical aspects of ecosystems in a compressive and relevant way. The inherent randomness and uncertainty in ecosystems is captured by using probabilistic strategies.

1 Introduction

Since nature is very complex, the perfect model that explains it will be complex too. A complex model is not practical or good to use, so we should obtain a simple and useful model that keeps the most important natural factors.

The P system presented in [1] gives good results in order to study the evolution of the ecosystem based on the Bearded Vulture in the Catalan Pyrenees in the short term, but it does not take into account neither important factors such as the population density or the feeding limitations, nor other species that coexist and compete for space and feeding with the Bearded Vulture. Besides, it was

G. Păun et al. (Eds.): WMC 2009, LNCS 5957, pp. 182–195, 2010.
© Springer-Verlag Berlin Heidelberg 2010

accepted at the said model that the population growth rate of the Bearded Vulture was constant.

In the Catalan Pyrenees, in the North-east of Spain, three vulture species inhabits sharing the geographic space and the existent food resources. In this work, we present a P system for modelling an ecosystem based on three vulture species and the prey species present from which scavengers obtain most of their energy requirements. Apart from adding two new predator species (the Egyptian Vulture *Neophron percnopterus* and Eurasian Griffon Vulture *Gyps fulvus*), we introduce new prey species (making a total of 13 species in comparison to the 5 species appearing at [1]) in the new model that provide feeding resources for the scavenger community. Besides, new rules are introduced to limit the maximum amount of animals that can be supported by the ecosystem as well as the amount of grass available for the herbivorous species. At the new model, it is considered that the population growth rate of the Bearded Vulture varies depending on the surface and orography of the system as well as on existing population. For a good management of the ecosystem, it is suitable to know the biomass every species leaves annually. For this reason, it is interesting to codify this information at the system output.

For the modelling of the ecosystem, we need the biological parameters that are obtained experimentally and from the literature and they quantify the biological basic processes of the species and the physical environment of the ecosystem. The processes modelled are reproduction, feeding and mortality, and the physical factors that have been considered are the geographical limitations.

It has been developed simulator of the model written in JAVA by using the specification language P-Lingua [4]. This simulator allows us to experimentally validate the model as well as study the ecosystem dynamics under different initial conditions.

The paper is organized as follows. Next section shows a formal framework to model ecosystems by means of probabilistic P systems, and a P system modelling of the above mentioned ecosystem is presented. In Section 3, we experimentally validate the model presented in this paper by using a P-lingua simulator [4] and we also compare it to the one presented in [1].

2   A Formal Framework to Model Ecosystems

In this section, we present a model of the ecosystem described above by means of probabilistic P systems.

First, we define the P systems based framework (probabilistic P systems), where additional features such as electrical charges which describe specific properties in a better way, are used.

Definition 1. *A probabilistic P system of degree $q \geq 1$ is a tuple*

$$\Pi = (\Gamma, \mu, \mathcal{M}_1, \ldots, \mathcal{M}_q, R, \{c_r\}_{r \in R}), \ where:$$

- Γ is the alphabet (finite and nonempty) of objects (the working alphabet);
- μ is a membrane structure (a rooted tree), consisting of q membranes, labelled by $1, 2, \ldots, q$. The skin membrane is labelled by 1. We also associate electrical charges with membranes from the set $\{0, +, -\}$, neutral, positive and negative;
- $\mathcal{M}_1, \ldots, \mathcal{M}_q$ are strings over Γ, describing the multisets of objects initially placed in the q regions of μ;
- R is a finite set of evolution rules. An evolution rule associated with the membrane labelled by i is of the form

$$r : \; u\,[\,v\,]_i^{\alpha} \xrightarrow{c_r} u'\,[\,v'\,]_i^{\alpha'}$$

where u, v, u', v' are multisets over Γ, $\alpha, \alpha' \in \{0, +, -\}$, $1 \le i \le q$, and c_r is a real number between 0 and 1. Besides, if $r_1, \ldots, r_t$ are rules whose left-hand side is $u\,[\,v\,]_i^{\alpha}$ then it must verify $\sum_{j=1}^{t} c_{r_j} = 1$, being c_{r_j} the probabilistic constant associated with rule r_j.

We denote by $[\,v\, \xrightarrow{c_r} v'\,]_i^{\alpha}$ rule $u\,[\,v\,]_i^{\alpha} \xrightarrow{c_r} u'\,[\,v'\,]_i^{\alpha'}$ in the case $u = u' = \lambda$, and $\alpha = \alpha'$. In the same way, we denote by $u\,[\,v\,]_i^{\alpha} \to u'\,[\,v'\,]_i^{\alpha'}$ rule $u\,[\,v\,]_i^{\alpha} \xrightarrow{c_r} u'\,[\,v'\,]_i^{\alpha'}$ in the case $c_r = 1$.

We assume that a global clock exists, marking the time for the whole system (for all compartments of the system); that is, all membranes and the application of all the rules are synchronized.

The multisets of objects present at any moment in the n *regions* of the system constitute the *configuration* of the system at that moment. Particularly, tuple $(\mathcal{M}_1, \ldots, \mathcal{M}_q)$ is the initial configuration of the system.

The P system can pass from one configuration to another by using rules from R as follows:

- A rule $u\,[\,v\,]_i^{\alpha} \xrightarrow{c_r} u'\,[\,v'\,]_i^{\alpha'}$ is applicable (with a probability c_r) to a membrane labelled by i, and with α as electrical charge, when multiset u is contained in the father of membrane i, and multiset v is contained in membrane i. When rule $u\,[\,v\,]_i^{\alpha} \xrightarrow{c_r} u'\,[\,v'\,]_i^{\alpha'}$ is applied, multiset u (resp. v) in the father of membrane i (resp. membrane i) is removed from that membrane and multiset u' (resp. v') is produced in it.
- The rules are applied in a *maximal consistent parallel* way, that is, all those rules of type $u_1\,[\,v_1\,]_i^{\alpha} \xrightarrow{c_r} u'_1\,[\,v'_1\,]_i^{\alpha'}$ and $u_2\,[\,v_2\,]_i^{\alpha} \xrightarrow{c_s} u'_2\,[\,v'_2\,]_i^{\alpha'}$ must be applied simultaneously in a maximal way.
- The constant c_r associated with rule r indicates the affinity of the rule for its application.

2.1   A P System Based Model of the Ecosystem

Let D be a natural number higher than 0, which will represent the number of years to be simulated in the evolution of the ecosystem. At the definition of a probabilistic P system modelling the ecosystem described at Section 1, $n = 17$

represents the different types of animals of the 13 species which compose the ecosystem under study. We considerer two types of animals for the Red Deer due to the fact that males are highly valued by hunters and this implies that the mortality rate of males ($i = 6$) is higher than that of females ($i = 5$). We also consider two types of animals, denoted by A (annual) and P (periodical), for domestic ones (except for horses) because some of them spend only six months in the mountain.

Next, we present a list of the constants associated with the rules where the corresponding meanings are specified (index i, $1 \leq i \leq n$, represents the type of animal).

- $g_{i,1}$: 1 for wild animals and 0 for domestic animals.
- $g_{i,2}$: proportion of time they remain in the mountain during the year.
- $g_{i,3}$: age at which adult size is reached. This is the age at which the animal eats like and adult does, and at which if the animal dies, the amount of biomass it leaves is similar to the total one left by an adult. Moreover, at this age it will have surpassed the critical early phase during which the mortality rate is high.
- $g_{i,4}$: age at which it starts to be fertile.
- $g_{i,5}$: age at which it stops being fertile.
- $g_{i,6}$: average life expectancy in the ecosystem.
- $g_{i,7}$: maximum density of the ecosystem.
- $g_{i,8}$: number of animals that survive after reaching maximum density of the ecosystem.
- $k_{i,1}$: proportion of females in the population (per one).
- $k_{i,2}$: fertility rate (proportion of fertile females that reproduce).
- $k_{i,3}$: number of descendants per each fertile female that reproduces.
- $k_{i,4}$: it is equal to 0 when the species go through a natural growth and it is equal to 1 when animals are nomadic (the Bearded Vulture moves from one place to another until it is 6–7 years old, when it settles down).
- $m_{i,1}$: natural mortality rate in the first years, $age < g_{i,3}$ (per one).
- $m_{i,2}$: mortality rate in adult animals, $age \geq g_{i,3}$ (per one).
- $m_{i,3}$: percentage of domestic animals belonging to non–stabilized populations which are withdrawn in the first years.
- $m_{i,4}$: is equal to 1 if the animal dies at the age of $g_{i,6}$ and it is not retired, and it is equal to 0 if the animal does not die at the age of $g_{i,6}$ but it is retired from the ecosystem.
- $f_{i,1}$: amount of bones from young animals when they die, $age < g_{i,3}$.
- $f_{i,2}$: amount of meat from young animals when they die, $age < g_{i,3}$.
- $f_{i,3}$: amount of bones from adult animals when they die, $age \geq g_{i,3}$.
- $f_{i,4}$: amount of meat from adult animals when they die, $age \geq g_{i,3}$.
- $f_{i,5}$: amount of bones necessary per year and animal (1 unit is equal 0.5 kg of bones).
- $f_{i,6}$: amount of grass necessary per year and animal.
- $f_{i,7}$: amount of meat necessary per year and animal.

The values of these constants have been obtained experimentally, except for $k_{i,4}$ and $m_{i,4}$ (see [2], [3], [5], [6] for details). Constants k, m and f are associated with reproduction, mortality and feeding rules, respectively. Constants g are associated with the remaining rules.

Let us consider the following probabilistic P system of degree 2 with (only) two electrical charges (neutral and positive)

$$\Pi_D = (\Gamma, \mu, \mathcal{M}_1, \mathcal{M}_2, R, \{c_r\}_{r\in R}), \text{ where:}$$

- $\Gamma = \{X_{ij}, Y_{ij}, V_{ij}, Z_{ij} : 1 \leq i \leq n, \ 0 \leq j \leq g_{i,6}\} \cup$
  $\{B, \ G, \ M, \ B', \ G', \ M', \ C, \ C'\} \cup \{h_s : 1 \leq s\} \cup$
  $\{H_i, \ H'_i, \ F_i, \ F'_i, \ T_i, \ a_i, \ b_{0i}, \ b_i, \ d_i, \ e_i : 1 \leq i \leq n\}$

is the working alphabet.

Symbols X, Y, V and Z represent the same animal but in different states. Index i is associated with the type of animal, index j is associated with their age, and $g_{i,6}$ is the average life expectancy. It also contains the auxiliary symbols $B, \ B'$, which represent bones, $M, \ M'$, which represent meat and $G, \ G'$, which represent the amount of grass available for the feeding of the animals in the ecosystem. Objects $H_i, \ H'_i$ represent the biomass of bones, and objects $F_i, \ F'_i$ represent the biomass of meat left by species i in different states. Object C enables the creation of objects $B', \ M'$ and G' which codify bones and meat (artificially added by human beings) as well as the grass generated by the ecosystem itself. Besides, object C produces objects C' which in turn generate object C allowing the beginning of a new cycle. At the P system design, different objects (i.e. $G, \ G'$) represent the same entity (in this case, grass) with the purpose of synchronizing the model. T_i is an object used for counting the existing animals of species i. If a species overcomes the maximum density, values will be regulated. Objects $b_{0i}, \ b_i$ and e_i allow us to control the maximum number of animals per species in the ecosystem. At the moment when a regulation takes place, object a_i allows us to eliminate the number of animals of species i that exceeds the maximum density. Object d_i is used to put under control domestic animals that are withdrawn from the ecosystem for their marketing.

- $\mu = [\ [\ ]_2\ ]_1$ is the membrane structure. We consider two regions, the skin and an inner membrane. The first region is important to control the densities of every species do not overcome the threshold of the ecosystem. Animals reproduce, feed and die in the inner membrane. For the sake of simplicity, neutral polarization will be omitted.

- $\mathcal{M}_1$ and $\mathcal{M}_2$ are strings over Γ, describing the multisets of objects initially placed in regions of μ (encoding the initial population and the initial food);
  - $\mathcal{M}_1 = \{b_{0i}, X_{ij}^{q_{ij}}, h_t^{q_{1j}} : 1 \leq i \leq n, \ 0 \leq j \leq g_{i,6}\}$, where q_{ij} indicates the number of animals of species i initially present in the ecosystem whose age is j, and $t = \max\{1, \lceil \frac{\sum_{j=8}^{21} q_{1j} - 6}{1.352} \rceil\}$. The mathematical expression give for t it was obtained using the lineal regression that appear in figure 1;
  - $\mathcal{M}_2 = \{C\}$

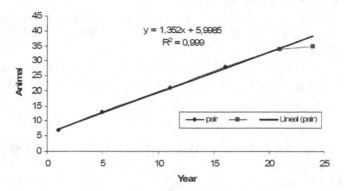

Fig. 1. Linear regression between numbers of pairs and years for the bearded vulture

- The set R of evolution rules consists of:
  - The first rule represents the contribution of energetic resources to the ecosystem at the beginning of each cycle and it is essential for the system to evolve. The second rule is useful to synchronize the process.
    $$r_0 \equiv [C \rightarrow B'^{\alpha} M'^{\beta} G'^{\gamma} C']_2^0,$$
    where α and β are the double of kilos of bones and meat that are externally introduced to the ecosystem, and γ is the amount of grass produced by the ecosystem.
    $$r_1 \equiv [b_{0,i} \rightarrow b_i]_1^0.$$
  - *Variation rules of the population.*
    We consider two cases due to the fact that in nomadic species the said variation is influenced by animals from other ecosystems.
    * Case 1. Non–nomadic species ($k_{i,4} = 0$).
      · Adult males:
      $$r_2 \equiv [X_{ij} \xrightarrow{(1-k_{i,1})\cdot(1-k_{i,4})} Y_{ij}]_1^0, \ 1 \leq i \leq n, \ g_{i,4} \leq j < g_{i,5}.$$
      · Adult females that reproduce:
      $$r_3 \equiv [X_{ij} \xrightarrow{k_{i,2}\cdot k_{i,1}\cdot(1-k_{i,4})} Y_{ij} Y_{i0}^{k_{i,3}}]_1^0, \ 1 \leq i \leq 4, \ g_{i,4} \leq j < g_{i,5}.$$
      $$r_4 \equiv [X_{ij} \xrightarrow{k_{i,2}\cdot k_{i,1}\cdot(1-k_{i,4})} Y_{ij} Y_{i0}^{k_{i,3}}]_1^0, \ 7 \leq i \leq n, \ g_{i,4} \leq j < g_{i,5}.$$
      $$r_5 \equiv [X_{5j} \xrightarrow{0.5\cdot k_{5,2}} Y_{5j} Y_{50}^{k_{i,3}}]_1^0, \ g_{5,4} \leq j < g_{5,5}.$$
      $$r_6 \equiv [X_{5j} \xrightarrow{0.5\cdot k_{5,2}} Y_{5j} Y_{60}^{k_{i,3}}]_1^0, \ g_{5,4} \leq j < g_{5,5}.$$
      · Adult females that do not reproduce:
      $$r_7 \equiv [X_{ij} \xrightarrow{(1-k_{i,2})\cdot k_{i,1}\cdot(1-k_{i,4})} Y_{ij}]_1^0, \ 1 \leq i \leq n, \ g_{i,4} \leq j < g_{i,5}.$$
      · Old females and males that do not reproduce:
      $$r_8 \equiv [X_{ij} \xrightarrow{1-k_{i,4}} Y_{ij}]_1^0, \ 1 \leq i \leq n, \ g_{i,5} \leq j \leq g_{i,6}.$$
      · Young animals that do not reproduce:
      $$r_9 \equiv [X_{ij} \xrightarrow{1-k_{i,4}} Y_{ij}]_1^0, \ 1 \leq i \leq n, \ 1 \leq j < g_{i,4}.$$

* Case 2. Nomadic species ($k_{i,4} = 1$).

$r_{10} \equiv [X_{1j}h_s \xrightarrow{v_s} Y_{1(g_{i,4}-1)}Y_{1j}h_{s+1}^2]_1^0$, $1 \leq i \leq n$, $g_{i,4} \leq j \leq g_{i,6}$, $t \leq s \leq D_1$, where $v_s = 1.352/(1.352s+6)$ and $D_1=\min\{21, D+t-1\}$.

$r_{11} \equiv [X_{1j}h_s \xrightarrow{0.01} Y_{1(g_{i,4}-1)}Y_{1j}h_{s+1}^2]_1^0$, $1 \leq i \leq n$, $g_{i,4} \leq j \leq g_{i,6}$, $D_3 \leq s \leq D_2$, where $D_2=\max\{21, D+t-1\}$ and $D_3=\max\{21, t\}$.

$r_{12} \equiv [X_{1j}h_s \xrightarrow{1-v_s} Y_{1j}h_{s+1}]_1^0$, $1 \leq i \leq n$, $g_{i,4} \leq j \leq g_{i,6}$, $t \leq s \leq D_1$.

$r_{13} \equiv [X_{1j}h_s \xrightarrow{0.99} Y_{1j}h_{s+1}]_1^0$, $1 \leq i \leq n$, $g_{i,4} \leq j \leq g_{i,6}$, $D_3 \leq s \leq D_2$.

- *Mortality rules.*
    * Young animals that survive:

    $r_{14} \equiv Y_{ij}[\ ]_2^0 \xrightarrow{1-m_{i,1}-m_{i,3}} [V_{ij}T_i]_2^+$, $1 \leq i \leq n$, $0 \leq j < g_{i,3}$.

    * Young animals that die:

    $r_{15} \equiv Y_{ij}[\ ]_2^0 \xrightarrow{m_{i,1}} [H_i'^{f_{i,1} \cdot g_{i,2}} F_i'^{f_{i,2} \cdot g_{i,2}} B'^{f_{i,1} \cdot g_{i,2}} M'^{f_{i,2} \cdot g_{i,2}}]_2^+$, $1 \leq i \leq n$, $0 \leq j < g_{i,3}$.

    * Young animals that are retired from the ecosystem:

    $r_{16} \equiv [Y_{ij} \xrightarrow{m_{i,3}} \lambda]_1^0$, $1 \leq i \leq n$, $0 \leq j < g_{i,3}$.

    * Adult animals that do not reach an average life expectancy and survive:

    $r_{17} \equiv Y_{ij}h_s^{k_{i,4}}[\ ]_2^0 \xrightarrow{1-m_{i,2}} [V_{ij}T_ih_s^{k_{i,4}}]_2^+$, $1 \leq i \leq n$, $g_{i,3} \leq j < g_{i,6}$, $t+1 \leq s \leq D+t$.

    * Adult animals that do not reach an average life expectancy and die:

    $r_{18} \equiv Y_{ij}h_s^{k_{i,4}}[\ ]_2^0 \xrightarrow{m_{i,2}} [H_i'^{f_{i,3} \cdot g_{i,2}} F_i'^{f_{i,4} \cdot g_{i,2}} B'^{f_{i,3} \cdot g_{i,2}} M'^{f_{i,4} \cdot g_{i,2}}$
    $V_{i,g_{i,4}-1}^{k_{i,4}} h_s^{k_{i,4}} T_i^{k_{i,4}}]_2^+$, $1 \leq i \leq n$, $g_{i,3} \leq j < g_{i,6}$, $t+1 \leq s \leq D+t$.

    * Animals that reach an average life expectancy and die in the ecosystem:

    $r_{19} \equiv Y_{ig_{i,6}}h_s^{k_{i,4}}[\ ]_2^0 \xrightarrow{c_{19}} [H_i'^{f_{i,3} \cdot g_{i,2}} F_i'^{f_{i,4} \cdot g_{i,2}} B'^{f_{i,3} \cdot g_{i,2}} M'^{f_{i,4} \cdot g_{i,2}}$
    $V_{i,g_{i,4}-1}^{k_{i,4}} h_s^{k_{i,4}} T_i^{k_{i,4}}]_2^+$, $1 \leq i \leq n$, being $c_{19} = k_{i,4} + (1 - k_{i,4}) \cdot (m_{i,4} + (1 - m_{i,4}) \cdot m_{i,2})$, $t+1 \leq s \leq D+t$.

    * Animals that reach an average life expectancy and are retired from the ecosystem:

    $r_{20} \equiv [Y_{ig_{i,6}}h_s^{k_{i,4}} \xrightarrow{(1-k_{i,4}) \cdot (1-m_{i,4}) \cdot (1-m_{i,2})} \lambda]_1$, $1 \leq i \leq n$, $t+1 \leq s \leq D+t$.

- *Density regulation rules.*
    * Creation of objects that are going to enable the control of the maximum number of animals in the ecosystem:

    $r_{21} \equiv b_i[\ ]_2^0 \to [b_i a_i^{\lceil 0,9*g_{i,7}\rceil} e_i^{\lceil 0,2*g_{i,7}\rceil}]_2^+$, $1 \leq i \leq n$.

    * Evaluation of the density of the different species in the ecosystem:

    $r_{22} \equiv [T_i^{g_{i,7}} a_i^{(g_{i,7}-g_{i,8})} \to \lambda]_2^+$, $1 \leq i \leq n$.

    * Generation of randomness in the number of animals:

    $r_{23} \equiv [e_i \xrightarrow{0,5} a_i]_2^+$, $1 \leq i \leq n$.

    $r_{24} \equiv [e_i \xrightarrow{0,5} \lambda]_2^+$, $1 \leq i \leq n$.

* Change of the names of the objects which represent animals:
  $r_{25} \equiv [V_{ij} \to Z_{ij}]_2^+,\ 1 \le i \le n,\ 0 \le j < g_{i,6}.$
* Change of the names of the objects which represent food resources:
  $r_{26} \equiv [G' \to G]_2^+.$
  $r_{27} \equiv [B' \to B]_2^+.$
  $r_{28} \equiv [M' \to M]_2^+.$
  $r_{29} \equiv [C' \to C]_2^+.$
  $r_{30} \equiv [H_i' \to H_i]_2^+,\ 1 \le i \le n.$
  $r_{31} \equiv [F_i' \to F_i]_2^+,\ 1 \le i \le n.$

- *Feeding rules.*
  $r_{32} \equiv [Z_{ij} h_s^{k_{i,4}} a_i B^{f_{i,5} \cdot g_{i,2}} G^{f_{i,6} \cdot g_{i,2}} M^{f_{i,7} \cdot g_{i,2}}]_2^+ \to X_{i(j+1)} h_s^{k_{1,4}} [\,]_2^0,\ 1 \le i \le n,\ 0 \le j \le g_{i,6},\ t+1 \le s \le D+t.$

- *Updating rules.*
  The purpose of the following rules is to make a balance at the end of the year. That is, the leftover food is not useful for the next year, so it is necessary to eliminate it. But if the amount of food is not enough, some animals die.

  * Elimination of the remaining bones, meat and grass:
    $r_{33} \equiv [G \to \lambda]_2^0.$
    $r_{34} \equiv [M \to \lambda]_2^0.$
    $r_{35} \equiv [B \to \lambda]_2^0.$
    $r_{36} \equiv [T_i \to \lambda]_2^0,\ 1 \le i \le n.$
    $r_{37} \equiv [a_i \to \lambda]_2^0,\ 1 \le i \le n.$
    $r_{38} \equiv [e_i \to \lambda]_2^0,\ 1 \le i \le n.$
    $r_{39} \equiv [b_i]_2^0 \to b_i[\,]_2^0,\ 1 \le i \le n.$
    $r_{40} \equiv [H_i]_2^0 \to H_i[\,]_2^0,\ 1 \le i \le n.$
    $r_{41} \equiv [F_i]_2^0 \to F_i[\,]_2^0,\ 1 \le i \le n.$

  * Young animals that die because of a lack of food:
    $r_{42} \equiv [Z_{ij} \xrightarrow{g_{i,1}} H_i'^{f_{i,1}} F_i'^{f_{i,2}} B'^{f_{i,1}} M'^{f_{i,2}}]_2^0,\ 1 \le i \le n,\ 0 \le j < g_{i,3}.$

    $r_{43} \equiv [Z_{ij}]_2^0 \xrightarrow{1-g_{i,1}} d_i[\,]_2^0,\ 1 \le i \le n,\ 0 \le j < g_{i,3}.$

  * Adult animals that die because of a lack of food:
    $r_{44} \equiv [Z_{ij} h_s^{k_{1,4}} \xrightarrow{g_{i,1}} H_i'^{f_{i,3}} F_i'^{f_{i,4}} B'^{f_{i,3}} M'^{f_{i,4}}]_2^0,\ 1 \le i \le n,\ g_{i,3} \le j \le g_{i,6},\ t+1 \le s \le D+t.$

    $r_{45} \equiv [Z_{ij} h_s^{k_{1,4}} \xrightarrow{1-g_{i,1}} \lambda]_2^0,\ 1 \le i \le n,\ g_{i,3} \le j \le g_{i,6},\ t+1 \le s \le D+t.$

  The purpose of these rules is to eliminate objects H and F associated with the quantity of biomass left by every species.
    $r_{46} \equiv [H_i \to \lambda]_1^0,\ 1 \le i \le n.$
    $r_{47} \equiv [F_i \to \lambda]_1^0,\ 1 \le i \le n.$

2.2   Structure of the P System Running

The model of the ecosystem presented in the previous Section includes new ingredients with the aim to overcome the limitations found at the model described in [1]. More specifically, the modifications made are the following:

– It has been added new species which have active roles in the ecosystem
  under study, although their roles are perhaps less relevant that those of the
  first species studied. These species are European mouflon, the wild boar, the
  horse, the goat and the cow. Besides, it has been included greedy species
  such as the Egyptian Vulture and the Griffon Vulture which compete with
  the Bearded Vulture.
– It is considered that the population growth rate of the Bearded Vulture
  varies depending on the surface and orography of the system as well as on
  the existing population.
– A new module has been added in order to regulate the population density
  of the ecosystem.
– The mortality module has been modified in order to consider that after an
  animal dies, in addition to the bones it leaves at the ecosystem, its meat
  serves as food for other animals.
– The feeding module has also been modified because the feeding resources for
  the species at the ecosystem have been modelled in this new approach. For
  this reason, new objects have been introduced representing, apart from the
  bones, the amount of meat and grass available at the ecosystem.

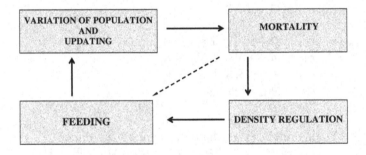

Fig. 2. Modules of the P system

In this model, a module devoted to control the density has been introduced.
From the point of view of the execution of the system, the module has been
incorporated between the Mortality and the Feeding modules. These are depicted
in Figure 2.

Let us recall that, objects X represent the different species along the execution of the reproduction module. Objects X evolve to objects Y (mortality
module) when they pass to the mortality module, and these objects Y evolve to
objects V (density module), together with objects T which represent the number of individuals per each species. Then, objects V evolve to objects Z (feeding
module). Objects T will allow the activation of the process of auto–regulation of
the ecosystem when the number of individuals of a species exceed the threshold
of maximum density, which is codified by objects a.

When a cycle is produced, all objects which are not associated with species
are eliminated, except the biomass generated by the animals that have died due
to the process of regulation.

3   Results and Discussions

The software tool used for the purposes of this paper is based on P-Lingua 2.0 [4]. P-Lingua is a new programming language able to define P systems of different types (from now on, frameworks). For instance, P-Lingua can define any P system within the probabilistic framework mentioned in this paper.

Next, we describe how to implement in P–Lingua the applicability of the rules to a given configuration.

(a) Rules are classified into disjoint classes so that all the rules belonging to a class have the same left–hand side.
(b) Let $\{r_1, \ldots, r_t\}$ be one of the said sets of rules. Let us suppose that the common left-hand side is $u\ [v]_i^\alpha$ and their respective probabilistic constants are $c_{r_1}, \ldots, c_{r_t}$. In order to determine how these rules are applied to a give configuration, we proceed as follows:

  - It is computed the greatest number N so that u^N appears in the father membrane of i and v^N appears in membrane i.
  - N random numbers x such that $0 \le x < 1$ are generated.
  - For each k, $1 \le k \le t$, let n_k be the amount of numbers generated belonging to interval $[\ \sum_{j=0}^{k-1} c_{r_j}\ ,\ \sum_{j=0}^{k} c_{r_j})$ (assuming that $c_{r_0} = 0$).
  - For each k, $1 \le k \le t$, rule r_k is applied n_k times.

P-Lingua 2.0 provides a JAVA library that defines algorithms in order to simulate P system computations for each supported framework, so we are using a common algorithm for all P systems within the probabilistic framework.

By defining the ecosystem model by a P system written in P-Lingua, it is possible to check, validate and improve the model in a flexible way, instead of developing a new "ad hoc" simulator for each new model.

The application has a friendly user-interface, which sits on the P-Lingua JAVA library, allowing the user to change the initial parameters of the ecosystem in an easy way without special knowledge about the P system or the initial multisets. The main objective is to make virtual experiments on the ecosystem.

The current version of this software is a prototype GPL licensed [8].

The model designed is experimentally validated by using the simulator previously described as well as the data from Table 1.

Table 1. Number of animals in the Catalan Pyrenees (1979–2009)

Specie	79	84	87	89	93	94	95	99	00	05	08	09
Bearded V.	-	7	-	13	-	-	21	-	28	34	35	-
Egyptian V.	-	-	29	-	34	-	-	-	40	-	66	-
Griffon V.	29	-	-	106	-	-	-	377	-	-	-	842
Pyrenean C.	-	-	-	-	-	9000	-	-	-	-	12000	-
Red deer	-	-	-	-	-	1000	-	-	-	-	5500	-
Fallow deer	-	-	-	-	-	600	-	-	-	-	1500	-
Roe deer	-	-	-	-	-	1000	-	-	-	-	10000	-

At the validation process, we have focused on the evolution of wild species populations. For that purpose, it has been validated the ecosystem dynamics for a period of 14 years, since 1994. The Bearded Vulture (respectively the Griffon and the Egyptian Vultures) populations at the initial year has been considered according to the data Table 1 by means of a logarithmic (respectively, exponential) regression (see Figure 3).

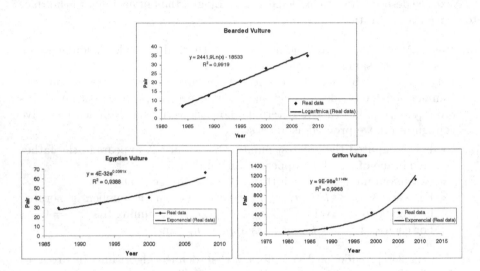

Fig. 3. Regression relationships between numbers of pairs and years

At the validation process, values obtained from the simulator running have been compared to those obtained experimentally. It is also worth noting that we have focused on the population dynamics at wild species from which there are only data about the initial (1994) and final (2008) years, except for scavengers birds which we have more information about (see Table 1 for details).

Bearing in mind the model designed is probabilistic, the ecosystem evolution throughout the period under study has been obtained by running the simulator for 100 times having the same input data. The simulator executions have allowed us to estimate the standard deviation and compute the population confidence intervals of the different species. The result presented in Figure 4 is the average of the 100 simulator executions.

Finally, we have compared the model presented in this work (we refer to it as model II) to the model presented in [1] (we refer to it as model I). For that purpose, we have used the simulator previously described studying the ecosystem evolution for a period of 10 years from 1994 on. Some of the results are shown at the Figure 5. Both models present good results until 2008, regarding experimental data (except for the Pyrenean Chamois). Nonetheless, at the simulations corresponding to the years later to 2008, it is noticed a great difference between models due to the fact that model I did not consider the regulation of the populations.

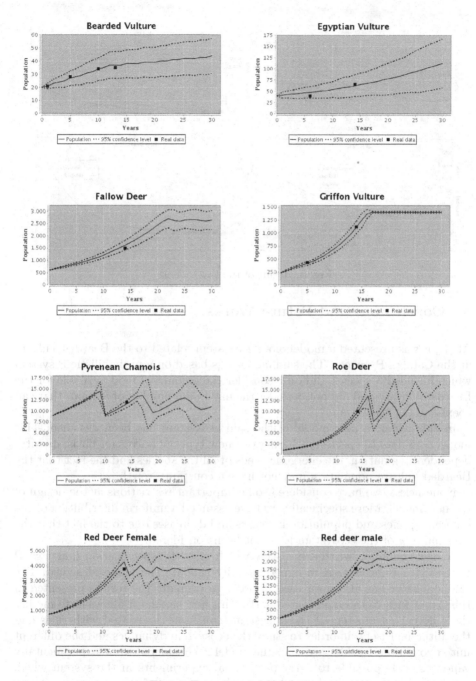

Fig. 4. Experimental Validation

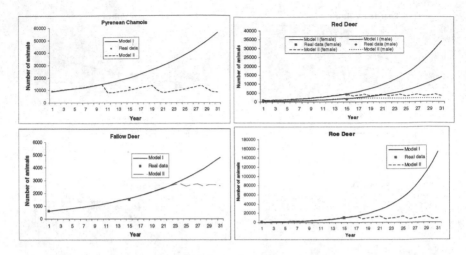

Fig. 5. Result of the two models

4   Conclusions and Future Works

At [1], it was presented a model of an ecosystem related to the Bearded Vulture at the Catalan Pyrenees. The said model was based on a probabilistic P system which included 5 species, did only consider the amount of food available for the Bearded Vulture and did not consider the maximum number of animals that can coexist in the ecosystem.

In this paper, a new model of the said ecosystem has been designed. This model considers 13 species, including two new types of scavenger birds, density-dependent regulation, the energetic needs of all the species and the fact that the Bearded Vulture population may not have a constant growth.

Nonetheless, we have considered some important restrictions at the design of the new model. More specifically, we have assumed a uniform distribution of the different species and population increases and decreases due to the fact that the external flow of the ecosystem have not been considered.

A new simulator written in JAVA which uses the specification language P Lingua [4] has been used to experimentally validate the model designed. The said simulator has also been used in order to compare the results presented in this paper with to those presented in [1]. This new simulator allows us to modify the different parameters of the P system (constants associated with rules and the initial multisets) in order to study the ecosystem dynamics and the different initial condition. In this way, once the model is considered to be experimentally validated, it is possible to carry out virtual experiments in the system which can provide hypotheses about the possible evolution of the ecosystem. These hypotheses, filtered by experts in a suitable way, can be useful for the ecologists when taking decisions which favour both the balance of the ecosystem and the preservation of the endangered species such as the Bearded Vulture.

In a future work, we hope to add new ingredients to this model which overcome the restrictions imposed on it that where previously referred to. For that purpose, we are studying the possibility of considering multienvironment P systems (see [7], for details) as a new modelling scenario. This will imply an important revision of the simulator and the searching of more efficient algorithms which simulate the running of the the probabilistic strategy.

Acknowledgement

The authors acknowledge the support of the project TIN2006–13425 of the Ministerio de Educación y Ciencia of Spain, cofinanced by FEDER funds, and the support of the Project of Excellence with *Investigador de Reconocida Valía* of the Junta de Andalucía, grant P08-TIC-04200.

References

1. Cardona, M., Colomer, M.A., Pérez–Jiménez, M.J., Sanuy, D., Margalida, A.: Modelling ecosystems using P Systems: The Bearded Vulture, a case of study. In: Corne, D.W., Frisco, P., Paun, G., Rozenberg, G., Salomaa, A. (eds.) WMC 2008. LNCS, vol. 5391, pp. 137–156. Springer, Heidelberg (2009)
2. Brown, C.J.: Population dynamics of the Bearded Vulture Gypaetus barbatus in southern Africa. African Journal of Ecology 35, 53–63 (1997)
3. Donázar, J.A.: Los buitres ibéricos: biología y conservación. In: Reyero, J.M. (ed.), Madrid, Spain (1993)
4. García–Quismondo, M., Gutiérrez–Escudero, R., Martínez, M.A., Orejuela, E., Pérez–Hurtado, I.: P–Lingua 2.0. A sofware framework for cell-like P systems. International Journal of Computers, Comunications and Control IV(3), 234–243 (2009)
5. Margalida, A., Bertran, J., Heredia, R.: Diet and food preferences of the endangered Bearded Vulture Gypaetus barbatus: a basis for their conservation. Ibis 151, 235–243 (2009)
6. Margalida, A., García, D., Cortés-Avizanda, A.: Factors influencing the breeding density of Bearded Vultures, Egyptian Vultures and Eurasian Griffon Vultures in Catalonia (NE Spain): management implications. Animal Biodiversity and Conservation 30(2), 189–200 (2007)
7. Romero, F.J., Pérez–Jiménez, M.J.: A model of the quorum sensing system in Vibrio Fischeri using P systems. Artificial Life 14(1), 95–109 (2008)
8. GPL license, http://www.gnu.org/copyleft/gpl.html

Metabolic P System Flux Regulation by Artificial Neural Networks

Alberto Castellini[1], Vincenzo Manca[1], and Yasuhiro Suzuki[2]

[1] Verona University, Dept. of Computer Science
Strada Le Grazie 15, 37134 Verona, Italy
{alberto.castellini,vincenzo.manca}@univr.it
[2] Nagoya University, Dept. of Complex Systems Science
Furo-cho, Chikusa-ku, Nagoya, 464-8601, Japan
ysuzuki@is.nagoya-u.ac.jp

Abstract. Metabolic P systems are an extension of P systems employed for modeling biochemical systems in a discrete and deterministic perspective. The generation of MP models from observed data of biochemical system dynamics is a hard problem which requires to solve several subproblems. Among them, flux tuners discovery aims to identify substances and parameters involved in tuning each reaction flux. In this paper we propose a new technique for discovering flux tuners by means of neural networks. This methodology, based on backpropagation with weight elimination for neural network training and on an heuristic algorithm for computing tuning indexes, has achieved encouraging results in a synthetic case study.

1 Introduction

Many kinds of models have been developed in order to provide new insight on chemically reacting systems, among them, ordinary differential equations (ODE) [31] represent a milestone for continuous and deterministic modeling, while models based on the Gillespie's algorithm [12] are widely used for discrete and stochastic modeling. A key point in the development of new modeling frameworks seems to be represented by the choice of the right abstraction level, since complex systems usually show different characteristics when viewed from different "distances". The majority of models now available seem to be either very low level (too detailed), or very high level (too coarse grain), while many biological systems seem to require an intermediate level of abstraction. The *executable biology* approach [9] suggests to employ *computational models*, namely, a new class of models that mimic natural phenomena by executing algorithm instructions, rather than using computer power to analyze mathematical relationships among the elements of biological systems.

Rewriting systems, in their basic form, consist of a set of terms and a set of rewriting rules stating how terms can be transformed. Many extensions of these systems have been applied to biological modeling, such as the well known *L systems* [17], developed in 1968 by the Hungarian theoretical biologist and botanist

G. Păun et al. (Eds.): WMC 2009, LNCS 5957, pp. 196–209, 2010.
© Springer-Verlag Berlin Heidelberg 2010

Lindenmayer to provide a formal description of the growth patterns of various types of algae. *P systems* [26,27], from the name of G. Păun who devised them in 1998, represent a novel computational model originated from the combination of multisets rewriting systems and membrane compartmentalization. This approach lends itself to be used as a computational model for biological systems, in which multisets of objects represent chemical elements, while rewriting rules and rewriting application strategies represent a kind of algorithm to be executed for mimicking phenomena under investigation.

Several extensions of P systems have been developed so far [7,28], some of them also coping with biological systems modeling [23,25,29,30]. In particular, *metabolic P systems*, or *MP systems*, suggest a deterministic strategy, based on the generalization of chemical laws, for computing the amount of objects transformed by rules at each computational step [18,19,21,22,23]. Equivalences between MP systems and, respectively, autonomous ODE [10] and Hybrid Functional Petri nets [3,4] have been recently proved, and several biological processes have been modeled by means of MP systems, such as the Lotka-Volterra dynamics [23], the mitotic cycles in early amphibian embryos [22] and the *lac* operon gene regulatory mechanism in glycolytic pathway [3]. These case studies show that, being intrinsically time-discrete and based on multiset rewriting, MP models are able to give a different viewpoint on biological processes respect to traditional ODE models. A software called MetaPlab has been also proposed [6,24,32] which enables the user to generate MP models by means of some useful graphical tools, and then to simulate their dynamics, to automatically estimate regulation functions and to perform many other tasks.

An MP system involves *i)* a set of *substances*, *ii)* a set of *parameters* (e.g., temperature, pH, etc.) and *iii)* a set of *reactions* each equipped with a corresponding *flux regulation function*. These functions compute reaction fluxes, that is, the amount of substances transformed by each reaction when the system is in a given state.

A crucial problem of MP model designing concerns the synthesis of flux regulation functions from observed time evolutions. In particular, the question is the following: "Given the time-series of substance concentrations and parameter values of a process observed every time interval τ, and given the stoichiometry of the system under investigation, which are the flux regulation functions that make an MP model evolve according with the observed dynamics?" The *log-gain theory* [18,19,20] supports a first step of the regulation function synthesis by enabling to deduce the time-series of flux values from the time-series of substances and parameters of an observed dynamics. Once flux time-series have been generated, regression techniques are used to discover the functions that compute these fluxes.

In [5] a new approach is proposed to the synthesis of MP regulation functions relying on artificial neural networks (ANNs) as universal function approximators [2], and employing both traditional and evolutionary algorithms [13] for learning these networks. Moreover, a plug-in tool for MetaPlab has been implemented to automate the learning stage. Here we extend this approach with a technique for weight elimination in ANNs [2,33] and an algorithm for identifying flux "tuners", namely, the set of substances and parameters actually involved in the regulation of each flux. In the next section we formally introduce MP systems and the problem of flux discovery, while Section 3 presents the usage of ANNs for flux regulation function synthesis. In Section 4 and 5 we report, respectively, the new technique for discovering flux tuners and an application of this technique to a simple case study.

2   MP Systems and MP Graphs

In MP systems *reactions* transform *substances*, *flux regulation maps* establish the amount of matter transformed by each reaction at each step, and *parameters*, which are not directly involved in reactions, affect with substances flux regulation maps. We refer to [20] for a formal definition of these systems, where also a detailed motivation of the principles underlying them is given.

The main intuition of MP dynamics is the *mass partition principle*, which expresses a discrete, deterministic and molar reading of metabolic transformations, as opposite to the infinitesimal deterministic and local perspective of the mass action principle of classical differential models. For our further discussion it is useful to focus on the following simple example. Let r_1, r_2, r_3, r_4 be the following set of reactions:

$$
\begin{aligned}
&r_1 : 2a + b \to c &\qquad &r_3 : b + c \to a \\
&r_2 : b \to c &\qquad &r_4 : a \to 2b
\end{aligned}
\tag{1}
$$

We consider the substances a, b, c along the time instants $i = 0, 1, 2, \ldots$ (for the sake of simplicity here we avoid to consider parameters) and $\Delta a[i], \Delta b[i], \Delta c[i]$ are the variations of a, b, c, respectively, at time i. The quantities $u_1[i], u_2[i], u_3[i], u_4[i]$, are the number of molar units transformed by reactions r_1, r_2, r_3, r_4, respectively, in the step from time i to time $i+1$. According to reactions (1) we get the following linear system at time i:

$$
\begin{aligned}
\Delta a[i] &= -2u_1[i] + u_3[i] - u_4[i] \\
\Delta b[i] &= -u_1[i] - u_2[i] - u_3[i] + 2u_4[i] \\
\Delta c[i] &= u_1[i] + u_2[i] - u_3[i]
\end{aligned}
\tag{2}
$$

which becomes, in vector notation:

$$
\Delta X[i] = \mathbb{A} \times U[i],
\tag{3}
$$

where $((\cdot)^T$ is the transpose operator)

$$
\Delta X[i] = (\Delta a[i], \Delta b[i], \Delta c[i])^T,
$$
$$
U[i] = (u_1[i], u_2[i], u_3[i], u_4[i])^T,
$$

$$
\mathbb{A} = (\mathbb{A}(x, r) | x \in X, r \in R) = \begin{pmatrix} -2 & 0 & 1 & -1 \\ -1 & -1 & -1 & 2 \\ 1 & 1 & -1 & 0 \end{pmatrix}.
$$

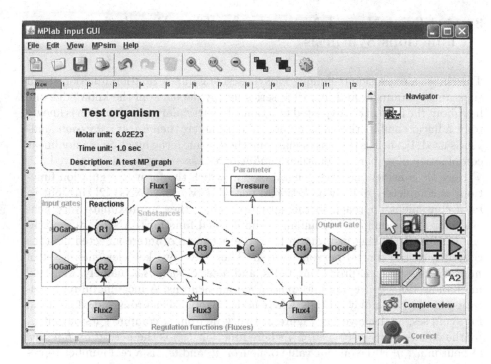

Fig. 1. An MP graph visualized by a graphical user interface of MetaPlab. Frame labels point out MP system elements in the MP graph representation. Substances, reactions and parameters describe the stoichiometry of the system, while fluxes regulate the dynamics.

The log-gain theory for MP systems [20] provides algebraic methods that, from a time-series of vectors $\Delta X[i]$, generates the time-series of $U[i]$. When $U[i]$ are known, we face the problem of discovering some functions $\varphi_1, \ldots, \varphi_m$ such that $(\varphi_1(X[i]), \varphi_2(X[i]), \ldots, \varphi_m(X[i]))^T = U[i]$. This problem of regulation maps discovery can be split into two subproblems: $i)$ selecting the variables of each φ_j, $ii)$ defining the right mathematical form of φ_j. In the following we propose some new methodologies, based on ANNs, for solving both these subproblems. As a graphical representation of MP systems we will employ bipartite graphs called *MP graphs* [22], which are shown in Figure 1). Substances, parameters, reactions and fluxes (e.g., respectively, A, *Pressure*, R_3 and $Flux_1$ in Figure 1) are depicted by different kind of nodes; stoichiometric (plain) arches connect reactant to reactions (e.g., $A \to R_3$) or reactions to products (e.g., $R_3 \to C$) and they possibly have labels denoting reaction stoichiometry, if it is different from '1' (e.g., label '2' on arch $R_3 \to C$); regulatory (dashed) arches having a black arrow link fluxes to the reaction they regulate (e.g., $Flux_1 \to R_1$); finally, regulatory (dashed) arches having a white arrow connect substances or parameters to the fluxes they regulate (e.g., $C \to Flux_1$). Notice that environment compartmentalization is not considered in the current version of the model but this feature will be topic of future work.

3   Artificial Neural Networks for Flux Regulation Functions Synthesis

The choice a regression technique for synthesizing flux regulation functions from substance, parameter and flux time-series deeply depends on the knowledge one has about the form of the expected functions. In particular, if the function is known to be a linear combination of its numerical parameters then *linear regression* analysis is used [1], such as the least squares method, while if the function is a nonlinear combination of its parameters then *nonlinear regression* analysis is employed.

Here we consider the very general case in which the form of regulation functions is completely unknown. Artificial neural networks (ANNs) [2] turn out to be a convenient approach in this situation, since they approximate very general maps just nonlinearly combining simple seed functions. ANNs are a mathematical model taking inspiration from the networks of interconnected neurons constituting the central nervous system. They have two key elements: a set of *neurons*, representing processing units, and a set of *synapses*, namely, weighted interconnections conveying information among neurons. A meaningful representation for ANNs employs graphs, where nodes symbolize neurons and edges stand for synapses, as displayed in Figure 2. Every neuron u_j computes its output y_j by the equation $y_j = f(\sum_i w_{ji} y_i)$, where function $f(\cdot)$ is the *activation function* of neuron u_j, y_i is the output value of neuron u_i and w_{ji} is a real number representing the weight of the synapse connecting u_i to u_j. Activation functions are usually nonlinear functions, such as the *logistic sigmoid*, $f(x) = \frac{1}{1+e^{-x}}$, or *tanh*, $f(x) = \frac{e^x - e^{-x}}{e^x + e^{-x}}$, but also other kind of function can be considered. A particular type of ANN we consider here are *feed-forward* neural networks, which have no feedback loops. In these networks, neurons are usually arranged in layers, where the *input-layer* receives input from the environment, the *output-layer* returns its output to the environment, and *hidden layers* process the information and pass it on through the network.

We employ ANNs for discovering flux regulation functions since they have a natural ability to represent both linear and nonlinear relationships between a set of input variables (e.g., substance and parameters) and a set of output variables (e.g., fluxes), and they are also able to learn these relationships from data sets. Moreover, it has been proved that ANNs having at least one hidden layer and sigmoid neurons are able to approximate any continuous functional mapping, if no limit is imposed on the number of hidden neurons [11].

Given an MP system with n substances, k parameters and m reactions we connect to it m neural networks, each having $n + k$ input neurons connected to substance and parameter nodes, and one output neuron linked to a specific flux node, as displayed in Figure 3. The number of hidden layers and hidden neurons should be tuned according to the complexity of the functions under investigation. As a rule of thumb, the more "complex" the regulation function, the higher the number of hidden layers and hidden neurons. If the complexity of the searched function is unknown, then different topologies should be tested, until a good approximation is found.

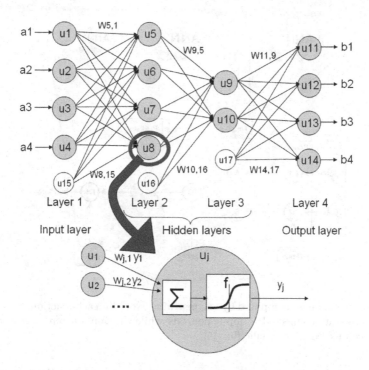

Fig. 2. A feed-forward neural network having four layers of neurons; an input layer with four input neurons and one bias neuron; two hidden layers with, respectively, four and two normal neurons, and one bias neuron; an output layer with four output neurons. Every neuron of layer i is connected to every (non-bias) neuron of layer $i+1$. Normal neurons (gray nodes) compute an activation function (usually sigmoid) of a weighted sum of their input. Bias neurons (white nodes) provide a constant unitary input.

Once the neural network topologies have been defined the information contained into a training set of observed data, has to be stored within synaptic weights. The process of weight tuning is called *training* and it is performed by the so called *learning algorithms*, namely, optimization techniques able to search for a set of weights which gives to the network a behavior defined by a set of examples, the *training set*. For MP flux regulation functions, a training set is represented by time-series of substances and parameters, generally collected by observations, and flux time-series computed by the log-gain method [18,19,20]. During the training stage, data are cyclically "observed" by neural networks which update their weight values at each training epoch (according to some learning rules) in order to minimize the square error between their outputs and the target outputs stored in the training set.

In [5] a Java software called *NeuralSynth* has been presented which trains feed-forward neural networks, within the MetaPlab suite [32], by means of four optimization algorithms, namely, *backpropagation* [2], *genetic algorithms* (GA) [13,35], *particle swarm optimization* (PSO) [14] and a *memetic algorithm* [15]. In

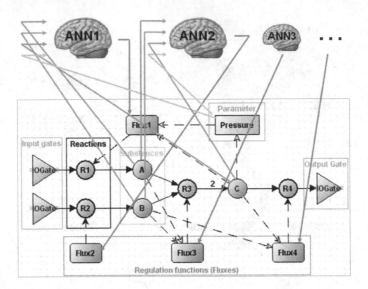

Fig. 3. MP system fluxes computed by one neural network for each reaction. Substances and parameters are connected to input neurons while the only output neuron of each network is connected to a specific flux [5].

that work the memetic algorithm has been proved to achieved the best performance in discovering regulation functions for an MP model of the mitotic cycle in early amphibian embryos.

4 Flux Tuners Discovery by Artificial Neural Networks

The problem we tackle in this section concerns the automatic discovery of flux tuners from observed data. In the following we will call *tuners* of a flux regulation function φ_i, the variables (i.e., substances and parameters) involved in the function [18]. Indeed, it is known that every reaction of a biochemical system transforms reactants into products with a rate depending on the instantaneous value of some substances and parameters. Discovering these elements provides key understanding about the system and it may suggest new experiments.

From regression theory it turns out that regulation functions should have as few independent variables as possible in order to give to MP models the best predictions capabilities [1]. This statement could sound a bit counterintuitive since it seems logical that, if a regulation function incorporates as many variables as possible, then its flux prediction should be more accurate. As a matter of fact, this is true only if the number of data points to be fitted has no limitations (which is not realistic), indeed, because of the *curse of dimensionality*, as the dimensionality of the fitting surface increases also the degrees of freedom of this surface increase, and the number of points needed to achieve a good fitting surface increases as well. Therefore, functions generated by regression methods have to be parsimonious in the number of independent variables in order to capture the systematic trend of data while avoiding uncertainty and overfitting typical of high-dimensional functions [1].

The methodology we present in the following for discovering flux tuners by means of neural networks, consists of two steps: *i)* application of the *weight elimination* technique [2,33], during the network training, for removing unnecessary synapse weights, *ii)* assignment, to each substance (parameter) of the MP system, of a *tuning index* for each flux, rating the propensity of the substance (parameter) to tune the flux itself.

4.1 Weight Elimination

Weight elimination [2,33] is a technique aiming to find a neural network which fits a specific training set by using the smallest number of weights. The hypothesis on which this method is based states that "if several networks fit the data equally well, then the network having the smallest number of weights will on average provide the best generalization", that is, it will get the best predictions for new data.

The idea is to add to the backpropagation cost function (usually a square error), a term which "counts" the number of weights, obtaining the new cost function [2]:

$$E = \sum_{k \in \mathcal{T}} (target_k - output_k)^2 + \lambda \sum_{i \in \mathcal{C}} \frac{w_i^2}{\hat{w}^2 + w_i^2}. \tag{4}$$

and then to minimize this function by means of backpropagation. The first term of Equation (4), called *performance term*, represents the square error between network output and target output over the entire training set $\mathcal{T}$. The second term, named *complexity term*, deals with the network size. Its sum, which extends over all the synapses $\mathcal{C}$, adds a penalty value close to unity (times λ) to each weight $w_i \in \mathbb{R}$ such that $|w_i| >> \hat{w}$, while it adds a penalty term approaching to zero to each weight w_i such that $|w_i| << \hat{w}$. The parameter $\lambda \in \mathbb{R}^+$ represents the relative importance of the network simplicity with respect to the network performance.

When the classical backpropagation learning algorithm is employed with the cost function of Equation (4), weights are updated at each step according to the gradient of both the performance and the complexity terms, thus a trade-off between a small fitting error and a small number of weights is found. In other words, the complexity term tends to "push" every weight to zero with a strength proportional to weight magnitudes and to λ, while the performance term keeps far from zero the weights actually needed to fit training data. Notice that, parameter λ is a sensitive factor in this procedure, since if it is too small, then the complexity term has no effect, while if it is too large then all the weights are driven to zero. Moreover, the value of λ usually changes depending on the problem. In [33] some heuristic rules are presented for dynamically tuning the value of λ during the training process in order to find a minimal network while achieving a desired level of performance on training data.

The weight-elimination technique has been implemented in the NeuralSynth plug-in [5]. The first step of our tuner discovery strategy can be performed by this software, so that, neural networks are trained on time-series data and, at the same time, their unnecessary weights are removed.

4.2   Assignment of Tuning Indexes

The second step of our strategy for tuner discovery involves the analysis of the neural networks achieved at the first step, with the aim to evaluate the sensibility of each flux to the variation of each substance and parameter. Given a trained (and minimized) neural network encoding a regulation function $\varphi(q)$, we assign to each input neuron x (which is connected to a substance or a parameter node according to the schema of Figure 3) a *tuning index*:

$$\xi(x) = \sum_{p \in path(x,o)} \prod_{w \in p} |w| \tag{5}$$

where $path(x, o)$ is the set of all paths from the input neuron x to the (only) output neuron o (connected to a flux node according to the schema of Figure 3), and each path $p \in path(x, o)$ is, in turn, the set of weights of synapses on the path from x to o. In other words, the tuning index $\xi(x)$ rates the "saliency" of the substance (parameter) connected to the input neuron x to tune the flux connected to the output neuron o. This index is computed by summing, for every path from the input neuron x to the output neuron o, the product of weights in the path. Similar techniques [16,34] have been already employed in several fields, such as in finance [8].

The idea behind this heuristic for computing tuning indexes is informally explained by means of Figure 4. In that picture, red thin arrows represent synapses having weights with small absolute values, green thick arrows stand for synapses having weights with large absolute values, and orange medium-thickness arrows represent synapses having weights with medium size absolute values. From Figure 4 it is evident that the contribution of a single path from the input neuron u_1 (related to substance A) to the output neuron u_9 (connected to flux F_1), is proportional to the product of the absolute values of weights on the path between u_1 and u_9. Moreover, the overall contribution of input A in tuning output F_1 is related to the sum of the contributions of every path. This is because each neuron computes a sigmoid function of the weighted sum of its inputs, as already described in Section 3.

Let us consider a simple example. On the left side of Figure 4, the contribution of path $u_1 \rightarrow u_5 \rightarrow u_9$, that is $|w_{5,1}| \cdot |w_{9,5}|$, is smaller than the contribution of path $u_1 \rightarrow u_6 \rightarrow u_9$, that is, $|w_{6,1}| \cdot |w_{9,6}|$, since $|w_{5,1}|$ and $|w_{9,5}|$ are smaller than $|w_{6,1}|$ and $|w_{9,6}|$. The tuning index of substance A with respect to flux F_1 is the sum $|w_{5,1}| \cdot |w_{9,5}| + |w_{6,1}| \cdot |w_{9,6}| + |w_{7,1}| \cdot |w_{9,7}|$. On the right side of the same picture it is shown that the contribution of substance B in tuning flux F_1 is almost insignificant, since the absolute values of all the weights on the paths between the input neuron u_2 (connected to B) and the output neuron u_9 have

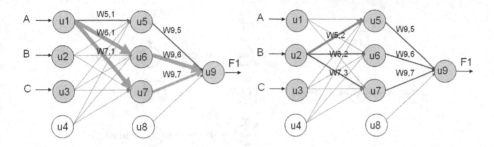

Fig. 4. Weight analysis of paths from the input neurons u_1 (on the left) and u_2 (on the right), to the output neuron u_9 for computing the tuning indexes of, respectively, substance A and B in respect of flux F_1.

small or medium sizes. Accordingly, the tuning index of substance A will be greater than the tuning index of substance B.

5   A Case Study: The Sirius Model

In this section we report some preliminary results of the application of the tuners discovery strategy explained above to a simple case study. The MP system we investigate, called Sirius, does not have any biological counterpart but its analysis is however interesting because of the oscillations it generates when specific regulation functions are employed. As displayed in Figure 5, Sirius has three substances, A, B and C, and five reactions $R_1, \ldots, R_5$. In [18] the following flux regulation functions have been manually generated:

$$F1 = \frac{k_1 a}{k_1 + k_2 c + k_4 b + k_a}$$

$$F2 = \frac{k_2 ac}{k_1 + k_2 c + k_4 b + k_a}$$

$$F3 = \frac{k_3 b}{k_3 + k_b} \tag{6}$$

$$F4 = \frac{k_4 ab}{k_1 + k_2 c + k_4 b + k_a}$$

$$F5 = \frac{k_5 c}{k_5 + k_c}$$

where $k_1 = k_3 = k_5 = 4$, $k_2 = k_4 = 0.02$, and $k_a = k_b = k_c = 100$. Notice that, functions F_1, F_2 and F_4 have the same denominator but the numerator of F_1 is characterized by tuner A, numerator of F_2 by tuners A and C, and numerator of F_4 is characterized by tuners A and B. On the other hand, functions F_3 and F_5 are characterized, respectively, by tuners B and C. The oscillatory dynamics generated by these functions, displayed in Figure 5, is featured by a very similar trend for substances B and C, which differ only in the first fifty steps.

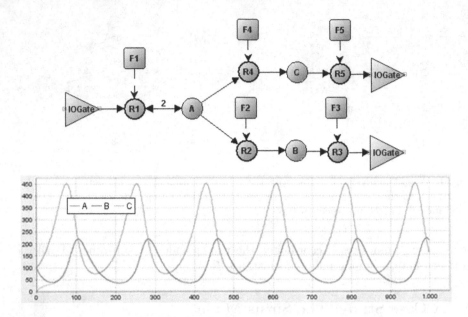

Fig. 5. On top: Sirius model. At the bottom: Sirius dynamics.

We have sampled the dynamics of Figure 5 in order to obtain three substance time-series (one for each substance), each having 1000 values, and we have computed the related five flux time-series (one for each flux) by the log-gain theory. Subsequently, these time-series have been employed to train five neural networks (one for each regulation function) by means of backpropagation with weight elimination. Specifically, substance values have been used as inputs and flux values as target outputs during the training process performed by the software Neural-Synth. We run the computation of the tuning indexes of each flux for five times and, subsequently, we have calculated the mean and the standard deviations of these indexes for each flux regulation function. The best results, reported in Table 1, have been achieved by employing $\lambda = 0.0001$ and $w_0 = 1.0$ for weight elimination and neural networks having one hidden layer with three neurons. This value of parameter w_0 tends to eliminate weights between (about) -5.0 and 5.0, which is consistent with the random initialization of neural network weights between -1.0 and 1.0. The parameter λ has been manually tuned for this case study but some heuristics [33] will be considered to dynamically tune its value during the training process. The network topology has been generated according to the complexity of the searched regulation function.

Let us analyze the results of Table 1. The first row reports the mean relative tuning indexes of flux F_1 and, in brackets, the standard deviation of the relative tuning indexes over the five tests performed. *Relative tuning indexes* are computed by normalizing tuning indexes (Equation 5) with respect to their sum. Value 0.918 in the first column, states that substance A have obtained a mean relative tuning index of 91.8% for flux F_1 over the five tests. Substances B and C,

Table 1. Mean tuning indexes and related standard deviations (in brackets) of substances A, B and C with respect to fluxes F_1, F_2, F_3, F_4, F_5. These results have been computed by performing five tests for each flux.

	A	**B**	**C**
F_1	0.918 (0.044)	0.043 (0.026)	0.038 (0.017)
F_2	0.336 (0.001)	0.301 (0.209)	0.362 (0.209)
F_3	0.018 (0.017)	0.971 (0.020)	0.010 (0.009)
F_4	0.337 (0.006)	0.525 (0.292)	0.136 (0.292)
F_5	0.020 (0.027)	0.084 (0.112)	0.895 (0.111)

respectively in the second and third columns, have achieved mean relative tuning indexes of 4.3% and 3.8%. This result agrees with the form of function F_1, by which dynamics data have been generated, indeed function F_1 is deeply related to substance A, which appears in the numerator of this function. By analyzing the third row of Table 1, concerning flux F_3, we observe that substance B, which appears in the numerator of function F_3, has achieved a mean relative tuning index of 97.1%, while substances A and C, that are not arguments of function F_3, have scored only 1.8% and 1.0%. Quite good results have been achieved for flux F_4 (in the forth row), indeed the variables appearing in its numerator, namely A and B, have scored mean relative tuning indexes of, respectively, 33.7% and 52.5% in contrast to the 13.6% scored by substance C. Flux F_5, in the last row of the table, has mean relative tuning indexes of 2.0% for A, 8.4% for B and 89.5% for C, according to the form of function F_5 which includes only substance C amomg its arguments. On the other hand, the result achieved for flux F_2 (in the second row) deserves further investigations, since the mean relative tuning indexes seem to be not enough informative. Their values (33.6 for A, 30.1 for B and 36.2 for C) are so close to each other that we cannot deduce A and C to be the only tuners for F_2 (as it appears in the numerator of function F_2). We believe that this problem can be due to the high similarity between the dynamics of substance B and C, which makes it difficult to distinguish between the two inputs. This is confirmed also by the high standard deviation values achieved for substances B and C in both fluxes F_2 and F_4, which points out a large variance in the relative tuning indexes computed over the five tests. The dynamics trend of the model obtained by this approach, displayed in [5], is very similar to the original one shown in Figure 5.

6   Conclusions and Future Work

In this paper we have presented a new technique, based on artificial neural networks, for discovering flux tuners within the framework of MP systems. This strategy involves a first training stage wherein each neural network learns a flux regulation function from observed time-series by means of backpropagation with weight elimination. Subsequently, for each flux a tuning index is associated to each substance and parameter of the MP system in order to evaluate its propensity to tune the flux. The technique has achieved encouraging results in a synthetic case study wherein data have been generated by known functions. Further work has to be done in order to get a stronger validation for real biological systems. Moreover, some heuristic techniques employed in this paper to learn neural networks (i.e., evolutionary and swarm optimization), will be directly employed for discovering flux tuners.

References

1. Aczel, A.D., Sounderpandian, J.: Complete Business Statistics. McGraw-Hill, New York (2006)
2. Bishop, C.M.: Neural Networks for Pattern Recognition. Oxford University Press, Oxford (1995)
3. Castellini, A., Franco, G., Manca, V.: Hybrid functional Petri nets as MP systems. Natural Computing 9121 (2009), doi:10.1007/s11047-009-9121-4
4. Castellini, A., Franco, G., Manca, V.: Toward a representation of hybrid functional Petri nets by MP systems. In: Suzuki, Y., et al. (eds.) Natural Computing. PICT, vol. 1, pp. 28–37. Springer, Japan (2009)
5. Castellini, A., Manca, V.: Learning regulation functions of metabolic systems by artificial neural networks. In: Proceedings of the Genetic and Evolutionary Computation Conference, GECCO 2009. ACM Publisher, New York (2009)
6. Castellini, A., Manca, V.: MetaPlab: A computational framework for metabolic P systems. In: Corne, D.W., Frisco, P., Paun, G., Rozenberg, G., Salomaa, A. (eds.) WMC 2008. LNCS, vol. 5391, pp. 157–168. Springer, Heidelberg (2009)
7. Ciobanu, G., Păun, G., Pérez-Jiménez, M.J. (eds.): Applications of Membrane Computing. Springer, Berlin (2006)
8. du Jardin, P.: Bankruptcy prediction and neural networks: the contribution of variable selection methods. In: Proceedings of ESTSP 2008, pp. 271–284 (2008)
9. Fisher, J., Henzinger, T.A.: Executable cell biology. Nature Biotechnology 25(11), 1239–1249 (2007)
10. Fontana, F., Manca, V.: Discrete solutions to differential equations by metabolic P systems. Theoretical Computer Science 372(2-3), 165–182 (2007)
11. Funahashi, K.: On the approximate realization of continuous mappings by neural networks. Neural Networks 2(3), 183–192 (1989)
12. Gillespie, D.T.: A general method for numerically simulating the stochastic time evolution of coupled chemical reactions. J. of Computational Physics 22, 403–434 (1976)
13. Holland, J.H.: Adaptation in Natural and Artificial Systems. University of Michigan Press, Ann Arbor (1975)
14. Kennedy, J., Eberhart, R.: Particle swarm optimization. In: Proc. IEEE Int. Conf. on Neural Networks, vol. 4, pp. 1942–1948 (1995)
15. Krasnogor, N., Smith, J.E.: A tutorial for competent memetic algorithms: model, taxonomy, and design issues. IEEE Trans. Evolutionary Computation 9(5), 474–488 (2005)
16. Leray, P., Gallinari, P.: Feature selection with neural networks. Behaviormetrika 26, 16–16 (1998)
17. Lindenmayer, A.: Mathematical models for cellular interactions in development I. Filaments with one-sided inputs. J. of Theoretical Biology 18(3), 280–299 (1968)

18. Manca, V.: The Metabolic algorithm: Principles and applications. Theoretical Computer Science 404, 142–157 (2008)
19. Manca, V.: Fundamentals of metabolic P systems. In: Păun, G., et al. (eds.) Handbook of Membrane Computing, ch. 16. Oxford University Press, Oxford (2009)
20. Manca, V.: Log-gain principles for metabolic P systems. In: Condon, A., et al. (eds.) Algorithmic Bioprocesses. Natural Computing Series, ch. 28. Springer, Heidelberg (2009)
21. Manca, V.: Metabolic P dynamics. In: Păun, G., et al. (eds.) Handbook of Membrane Computing, ch. 17. Oxford University Press, Oxford (2009)
22. Manca, V., Bianco, L.: Biological networks in metabolic P systems. BioSystems 91(3), 489–498 (2008)
23. Manca, V., Bianco, L., Fontana, F.: Evolutions and oscillations of P systems: Applications to biochemical phenomena. In: Mauri, G., Păun, G., Jesús Pérez-Jímenez, M., Rozenberg, G., Salomaa, A. (eds.) WMC 2004. LNCS, vol. 3365, pp. 63–84. Springer, Heidelberg (2005)
24. Manca, V., Castellini, A., Franco, G., Marchetti, L., Pagliarini, R.: Metaplab 1.1 user guide (2009), http://mplab.scienze.univr.it
25. Pérez-Jiménez, M.J., Romero-Campero, F.J.: P systems: a new computational modelling tool for systems biology. In: Priami, C., Plotkin, G. (eds.) Transactions on Computational Systems Biology VI. LNCS (LNBI), vol. 4220, pp. 176–197. Springer, Heidelberg (2006)
26. Păun, G.: Computing with membranes. Journal of Computer and System Sciences 61(1), 108–143 (2000)
27. Păun, G.: Membrane Computing. An Introduction. Springer, Berlin (2002)
28. The P Systems Web Site, http://ppage.psystems.eu/
29. Suzuki, Y., Fujiwara, Y., Takabayashi, J., Tanaka, H.: Artificial life applications of a class of P systems: Abstract rewriting systems on multisets. In: Calude, C.S., Pun, G., Rozenberg, G., Salomaa, A. (eds.) Multiset Processing. LNCS, vol. 2235, pp. 299–346. Springer, Heidelberg (2001)
30. Suzuki, Y., Tanaka, H.: Modeling p53 signaling pathways by using multiset processing. In: [7], pp. 203–214
31. Voit, E.O.: Computational Analysis of Biochemical Systems: A Practical Guide for Biochemists and Molecular Biologists. Cambridge University Press, Cambridge (2000)
32. MetaPlab website, http://mplab.scienze.univr.it
33. Weigend, A.S., Rumelhart, D.E., Huberman, B.A.: Generalization by weight-elimination with application to forecasting. In: Lippmann, R., et al. (eds.) NIPS, pp. 875–882. Morgan Kaufmann, San Francisco (1990)
34. Yacoub, M., Bennani, Y.: HVS: A heuristic for variable selection in multilayer artificial neural network classifier. In: Proc. of ANNIE 1997, pp. 527–532 (1997)
35. Yao, X.: Evolving artificial neural networks. Proceedings of the IEEE 87(9), 1423–1447 (1999)

A Novel Variant of P Systems for the Modelling and Simulation of Biochemical Systems

Paolo Cazzaniga[1], Giancarlo Mauri[1], Luciano Milanesi[2],
Ettore Mosca[2], and Dario Pescini[1]

[1] Università degli Studi di Milano-Bicocca
Dipartimento di Informatica, Sistemistica e Comunicazione
Viale Sarca 336, 20126 Milano, Italy
{cazzaniga,mauri,pescini}@disco.unimib.it
[2] Consiglio Nazionale Ricerche, Istituto Tecnologie Biomediche
Via Fratelli Cervi 93, 20090 Segrate (MI), Italy
{ettore.mosca,luciano.milanesi}@itb.cnr.it

Abstract. In the last decade, different computing paradigms and modelling frameworks for the description and simulation of biochemical systems have been proposed. Here, we consider membrane systems, in particular, tissue P systems and τ-DPP, for the development of a novel variant of membrane systems with sizes associated to the volumes involved in the structure and to the molecular species occurring inside the system. Moreover, this variant allows the communication of objects among non adjacent membranes arranged in a hybrid structure, that is, organised in a tissue-like fashion where nodes can have a complex internal structure. The features presented in the new variant of P systems can be used to describe, among others, reaction-diffusion systems, where molecules are involved both in chemical reactions and diffusive processes, and their movements depend on the free space of the volumes; or systems where exist privileged pathways between membranes, which are inspired by the role of microtubule in protein transport within the intracellular space.

1 Introduction

Membrane systems [17], also known as P systems, are one of the computation models inspired by the structure and the functioning of living cells presented in the recent years. The basic model consists of a hierarchical structure composed by several membranes, embedded into a main membrane called the *skin*. Membranes divide the space into *regions*, that contain *objects* (represented by symbols over an alphabet) and *evolution rules*.

The current variants of P systems used in the modelling of biochemical systems provide a description where membranes can contain up to an infinite number of molecules because the sizes of the structure components and of the objects involved are not considered. Moreover, the communication channels are limited to adjacent membranes. In particular, in the framework of tree-like P systems, the communication is permitted from/to a membrane to/from another

G. Păun et al. (Eds.): WMC 2009, LNCS 5957, pp. 210–226, 2010.

one immediately inside or outside the first one. On the other hand, working with tissue P systems (or tP systems), communication of objects is achieved using the "synapses" defined among nodes (recently, different variants on P systems have been introduced to better model economic or socio-economic processes, databases, operating systems, etc., in which different communication strategies are used, like *Hyperdag P systems* [16], or variants like that presented in [19]). In addition, either variant of P systems use only a tree-like, or a tissue-like structure, while hybrid structures are not considered. For instance, the description of tissues where nodes have a complex internal structure or tree-like systems with membranes enclosing a tissue, have never been defined.

In this paper, we present a novel variant of P systems where we exploit tP systems [15] to describe the topological organisation of the membranes and to denote the possible communication channels of the system. Furthermore, for the description of the dynamics, we consider τ-DPP, presented in [7]. Within the framework of τ-DPP, the probabilities are associated to the rules, following the method introduced by Gillespie in [10]. In particular, τ-DPP extends the tau-leaping procedure [5] in order to quantitatively simulate the behaviour of multi-volume biological and chemical systems.

Starting from the structure of tP systems and the description of the dynamics provided by τ-DPP, we introduce a variant of tP systems, called $S\tau$-DPP, in which we associate a measure to membranes and objects, representing respectively, the "size" of the volume where the computation occurs and the volume occupied by objects. Both the size of membranes and objects are useful to describe any real system where it is important to avoid the infinite accumulation of objects inside the system membranes, which is very important in chemical system and cannot be achieved by simply bounding the "capacity" of the membranes or limiting the maximum number of each kind object allowed inside a particular volume, since each object can have a different size.

In the framework of $S\tau$-DPP, the system's structure is independent from the communication channels between membranes, hence, two different graphs are used in the description: the first one denotes the membrane structure, while the second graph specifies the connections between membranes which allow the communication of objects. Furthermore, the membrane structure can be hybrid (since it combines tree-like P systems with tissue P systems), and the communication can be performed between non adjacent membranes, to denote privileged pathways between membranes. This characteristic of $S\tau$-DPP takes inspiration from a specific component of living cells, called *microtubule* [24], with the aim of reproducing its role as intracellular "highway" for the transport of other cellular components, such as vesicles and proteins.

The paper is organised as follows: in Section 2 we recall the basic notions of P systems, tP systems and τ-DPP variants; in Section 3 we introduce the $S\tau$-DPP variant; in Section 4 we present a test case: a biochemical system with preferential communication pathways; finally in Section 5 we discuss the modelling power of our variant of P systems and we conclude with some possible developments for this work.

2   Membrane Systems

In this section we describe the framework of membrane systems [18], recalling their basic notions and definitions. We then present tissue P systems, a variant consisting of a set of several cells connected through protein channels [15]. Finally, we describe τ-DPP, a computational method introduced in [7], used to describe and perform stochastic simulations of multi-volume biological and chemical systems.

2.1   Basic Notions of P Systems

P systems, or membrane systems, have been introduced in [17] as a class of unconventional computing devices of distributed, parallel and nondeterministic type, inspired by the compartmental structure and the functioning of living cells.

In order to define a basic P system, three main parts need to be introduced: the *membrane structure*, the *objects* and the *rules*.

The *membrane structure* defines the topological and hierarchical organisation of a system consisting of distinct compartments. The definition of membrane structure is given through a set of membranes with a distinct label (usually numbers), hierarchically organised inside a unique membrane, named *skin membrane*. Among others, a representation of a membrane structure is given by using a string of square parentheses.

In particular, each membrane identifies a *region*, delimited by the membrane itself and any other adjacent membrane possibly present inside it. The number of membranes in a membrane structure is called the *degree* of the P system. The whole space outside the skin membrane is called the *environment*.

The internal state of a P system is described by the *objects* (represented by symbols taken from an alphabet V) occurring inside the membranes. In order to denote the presence of multiple copies of the same object inside a membrane, multisets are usually used.

The objects inside the membranes of a P system are transformed by means of *evolution rules*. These are multiset rewriting rules of the form $r_i : u \to v$, where u and v are multisets of objects. The meaning of the generic rule i is that the multiset u is modified into the multiset v.

Moreover, it is possible to associate a target to v, representing the membrane where the multiset v is placed when the rule is applied. There are three different types of target. If the target is *here*, then the object remains in the region where the rule is executed (usually, this target label is omitted in the systems description). If the target is *out*, then the object is sent out from the membrane containing the rule and placed to the outer region (the environment in the case of skin membrane). Finally, if the target is in_j, where j is a label of a membrane, then the object is sent into the membrane labelled with j. It is possible to apply this kind of rule, only if the membrane j is placed immediately inside the membrane where the rule is executed.

Starting from an initial configuration (described by a membrane structure containing a certain number of objects and a fixed set of rules), and letting the system evolve, a computation is obtained. A universal clock is assumed to exist: at each step, all rules in all regions are simultaneously applied to all objects which can be the subjects of evolution rules. So doing, the rules are applied in a maximal parallel manner, hence the membranes evolve simultaneously. If no further rule can be applied, the computation halts. The result of a computation is the multiset of objects contained into a previously specified *output membrane* or the environment.

For a complete and extensive overview of P systems, we refer the reader to [18], and to the P Systems Web Page (http://ppage.psystems.eu).

2.2 tP Systems

The basic definition of P systems consists of a membrane structure organised in a tree-like structure. In [15], *tP systems* have been defined to describe a tissue-like architecture, where cells are placed in the nodes of a (directed) graph, and objects are communicated along the edges of the graph. These communication channels are called synapses. Moreover, the communication of objects can be achieved both in a replicative and non-replicative manner, that is, the objects are sent to all the adjacent cells or to only one adjacent cell, respectively.

In general, the structure of a tP system is composed by elementary membranes, namely, each node of the system is represented by an elementary membrane. Furthermore, the communication of objects is allowed, as in standard P systems, only to/from adjacent membranes.

Tissue P systems have been further elaborated, for example in [9] and [20], with recent results about both theoretical properties [1] and applications [13]. The variants of tP systems considered in the literature essentially differ in the mechanisms used to communicate objects between cells. For instance, particular sets of communication rules (i.e., symport and antiport rules) can be assigned to the edges of the graph that defines the structure of the tissue, in order to model the existence of communication channels among the cells [12,9].

Alternatively, there are evolution-communication tP systems (adopting the terminology introduced in [6]), where the objects produced by particular transformations occurring inside the cells are nondeterministically propagated from one place to another one [14,2].

2.3 τ-DPP

We recall now the basic definition of the stochastic simulation technique called τ-DPP [7], where the probabilities are associated to the rules, following the method introduced by Gillespie in [10]. The aim of τ-DPP is to extend the single-volume algorithm of tau-leaping [5], in order to simulate multi-volume systems, where the distinct volumes are arranged according to a specified hierarchy. The structure of the system is required to be kept fixed during the evolution (this requirement is satisfied by the variants of membrane system we consider here). Hence, the spatial arrangement of P system is exploited in the τ-DPP description. In particular, τ-DPP has been defined starting from a variant of P systems called dynamical probabilistic P systems (DPP). DPP, presented in [23], exploit the membrane structure of P systems and associate probabilities with the rules, such values vary (dynamically), according to a prescribed strategy, during the evolution of the system. They have been introduced to take into account the stochasticity of the modelled systems and to probe different levels of parallelism of the rules executions. For the formal definitions of DPP and examples of simulated systems, we refer the reader to [22,21,4,3].

There is a difference between these two membrane systems variants: DPP provides only a qualitative description of the analysed system, that is, "time" is not associated to the evolution steps, while τ-DPP is able to give a quantitative description tracing the time-stream of the evolution.

The τ-DPP approach is designed to share a common time increment among all membranes, used to extract the rules that will be executed in each compartment (at each step). This improvement is achieved using, inside the membranes of τ-DPP, a modified tau-leaping algorithm, which gives the possibility to simulate the time evolution of every volume as well as that of the entire system.

The internal behaviour of the membranes is therefore described by means of a modified tau-leaping procedure. The original method, first introduced in [11], is based on the stochastic simulation algorithm (SSA) presented in [10]. These approaches are used to describe the behaviour of chemical systems, computing the probabilities of the reactions placed inside the system and the length of the step (at each iteration), according to the current system state. While SSA is proved to be equivalent to the Chemical Master Equation (CME), therefore it provides the exact behaviour of the system, the tau-leaping method describes an approximated behaviour with respect to the CME, but it is faster for what concerns the computational time required.

To describe the correct behaviour of the whole system, all the volumes evolve in parallel, through a strategy used to compute the probabilities of the rules (and then, to select the rules that will be executed), and to choose the "common" time increment that will be used to update the system state. The method applied for the selection of the time step length is the following. Each membrane independently computes a candidate time increment (exploiting the tau-leaping procedure), based on its internal state. The smallest time increment among all membranes is then selected and used to describe the evolution of the whole system, during the current iteration. Since all volumes *locally* evolve according to the same time increment, τ-DPP is able to correctly work out the *global* dynamics of the system. Moreover, using the "common" time increment inside the membranes, it is possible to manage the communication of objects among them. This is achieved because the volumes are naturally *synchronised* at the end of each iterative step, when all the rules are executed.

3   Sτ-DPP

Sτ-DPP is a membrane system variant based on the structure definition of basic tP systems, and on the dynamics description of τ-DPP. Sτ-DPP has some additional features: nodes (arranged in a tissue-like fashion) can have a complex structure hierarchically organised in a tree-like structure; moreover, a measure denoting the size of membranes and objects is considered, and the rules defined inside each membrane will be enabled only in the case there is sufficient space, for instance, to "create" new objects or to receive objects from other membranes. The sizes considered here can be used in the modelling of biochemical systems where diffusive processes play an important role in the system dynamics and it is important to avoid the unlimited accumulation of objects in a region of finite size.

In order to correctly describe the hierarchy of complex nodes of the system we first need a directed graph representing the topology of the membranes. In particular, undirected edges indicate that the two membranes are placed on the same level (as in the first definition of tP systems). On the other hand, directed edges denote that the target membrane is contained inside the source membrane.

Another directed graph is needed to represent the communication channels among the membranes. Clearly, the arrows of the edges indicate the direction of the (permitted) flow of objects among membranes. Note that, the communication graph can contain edges which are not indicated inside the structure graph. The meaning of these particular edges is to represent communication channels that connect non adjacent membranes. Thanks to these arcs it is possible to create privileged pathways of communication between membranes.

The features proposed in the framework of S_T-DPP can be exploited to represent (among the other real life systems) reaction-diffusion systems [8], mathematical models which capture the dynamics of a set of substances involved in a number of chemical reactions, considering both the temporal and spatial dimensions. In this case, the membrane structure can be used to represent a reaction volume as a sum of a number of finite size subvolumes and the communication graph will describe the diffusion among the considered regions.

In what follows, we will refer to membranes of volumes without distinction.

3.1   Definition

A tP systems with dimension associated with objects and membranes is defined as

$$\Pi = (\mathcal{V}, \mathcal{T}_G, \mathcal{C}_G, \mathcal{S}, \mathcal{M}, \mathcal{R}, \mathcal{C}, \mathcal{D}_X, \mathcal{D}_V), \text{ where:}$$

- $\mathcal{V} = \{V_0, \dots, V_N\}$ is the set of the volumes V_i of the system, $N \in \mathbb{N}$;
- $\mathcal{T}_G = (\mathcal{V}, A_T)$ is a graph representing the topological arrangement of the volumes in $\mathcal{V}$ and $A_T = A_T^u \cup A_T^d$ is the set of the arcs (V_l, V_k) which describes the arrangement of volumes. In particular, A_T^u is the subset of undirected arcs representing the connections between membranes placed at the same level in the membrane structure. On the contrary, A_T^d is the set of directed arcs (V_l, V_k) (where $V_k \in V_l$) which specifies the inclusion relations between volumes. We also define the set of the volumes enclosed in V_i as $a_T(V_i) = \{V_l \mid V_l \in \mathcal{V}, (V_i, V_l) \in A_T^d\}$;

– $C_G = (V, A_C)$ is a directed graph representing the connections (communication channels) among the volumes in V. A_C is the set of directed arcs (V_l, V_k) (from V_l to V_k) which denote the presence of communication channels between volumes and specify in which directions the flow of objects is allowed;

– $S = \{X_1, \ldots, X_M\}$ is the set of molecular species, $M \in \mathbb{N}$, that is, the alphabet of the system;

– $\mathcal{M} = \{M_0, \ldots, M_N\}$, is the set of the multisets occurring inside the membranes $V_0, \ldots, V_N$, representing the internal state of the volumes. The multiset M_i ($0 \leq i \leq N$) is defined over S^*;

– $\mathcal{R} = \{R_0, \ldots, R_N\}$ is the set of the sets of rules defined in volumes $V_0, \ldots, V_N$, respectively. A rule can be of internal or of communication type (as described below);

– $\mathcal{C} = \{C_0, \ldots, C_N\}$ is the set of the sets of stochastic constants associated to the rules defined in volumes $V_0, \ldots, V_N$.

– $\mathcal{D}_X = \{D_{X_1}, \ldots, D_{X_M}\}$, with $D_{X_j} \in \mathbb{R}^+$, is the set of the sizes of the molecular species $X_1, \ldots, X_M$, respectively.

– $\mathcal{D}_V = \{D_{V_0}, \ldots, D_{V_N}\}$, with $D_{V_i} \in \mathbb{R}^+$, is the set of the sizes of the volume $V_0, \ldots, V_N$, respectively.

The multiset M_i, describing the state of volume V_i ($i = 0, \ldots, N$), is defined as $M_i = (m_0, \ldots, m_M)$ where m_j denotes the number of molecules of the species X_j occurring inside V_i ($j = 0, \ldots, M$).

Given the internal state M_i of a membrane V_i together with the species volumes in $\mathcal{D}_X$, it is possible to define the *occupied volume* in V_i as:

$$O(V_i) = \sum_{j=1}^{M}(m_j \cdot D_{X_j}) + \sum_{V_l \in a_T(V_i)} D_{V_l} \tag{1}$$

Hence, it is possible to define the value of the *free space* in V_i as:

$$F(V_i) = D_{V_i} - O(V_i) \tag{2}$$

Note that, at each rule execution, the free space value has to be updated as $F(V_i) = F(V_i) - \sum_{j=1}^{M}(\beta_j - \alpha_j) \cdot D_{X_j}$, where α_j and β_j are the stoichiometric coefficients of the chemical species occurring in the executed rule.

The sets $R_0, \ldots, R_N$ define the rules occurring inside the membranes of the system. There are two different kind of rules which can be defined inside the volumes V_i: internal and communication rules. Internal rules are used to modify (evolve) the objects involved in their left-hand sides; communication rules send to other membranes the objects occurring in their left-hand sides without modifying them.

Internal rules have the general form $\alpha_1 X_1 + \alpha_2 X_2 + \cdots + \alpha_M X_M \to \beta_1 X_1 + \beta_2 X_2 + \cdots + \beta_M X_M$. Moreover, an internal rule is enabled inside V_i if $F(V_i) - \sum_{j=1}^{M}(\beta_j - \alpha_j) \cdot D_{X_j} \geq 0$. On the contrary, a communication rule, having the general form $\alpha_1 X_1 + \alpha_2 X_2 + \cdots + \alpha_M X_M \to (\beta_{1,1} X_1 + \cdots + \beta_{M,1} X_M, tgt_1) + (\beta_{1,2} X_1 + \cdots + \beta_{M,2} X_M, tgt_2) + \cdots + (\beta_{1,N} X_1 + \cdots + \beta_{M,N} X_M, tgt_N)$ is enabled inside membrane V_i if, for each volume V_{tgt_k}, $F(V_{tgt_k}) - \sum_{j=1}^{M} \beta_{j,k} D_{X_j} \geq 0$. Note that, communication rules send objects to target volumes which are always different from the source volume.

The sets of stochastic constants $C_0, \ldots, C_N$, associated to the sets of rules $R_0, \ldots, R_N$, are needed to compute the probabilities of the rule applications (also called propensity functions), along with a combinatorial function depending on the left-hand side of the rule [10].

In order to obtain a correct description of the system dynamics, we need to check if a rule r_μ (internal or communicating) is applicable. Therefore, we need to compute the effect of a rule on the free space of the volume affected by the rule. It is clear that a rule can be executed only if the free space of the volume, after the rule application, is greater or equal to zero. The rule applicability is computed differently for internal and communication rules. Given an internal rule occurring inside volume V_i, we need to check if:

$$F(V_i) - \sum_{j=1}^{M} (\beta_j - \alpha_j) \cdot D_{X_j} \geq 0$$

For what concerns a communication rule r_μ, we need to check the free space of the targets indicated by the rule:

$$\forall \ tgt_l \ \text{of} \ r_\mu, \ F(V_{tgt_l}) - \sum_{j=1}^{M} (\beta_j \cdot D_{X_j}) \geq 0$$

where the values β_j are the stoichiometric coefficients of the molecular species associated with V_{tgt_l}.

Note that, using a modified version of the tau-leaping algorithm to describe the behaviour of the system, at each iteration step, a number of rules is applied in parallel. Hence, the applicability of the parallel execution of the rules has to be verified in order to update the state of the system.

3.2   The Algorithm

We now describe the algorithm used to simulate the evolution of the entire system. Each step is executed *independently* and *in parallel* within each volume V_i ($i = 0, \ldots, N$) of the system. In the following description, the algorithm execution naturally proceeds according to the order of instructions, when not otherwise specified by means of "go to" commands.

Step 1. Initialisation: load the description of volume V_i, which consists in the initial quantities of all object types, the set of rules and their respective stochastic constants, the volume and the objects dimensions.

Step 2. Compute the initial free space of the volume V_i using Equation 2.

Step 3. Compute the propensity function a_μ of each rule $r_\mu \in R_i$, where $\mu = 1, \ldots, l$, and evaluate the sum of all the propensity functions in V_i, $a_0 = \sum_{\mu=1}^{l} a_\mu$. If $a_0 = 0$, then go to *step 4*, otherwise go to *step 6*.

Step 4. Set τ_i, the length of the step increment in volume V_i, to ∞.

Step 5. Wait for the communication of the smallest time increment $\tau_{min} = \min\{\tau_0, \ldots, \tau_N\}$ among those generated independently inside all volumes $V_0, \ldots, V_N$, during the current iteration, then **go to** *step 14*.

Step 6. Generate the step size τ_i according to the internal state, and select the way to proceed in the current iteration (i.e. SSA-like evolution, tau-leaping evolution with non-critical reactions only, or tau-leaping evolution with non-critical reactions and one critical reaction), using the selection procedure defined in [5].

Step 7. Wait for the communication of the smallest time increment $\tau_{min} = \min\{\tau_0, \ldots, \tau_N\}$ among those generated independently inside all volumes, during the current iteration.

Step 8. According to the evolution strategy of the current iteration:
- if the evolution is SSA-like and the value $\tau_i = \tau_{SSA}$ generated inside the volume is greater than τ_{min}, then **go to** *step 9*;
- if the evolution is SSA-like and $\tau_i = \tau_{SSA}$ is equal to τ_{min}, then **go to** *step 12*;
- if the evolution is tau-leaping with non-critical reactions plus one critical reaction, and $\tau_i = \tau_{nc1c}$ is equal to τ_{min}, then **go to** *step 13*;
- if the evolution is tau-leaping with non-critical reactions plus one critical reaction and $\tau_i = \tau_{nc1c}$ is greater than τ_{min}, then **go to** *step 14*;
- if the evolution is tau-leaping with non-critical reactions only ($\tau_i = \tau_{nc}$), then **go to** *step 14*.

Step 9. Compute $\tau_{SSA} = \tau_{SSA} - \tau_{min}$.

Step 10. Wait for possible communication of objects from other volumes, by means of communication rules. If some object is received, then **go to** *step 16*, otherwise **go to** *step 11*.

Step 11. Set $\tau_i = \tau_{SSA}$ for the next iteration, then **go to** *step 7*.

Step 12. Using the SSA strategy [10], extract the rule that will be applied in the current iteration, then **go to** *step 15*.

Step 13. Extract the critical rule that will be applied in the current iteration.

Step 14. Extract the set of non-critical rules that will be applied in the current iteration.

Step 15. Check if the execution of the selected rules (considering all the volumes) leads to an unfeasible state, namely, there are negative amounts of molecules, or if there is not enough space either inside the volume V_i (for internal rules) or inside the target volumes (for communication rules). If one of these conditions is satisfied, reduce τ_{min} by half and send the new value to the other membranes, then **go to** *step 8*.

Step 16. If a new value of τ_{min} reduced by half is received, then **go to** *step 8*, otherwise **go to** *step 17*.

Step 17. Update the internal state by applying the extracted rules (both internal and communication) to modify the current number of objects, then check for objects (possibly) received from the other volumes, and finally update the value of the free space $F(V_i)$.

Step 18. If the termination criteria is satisfied, then finish, otherwise **go to** *step 3*.

The algorithm described above is based on the τ-DPP procedure presented in [7], this new version is obtained by considering the size of the objects and membranes and checking if the execution of the selected rules leads to unfeasible states of the system. The other features introduced in Sτ-DPP regarding the possibility to send objects to non-adjacent membranes and the definition of hybrid membrane structure do not affect the simulation procedure, since they are implicitly considered in the algorithm.

The algorithm begins by loading the initial conditions of the membrane. The next operation consists in the calculation of the free space of the volume and in the computation of the propensity functions (and their sum a_0) in order to check if, inside the membrane, it is possible to execute some reaction. If the sum of the propensity functions is zero, then the value of τ is set to ∞ and the membrane waits for the communication of the smallest τ computed among the other membranes (τ_{min}) in order to synchronise with them; then, it checks if it is the target of some communication rule applied inside the other volumes. These operations are needed in order to properly update the internal state of the membrane.

On the other hand, if the sum of the propensity functions is greater than zero, the membrane will compute a τ value based only on its internal state, following the first part of the original tau-leaping procedure [5]. Besides this operation, the membrane selects the kind of evolution for the current iteration (like the computation of τ, this procedure is executed independently from the other volumes).

The algorithm proceeds to *step 7*, where the membrane receives the smallest τ value computed by the volumes. This will be the common value used to update the state of the entire system. It is necessary to proceed inside every membrane using the same time increment, in order to manage the communication of objects.

At this stage, the membrane knows the length of the time step and the kind of evolution to perform. The next step consists in the extraction of the rules that will be applied in the current iteration. In order to properly extract the rules, several conditions need to be checked.

In the case the membrane is evolving using the SSA strategy: if τ_{min} is the value generated inside itself, then it is possible to extract the rule, otherwise the execution of the rule is not allowed, because the step is "too short". In the next stage, the membrane verifies for possible incoming objects, to update its internal state according to the communication rules (possibly) executed inside other regions. Finally, if its state is changed (according to some internal or communication rule), then the membrane, in the successive iteration, will compute a new value of τ. On the contrary, the value of the time increment will be the result of the application of *step 9*.

If the evolution strategy corresponds to a tau-leaping step with the application of a set of non-critical reactions and one critical reaction, the algorithm verifies if the value of τ computed by the membrane is equal to τ_{min}. If this is true, the membrane selects the set of non-critical reactions to execute as well as the critical reaction. The execution of the critical reaction is allowed because, here τ_{min}

represents the time needed to execute it. Otherwise, the application of the critical reaction is forbidden and the membrane will execute non-critical reactions only.

If the membrane is following the tau-leaping strategy with the execution of non-critical reactions only, τ_{min} is used to extract the rules (from the set of non-critical) to apply in the current iteration.

In the next step, the algorithm checks if the execution of the rules selected inside all volumes of the system leads to negative amounts of the molecular quantities or if the entire set of rules is enabled, that is, the effects of the rules application result in positive values of the free space of each volume. If these conditions are not satisfied, then the set of selected rules cannot be executed, therefore, the value of τ is reduced by half and the algorithm goes back to *step 8* in order to select a new (possibly smaller) set of rules. On the contrary, if the conditions on the set of rules are satisfied, then the system can be updated. Here, every membrane executes the selected rules and updates its state and free space according to both internal and communication rules. This step is executed in parallel inside every membrane, therefore it is possible to correctly manage the "passage" of objects and to synchronise the volumes.

The last step checks if the termination criterion is satisfied in order to stop the simulation. Here, conditions for the termination of the execution are related to the time of the simulation, to the number of iteration executed or to the absence of free space.

4 A Test Case for Sτ-DPP

In this section we present a system with preferential communication pathways. In particular, we define a model in which it is present a so called *microtubule* [24]. The microtubule is a sort of intracellular "highway" for the transport of other cellular components, such as vesicles and proteins. In order to define a preferential pathway, we exploit the communication between non adjacent membranes to move objects in particular directions. We consider the membrane system Υ represented in Figure 1, where:

- $\mathcal{V} = \{V_0, \ldots, V_7\}$;
- $\mathcal{T}_G = (\mathcal{V}, A_{\mathcal{T}})$, where $A_{\mathcal{T}}^u = \{(V_2, V_3), (V_4, V_5), (V_6, V_7)\}$, and $A_{\mathcal{T}}^d = \{(V_0, V_1), (V_1, V_2), (V_1, V_3), (V_3, V_4), (V_3, V_5), (V_5, V_6), (V_5, V_7)\}$;
- $\mathcal{C}_G = (\mathcal{V}, A_C)$, $A_C = \{(V_0, V_1), (V_1, V_0), (V_1, V_2), (V_1, V_3), (V_3, V_1), (V_3, V_2), (V_3, V_4), (V_3, V_5), (V_5, V_3), (V_5, V_4), (V_5, V_6), (V_5, V_7), (V_7, V_5), (V_6, V_7)\}$;
- $\mathcal{S} = \{X_1, X_2\}$;
- $\mathcal{M} = \{M_0, \ldots, M_7\}$, $M_0 = \{X_1^{10^5}, X_2^{10^5}\}$, $M_3 = M_4 = \ldots = M_7 = \emptyset$;
- $\mathcal{R} = \{R_0, \ldots, R_7\}$, $R_0 = \{r_{0,0}, r_{0,1}\}$, $R_1 = \{r_{1,0}, \ldots, r_{1,4}\}$, $R_2 = \{r_{2,0}\}$, $R_3 = \{r_{3,0}, \ldots, r_{3,2}, \ldots, r_{3,4}\}$, $R_4 = \{r_{4,0}\}$, $R_5 = \{r_{5,0}, \ldots, r_{5,4}\}$, $R_6 = \{r_{6,0}\}$, $R_7 = \{r_{7,0}, \ldots, r_{7,4}\}$;
- $\mathcal{C} = \{C_0, \ldots, C_7\}$, where the value of all the stochastic constants is set to 1;
- $\mathcal{D}_X = \{1, 1\}$;
- $\mathcal{D}_V = \{10^6, 11 \cdot 10^4, 10^4, 10^6, 8 \cdot 10^4, 5 \cdot 10^4, 10^4, 2 \cdot 10^4\}$.

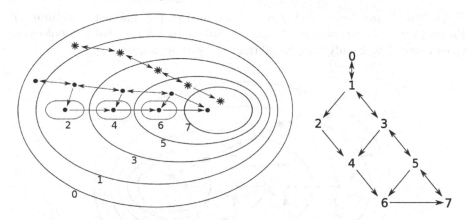

Fig. 1. The membrane system Υ with communication channels between non adjacent membranes, in the graphical representation (left side), the molecular species X_1 is indicated as a dot, while X_2 is denoted as a star, the arrows indicate the allowed flow of information of the two molecular species. On the right side, the communication graph $\mathcal{C}_G$, is reported.

Υ represents a simplified version of a "cell" in which the "movement" of molecules X_1 and X_2 from the "extracellular space" (membrane V_0) to the "nucleus" (membrane V_7), passing through nested regions of the "cytoplasm", (membranes V_1, V_3, V_5), or through a "microtubule" (membranes V_2, V_4, V_6), is described. The rules listed in Table 1 describe the diffusive processes (through the membranes of the system) related to X_1 and X_2. Note that, only X_1 molecules can "enter" into the microtubule regions, and then, they can move only towards membrane V_7 (the nucleus). On the contrary, in the other regions, that is, outside the microtubule, the diffusion is enabled in every direction for both molecules X_1 and X_2.

Hereafter we report the results obtained by simulating the membrane system Υ. In the first set of simulations we compared the behaviour of the system as

Table 1. Rules of the membrane system Υ. The stochastic constants associated to the rules are all set to 1.

Reaction	Reaction
$r_{0,0} : X_1 \rightarrow (X_1, 1)$	$r_{3,2} : X_1 \rightarrow (X_1, 4)$
$r_{0,1} : X_2 \rightarrow (X_2, 1)$	$r_{3,3} : X_1 \rightarrow (X_1, 5)$
$r_{1,0} : X_1 \rightarrow (X_1, 0)$	$r_{3,4} : X_2 \rightarrow (X_2, 5)$
$r_{1,1} : X_2 \rightarrow (X_2, 0)$	$r_{4,0} : X_1 \rightarrow (X_1, 6)$
$r_{1,2} : X_1 \rightarrow (X_1, 2)$	$r_{5,1} : X_1 \rightarrow (X_1, 3)$
$r_{1,3} : X_1 \rightarrow (X_1, 3)$	$r_{5,2} : X_2 \rightarrow (X_2, 3)$
$r_{1,4} : X_2 \rightarrow (X_2, 3)$	$r_{5,3} : X_1 \rightarrow (X_1, 6)$
$r_{2,0} : X_1 \rightarrow (X_1, 4)$	$r_{5,4} : X_1 \rightarrow (X_1, 7)$
$r_{3,0} : X_1 \rightarrow (X_1, 1)$	$r_{5_5} : X_2 \rightarrow (X_2, 7)$
$r_{3,1} : X_2 \rightarrow (X_2, 1)$	$r_{6,0} : X_1 \rightarrow (X_1, 7)$

reported in Figure 1 with the behaviour obtained with a different configuration (Figure 2), namely, we "sealed" the microtubule in such a way that the molecules X_1 can enter into it only from membrane V_1 (into membrane V_2).

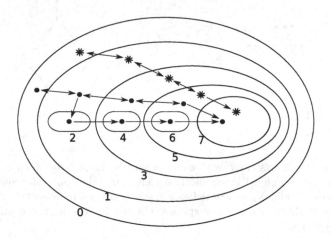

Fig. 2. The membrane system Υ with "sealed" microtubule. This configuration allows the movement of molecules X_1 into the microtubule only from membrane V_1.

In Figure 3, the dynamics of membrane V_0 and V_7 is shown. As expected, the diffusion of molecules X_2 (solid line) is slower than that of molecules X_1 in both configurations: with the microtubule accessible from any membrane (dots) or with the "sealed" microtubule (dashed line).

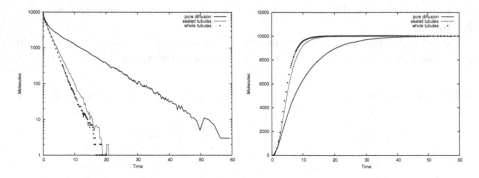

Fig. 3. Dynamics of membrane V_0 (left) and V_7 (right). In both graphs, the dynamics of X_2 (solid line) and X_1 with the completely open microtubule (dots) and the sealed microtubule (dashed line), is reported.

Figure 4 shows the dynamics of membranes V_1, V_2, V_3 and V_6 of the system Υ. Inside membrane V_1 and V_3 (top left and right of Figure 4) there is a greater amount of X_2 with respect to molecules X_1, since the latter can diffuse towards membrane V_7 passing through the microtubule. On the other hand, inside the membranes which compose the microtubule, the dynamics obtained with the two different configurations are comparable for what concerns membrane V_2; in the other membranes, and in the configuration with sealed microtubule, the quantity of X_1 is lower, since this kind of molecules cannot enter into the microtubule from any membrane of the cytoplasm and they therefore move through membranes V_3 and V_5, resulting in a slower diffusion towards membrane V_7.

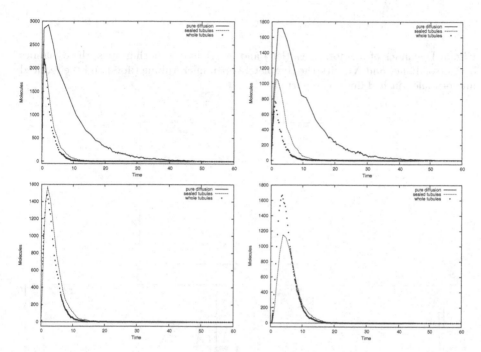

Fig. 4. Dynamics of membrane V_1 and V_3 (top left and right) and membrane V_2 and V_6 (bottom left and right). In the graphs, the dynamics of X_2 (solid line) and X_1 with the completely open microtubule (dots) and the sealed microtubule (dashed line), is reported. Clearly, the quantity of X_2 inside membranes V_2 and V_6 is always zero because this molecular species cannot enter into the microtubule.

In the second set of simulations, we considered a configuration where the size of one membrane of the microtubule (namely, membrane V_4) is much more smaller than the size used in the initial configuration. In particular the size of membrane V_4 has been modified from $8 \cdot 10^4$ to 100.

The results of the simulations show that the diffusion towards the nucleus V_7 of the cell is slower in the configuration with the bottleneck represented by the reduced size of membrane V_4 (Figure 5).

Figure 6 shows the dynamics of the membranes representing the microtubule: V_2 (top), V_4 (bottom left) and V_6 (bottom right). As expected, the reduced size

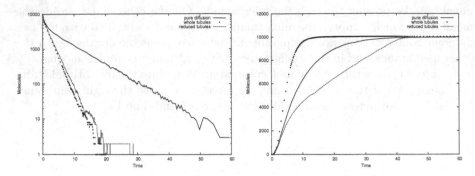

Fig. 5. Dynamics of membrane V_0 (left) and V_7 (right). In both graphs, the dynamics of X_2 (solid line) and X_1 with the completely open microtubule (dots) and the reduced microtubule (dashed line), is reported.

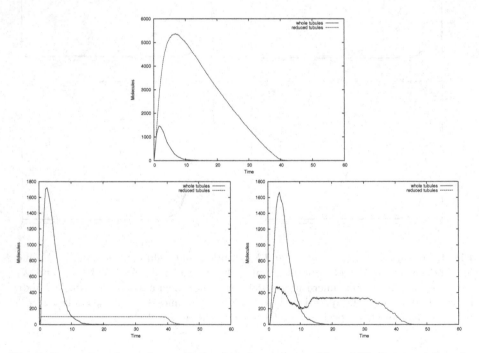

Fig. 6. Dynamics of membrane V_2 (top) and membrane V_4 and V_6 (bottom left and right). In the graphs, the dynamics of X_1 with the completely open microtubule (solid line) and the reduced microtubule (dashed line), is reported.

of membrane V_4 leads to an accumulation of X_1 inside membrane V_2 since the molecules enter into this membrane but they cannot move because membrane V_4 is full. As a result, also the quantity of molecules X_1 inside membrane V_6 is low, because of the slower communication of molecules from membrane V_4.

5 Discussion and Future Developments

In this paper we presented a new variant of P systems, called $S\tau$-DPP, inspired to tP systems and τ-DPP. The novel properties of $S\tau$-DPP consist in the representation of the membranes structure and the communication between membranes of the system by using two distinct graphs, the possibility to define tissue-like structure where nodes have a complex internal architecture, the association of a size to objects and membranes and the consequent handling of the free space during the system evolution with a new version of the τ-DPP simulation technique.

The introduction of the new properties enables the formalism to be used to model a number of real systems in which, first of all, the unlimited accumulation of objects within membranes is not possible or, in other words, in which the free space of the volumes is a critical resource for the system dynamics.

Moreover, the use of two distinct graphs for describing the membranes structure and the communication between membranes provides a formalism with a strong expressive power, in which it is possible to have communication channels between membranes that are not adjacent. This feature allows the creation of preferential paths of communication, as it has been shown in the test case, where the role of microtubules in the protein transport within cells has been reproduced.

As a future improvement of this work, we plan to better characterise and study the role of space occupation and diffusion of molecules between the volumes of the modelled systems. Furthermore, the simulation algorithm can be optimised in order to obtain a more efficient procedure and, otherwise, alternative strategies to select the rules and to verify their applicability can be tested.

Another possible extension of this work consists in studying the computational power of $S\tau$-DPP in order to prove if it is computationally (Turing) complete.

Acknowledgement

This work has been supported by the NET2DRUG, EGEE-III, BBMRI, EDGE European projects, by the MIUR FIRB LITBIO (RBLA0332RH), ITALBIONET (RBPR05ZK2Z), BIOPOPGEN (RBIN064YAT), CNR-BIOINFORMATICS initiatives, and by the project FAR-08 "Modelli di calcolo naturale e applicazioni".

References

1. Alhazov, A., Freund, R., Oswald, M.: Cell/symbol complexity of tissue p systems with symport/antiport rules. Int. J. Found. Comput. Sci. 17(1), 3–25 (2006)
2. Bernardini, F., Gheorghe, M.: Cell communication in tissue p systems: universality results. Soft Comput. 9(9), 640–649 (2005)
3. Besozzi, D., Cazzaniga, P., Pescini, D., Mauri, G.: Seasonal variance in p system models for metapopulations. Progress in Natural Science 17, 392–400 (2007)
4. Besozzi, D., Cazzaniga, P., Pescini, D., Mauri, G.: Modelling metapopulations with stochastic membrane systems. Biosystems 91(3), 499–514 (2008)

5. Cao, Y., Gillespie, D.T., Petzold, L.R.: Efficient step size selection for the tau-leaping simulation method. J. Chem. Phys. 124(4), 44109 (2006)
6. Cavaliere, M.: Evolution-communication p systems. In: Păun, G., Rozenberg, G., Salomaa, A., Zandron, C. (eds.) WMC 2002. LNCS, vol. 2597, pp. 134–145. Springer, Heidelberg (2003)
7. Cazzaniga, P., Pescini, D., Besozzi, D., Mauri, G.: Tau leaping stochastic simulation method in p systems. In: Hoogeboom, H.J., Păun, G., Rozenberg, G., Salomaa, A. (eds.) WMC 2006. LNCS, vol. 4361, pp. 298–313. Springer, Heidelberg (2006)
8. De Wit, A.: Spatial patterns and spatiotemporal dynamics in chemical systems. Adv. Chem. Phys. 109, 435–513 (1999)
9. Freund, R., Păun, G., Pérez-Jiménez, M.J.: Tissue p systems with channel states. Theoretical Computer Science 330(1), 101–116 (2005)
10. Gillespie, D.T.: Exact stochastic simulation of coupled chemical reactions. The Journal of Physical Chemistry 81(25), 2340–2361 (1977)
11. Gillespie, D.T.: Approximate accelerated stochastic simulation of chemically reacting systems. The Journal of Chemical Physics 115, 1716–1733 (2001)
12. Ionescu, M., Martín-Vide, C., Păun, A., Păun, G.: Unexpected universality results for three classes of P systems with symport/antiport. Natural Computing: an international journal 2(4), 337–348 (2003)
13. Marion, O.: Independent agents in a globalized world modelled by tissue P systems. Artificial Life and Robotics 11(2), 171–174 (2007)
14. Martín-Vide, C., Păun, G., Pazos, J., Rodríguez-Patón, A.: Tissue p systems. Theoretical Computer Science 296(2), 295–326 (2003)
15. Martín-Vide, C., Pazos, J., Păun, G., Rodríguez-Patón, A.: A new class of symbolic abstract neural nets: Tissue p systems. In: Ibarra, O.H., Zhang, L. (eds.) COCOON 2002. LNCS, vol. 2387, pp. 290–299. Springer, Heidelberg (2002)
16. Nicolescu, R., Dinneen, M., Kim, Y.: Structured modeling with hyperdag P systems. In: Proc. 7th Brainstorming week on Membrane Computing, vol. II, pp. 85–108 (2009)
17. Păun, G.: Computing with membranes. Journal of Computer and System Sciences 61, 108–143 (1998)
18. Păun, G.: Membrane Computing. An Introduction. Springer, Berlin (2002)
19. Păun, G., Păun, R.A.: Membrane computing as a framework for modeling economic processes. In: International Symposium on Symbolic and Numeric Algorithms for Scientific Computing, pp. 11–18 (2005)
20. Păun, G., Sakakibara, Y., Yokomori, T.: P systems on graphs of restricted forms. Publicationes Mathematicae Debrecen 60, 635–660 (2002)
21. Pescini, D., Besozzi, D., Mauri, G.: Investigating local evolutions in dynamical probabilistic p systems. In: Proc. Seventh International Symposium on Symbolic and Numeric Algorithms for Scientific Computing SYNASC 2005, p. 440 (2005)
22. Pescini, D., Besozzi, D., Mauri, G., Zandron, C.: Dynamical probabilistic p systems. International Journal of Foundations of Computer Science 17, 183–204 (2006)
23. Pescini, D., Besozzi, D., Zandron, C., Mauri, G.: Analysis and simulation of dynamics in probabilistic p systems. In: Carbone, A., Pierce, N.A. (eds.) DNA 2005. LNCS, vol. 3892, pp. 236–247. Springer, Heidelberg (2006)
24. Pouton, C.W., Wagstaff, K.M., Roth, D.M., Moseley, G.W., Jans, D.A.: Targeted delivery to the nucleus. Adv. Drug Deliv. Rev. 59(8), 698–717 (2007)

Implementing P Systems Parallelism
by Means of GPUs

Jose M. Cecilia[2], José M. García[2], Ginés D. Guerrero[2],
Miguel A. Martínez–del–Amor[1], Ignacio Pérez–Hurtado[1],
and Mario J. Pérez–Jiménez[1]

[1] Research Group on Natural Computing
Department of Computer Science and Artificial Intelligence
University of Sevilla
Avda. Reina Mercedes s/n, 41012 Sevilla, Spain
{mdelamor,perezh,marper}@us.es
[2] Grupo de Arquitectura y Computación Paralela
Dpto. Ingeniería y Tecnología de Computadores
Universidad de Murcia
Campus de Espinardo, 30100 Murcia, Spain
{chema,jmgarcia,gines.guerrero}@ditec.um.es

Abstract. Software development for Membrane Computing is growing
up yielding new applications. Nowadays, the efficiency of P systems simulators have become a critical point when working with instances of large
size. The newest generation of GPUs (Graphics Processing Units) provide a massively parallel framework to compute general purpose computations. We present GPUs as an alternative to obtain better performance
in the simulation of P systems and we illustrate it by giving a solution
to the N-Queens problem as an example.

1 Introduction

Membrane Computing is an emerging branch within Natural Computing that
was introduced by Gh. Păun [24]. The main idea is to consider biochemical
processes taking place inside living cells from a computational point of view,
in a way that gives us a new nondeterministic model of computation by using
cellular machines.

Up to now, it has not been possible to have implementations neither *in vivo*
nor *in vitro* of P systems, so the computation and analysis of these devices are
performed by simulators. Therefore, P systems simulators are tools that help
the researchers to extract results from models. Since P systems was presented,
many software applications have been produced [11]. These simulators have to
be as much efficient as possible when handling large problem sizes. Thus, the
massively parallel nature of P systems computations points out to look for a
massively parallel technology where the simulator can run efficiently.

Parallel computation on clusters is the traditional environment to speedup parallel applications. Particularly, many simulators of P systems have been

G. Păun et al. (Eds.): WMC 2009, LNCS 5957, pp. 227–241, 2010.

designed for clusters of computers [4]. However, this computation is relatively expensive and it is available for organizations that have enough resources to buy and maintain those clusters. Nowadays, there are other cheaper solutions in the computer market that also provides parallel environments. Among these solutions, the newest generation of graphics processor units (GPUs) are massively parallel processors which allow to develop a wide range of parallel applications. We also recall that other parallel computing platforms for P systems simulators are being investigated, such as special hardware circuits [6] and FPGAs [20].

GPUs can support several thousand of concurrent threads providing a massively parallel environment where parallel applications can obtain huge performance [14][17][29]. Current Nvidia's GPUs, for example, contain up to 240 scalar processing elements per chip [16], they are programmed using C and CUDA [32][21], and they have low cost compared with a cluster of computers.

In this paper, we use CUDA as parallel programming environment for P systems simulator in order to speedup the simulation. The input of the simulator is a P system which is defined by using the P-Lingua [5] programming language, and the output is a detailed list of information of every configuration of the computation. The simulation is divided in two main stages: *selection stage* and *execution stage*. At this stage of development, the simulator simulates recognizer P systems with active membranes, the *selection stage* is executed on the GPU and the *execution stage* is executed on the CPU.

The rest of the paper is structured as follows. In Section 2 several definitions and concepts are given for a correct understanding of the paper. Section 3 introduces the Compute Unified Device Architecture (CUDA) and some concepts of programming on GPUs are specified. In Section 4 we explain the design of the simulator. In Section 5 we implement a solution to the N-Queens problem using the simulator and P-Lingua. Finally, in Section 6 we show some results and compare them with the sequential version of the simulator. The paper ends with some conclusions and ideas for future work in Section 7.

2   Preliminaries

Polynomial time solutions to **NP**-complete problems in Membrane Computing are achieved by trading time for space. This is inspired by the capability of cells to produce an exponential number of new membranes in polynomial time. There are many ways a living cell can produce new membranes: *mitosis* (cell division), *autopoiesis* (membrane creation), *gemmation*, etc. Following these inspirations a number of different models of P systems has arisen, and many of them proved to be computational completeness (they are equivalent in power to Turing machines).

In this paper we focus on the model of *P systems with active membranes*. It is one of the most studied models in Membrane Computing and one of the first models presented by Gh. Păun [25]. P systems with active membranes is formed by a membrane structure, where a label and a polarization is associated to each membrane. In this model, every elementary membrane is able to divide itself by reproducing its content into a new membrane.

Here we provide a short recall of its features (see [25] for details). The model of P system with active membranes is a construct of the form $\Pi = (O, H, \mu, \omega_1, \ldots, \omega_m, R)$, where $m \geq 1$ is the initial degree of the system; O is the alphabet of *objects*, H is a finite set of *labels* for membranes; μ is a membrane structure (a rooted tree), consisting of m membranes injectively labelled with elements of H, $\omega_1, \ldots, \omega_m$ are strings over O, describing the *multisets of objects* placed in the m regions of μ; and R is a finite set of *rules*, where each rule is of one of the following forms:

(a) $[a \rightarrow v]_h^\alpha$ where $h \in H$, $\alpha \in \{+, -, 0\}$ (electrical charges), $a \in O$ and v is a string over O describing a multiset of objects associated with membranes and depending on the label and the charge of the membranes (*evolution rules*).

(b) $a\,[\,]_h^\alpha \rightarrow [b]_h^\beta$ where $h \in H$, $\alpha, \beta \in \{+, -, 0\}$, $a, b \in O$ (*send-in communication rules*). An object is introduced in the membrane, possibly modified, and the initial charge α is changed to β.

(c) $[a]_h^\alpha \rightarrow [\,]_h^\beta b$ where $h \in H$, $\alpha, \beta \in \{+, -, 0\}$, $a, b \in O$ (*send-out communication rules*). An object is sent out of the membrane, possibly modified, and the initial charge α is changed to β.

(d) $[a]_h^\alpha \rightarrow b$ where $h \in H$, $\alpha \in \{+, -, 0\}$, $a, b \in O$ (*dissolution rules*). A membrane with a specific charge is dissolved in reaction with a (possibly modified) object.

(e) $[a]_h^\alpha \rightarrow [b]_h^\beta [c]_h^\gamma$ where $h \in H, \alpha, \beta, \gamma \in \{+, -, 0\}$, $a, b, c \in O$ (*division rules*). A membrane is divided into two membranes. The objects inside the membrane are replicated, except for a, that may be modified in each membrane.

Rules are applied according to the following principles:

- All the elements which are not involved in any of the operations to be applied remain unchanged.
- Rules associated with label h are used for all membranes with this label, no matter whether the membrane is an initial one or whether it was generated by division during the computation.
- Rules from (a) to (e) are used as usual in the framework of membrane computing, i.e., in a maximal parallel way. In one step, each object in a membrane can only be used by at most one rule (non-deterministically chosen), but any object which can evolve by a rule must do it (with the restrictions indicated below).
- Rules (b) to (e) cannot be applied simultaneously in a membrane in one computation step.
- An object a in a membrane labelled with h and with charge α can trigger a division, yielding two membranes with label h, one of them having charge β and the other one having charge γ. Note that all the contents present before the division, except for object a, can be the subject of rules in parallel with the division. In this case we consider that in a single step two processes take place: "first" the contents are affected by the rules applied to them, and "after that" the results are replicated into the two new membranes.

– If a membrane is dissolved, its content (multiset and interior membranes) becomes part of the immediately external one. The skin is never dissolved neither divided.

Note that P systems can be seen as devices with two levels of parallelism: among membranes (every membrane works independently, with the exception of when there are communication across them) and among objects inside a membrane (the rules are applied to the existing multiset of objects in a maximal parallel way).

Recognizer P systems were introduced in [26], and constitute the natural framework to study the solvability of decision problems. The data representing an instance of the problem has to be provided to the P system to compute the appropriate answer. This is done by codifying each instance as a multiset placed in an *input membrane*. The output of the computation, *yes* or *no*, is sent to the environment in every halting configuration.

Furthermore, the act of simulating something generally entails representing certain key characteristics or behaviours of some physical, or abstract, system. However, an emulation tool duplicates the functions of one system by using a different system, so that the second system behaves like (and appears to be) the first system. With the current technology, we can not emulate the functionality of a cellular machine by using a conventional computer to solve **NP**-complete problems in polynomial time, but we can simulate these cellular machines, not necessarily in polynomial time, in order to aid researchers. However, depending on the underlying technology where the simulator is executed, the simulations can take too much time.

The technology used for this work is called CUDA (Compute Unified Device Architecture). CUDA is a co-designed hardware and software solution to make easier developing general-purpose applications on the Graphics Processor Unit (GPU) [34]. GPUs, that are one of the main components of traditional computers, originally were specialized for math-intensive, highly parallel computation which is the nature of graphics applications. These characteristics of the GPU were very attractive to accelerate scientific applications which have massively parallel computations. However, the problem was the way to program general purpose applications on the GPU. This way involved to deal with GPUs designed for video games, so they have had to tune their applications using programming idioms tied to computer graphics, programing environment tightly constrained, etc [17] [14]. The CUDA extensions developed by Nvidia provides an easier environment to program general-purpose applications onto the GPU, because it is based on ANSI C, supported by several keywords and constructs. ANSI C is the standard published by the American National Standards Institute (ANSI) for the C programming language, which is one of the most used.

P systems devices are massively parallel, what fits, in a similar way, into massively parallel nature of the GPUs with thousands of threads running in parallel.

These threads are units of execution which execute the same code concurrently on different pieces of data.

3   Graphics Processing Unit

Driven by the video games market, programmable GPUs (Graphics Processing Units) have evolved into a highly parallel, multithreaded, manycore processor. They were designed to accelerate graphics applications, which transform three-dimensional data (coordinates of triangle vertices) into pixels that are displayed on a screen, using for this task programming interfaces such as OpenGL and DirectX. The massively parallel nature of graphics applications and its arithmetic intensity leads the researches to explore more general non-graphics applications onto the GPU, creating a new programming field called GPGPU (General-Purpose on GPUs).

GPUs have become an inexpensive and readily available single-chip massively parallel system. However, GPGPU programmers had to deal with the limitations and difficulties of constrained graphics primitives to compute their non-graphics computations. The emergence of Compute Unified Device Architecture (CUDA) [34] programming model, proposed by Nvidia Corporation in 2007, has helped to develop highly-parallel applications onto the GPU easier than it was before. CUDA allows GPGPU programmers to develop their applications in a more familiar environment by using C/C++ programming language, with some extensions to manipulate special aspects of the GPU. Moreover, Nvidia consolidated this trend launching a line of GPUs optimized for general purpose computations called TESLA [16].

In this work we use a Tesla C1060 graphics processor unit (GPU) from Nvidia as hardware target for its study. This section introduces the Tesla C1060 computing architecture. In addition, it analyses the threading model of Tesla architectures, and also the most important issues in the CUDA programming environment.

3.1   Tesla C1060 Base Microarchitecture

The Tesla C1060 [16] is based on a scalable processor array which has 240 streaming-processor (SP) cores organised as 30 streaming multiprocessor (SM). The applications start at the host side (the CPU) which communicates with the device side (the GPU) through a PCI-Express x16 bus (see the top of figure 1).

The SM is the processing unit, and it is unified graphics and computing multiprocessor. Every SM contains eight SPs arithmetic cores, one double precision unit, 16-Kbyte read/write shared memory, a set of 16384 registers, and access to the off-chip memory (global/local memory). The access to shared memory is very cheap, however, the access to the off-chip memory has low performance because it is out of the chip, as it is shown on figure 1. In addition, table 1 shows all memories available on the GPU and also the cost to access them.

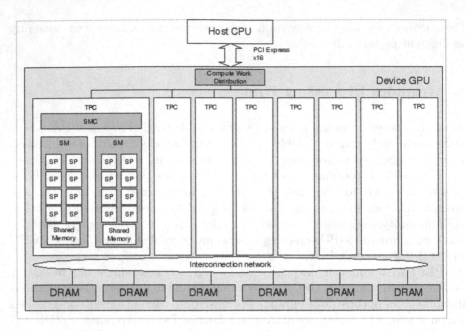

Fig. 1. Tesla C1060 GPU with 240 SPs: Streamming Processors, organised in 30 SMs: Streamming Multiprocessors

Table 1. Memory System on the Tesla C1060

Memory	Location	Size	Latency	Access
Registers	On-Chip	16384 32-bits Registers per SM	$\simeq 0$ cycles	R/W
Shared Memory	On-Chip	16 KB per SM	$\simeq registers$	R/W
Constant	On-Chip	64 KB	$\simeq registers$	R
Texture	On-Chip	Up to Global	> 100 cycles	R
Local	Off-Chip	4 GB	400-600 cycles	R/W
Global	Off-Chip	4 GB	400-600 cycles	R/W

3.2   Parallel Computing with CUDA

The GPU is seen as a cooprocessor that executes data-parallel *kernel* functions. The user creates a program encompassing CPU code (Host code) and GPU code (Kernel code). They are separated and compiled by *nvcc* (Nvidia's compiler for CUDA code) as shown in figure 2.

Firstly, the host code is responsible for transfering data from the main memory (RAM or host memory) to the GPU memory (device memory), using CUDA instructions, such as *cudamemcpy*. Moreover, the host code has to state the number of threads executing the kernel function and the organization of them. Threads execute the kernel code, and they are organized into a three-level hierarchy as it is shown in figure 3. At the highest level, each kernel creates a single grid that consists of many thread blocks. Each thread block can contain up to 512

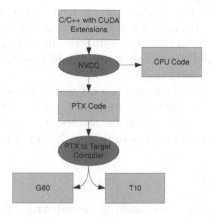

Fig. 2. Nvcc compilation process

threads, which can share data through Shared Memory and can perform barrier synchronization by invoking the –*syncthreads* primitive [31]. Besides, thread blocks can not perform synchronization. The synchronization across blocks can only be obtained by terminating the kernel.

Furthermore, the host code calls the kernel function like a C function by passing parameters if it is needed, and also by specifying the number of threads per block and the number of blocks making up the grid. Each block within the grid has their own identifier [22]. This identifier can be one, two or three

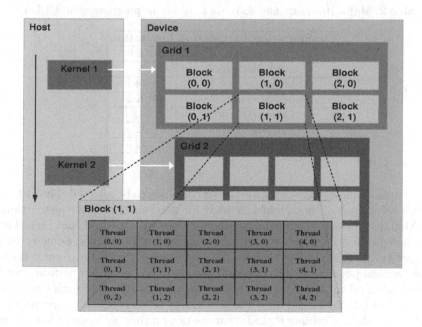

Fig. 3. Thread organization in CUDA programming model

dimensions depending on how the programmer has declared the grid, accessed via .x, .y, and .z index fields. Each thread within the block have their own identifier which can be one, two or three dimensions as well. Combining thread and block identifiers, the threads can access to different data address, and also select the work that they have to do.

The kernel code is specified through the key word __global__ and the syntax is: __global__ kernelName <<< dimGrid, dimBlock >>> (...parameter list...) where dimGrid and dimBlock are three-elements vectors that specify the dimensions of the grid in blocks and the dimensions of the blocks in threads, respectively [21].

3.3   Threading Model

A SM is a hardware device specifically designed with multithreaded capabilities. Each SM manages and executes up to 1024 threads in hardware with zero scheduling overhead. Each thread has its own thread execution state and can execute an independent code path. The SMs execute threads in a Single-Instruction Multiple-Thread (SIMT) fashion [16]. Basically, in the SIMT model all the threads execute the same instruction on different piece of data. The SMs create, manage, schedule and execute threads in groups of 32 threads. This set of 32 threads is called *Warp*. Each SM can handle up to 32 Warps (1024 threads in total, see table 2). Individual threads of the same Warp must be of the same type and start together at the same program address, but they are free to branch and execute independently.

Table 2. Major Hardware and Software Limitations programing on CUDA

Configuration Parameters	Limitation
Threads/SM	1024
Thread Blocks/SM	8
32-bit Registers/SM	16384
Shared Memory/SM	16KB
Threads/Block	512
Threads/Warp	32
Warps/SM	32

The execution flow begins with a set of Warps ready to be selected. The instruction unit selects one of them, which is ready for issue and executing instructions. The SM maps all the threads in an active Warp per SP core, and each thread executes independently with its own instructions and register state. Some threads of the active Warp can be inactive due to branching or predication, and it is also another critical point in the optimisation process. The maximum performance is achieved when all the threads in an active Warp takes the same path (the same execution flow). If the threads of a Warp diverge, the Warp serially executes each branch path taken, disabling threads that are not on that path, and when all the paths complete, the threads reconverge to the original execution path.

4 Design of the Simulator for Recognizer P Systems

In this section we briefly describe the simulator of recognizer P systems with active membranes, elementary division and polarization. Firstly, we explain the previous work that we have done in order to prepare the development of the parallel simulator on the GPU. Then, we introduce the algorithm design in the CUDA programming language, and finally, we finish with our simulator's design.

4.1 Design of the Baseline Simulator

As previously mentioned, CUDA programming model is based on C/C++ language. Therefore, the first recommended step when developing applications in CUDA is to start from a baseline algorithm written in C++, where some parts can be susceptible to be parallelized on the GPU.

In this work, we have based on the simulator for P systems with active membranes developed in PLinguaCore [5]. This sequential (or single-threaded) simulator is programmed in JAVA, so the first step was to translate the code to C++.

The simulator is executed into two main stages: *selection stage* and *execution stage*. The *selection stage* consists of the search for the rules to be executed in each membrane. Once the rules have been selected, the *execution stage* consists of the execution of these rules.

The input data for the *selection stage* consists of the description of the membranes with their multisets (strings over the working alphabet O, labels associated with the membrane in H, etc.), and the set of rules R to be selected. The output data of this stage is the set of selected rules. Only the *execution stage* changes the information of the configuration. It is the reason because *execution stage* needs synchronization when accessing to the membrane structure and the multisets. At this point of implementation, we have parallelized the *selection stage* on the GPU, and the *execution stage* is still executed on the CPU because of the synchronization problem.

We also have developed an adapted sequential simulator for the CPU (called *fast sequential* simulator), which has the same constraints as the CUDA simulator explained in the next subsections to make a fair comparison among them. This simulator achieves much better performance than the original sequential simulator.

4.2 Algorithm Design in CUDA

Whenever we design algorithms in the CUDA programming model, our main effort is dividing the required work into processing pieces, which have to be processed by TB thread blocks of T threads each. Using a thread block size of $T=256$, it is empirically determined to obtain the overall best performance on the Tesla C1060 [28]. Each thread block access to one different set of input data, and assigns a single or small constant number of input elements to each thread.

Each thread block can be considered independent to the other, and it is at this level at which internal communication (among threads) is cheap using explicit barriers to synchronize, and external communication (among blocks) becomes expensive, since global synchronization only can be achieved by the barrier implicit between successive kernel calls. The need of global synchronization in our designs requires successive kernel calls even to the same kernel.

4.3 Design of the Parallel Simulator

In our design, we identify each membrane as a thread block where each thread represents at least an element of the alphabet O. Each thread block runs in parallel looking for the set of rules that has to select for its membrane, and each individual thread is responsible for selecting the rules associated with the object that it represents (each thread selects the rules that need to be executed by using the represented object).

As result of the *execution stage*, the membranes can vary including news elements, dissolving membranes, dividing membranes, etc. Therefore, we have to modify the input data for the *selection stage* with the newest structure of membranes, and then call the selection again. It is an iterative process until a halting configuration is reached.

Finally, our simulator presents some limitations, constrained by some peculiarities in the CUDA programming model. The main limitations are showed in table 3, and the following stand out among them: it can handle only two levels of membrane hierarchy for simplicity in synchronization (the skin and the rest of elementary membranes), which is enough for solving lots of **NP**-complete problems; and the number of objects in the alphabet must be divisible by a number smaller than 512 (the maximum thread block size), in order to distribute the objects among the threads equally.

Table 3. Main limitations in the parallel simulator

Parameter	Limitation
Levels of membrane hierarchy	2
Maximum alphabet size	65535
Maximum label set size	65535
Maximum multiplicity of an object in an elementary membrane	65535
Alphabet size	Divisible by a number smaller than 512

5 A Case Study: Implementing a Solution to the N-Queens Problem

In this section, we briefly present a solution to the **N-Queens** problem by means of P systems, given by Miguel A. Gutiérrez–Naranjo et al in [10]. This family of P systems is our case study for the performance analysis of our simulator.

5.1   A Family of P Systems for Solving the N-Queens Problem

The **N-Queens** problem can be expressed as a formula in conjunctive normal form, in such way that one truth assignment of the formula is considered as a solution of the puzzle. A family of recognizer P systems for the SAT problem [27] can state whether exists a solution to the formula or not sending *yes* or *no* to the environment.

However, the *yes* ot *no* answer from the recognizer P system is not enough because it is also important to know the solutions. Besides, the system needs to give us the way to encode the state of the **N-Queens** problem.

The P system designed for solving the **N-Queens** problem is a modification of the P system for the SAT problem. It is an uniform family of deterministic recognizer P system which solves SAT as a decision problem (i.e., the P system sends *yes* or *no* to the environment in the last computation step), but also stores the truth assignments that makes true the formula encoded in the elementary membranes of the halting configuration.

5.2   Implementation

P-Lingua 1.0 [5] is a programming language useful for defining P system models with active membranes. We use P-Lingua to encode a solution to the **N-Queens** problem, and also to generate a file that our simulator can use as input. Figure 4 shows the P-Lingua process to generate the input for our simulator.

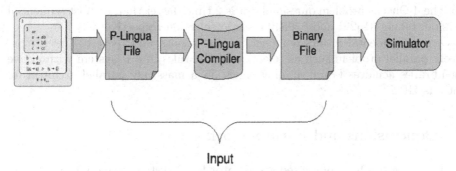

Input

Fig. 4. Generation of the simulator's input

P-Lingua 2.0 [7] is able to translates a P system written in P-Lingua language into a binary file. A binary file is a file whose information is encoded in Bytes and bits (not understandable by humans like plain text), which is suitable for trying to compress the data. This binary file contains all the information of the P system (Alphabet, Labels, Rules, . . .) which is the input of our simulator.

In our tests, we use the P system for solving the 3-Queens and 4-Queens problems. The former creates 512 membranes and up to 1883 different objects. The latter creates 65536 membranes and up to 8120 different objects, and now the simulator can handle it because we have decreased the memory requirement by the simulator in [18]. On one hand, the P system for 5-Queens needs to generate 33554432 membranes and 25574 objects, what leads in a memory space limitation (requires up to 1.5TB). On the other hand, we point out that 2-Queens is a system with only 4 membranes, what are not enough for exploiting the parallelism in P systems.

6   Performance Analysis

We now examine the experimental performance of our simulator. Our performance test are based on the solutions to 3-Queens and 4-Queens problems previously explained in 5.2. We report the *selection stage* time which is executed on the GPU, and compare it with the *selection stage* for the fast sequential code. We do not include the cost of transferring input (and output) data from (and to) host CPU memory across the PCI-Express bus to the GPU's on board memory, which negatively affects to the overall simulation time. Selection is one building block of a larger-scale computation. Our aim is to get a full implementation of the simulator on the GPU. In such case, the transfers across PCI-Express bus will be close to zero.

We have used the Nvidia GPU Tesla C1060 which has 240 execution cores and 4GB of device memory, plugged in a computer server with a Intel Core2 Quad CPU and 8GB of RAM, using the 32bits ubuntu server as Operating System.

The *selection stage* on the GPU takes about *171 msec* for the 3-Queens. So it is 2.7 times faster than the *selection stage* on the CPU which takes *465 msec*. For the 4-Queens problem our simulator is 2 times faster than the fast sequential version, taking 315291 and 629849 msec in selection respectively.

Our experimental results demonstrate the results we expect to see: a massively parallel problem such as selection of the rules in a P system with active membranes achieves faster running times on a massively parallel architecture such as GPU.

7   Conclusions and Future Work

In this paper, we have presented a simulator for P systems using CUDA. P system computations have a double parallel nature. The first level of parallelism is presented by the objects inside the membranes, and the second one is presented between membranes. Hence, we have simulated these P systems in a platform which provides those levels of parallelism. This platform is the GPU, with parallelism between thread blocks and threads. Besides, we have used a programming language called P-Lingua to encode P systems as input for our simulator. This tool helped us to encode the P system for solving the N-Queens problem in order to test our simulator.

Using the power and parallelism that provides the GPU to simulate P systems with active membranes is a new concept in the development of applications for membrane computing. Even the GPU is not a cellular machine, its features help the researches to accelerate their simulations allowing the consolidation of the cellular machines as alternative to traditional machines.

The first version of the simulator is presented for recognizer P systems with active membranes, elementary division and polarization, specifically, we have developed the *selection stage* of the simulator on the GPU. In forthcoming versions, we will include the execution version on the GPU. This issue allows a completely parallel execution on the GPU, avoiding CPU-GPU transfers in every step, which degrades system performance.

Moreover, we are working to obtain fully simulation of P systems with active membranes, deleting the limitations showed in table 3. Besides, we will include new funcionality in the simulator like not elementary division. Our aim is to develop a framework of P systems simulators running on the GPU, so we will study the simulation of other P systems models by using this parallel architecture.

It is also important to point out that this simulator is limited by the resources available on the GPU as well as the CPU (RAM, Device Memory, CPU, GPU). They limit the size of the instances of **NP**-complete problems whose solutions can be successfully simulated. Although developing general purpose programs on the GPU is easier than several years ago with tools such as CUDA, to extract the maximum performance of the GPU is still hard, so we need to make a deep analysis to obtain the maximum performance available for our simulator. For instance, in the following versions of the simulator we will reduce the memory requirements in order to simulate bigger instances of **NP**-complete problems and avoid idle threads, by deleting objects with zero multiplicity. For this task we can use spare matrix in our simulator's design.

The massively parallel environment that provides the GPUs is good enough for the simulator, however, we need to go beyond. The newest cluster of GPUs provides a higher massively parallel environment, so we will attempt to scale to those systems to obtain better performance in our simulated codes.

Finally, we will study new simulation algorithms based on algebra, and the adaptation of the design of P systems to the constraints of the GPU to make faster simulations. Furthermore, it would be interesting to avoid the brute force algorithms in P system computations, and start to design heuristics in the design of membrane solutions (i.e., avoiding membrane division as possible).

Acknowledgement

The first three authors acknowledge the support of the project from the Fundación Séneca (Agencia Regional de Ciencia y Tecnología, Región de Murcia) under grant 00001/CS/2007, and also by the Spanish MEC and European Commission FEDER under grant CSD2006-00046. The last three authors acknowledge the support of the project TIN2006–13425 of the Ministerio de Educación y Ciencia of Spain, cofinanced by FEDER funds, and the support of the "Proyecto de Excelencia con Investigador de Reconocida Valía" of the Junta de Andalucía under grant P08-TIC04200.

References

1. Alhazov, A., Pérez–Jiménez, M.J.: Uniform solution of QSAT using polarizationless active membranes. In: Durand-Lose, J., Margenstern, M. (eds.) MCU 2007. LNCS, vol. 4664, pp. 122–133. Springer, Heidelberg (2007)
2. Buck, I., Foley, T., Horn, D., Sugerman, J., Fatahalian, K., Houston, M., Hanrahan, P.: Brook for GPUs: stream computing on graphics hardware. In: SIGGRAPH 2004, pp. 777–786. ACM Press, New York (2004)
3. Ciobanu, G., Pérez–Jiménez, M.J., Păun, G. (eds.): Applications of membrane computing. Springer, Heidelberg (2006)
4. Ciobanu, G., Wenyuan, G.: P systems running on a cluster of computers. LNCS, vol. 2993, pp. 123–139. Springer, Heidelberg (2004)
5. Díaz–Pernil, D., Pérez–Hurtado, I., Pérez–Jiménez, M.J., Riscos–Núñez, A.: A P-Lingua programming environment for Membrane Computing. In: Corne, D.W., Frisco, P., Paun, G., Rozenberg, G., Salomaa, A. (eds.) WMC 2008. LNCS, vol. 5391, pp. 187–203. Springer, Heidelberg (2009)
6. Fernández, L., Martínez, V.J., Arroyo, F., Mingo, L.F.: A hardware circuit for selecting active rules in transition P systems. In: Proceedings of the Seventh International Symposium on Symbolic and Numeric Algorithms for Scientific Computing, p. 415 (2005)
7. García–Quismondo, M., Gutiérrez–Escudero, R., Martínez–del–Amor, M.A., Orejuela, E., Pérez–Hurtado, I.: P–Lingua 2.0. A software framework for cell-like P systems. Intern. J. Computers, Communications and Control IV(3), 234–243 (2009)
8. Garland, M., Grand, S.L., Nickolls, J., Anderson, J., Hardwick, J., Morton, S., Phillips, E., Zhang, Y., Volkov, V.: Parallel computing experiences with CUDA. IEEE Micro 28(4), 13–27 (2008)
9. Govindaraju, N.K., Manocha, D.: Cache–efficient numerical algorithms using graphics hardware. Parallel Computing 33(10-11), 663–684 (2007)
10. Gutiérrez–Naranjo, M.A., Martínez–del–Amor, M.A., Pérez–Hurtado, I., Pérez–Jiménez, M.J.: Solving the N–queens puzzle with P systems. In: Proc. 7th Brainstorming Week on Membrane Computing, vol. I, pp. 199–210 (2009)
11. Gutiérrez–Naranjo, M.A., Pérez–Jiménez, M.J., Riscos–Núñez, A.: Available membrane computing software. In: Applications of Membrane Computing, ch. 15, pp. 411–436. Springer, Heidelberg (2006)
12. Gutiérrez–Naranjo, M.A., Pérez–Jiménez, M.J., Riscos–Núñez, A.: Towards a programming language in cellular computing. Electronic Notes in Theoretical Computer Science 123, 93–110 (2005)
13. Harris, M., Sengupta, S., Owens, J.D.: Parallel prefix sum (Scan) with CUDA. GPU Gems 3 (2007)
14. Hartley, T.D., Catalyurek, U., Ruiz, A., Igual, F., Mayo, R., Ujaldon, M.: Biomedical image analysis on a cooperative cluster of GPUs and multicores. In: ICS 2008: Proce. 22nd annual international conference on Supercomputing, pp. 15–25. ACM, New York (2008)
15. Lam, M.D., Rothberg, E.E., Wolf, M.E.: The cache performance and optimizations of blocked algorithms. In: ASPLOS-IV: Proceedings of the fourth international conference on Architectural support for programming languages and operating systems, pp. 63–74. ACM, New York (1991)
16. Lindholm, E., Nickolls, J., Oberman, S., Montrym, J.: Nvidia Tesla. A unified graphics and computing architecture. IEEE Micro 28(2), 39–55 (2008)

17. Mark, W.R., Glanville, R.S., Akeley, K., Kilgard, M.J.: Cg – a system for programming graphics hardware in a C–like language. In: SIGGRAPH 2003, pp. 896–907. ACM, New York (2003)
18. Martínez–del–Amor, M.A., Pérez–Hurtado, I., Pérez–Jiménez, M.J., Cecilia, J.M., Guerrero, G.D., García, J.M.: Simulation of Recognizer P Systems by using Manycore GPUs. In: Proc. 7th Brainstorming Week on Membrane Computing, vol. II, pp. 45–58 (2009)
19. Michalakes, J., Vachharajani, M.: GPU acceleration of numerical weather prediction. In: IPDPS, pp. 1–7 (2008)
20. Nguyen, V., Kearney, D., Gioiosa, G.: An algorithm for non-deterministic object distribution in P systems and its implementation in hardware. In: Corne, D.W., Frisco, P., Paun, G., Rozenberg, G., Salomaa, A. (eds.) WMC 2008. LNCS, vol. 5391, pp. 325–354. Springer, Heidelberg (2009)
21. Nickolls, J., Buck, I., Garland, M., Skadron, K.: Scalable parallel programming with CUDA. Queue 6(2), 40–53 (2008)
22. Owens, J.D., Houston, M., Luebke, D., Green, S., Stone, J.E., Phillips, J.C.: Gpu computing. Proceedings of the IEEE 96(5), 879–899 (2008)
23. Owens, J.D., Luebke, D., Govindaraju, N., Harris, M., Krger, J., Lefohn, A.E., Purcell, T.J.: A survey of general–purpose computation on graphics hardware. Computer Graphics Forum 26(1), 80–113 (2007)
24. Păun, G.: Computing with membranes. Journal of Computer and System Sciences 61(1), 108–143 (2000); Turku Center for Computer Science-TUCS Report No 208
25. Păun, G.: Membrane Computing, An introduction. Springer, Berlín (2002)
26. Pérez–Jiménez, M.J., Romero–Jiménez, A., Sancho–Caparrini, F.: Complexity classes in models of cellular computing with membranes. Natural Computing 2(3), 265–285 (2003)
27. Pérez–Jiménez, M.J., Romero–Jiménez, A., Sancho–Caparrini, F.: A polynomial complexity class in P systems using membrane division. Journal of Automata, Languages and Combinatorics 11(4), 423–434 (2006)
28. Satish, N., Harris, M., Garland, M.: Designing efficient sorting algorithms for manycore GPUs. To Appear in Proc. 23rd IEEE International Parallel and Distributed Processing Symposium (2009)
29. Ruiz, A., Ujaldon, M., Andrades, J.A., Becerra, J., Huang, K., Pan, T., Saltz, J.H.: The GPU on biomedical image processing for color and phenotype analysis. In: BIBE, pp. 1124–1128 (2007)
30. Ryoo, S., Rodrigues, C., Baghsorkhi, S., Stone, S., Kirk, D., Mei Hwu, W.: Optimization principles and application performance evaluation of a multithreaded GPU using CUDA. In: Proc. 13th ACM SIGPLAN Symposium on Principles and Practice of Parallel Programming, pp. 73–82 (2008)
31. Ryoo, S., Rodrigues, C.I., Stone, S.S., Stratton, J.A., Ueng, S.-Z., Baghsorkhi, S.S., Hwu, W.W.: Program optimization carving for GPU computing. J. Parallel Distrib. Comput. 68(10), 1389–1401 (2008)
32. Nvidia CUDA Programming Guide 2.0. (2008),
   http://developer.download.nvidia.com/compute/cuda/2_0/docs/
   NVIDIA_CUDA_Programming_Guide_2.0.pdf
33. GPGPU organization. World Wide Web electronic publication,
   http://www.gpgpu.org
34. Nvidia CUDA. World Wide Web electronic publication,
   http://www.nvidia.com/cuda

Regulation and Covering Problems in MP Systems

Giuditta Franco, Vincenzo Manca, and Roberto Pagliarini

Verona University
Computer Science Department
Strada Le Grazie 15, 37134 Verona, Italy
{giuditta.franco,vincenzo.manca,roberto.pagliarini}@univr.it

Abstract. The study of efficient methods to deduce fluxes of biological reactions, by starting from experimental data, is necessary to understand metabolic dynamics, and is a central issue in systems biology. In this paper we report some initial results, together with related open problems, regarding the efficient computation of regulation fluxes in metabolic P systems. By means of Log-gain theory the system dynamics can be linearized, in such a way to be described by a recurrence equations system, of which we point out a few algebraic properties, involving covering problems.

1 Introduction

Since their first introduction [14], P systems have been widely investigated in the framework of formal language theory as innovative compartmentalized multiset rewriting systems [15], and different variants of them have been analyzed along with their computational power (for a complete list of references, see http://ppage.psystems.eu/). Although they were originally introduced as computational models, their biologically inspired structure and functioning, together with their feasibility as models of cellular and biomolecular processes, turned out to be a widely applicable modeling technique in several domains, including medicine (for immunological processes [6], and cellular tissue healing [5]), economics [16], linguistics and computer science (computer graphics, cryptography, approximate solutions to optimization problems) [4], and, of course, biology (for mechanosensitive channels [1], respiration in bacteria, photosynthesis, the protein kinase C activation [3]).

The intent to employ multiset processing in the compartmentalized framework provided by cell-like or tissue-like membrane structures in order to study real biological systems is nowadays vividly pursued, along with variants of P systems enriched with several other features, usually inspired by biology [2]. An important aspect in modeling biological reactions by rewriting rules was a thorough study of the rule application strategy [7], since the traditional nondeterministic maximally parallel way seemed to be not realistic enough. Along this recent more applicative trend, a body of research is focused on the modeling of

G. Păun et al. (Eds.): WMC 2009, LNCS 5957, pp. 242–251, 2010.
© Springer-Verlag Berlin Heidelberg 2010

metabolisms, where the main interest is devoted to the molecular reactions transforming matter rather than to the biochemicals distribution and coordination in compartments.

Metabolic P systems have been introduced as mono-membrane multiset rewriting grammars, whose rules are regulated by specific functions [8,10]. The aim is to control the matter transformation in a reactor by means of rules whose *fluxes* dynamically depend on the objects population concentration. This new strategy of rule application was inspired by real 'metabolic reactions', and it seems to lead multiset based computing towards interesting simulations of biological processes, such as complex oscillations [11], the mitotic cycle [3] and the non-photochemical quenching phenomenon [12]. Overall, a new way to observe the evolution rules of a system reproducing a metabolic reaction was proposed. Indeed, since the application of every rule changes the relative amounts of reacting substances, it was enforced that such quantities influence the reactivity of the rules in a way that their application depends on the current substances (and parameters) concentration, as it normally happens in biochemical phenomena. A simulator (named MetaPlab) applying evolution rules with this strategy has been developed, and employed to simulate several biological processes (it may be downloaded from the website `mplab.sci.univr.it`, where the reader may find also several references).

Before entering in more technical details, let us discuss a few other substantial (for modeling purposes) differences which have been introduced by metabolic P systems with respect to traditional membrane systems.

P systems are traditionally organized in a way that their evolution is synchronous, i.e., a global clock triggers the production of new symbols inside all membranes. In principle, one may try to increase the granularity of a P system in order to obtain fine-grained sequences of transitions, then consider the trajectories described by these sequences, and this description would be as accurate as fine the granularity of the P system is. In practice, it is likely that the desired granularity is obtained by adding auxiliary symbols or priority constraints in the system, to form (sometimes complex) priority relationships for the rewriting rules [6]. As a matter of fact, P systems do not provide tools for controling the resolution of the observation of intermediate states, and they are better suited to model a process as a sequence of "snapshots". With metabolic P systems instead, one assumes an *a priori* choice of the time interval τ, between consecutive observation instants, that depends on the macroscopic level at which considering the dynamics of the biological system. The flux values (also called reaction units) are computed according to the chosen observation granularity.

In metabolic P systems rules are obviously global and not compartmentalized, and the environment changes are taken into account by the fluxes associated to reactions. The state, on which reaction units depend, is given both by the value of some magnitudes, called *parameters*, which can influence the reactions (e.g., temperature and pressure), and by the amount of the *substances* inside the system. Some distinction between matter and not matter is fundamental to study metabolic processes, and the idea of considering *parameters* as elements of the system different from metabolic substrates, and having their own evolution, is new of metabolic P systems. Nevertheless, some similar ideas were formalized in the context of membrane systems, by means of *promoters* and *inhibitors*, that are respectively permitting and forbidding objects associated to regions, modeling the chemicals in the cell that, while supporting or forbidding certain reactions, can separately evolve, in parallel with the chemicals involved in the reactions [4]. Finally, we would like to emphasize that the approach of modeling by metabolic P systems assumes a novel perspective, by considering the rules only as matter transformation reactions rather than precise molecular interactions. The search of fluxes is therefore aimed at designing a model of the observed macroscopic reality with respect to the abstract transformations one has assumed, and it is different from the parameter (or rate) estimation typically studied in systems biology, even in the framework of membrane systems [17].

In the next section the problem we tackle is framed, after a brief introduction to MP systems and to Log-gain theory [9], specifically devised for them. A few results are reported in the third section, while a last section about open problems and ongoing work concludes the paper.

2   Framing the Problem

An MP system is completely specified by: *i)* m reactions, *ii)* m corresponding flux regulation functions, *iii)* n substances, which are the elements transformed by reactions, and their initial values, *iv)* k parameters, which are arguments (beside substances) of flux regulation functions, and *v)* k parameter evolution functions.

We assume $m > n$ (more rules than substances), as it realistically happens in biochemical systems. A few examples are given by the following protein-protein interaction networks: Ito (yeast) has 8868 known interactions among 3280 proteins, Giot (Drosophila) has 4780 known interactions among 4679 proteins, and Li (C. elegans) has 5534 known interactions among 3024 proteins, and by the following bacterial metabolic networks: Wolbachia pipientis has 8128 interactions over 2100 genes, S. enterica has 13309 interactions over 3717 genes, R. felis has 6966 interactions over 2062 genes, and A. phagocytophilum has 7924 over 2056 genes. About our hypothesis, by a more abstract perspective, we observe that since usually each metabolic reaction transforms few substances, in the case $m \leq n$ we would have a scarce competition for substances among the rules, and the interaction system would be not so interesting to analyze. Furthermore, as we will discuss, the problem we are going to describe would be just not so significant from an algebraic viewpoint. The number k of parameters instead, has no relationship with m and n, as it just represents the sensitivity of the system to the environment (parameters are internal or external controlling variables which somehow affect the system functioning).

A *state* Z is an $\mathbb{R}^{n+k}$ vector (reporting the current amounts of substances and parameters), while each rule r_j (with $j = 1, \ldots, m$) having some of the n substances as substrates and some as products, is associated to a couple of $\mathbb{R}^n$

vectors (r_j^-, r_j^+) (one of which possibly null), reporting the substance quantities respectively occurring in the premise and in the consequence of r_j. As an instance, we might have a system Q with three substances $\{a, b, c\}$, two parameters $\{v, w\}$ which values evolve according with their own function vector $(f_v(i), f_w(i))$, for $i \in \mathbb{N}$, and four rules

$$r_1 : ab \to aa, \ r_2 : bcc \to a, \ r_3 : ac \to \lambda, \ r_4 : abc \to bb.$$

The reactions respectively correspond to the vector couples:

$$(r_1^-, r_1^+) = ((1, 1, 0), (2, 0, 0)), \ (r_2^-, r_2^+) = ((0, 1, 2), (1, 0, 0)),$$
$$(r_3^-, r_3^+) = ((1, 0, 1), (0, 0, 0)), \ (r_4^-, r_4^+) = ((1, 1, 1), (0, 2, 0)).$$

Furthermore, four (one for each rule) flux regulation functions may be given, defined on $\mathbb{R}^5$ and having values in $\mathbb{R}$, in order to have the *fluxes* u_1, u_2, u_3, u_4, respectively associated to the rules.

There are a couple of features to point out when dealing with metabolic rules r. One is the *activation substrate* (that is, how many units of substrate are necessary in order that the rule be applied), given by the vector r^-, and the other one is the effect of the rule application, given by $r^+ - r^-$. This last vector gives the biochemical balance due to the application of the rule r, that is, how much of each substance was either consumed or produced. For example, in the above rule r_4, we need to have all u_4 units of a, u_4 units of b and u_4 units of c to activate the rule (i.e., to be able to apply the rule), while the rule effectively producing u_4 of b and consuming u_4 of a and of c. Of course, in cases of no substance production (as it is for r_3 in the example), the activation and the consumption of the rule coincide.

We call *stoichiometric matrix R*, the $(n \times m)$-dimensional matrix formed by the vectors $r^+ - r^-$, for every rule r, disposed according to a prefixed order. For example, in the example above, we have

$$R = \begin{pmatrix} 1 & 1 & -1 & -1 \\ -1 & -1 & 0 & 1 \\ 0 & -2 & -1 & -1 \end{pmatrix}. \tag{1}$$

The stoichiometric matrix is assumed to have maximal rank, as it is the case in our example. Should we have one row linearly dependent on the others, we could delete it (together with the corresponding substance in the system, as studying its dynamics would be not meaningful), and reset the whole system with the remaining substances (we newly say n) and the corresponding $n \times m$ stoichiometric matrix (having full rank, after a certain number of iterations of this procedure).

Analogously, the *activation matrix A* is formed by the vectors r^-, and for the example above we have:

$$A = \begin{pmatrix} 1 & 0 & 1 & 1 \\ 1 & 1 & 0 & 1 \\ 0 & 2 & 1 & 1 \end{pmatrix}.$$

The dynamics of a metabolic P system is given by both the evolution of parameters, according to their laws, and by the evolution of the vector X of substances, ruled by the following recurrence n-equations system [10] (where $\times$ denotes the ordinary matrix product and i the discrete instant of time):

$$X[i+1] = R \times U[i] + X[i]. \tag{2}$$

By considering $U[i]$ as the unknown vector (and X[i], X[i+1] given by observed numerical data), the linear system (2) (called ADA for "Avogadro and Dalton Action" [10]) has infinite solutions, as the number n of equations is usually smaller than the number m of variables (should we have the case $m \leq n$, from an algebraic point of view there would be no problem to eventually solve the system or figure out if there is not any solution).

In [9] the Log-gain theory was developed to design an MP model from observation experimental data, that is, to deduce the MP regulation fluxes from temporal series of the substances. From an algebraic viewpoint, such a theory provides us with other m equations and other n variables, that can be added to the ADA system (2) in order to obtain an $n + m$ equations system univocally solvable.

According to the simplest formulation of this theory, given a number of observation steps (at a specified time interval τ), and the corresponding time series of the observed states of a real metabolic system (with an assumed stoichiometry), the relative variations of any reaction flux u_j of the rule $r_j : \alpha_j \to \beta_j$ ($j = 1, \ldots, m$) is the sum of the relative variations of the reactants (i.e., the substances occurring in α_j), apart of some error p_j, which is introduced as a variable of the system:

$$\frac{u_j[i+1] - u_j[i]}{u_j[i]} = \sum_{a \in \alpha_j} \frac{a[i+1] - a[i]}{a[i]} + p_j.$$

We denote with $P[i]$ the m-dimensional vector of p variables, called *reaction offsets* [9], that is, of the errors introduced in the log-gain approximations of fluxes at step i. Furthermore, we denote with $Lg(U[i])$ the m-dimensional vector of relative fluctuations, that is ($\frac{u_j[i+1] - u_j[i]}{u_j[i]} \mid j = 1, \ldots, m$), for any $i \in \mathbb{N}$. Analogously $Lg(X[i])$ and $Lg(Z[i])$ are the vectors of relative fluctuations respectively of substances and of both substances and parameters. Therefore, in formal terms, the $m + n$ equations system we want to solve (in order to find the vector U[i+1]) is

$$\begin{cases} Lg(U[i]) = B \times Lg(X[i]) + C \cdot P[i+1] \\ R \times U[i+1] = X[i+2] - X[i+1] \end{cases} \tag{3}$$

where B is a $(m \times n)$-dimensional boolean matrix selecting, by matrix product, the reactants for each reaction, and C is an m-dimensional boolean vector selecting, by entrywise product[1], only n of the m reaction offsets (hence that are n other unknowns in the system, besides the m fluxes).

---

[1] For two matrices A and B having the same dimensions, the Schur product $C = A \cdot B$ is entrywise defined as $C_{ij} = A_{ij} \cdot B_{ij}$.

According to a more general formulation of the Log-gain theory [9], the relative variations of any reaction flux u_j of the rule $r_j : \alpha_j \rightarrow \beta_j$ $(j = 1, \ldots, m)$ is the sum of the relative variations of its *tuners*, which are both the substances (including the rule reactants) and the parameters which influence the reaction r_j. In this general case, the system (3) to solve becomes

$$\begin{cases} Lg(U[i]) = B^\star \times Lg(Z[i]) + C \cdot P[i+1] \\ R \times U[i+1] = X[i+2] - X[i+1] \end{cases} \tag{4}$$

where $B^\star$ is an $m \times (n + k)$-dimensional boolean matrix selecting, by matrix product, the tuners for each reaction.

Since in our recurrence systems we are assuming to know the fluxes $U[i]$ computed at the previous step, the reader could wonder about the value of the reaction fluxes at the initial observation step. There exists a heuristic algorithm to estimate it [13], by evaluating, along with few initial steps of observation, how much of each substance is necessary to activate the first evolution step.

3   A Few Results

Given an MP system of substances $\{x_1, \ldots, x_n\}$ and rules $\{r_1, \ldots, r_m\}$, $R(x)$ is defined as the set of all the rules involving x either as a reactant or as a product. A set R_0 of n rules is called *covering set* if $R(x) \cap R_0 \neq \emptyset$ for any substance x. Consequently, the boolean vector C from the system (3), if selecting offsets of rules of a covering set, is called *covering vector*.

Proposition 1. *Any set of n linearly independent rules is a covering.*

In fact, if we look at the n columns of the stoichiometric matrix, corresponding to n prefixed linearly independent rules, they cannot show a null row (otherwise they would be not linearly independent), then any substance (which corresponds to a row) is involved by at least one of the prefixed rules, and this implies they form a covering set.

Let us recall here that a set of n linearly independent rules always exists, because the stoichiometric matrix R (which columns are represented by the rules) is assumed to have maximal rank.

We observe that the system (3) may be rewritten as

$$\begin{cases} U[i+1] - C \cdot U[i] \cdot P[i+1] = (B \times Lg(X[i])) \cdot U[i] + U[i] \\ R \times U[i+1] = X[i+2] - X[i+1] \end{cases} \tag{5}$$

More interestingly, system (5) may be transformed in another one (6), computing the same flux values U, by applying a time constant block matrix in each step. The idea underlying this algebraic manipulation is to change the "fake" variables of the system from the m-dimensional vector $P[i+1]$ to the n-dimensional vector $W[i+1]$, obtained by taking the n non-null components of $C \cdot U[i] \cdot P[i+1]$. Hence, if we consider as a variable the vector $\begin{pmatrix} U[i+1] \\ W[i+1] \end{pmatrix}$ rather than $\begin{pmatrix} U[i+1] \\ P[i+1] \end{pmatrix}$, we get the following system

$$\begin{pmatrix} I_m & G \\ R & O_n \end{pmatrix} \times \begin{pmatrix} U[i+1] \\ W[i+1] \end{pmatrix} = \begin{pmatrix} (B \times Lg(X[i])) \cdot U[i] + U[i] \\ X[i+2] - X[i+1] \end{pmatrix} \qquad (6)$$

where I_m is the identity matrix of dimension m, O_n is the null square matrix of dimension n, R is the stoichiometric matrix, and G is an $(m \times n)$-dimensional boolean matrix, called *covering matrix*, such that each column has exactly one non-null element and the sum of the first h columns (for any $h = 1, \ldots, n$) coincide with the covering vector C in its first components containing h ones. In other words, if the non-null components of C are $j_1, j_2, \ldots, j_n$, then the non-null components of the corresponding covering matrix G are $(j_1, 1), (j_2, 2), \ldots, (j_n, n)$.

As a conclusion, we can deduce the following claim:

Proposition 2. *Systems (5) and (6) provide same solutions for $U[i+1]$.*

The systems (5) and (6) are equivalent with regard to the first m components of the unknown vector to compute (i.e., the fluxes we are looking for), because it holds that $G \times W[i+1] = C \cdot U[i] \cdot P[i+1]$. Formulating the system (6) has been helpful to perform fast and efficient computations for our simulations. Indeed, with it we do not need to compute the matrix at every computational step as for the system (3), but just once, and the blockwise matrix product can be easily performed by involving operations only on the submatrices.

Let us see all of this on the example Q introduced in the previous section. In the stoichiometric matrix R reported in (1), one can verify that $R_0 = \{r_1, r_2, r_3\}$ is a covering set. Then the 7×7 system (6) to solve is

$$\begin{pmatrix} I_4 & G \\ R & O_3 \end{pmatrix} \times \begin{pmatrix} U[i+1] \\ W[i+1] \end{pmatrix} = \begin{pmatrix} (B \times (LgX[i])) \cdot U[i] + U[i] \\ X[i+2] - X[i+1] \end{pmatrix}$$

where $G = \begin{pmatrix} 1 & 0 & 0 \\ 0 & 1 & 0 \\ 0 & 0 & 1 \\ 0 & 0 & 0 \end{pmatrix}$, $B = \begin{pmatrix} 1 & 1 & 0 \\ 0 & 1 & 1 \\ 1 & 0 & 1 \\ 1 & 1 & 1 \end{pmatrix}$, and $LgX[i] = \begin{pmatrix} Lg(a[i]) \\ Lg(b[i]) \\ Lg(c[i]) \end{pmatrix}$.

As a relevant consequence of the propositions above, we have the following result.

Theorem 1. *The system (6) is univocally solvable if and only if G corresponds to a linearly independent covering.*

It is enough to show that the matrix $\mathcal{N} = \begin{pmatrix} I_m & G \\ R & O_n \end{pmatrix}$ of the linear system (6) has a non-null determinant. Since $\mathcal{N}$ may be decomposed as the product of square matrices as in the following

$$\begin{pmatrix} I_m & G \\ R & O_n \end{pmatrix} = \begin{pmatrix} I_m & O_{m \times n} \\ R & I_n \end{pmatrix} \times \begin{pmatrix} I_m & G \\ O_{n \times m} & -R \times G \end{pmatrix}$$

then, $det(\mathcal{N}) = -det(R \times G)$ which is not null if and only if G corresponds to a linearly independent covering.

Finally, we would like to point out that:

Proposition 3. *For any C corresponding to a linearly independent covering, the values of U, obtained as solution of the systems (3) and (4), are equal, if the fluxes of the non-covered rules are assumed to depend only on their reactants.*

In more technical terms, along the rows corresponding to the zero components of $C \cdot P$, it holds that $B^\star$ coincides with B in the first n components and has the other k ones equal to zero. In other words, the solution U of the system (3) does not change if the offset covered rules have log-gains of the fluxes given by the sum of log-gains not only of the reactants, but also of other elements (substances or parameters).

In order to prove this, once we have chosen a linearly independent covering R_0, we arrange the rules of the system according to an order which disposes first the rules of the covering and then the others, so that the stoichiometric matrix R has the first n columns corresponding to the vectors $r^+ - r^-$, for the rules $r \in R_0$, and the others to the vectors $r^+ - r^-$, for $r \notin R_0$. We denote this form of the stoichiometric matrix as $R = (R_0 \ R_1)$, where R_0 is an $n \times n$ (and R_1 an $n \times (m-n)$) submatrix. The vectors $U = (U_0 U_1)$, $C = (C_0 C_1)$, and $P = (P_0 P_1)$ are arranged consistently, with C_0 having all the components equal to one and C_1 being an $(m-n)$-dimensional null vector. Namely, in the system (4), $B^\star = \begin{pmatrix} B_0^\star \\ B_1^\star \end{pmatrix}$, where $B_0^\star$ is a $n \times (n+k)$ boolean matrix selecting the tuners of each reaction in the matrix R_0, and $B_1^\star$ is an $(m-n) \times (n+k)$ boolean matrix selecting the tuners of the reactions in R_1, and if we consider the system (4) in its reformulation (5), we have

$$\begin{cases} U_0[i+1] = ((B_0^\star \times Lg(Z[i])) + 1 + C_0 \cdot P_0[i+1]) \cdot U_0[i] \\ U_1[i+1] = ((B_1^\star \times Lg(Z[i])) + 1 + C_1 \cdot P_1[i+1]) \cdot U_1[i] \qquad (7) \\ R_0 \times U_0[i+1] = X[i+2] - X[i+1] - R_1 \times U_1[i+1]. \end{cases}$$

Our hypothesis means that, if $B = \begin{pmatrix} B_0 \\ B_1 \end{pmatrix}$, with B_0 an $n \times n$ (and B_1 an $(m-n) \times n$)-dimensional matrix, then $B_1^\star = (B_1 \ O_{(m-n) \times k})$.

Under this assumption, since C_1 is a null vector and we know the vectors $Lg(Z[i])$ and $U[i]$, from the second equation of system (7) we deduce that

$$U_1[i+1] = (B_1^\star \times Lg(S[i])) \cdot U_1[i] + U_1[i] = (B_1 \times Lg(X[i])) \cdot U_1[i] + U_1[i]$$

which is evidently the U_1 solution of both systems (3) and (4).

On the other hand, since R_0 has a non-null determinant (because C is a linearly independent covering), the third equation of system (7) has a unique solution for U_0 (fluxes of covered rules), which does not depend on the matrix B_0 but only on the above computed $U_1[i+1]$. The matrix B_0 indeed selects only reactants for each rule of the covering R_0, while here we get the same values for both U_0 and U_1 even if there is $B_0^\star$ involved (by the first equation of the system (7)).

At this point, it is quite straightforward to see the general formulation of the previous result, that is

Theorem 2. *The system (4) (or (7)) has the same flux values, by keeping constant the choice of the tuners for the non-covered rules (i.e., $B_1^\star$), and arbitrarily modifying the choice of the set of tuners for the covered rules.*

This last remark points out the importance of the covering choice (i.e., the vector C): one should essentially select which are the rules which are not "crucial" for the dynamics (i.e. $B_0^\star$), because it does not really matter which are the substances or parameters which affect them.

An interesting consequence of our analysis is that the general linear system (4) to compute the m fluxes of a metabolic P system may be equivalently reduced to a minimal one, having m equations, given by

$$\begin{cases} Lg(\hat{U}[i]) = B^\star \times Lg(Z[i]) \\ R \times U[i+1] = X[i+2] - X[i+1] \end{cases}$$

where $B^\star$ is an $(m - n) \times (n + k)$-dimensional boolean matrix selecting, by matrix product, the tuners for each non-covered reaction of R, and $Lg(\hat{U}[i])$ is the subvector of $Lg(U[i])$ given by the relative fluctuations of the $m - n$ fluxes corresponding to the non-covered rules.

4   Future Work

Along the results presented in this paper, MP systems clearly give an exciting connection between linear algebra and rewriting rules, especially those covering all substances transformed within a metabolic system, with several facets that require further research. What linear independence of rules means in terms of biological dynamics, and what in terms of formal rewriting systems? As one of the referees observed, "it would be interesting to determine the meaning of linear independence of rules in the frameworks of both biology and rewriting P systems".

In order to compute fluxes of MP systems, that need to be positive in order rules be applied, it is still not clear which would be the choice of a "good" covering (among the linearly independent ones). Other similarly interesting problems could be outlined if we consider the covering set composed by rules involving all the substances *only* along their premises. A new formulation of our problem would replace the stoichiometric matrix R with the activation matrix A, and it would be interesting to investigate conditions we should have on the data to guarantee a "correct" (i.e., with positive fluxes) biological dynamics.

References

1. Ardelean, I., Besozzi, D., Garzon, M.H., Mauri, G., Roy, S.: P System Models for Mechanosensitive Channels. In: [4], ch. 2, pp. 43–80
2. Besozzi, D., Cazzaniga, P., Mauri, G., Pescini, D.: Modelling metapopulations with stochastic membrane systems. Biosystems 91, 499–514 (2008)

3. Bianco, L., Fontana, F., Franco, G., Manca, V.: P systems for biological dynamics. In: [4], ch. 3, pp. 81–126
4. Ciobanu, G., Pérez-Jiménez, M.J., Păun, G. (eds.): Applications of Membrane Computing. Springer, Berlin (2006)
5. Franco, G., Jonoska, N., Osborn, B., Plaas, A.: Knee joint injury and repair modeled by membrane systems. BioSystems 91, 473–488 (2008)
6. Franco, G., Manca, V.: A membrane system for the leukocyte selective recruitment. In: Martín-Vide, C., Mauri, G., Păun, G., Rozenberg, G., Salomaa, A. (eds.) WMC 2003. LNCS, vol. 2933, pp. 181–190. Springer, Heidelberg (2004)
7. Gutiérrez-Naranjo, M.A., Pérez-Jiménez, M.J., Riscos-Núñez, A.: On the degree of parallelism in membrane systems. Theoretical Computer Science 372(2-3), 183–195 (2007)
8. Manca, V.: Fundamentals of metabolic P systems. In: Păun, G., et al. (eds.) Handbook of Membrane Computing, ch. 16. Oxford University Press, Oxford (2009)
9. Manca, V.: Log-gain principles for metabolic P systems. In: Condon, A., et al. (eds.) Algorithmic Bioprocesses, ch. 28. Springer, Berlin (2009)
10. Manca, V.: The metabolic algorithm for P systems: Principles and applications. Theoretical Computer Science 404, 142–157 (2008)
11. Manca, V., Bianco, L., Fontana, F.: Evolution and oscillation in P systems: Applications to biological phenomena. In: Mauri, G., Păun, G., Jesús Pérez-Jímenez, M., Rozenberg, G., Salomaa, A. (eds.) WMC 2004. LNCS, vol. 3365, pp. 63–84. Springer, Heidelberg (2005)
12. Manca, V., Pagliarini, R., Zorzan, S.: A photosynthetic process modelled by a metabolic P system. Natural Computing (to appear, 2009), doi:10.1007/s11047-008-9104-x
13. Pagliarini, R., Franco, G., Manca, V.: An algorithm for initial fluxes of metabolic P systems. Int. J. of Computers, Communications & Control IV(3), 263–272 (2009)
14. Păun, Gh.: Computing with membranes. Journal of Computer and System Sciences 61(1), 108–143 (2000); Turku Center for Computer Science-TUCS Report 208, November 1998, www.tucs.fi
15. Păun, Gh.: Membrane Computing: An Introduction. Springer, Berlin (2002)
16. Păun, Gh., Păun, R.A.: Membrane computing as a framework for modeling economic processes. In: Proc. Seventh International Symposium on Symbolic and Numeric Algorithms for Scientific Computing, Timişoara, pp. 11–18 (2005)
17. Romero-Campero, F.J., Cao, H., Camara, M., Krasnogor, N.: Structure and parameter estimation for cell systems biology models. In: Proc. of the Genetic and Evolutionary Computation Conference, Atlanta, USA, July 12-16, pp. 331–338 (2008)

(Tissue) P Systems
with Hybrid Transition Modes

Rudolf Freund and Marian Kogler

Faculty of Informatics, Vienna University of Technology
Favoritenstr. 9, 1040 Vienna, Austria
{rudi,marian}@emcc.at

Abstract. In addition to the maximally parallel transition mode used from the beginning in the area of membrane computing, many other transition modes for (tissue) P systems have been investigated since then. In this paper we consider (tissue) P systems with hybrid transition modes where each set of a covering of the whole set of rules may work in a different transition mode in a first level and all partitions of rules work together at a (second) level of the whole system on the current configuration in a maximally parallel way. With all partitions of noncooperative rules working in the maximally parallel mode, we obtain a characterization of Parikh sets of ET0L-languages, whereas with hybrid systems with the partitions either working in the maximally parallel and in the $= 1$-mode or with all partitions working in the $= 1$-mode we can simulate catalytic or purely catalytic P systems, respectively, thus obtaining computational completeness.

1 Introduction

In the original model of P systems introduced as membrane systems by Gh. Păun (see [6], [12]), the objects evolve in a hierarchical membrane structure; in tissue P systems, for example considered by Gh. Păun, T. Yokomori, and Y. Sakakibara in [15] and by R. Freund, Gh. Păun, and M.J. Pérez-Jiménez in [8], the cells communicate within an arbitrary graph topology. The maximally parallel transition mode was not only used in the original model of membrane systems, but then also in many variants of P systems and tissue P systems investigated during the last decade. Rather recently several new transition modes for P systems and tissue P systems have been introduced and investigated, for example, the sequential and the asynchronous transition mode as well as the minimally parallel transition mode (see [3]) and the k-bounded minimally parallel transition mode (see [10]). In [9], a formal framework for (tissue) P systems capturing the formal features of these transition modes was developed, based on a general model of membrane systems as a collection of interacting cells containing multisets of objects (compare with the models of networks of cells as discussed in [1] and networks of language processors as considered in [4]). In this paper we consider coverings of the rule set with each partition being equipped with its own transition mode – which may not only be the transition modes usually

G. Păun et al. (Eds.): WMC 2009, LNCS 5957, pp. 252–263, 2010.
© Springer-Verlag Berlin Heidelberg 2010

considered in the area of P systems as the maximally parallel mode, but also modes well known from the area of grammar systems (e.g., see [5]) as the $= k$, $\leq k$, and the $\geq k$ modes for $k \geq 1$. A multiset of rules to be applied to a given configuration is composed from a multiset of rules from each partition working in the corresponding transition mode on a suitable partitioning of the objects in the underlying configuration.

The rest of this paper is organized as follows: In the second section, well-known definitions and notions are recalled. In the next section, we explain our general model of tissue P systems with hybrid transition modes and give some illustrative examples in the succeeding section. A characterization of the Parikh sets of ET0L-languages by tissue P systems with all partitions working in the maximally parallel transition mode is shown in the fourth section. In the fifth section, we establish some results on computational completeness by showing how catalytic P systems and purely catalytic P systems can be simulated by tissue P systems where one partition works in the maximally parallel mode and all the others in the $= 1$-mode and by tissue P systems where all partitions work in the $= 1$-mode, respectively. A short summary concludes the paper.

2 Preliminaries

We recall some of the notions and the notations used in the following as in [10]; for elements of formal language theory, we refer to [14].

Let V be a (finite) alphabet; then V^* is the set of all strings over V, and $V^+ = V^* - \{\lambda\}$ where λ denotes the empty string. RE, REG ($RE(T)$, $REG(T)$) denote the families of recursively enumerable and regular languages (over the alphabet T), respectively. For any family of string languages F, PsF denotes the family of Parikh sets of languages from F. By $\mathbb{N}$ we denote the set of all non-negative integers, by $\mathbb{N}^k$ the set of all vectors with components of non-negative integers. In the following, we will not distinguish between NRE, which coincides with $PsRE(\{a\})$, and $RE(\{a\})$.

Let V be a (finite) set, $V = \{a_1, ..., a_k\}$. A *finite multiset* M over V is a mapping $M : V \longrightarrow \mathbb{N}$, i.e., for each $a \in V$, $M(a)$ specifies the number of occurrences of a in M. The size of the multiset M is $|M| = \sum_{a \in V} M(a)$. A multiset M over V can also be represented by any string x that contains exactly $M(a_i)$ symbols a_i for all $1 \leq i \leq k$, e.g., by $a_1^{M(a_1)}...a_k^{M(a_k)}$. The set of all finite multisets over the set V is denoted by $\langle V, \mathbb{N} \rangle$.

Throughout the rest of the paper, we will not distinguish between a multiset from $\langle V, \mathbb{N} \rangle$ and its representation by a string over V containing the corresponding number of each symbol.

An ET0L system is a construct $G = (V, T, w, P_1, ..., P_m)$, $m \geq 1$, where V is an alphabet, $T \subseteq V$ is the terminal alphabet, $w \in V^*$ is the *axiom*, and P_i, $1 \leq i \leq m$, are finite sets of rules (*tables*) of noncooperative rules over V of the form $a \rightarrow x$. In a derivation step, all the symbols present in the current sentential form are rewritten using one table. The language generated by G, denoted by $L(G)$, consists of all the strings over T which can be generated in this way when starting from w. An ET0L system with only one table is called an E0L system. By $E0L$ and $ET0L$ we denote the families of languages generated by E0L systems and ET0L systems, respectively. It is known from [14] that $CF \subset E0L \subset ET0L \subset CS$, with CF being the family of context-free languages and CS being the family of context-sensitive languages. The corresponding families of sets of (vectors of) non-negative integers are denoted by XCF, $XE0L$, $XET0L$, and XCS, respectively, with $X \in \{N, Ps\}$.

A *register machine* is a construct $M = (n, B, l_0, l_h, I)$, where n is the number of registers, B is a set of instruction labels, l_0 is the start label, l_h is the halt label (assigned to HALT only), and I is a set of instructions of the following forms:

- $l_i : (\text{ADD}(r), l_j, l_k)$ add 1 to register r, and then go to one of the instructions labeled by l_j and l_k, non-deterministically chosen;
- $l_i : (\text{SUB}(r), l_j, l_k)$ if register r is non-empty (non-zero), then subtract 1 from it and go to the instruction labeled by l_j, otherwise go to the instruction labeled by l_k;
- $l_h : \text{HALT}$ the halt instruction.

A register machine M generates a set $N(M)$ of natural numbers in the following way: start with the instruction labeled by l_0, with all registers being empty, and proceed to apply instructions as indicated by the labels and by the contents of the registers. If we reach the HALT instruction, then the number stored at that time in register 1 is taken into $N(M)$. It is known (e.g., see [11]) that in this way we can compute every recursively enumerable set of natural numbers even with only three registers, where the first one is never decremented.

3   Networks of Cells

In this section we consider membrane systems as a collection of interacting cells containing multisets of objects like in [1] and [9]. For an introduction to the area of membrane computing, we refer the interested reader to the monograph [13], the actual state of the art can be seen in the web [16].

Definition 1. *A network of cells of degree $n \geq 1$ is a construct*

$$\Pi = (n, V, w, i_0, R) \quad \text{where}$$

1. *n is the number of cells;*
2. *V is a (finite) alphabet;*
3. *$w = (w_1, \ldots, w_n)$ where $w_i \in \langle V, \mathbb{N} \rangle$, for all $1 \leq i \leq n$, is the multiset initially associated to cell i;*
4. *i_0, $1 \leq i_0 \leq n$, is the output cell;*
5. *R is a finite set of rules of the form $X \to Y$ where $X = (x_1, \ldots, x_n)$, $Y = (y_1, \ldots, y_n)$, with $x_i, y_i \in \langle V, \mathbb{N} \rangle$, $1 \leq i \leq n$, are vectors of multisets over V. We will also use the notation*

$$(x_1, 1) \ldots (x_n, n) \to (y_1, 1) \ldots (y_n, n)$$

for a rule $X \to Y$.

A network of cells consists of n cells, numbered from 1 to n, that contain multisets of objects over V; initially cell i contains w_i. A *configuration* C of Π is an n-tuple of multisets over V $(u_1, \ldots, u_n)$; the *initial configuration* of Π, C_0, is described by w, i.e., $C_0 = w = (w_1, \ldots, w_n)$. Cells can interact with each other by means of the rules in R. The application of a rule

$$(x_1, 1) \ldots (x_n, n) \to (y_1, 1) \ldots (y_n, n)$$

means rewriting objects x_i from cells i into objects y_j in cells j, $1 \le i, j \le n$. A rule is called *noncooperative* if it is of the form $(a, i) \to (y_1, 1) \ldots (y_n, n)$ with $a \in V$.

The set of all multisets of rules *applicable* to C is denoted by $Appl\,(\Pi, C)$ (a procedural algorithm how to obtain $Appl\,(\Pi, C)$ is described in [9]).

We now consider a covering of R by subsets R_1 to R_h, $h \ge 1$. Usually, this covering may be a partitioning of R coinciding with a specific assignment of the rules to the cells, yet in this paper we do not restrict ourselves to such a constraint, but allow the rule sets R_1 to R_h to be working on arbitrary cells. For any multiset of rules R' containing rules from a set of rules R, we define $\|R'\|$ to be the number of rules in R'.

For the specific *transition modes* used for the subsets of rules R_j to be defined in the following, we consider the subsystems

$$\Pi_j = (n, V, w, i_0, R_j)\,.$$

The selection of multisets of rules from R_j, $1 \le j \le h$, applicable to a configuration C has to be a specific subset of $Appl\,(\Pi_j, C)$; for the transition mode ϑ, the selection of multisets of rules applicable to a configuration C is denoted by $Appl\,(\Pi_j, C, \vartheta)$. In contrast to the transition modes usually considered in the area of P systems as the asynchronous and the sequential mode, we also define some more general variants well known from the area of grammar systems (e.g., see [5]) as the derivation modes $= k$, $\ge k$, $\le k$ for $k \ge 1$.

Definition 2. *For the transition mode (Δk) with $\Delta \in \{=, \le, \ge\}$,*

$$Appl\,(\Pi_j, C, \Delta k) = \{R' \mid R' \in Appl\,(\Pi_j, C) \text{ and } \|R'\| \, \Delta k\}\,.$$

The *asynchronous* transition mode *(asyn)* with

$$Appl\,(\Pi_j, C, asyn) = Appl\,(\Pi_j, C)$$

is the special case of the transition mode Δk with Δk being equal to ≥ 1, i.e., in fact there are no particular restrictions on the multisets of rules applicable to C.

The *sequential* transition mode *(sequ)* with

$$Appl\,(\Pi_j, C, sequ) = \{R' \mid R' \in Appl\,(\Pi_j, C) \text{ and } \|R'\| = 1\}$$

is the special case of the transition mode Δk with Δk being equal to $= 1$, i.e., every multiset of rules $R' \in Appl\,(\Pi_j, C, sequ)$ has size 1.

The transition mode considered in the area of P systems from the beginning is the *maximally parallel* transition mode where we only select multisets of rules R' that are not extensible, i.e., there is no other multiset of rules $R'' \supsetneq R'$ applicable to C.

Definition 3. *For the* maximally parallel *transition mode (max),*

$$Appl\,(\Pi_j, C, max) = \{R' \mid R' \in Appl\,(\Pi_j, C) \text{ and there is}$$
$$no \; R'' \in Appl\,(\Pi_j, C) \text{ with } R'' \supsetneq R'\}.$$

Based on these transition modes for the partitions of rules R_j, we now are able to define a *network of cells with hybrid transition modes* as follows:

Definition 4. *A network of cells with hybrid transition modes of degree $n \geq 1$, in the following also called* tissue P system (with hybrid transition modes) *of degree $n \geq 1$, is a construct*

$$\Pi = (n, V, w, i_0, R, (R_1, \alpha_1), \dots, (R_h, \alpha_h)) \quad \text{where}$$

1. *(n, V, w, i_0, R) is a network of cells of degree n;*
2. *$R_1, \dots, R_h$ is a covering of R and the α_j, $1 \leq j \leq h$, are the transition modes assigned to the corresponding partitions of rules R_j.*

Based on the transition modes of the partitions R_j, we now can define how to obtain a next configuration from a given one in the whole system Π by applying in a maximally parallel way an applicable multiset of rules consisting of multisets of rules from the R_j each of those applied in the respective transition mode:

Definition 5. *Given a configuration C of Π, we non-deterministically choose a partition R_{j_1} and try to apply it; if this is not possible, we just continue with non-deterministically choosing another partition R_{j_2}; if we are able to apply R_{j_1} in the corresponding transition mode α_{j_1} with using a multiset of rules R'_{j_1}, we mark the objects affected by doing that and continue with non-deterministically choosing another partition R_{j_2} then being to be applied to a configuration not containing the objects marked for being used with the rules from R'_{j_1}. We continue with the same algorithm as for R_{j_1} eventually marking objects to be used with a multiset of rules R'_{j_2}, etc. In sum, we obtain a multiset of rules R' to be applied to C as the union of the multisets of rules R'_{j_m} constructed by the algorithm described above. The result of the transition step from the configuration C with applying R' is the configuration $Apply\,(\Pi, C, R')$, and we also write $C \Longrightarrow_\Pi C'$. The reflexive and transitive closure of the transition relation $\Longrightarrow_\Pi$ is denoted by $\Longrightarrow_\Pi^*$; if n transition steps take place, we write $\Longrightarrow_\Pi^n$ for $n \geq 0$.*

Definition 6. *A computation in a network of cells with hybrid transition modes Π starts with the initial configuration $C_0 = w$ and continues with transition steps as defined above. It is called* successful *if we reach a configuration C to which no partition R_j can be applied with respect to the transition mode α_j anymore (we also say that the computation halts).*

Definition 7. *As the results of halting computations we take the Parikh vectors or numbers of objects in the specified output cell i_0. The set of results of all computations then is denoted by $X(\Pi)$ with $X \in \{Ps, N\}$.*
We shall use the notation

$$XO_m h_h t P_n (\vartheta) \ [parameters\ for\ rules]$$

with $X \in \{Ps, N\}$ to denote the family of sets of Parikh vectors (Ps) and natural numbers (N), respectively, generated by tissue P systems Π of the form

$$(n', V, w, i_0, R, (R_1, \alpha_1), \ldots, (R_{h'}, \alpha_{h'}))$$

with $n' \leq n$, $|V| \leq m$, $h' \leq h$, and $\cup_{j=1}^{h} \{\alpha_j\} \subseteq \vartheta$ (ϑ contains the allowed transition modes); the parameters for rules *describe the specific features of the rules in R. If any of the parameters n, m, and h is unbounded, we replace it by $*$.*

4 Examples

As a first example, we construct a tissue P system with one cell initially containing two symbols a and two sets of rules each of them containing one rule affecting the symbol a using eventually different transition modes:

Example 1. Let

$$\Pi = (1, \{a\}, aa, 1, P_1 \cup P_2, (P_1, \alpha_1), (P_2, \alpha_2))$$

where $P_1 = \{a \to b\}$ and $P_2 = \{a \to c\}$. We now consider the results of computations in this tissue P system with different transition modes α_1 and α_2:

- α_1 and α_2 both are $= 1$: both the rule in P_1 and the rule in P_2 are applied exactly once, no matter which partition we choose first to be applied, i.e., $aa \Longrightarrow_{\Pi} bc$; hence, the result is bc.
- α_1 and α_2 both are max: recall that the transition modes of the rule sets do not take into account the rules in other rule sets, so both P_i try to apply their own rule twice. This conflict is solved in a non-deterministic way, i.e., $aa \Longrightarrow_{\Pi_1} bb$ or $aa \Longrightarrow_{\Pi_2} cc$; hence, the results are bb, cc.
- α_1 and α_2 both are ≥ 1: the rules in P_i are applied either once or twice. If the rule from each set is only applied once, we have a similar situation as before when using the transition mode $= 1$. If one or both sets attempt to apply their own rule twice, a conflict arises which is solved in a non-deterministic way. Thus, the result set is the union of the result sets considered in the cases $= 1$ and max, i.e., $\{bc, bb, cc\}$.
- α_1 is $= 1$, α_2 is ≥ 1: as before, yet we do not have to consider the case that the rule in P_1 is applied twice. Therefore, the result set is $\{bc, cc\}$.

– α_1 is $= 1$, α_2 is max: the conflict is solved by non-deterministically choosing to execute the rule in P_2 in a maximally parallel way thus consuming all symbols a before trying to execute the rule in P_1 (which then fails, as no symbol a is left) or else to execute P_1 before P_2 (resulting in one symbol a being transformed to b and one symbol a being transformed to c). This yields the same result set as in the case before ($\{bc, cc\}$).
– α_1 is ≥ 1, α_2 is max: P_1 and P_2 conflict with respect to either one symbol (if the rule in the partition chosen first is applied only once) or to both symbols (if it is applied twice). If the conflict arises with respect to one symbol, the conflict resolution yields $\{bc, cc\}$; otherwise, as in the case when α_1 and α_2 both are max, the results are bb, cc. The set of all possible computation results thus is the union of both cases, i.e., $\{bc, bb, cc\}$.

Usually, with only taking results from halting computations and using the maximally parallel transition mode without using a covering of the rule set R, with noncooperative rules it is not possible to generate sets like $\{a^{2^n} \mid n \geq 0\}$ (compare with the results established in [2], where the variant of unconditional halting was used instead, i.e., the results were taken in every computation step). As the following example shows, such sets can easily be obtained with specific partitions of noncooperative rules all of them working in the maximally parallel transition mode:

Example 2. Consider the tissue P system (of degree 1)

$$\Pi = (1, \{a, b\}, b, 1, P_1 \cup P_2, (P_1, max), (P_2, max))$$

with $P_1 = \{b \to bb\}$ and $P_2 = \{b \to a\}$. As elaborated in the previous example, we can either apply $b \to bb$ OR $b \to a$ in a maximally parallel way, but not mix both rules. Hence, as long as we apply P_1 in the maximally parallel mode, in each transition step we double the number of objects b. As soon as we choose to apply P_2 in the maximally parallel mode, the computation comes to an end yielding a^{2^n} for some $n \geq 0$, i.e.,

$$b \Longrightarrow_\Pi^n b^{2^n} \Longrightarrow_\Pi a^{2^n},$$

hence, $X(\Pi) = \{a^{2^n} \mid n \geq 0\}$ with $X \in \{Ps, N\}$.

5   Characterization of *ET0L*

In this section we show that tissue P systems with all partitions (of noncooperative rules) working in the maximally parallel transition mode exactly yield the Parikh sets of ET0L-languages.

Theorem 1. $PsET0L = PsO_*h_*tP_n(\{max\}) [noncoop]$ *for all* $n \geq 1$.

Proof. We first show $PsET0L \supseteq PsO_*h_*tP_*(\{max\}) [noncoop]$. Let

$$\Pi = (n, V, w, i_0, R, (R_1, max), \dots, (R_h, max))$$

be a tissue P system with hybrid transition modes with all partitions working in the max-mode. We first observe that an object a from V in the cell m, $1 \leq m \leq n$, can be represented as a new symbol (a, m). Hence, in the ET0L-system

$$G = (V', T, w', P_1, \ldots, P_d, P_f)$$

simulating Π, we take $T = V$ and $V' = V'' \cup V \cup \{\#\}$ with

$$V'' = \{(a, m) \mid a \in V, 1 \leq m \leq n\}.$$

In the axiom w', every symbol a in cell m is represented as the new symbol (a, m). Observe that a noncooperative rule

$$(a, i) \rightarrow (y_1, 1) \ldots (y_n, n)$$

can also be written as

$$(a, i) \rightarrow (y_{1,1}, 1) \ldots (y_{1,d_1}, 1) \ldots (y_{n,1}, 1) \ldots (y_{n,d_n}, n)$$

where all $y_{i,j}$ are objects from V and in that way can just be considered as a pure context-free rule over V''.

For every sequence of partitions $l = \langle R'_1, \ldots, R'_h \rangle$ such that $\{R'_1, \ldots, R'_h\} = \{R_1, \ldots, R_h\}$, we now construct a table P_l for G as follows:

$P_l := \{x \rightarrow x \mid x \in V', x \neq y$ for all rules $y \rightarrow v$ in $\cup_{i=1}^{h} R_i\}$;
for $i = 1$ to h do
    begin
       $R''_i := \{x \rightarrow w \mid x \rightarrow w \in R'_i$ and $x \neq y$ for all rules $y \rightarrow v$ in $P_l\}$;
       $P_l := P_l \cup R''_i$
    end

As all partitions work in the max-mode, a partition applied first consumes all objects for which it has suitable rules. Finally, to fulfill the completeness condition for symbols usually required in the area of Lindenmayer systems, we have added unit rules $a \rightarrow a$ for all objects not affected by the rule sets $R_1, \ldots, R_h$. In that way, one transition step in Π with using a multiset of rules marking the objects in the underlying configuration according to the sequence of partitions $\langle R'_1, \ldots, R'_h \rangle$ exactly corresponds with an application of the table P_l in G. To extract the terminal configurations, we have to guarantee that no rule from $\cup_{i=1}^{h} R_i$ can be applied anymore (they are projected on the trap symbol $\#$) and project the symbols (a, i_0) from the output membrane to the terminal symbols a, which is accomplished by the final table

$P_f := \{x \rightarrow \# \mid x \in V''$ for some rule $x \rightarrow v$ in $\cup_{i=1}^{h} R_i\} \cup \{\# \rightarrow \#\}$
   $\cup \{(a, j) \rightarrow \lambda \mid a \in V, j \neq i_0$ and there is no rule $(a, j) \rightarrow v$ in $\cup_{i=1}^{h} R_i\}$
   $\cup \{(a, i_0) \rightarrow a \mid a \in V$ and there is no rule $(a, i_0) \rightarrow v$ in $\cup_{i=1}^{h} R_i\}.$

We now show the inclusion

$$PsET0L \subseteq PsO_* h_* tP_1 (\{max\}) [noncoop].$$

Let $G = (V, T, w, P_1, \ldots, P_n)$ be an ET0L-system. Then we construct the equivalent tissue P system with only one cell and $n + 2$ partitions all of them working in the maximally parallel mode

$$\Pi = (1, V \cup T' \cup \{\#\}, h(w), 1, R, (R_1, max), \ldots, (R_{n+2}, max))$$

as follows:

The renaming homomorphism $h : V \to (V - T) \cup T'$ is defined by $h(a) = a$ for $a \in V - T$ and $h(a) = a'$ for $a \in T$. Then we simply define $R_i = h(P_i)$ for $1 \leq i \leq n$, i.e., in all rules we replace every terminal symbol a from T by its primed version a'. If G has produced a terminal multiset, then Π should stop with yielding the same result, which is accomplished by applying the partition

$$R_{n+1} = \{a' \to a \mid a \in T\} \cup \{x \to \# \mid x \in V - T\};$$

if the terminating rule set R_{n+1} is applied while objects from $V - T$ are still present, trap symbols $\#$ are generated, which causes a non-terminating computation in Π because of the partition $R_{n+2} = \{\# \to \#\}$. We remark that the simulation works just as well when exactly one partition is selected non-deterministically from the set of partitions where at least one rule is applicable instead of using the algorithm outlined in Definition 5. These observations conclude the proof. □

6   Simulation of (Purely) Catalytic P Systems and Computational Completeness

Membrane systems with catalytic rules were already defined in the original paper of Gheorghe Păun (see [12]), but used together with other noncooperative rules. In the notations of this paper, a *noncooperative rule* is of the form $(a, i) \to (y_1, 1) \ldots (y_n, n)$, and a *catalytic rule* is of the form

$$(c, i)(a, i) \to (c, i)(y_1, 1) \ldots (y_n, n)$$

where c is from a distinguished subset $V_C \subset V$ such that in all rules – noncooperative rules *(noncoop)* and catalytic rules *(cat)* of the whole system – the y_i are from $(V - V_C)^*$ and the symbols a are from $(V - V_C)$.

A *catalytic tissue P system*

$$\Pi = (n, V, C, w, i_0, R, (R, max))$$

can be interpreted as a tissue P system with hybrid transition modes where the single rule set works in the maximally parallel transition mode and the rules are noncooperative rules and catalytic rules. If all rules in R are catalytic ones, such a system is called *purely catalytic*. We recall the fact that in the original variant of catalytic P systems, the connection graph of the catalytic tissue P system as defined above is a tree. By $XO_mC_ktP_n[cat]$ $(XO_mC_ktP_n[pcat])$ with $X \in \{Ps, N\}$ we denote the family of sets of Parikh vectors (Ps) and natural numbers (N), respectively, generated by (purely) catalytic tissue P systems of the form $(n', V, C, w, i_0, R, (R, max))$ with $n' \leq n$, $|V| \leq m$, and $|C| \leq k$. If any of the parameters n, m, and h is unbounded, we replace it by $*$.

We now show that catalytic tissue P systems can be simulated by tissue P systems with hybrid transition modes using the maximally parallel transition mode for one partition and the $= 1$-mode for all other partitions of noncooperative rules:

Theorem 2. $XO_m C_k t P_n [cat] \subseteq XO_m h_{k+1} t P_n (\{max, = 1\}) [noncoop]$ *for* $X \in \{Ps, N\}$ *and all natural numbers* m, k, *and* n.

Proof. Let $\Pi = (n, V, C, w, i_0, R, (R, max))$ be a catalytic tissue P system with n cells. Then we construct an equivalent tissue P system with hybrid transition modes Π' as follows:

$$\Pi' = (n, V, w, i_0, R, (R_1, = 1), \ldots, (R_k, = 1), (R_{k+1}, max))$$

where, for $C = \{c_j \mid 1 \leq j \leq k\}$,

$$R_j = \{r \mid r \in R \text{ and } r = (c_j, i)(a, i) \to (c_j, i)(y_1, 1) \ldots (y_n, n)\}$$

for $1 \leq j \leq k$ and $R_{k+1} = R - \cup_{j=1}^{k} R_j$. For each catalyst c_j, the catalytic rules involving c_j form the partition R_j, from which at most one rule can be taken in any transition step, i.e., the R_j, $1 \leq j \leq k$, are combined with the $= 1$-mode, and the remaining noncooperative rules from R are collected in R_{k+1} and used in the max-mode. The equivalence of the systems Π' and Π immediately follows from the definition of the respective transition modes and the resulting transitions in these systems. $\square$

From the proof of the preceding theorem, we immediately infer the following result for purely catalytic tissue P systems:

Theorem 3. *For* $X \in \{Ps, N\}$ *and all natural numbers* m, k, *and* n,

$$XO_m C_k t P_n [pcat] \subseteq XO_m h_k t P_n (\{= 1\}) [noncoop].$$

In [7] it was shown that only three catalysts are sufficient in one cell, using only catalytic rules with the maximally parallel transition mode, to generate any recursively enumerable set of natural numbers. Hence, by showing that (tissue) P systems with purely catalytic rules working in the maximally parallel transition mode can be considered as tissue P systems with each set of noncooperative rules working in the $= 1$-mode corresponding to the set of rules related with a specific catalyst, thus covering the whole rule set for the single cell, we obtain the interesting result that in this case we get a characterization of the recursively enumerable sets of natural numbers by using only noncooperative rules (in fact, this covering replaces the use of the catalysts). In sum, from Theorem 2 and Corollary 3 and the results from [7] we obtain the following result showing computational completeness for tissue P systems with hybrid transition modes:

Corollary 1. $NRE = NO_*h_3tP_1 (\{= 1\})\,[noncoop]$
$$NO_*h_3tP_1 (\{max, = 1\})\,[noncoop]\,.$$

We mention that the $= 1$ mode in any case can be replaced by the ≤ 1-mode which immediately follows from the definition of the respective transition modes. Moreover, having the partitions working in the $= 1$-mode on the first level and using maximal parallelism on the second level of the whole system corresponds with the min_1 transition mode as introduced in [10] - this min_1 transition mode forces to take exactly one rule or zero rules from each partition into an applicable multiset of rules in such a way that no rule from a partition not yet considered could be added. Hence, the result of Theorem 1 directly follows from the results proved in [7] in the same way as shown in [10] for the min_1 transition mode. From the proof of Theorem 2 and the results proved in [7], also the following general computational completeness results for tissue P systems with hybrid transition modes follow:

Theorem 4. *For* $X \in \{Ps, N\}$,

$$XRE = XO_*h_*tP_1 (\{= 1\})\,[noncoop]$$
$$XO_*h_*tP_1 (\{max, = 1\})\,[noncoop]\,.$$

7   Summary

In this paper we have introduced tissue P systems with hybrid transition modes. With noncooperative rules as well as with the maximally parallel transition mode for all partitions, we obtain a characterization of the extended tabled Lindenmayer systems, whereas with the $= 1$-mode for three partitions or with the $= 1$-mode for two partitions and the maximally parallel transition mode for one partition we already are able to generate any recursively enumerable set of natural numbers. As for (purely) catalytic P systems, the descriptional complexity, especially with respect to the number of partitions, of tissue P systems with hybrid transition modes able to generate any recursively enumerable set of (vectors of) natural numbers remains as a challenge for future research.

References

1. Bernardini, F., Gheorghe, M., Margenstern, M., Verlan, S.: Networks of cells and Petri nets. In: Gutiérrez-Naranjo, M.A., et al. (eds.) Proc. Fifth Brainstorming Week on Membrane Computing, Sevilla, pp. 33–62 (2007)
2. Beyreder, M., Freund, R.: Membrane systems using noncooperative rules with unconditional halting. In: Corne, D.W., Frisco, P., Paun, G., Rozenberg, G., Salomaa, A. (eds.) WMC 2008. LNCS, vol. 5391, pp. 129–136. Springer, Heidelberg (2009)
3. Ciobanu, G., Pan, L., Păun, Gh., Pérez-Jiménez, M.J.: P systems with minimal parallelism. Theoretical Computer Science 378(1), 117–130 (2007)
4. Csuhaj-Varjú, E.: Networks of language processors. Current Trends in Theoretical Computer Science, 771–790 (2001)

5. Csuhaj-Varjú, E., Dessow, J., Kelemen, J., Păun, Gh.: Grammar Systems: A Grammatical Approach to Distribution and Cooperation. Gordon and Breach Science Publishers, Amsterdam (1994)
6. Dassow, J., Păun, Gh.: On the power of membrane computing. Journal of Universal Computer Science 5(2), 33–49 (1999)
7. Freund, R., Kari, L., Oswald, M., Sosík, P.: Computationally universal P systems without priorities: two catalysts are sufficient. Theoretical Computer Science 330, 251–266 (2005)
8. Freund, R., Păun, Gh., Pérez-Jiménez, M.J.: Tissue-like P systems with channel states. Theoretical Computer Science 330, 101–116 (2005)
9. Freund, R., Verlan, S.: A formal framework for P systems. In: Eleftherakis, G., Kefalas, P., Păun, Gh. (eds.) Pre-proceedings of Membrane Computing, International Workshop – WMC8, Thessaloniki, Greece, pp. 317–330 (2007)
10. Freund, R., Verlan, S.: (Tissue) P systems working in the k-restricted minimally parallel derivation mode. In: Csuhaj-Varjú, E., et al. (eds.) Proceedings of the International Workshop on Computing with Biomolecules, Österreichische Computer Gesellschaft, pp. 43–52 (2008)
11. Minsky, M.L.: Computation – Finite and Infinite Machines. Prentice Hall, Englewood Cliffs (1967)
12. Păun, Gh.: Computing with membranes. J. of Computer and System Sciences 61(1), 108–143 (2000); TUCS Research Report 208 (1998), http://www.tucs.fi
13. Păun, Gh.: Membrane Computing. An Introduction. Springer, Berlin (2002)
14. Rozenberg, G., Salomaa, A. (eds.): Handbook of Formal Languages, 3 vols. Springer, Berlin (1997)
15. Păun, Gh., Sakakibara, Y., Yokomori, T.: P systems on graphs of restricted forms. Publicationes Matimaticae 60, 635–660 (2002)
16. The P Systems web page, http://ppage.psystems.eu

An Overview of P-Lingua 2.0

Manuel García-Quismondo, Rosa Gutiérrez-Escudero, Ignacio Pérez-Hurtado,
Mario J. Pérez-Jiménez, and Agustín Riscos-Núñez

Research Group on Natural Computing
Department of Computer Science and Artificial Intelligence
University of Sevilla
Avda. Reina Mercedes s/n, 41012, Sevilla, Spain
mangarfer2@alum.us.es, {rgutierrez,perezh,marper,ariscosn}@us.es

Abstract. P–Lingua is a programming language for membrane computing which aims to be a standard to define P systems. In order to implement this idea, a Java library called pLinguaCore has been developed as a software framework for cell–like P systems. It is able to handle input files (either in XML or in P–Lingua format) defining P systems from a number of different cell–like P system models. Moreover, the library includes several built–in simulators for each supported model. For the sake of software portability, pLinguaCore can export a P system definition to any convenient output format (currently XML and binary formats are available). This software is not a closed product, but it can be extended to accept new input or output formats and also new models or simulators.

The term P–Lingua 2.0 refers to the software package consisting of the above mentioned library together with a user interface called pLinguaPlugin (more details can be found at http://www.p-lingua.org).

Finally, in order to illustrate the software, this paper includes an application using pLinguaCore for describing and simulating ecosystems by means of P systems.

1 Introduction

The initial definition of a *membrane system* as a computing device, introduced by Gh. Păun [14], can be interpreted as a flexible and general framework. Indeed, a large number of different models have been defined and investigated in the area: P systems with symport/antiport rules, with active membranes, with probabilistic rules, etc. There were some attempts to establish a common formalization covering most of the existing models (see e.g. [5]), but the membrane computing community is still using specific syntax and semantics depending on the model they work with.

Each model displays characteristic semantic constraints that determine the way in which rules are applied. Hence, the need for software simulators capable of taking into account different scenarios when simulating P system computations comes to the fore. Moreover, simulators have to precisely define the specific P system that is to be simulated. Along this paper, the term *simulator input* will

G. Păun et al. (Eds.): WMC 2009, LNCS 5957, pp. 264–288, 2010.

be used to refer to the definition (on a text file) of the P system to be simulated. One approach to implement the simulators input could be defining a specific input file format for each simulator. Nevertheless, this approach would require a great redundant effort. A second approach could be to standardize the simulator input, so all simulators need to process inputs specified in the same format. These two approaches raise up a trade-off: On the one hand, specific simulator inputs could be defined in a more straightforward way, as the used format is closer to the P system features to simulate. On the other hand, although the latter approach involves analyzing different P systems and models to develop a standard format, there is no need to develop completely a new simulator every time a new P system should be simulated, as it is possible to use a common software library in order to parse the standard input format. Moreover, users would not have to learn a new input format every time they use a different simulator and would not need to change the way to specify P systems which need to be simulated every time they move on to another model, as they would keep on using the standard input format.

This second approach is the one considered in P–Lingua project, a programming language whose first version, presented in [3], is able to define P systems within the active membrane P system model with division rules. The authors also provide software tools for compilation, simulation and debug tasks.

As P–Lingua is intended to become a standard for P systems definition, it should also consider other models. At the current stage, P–Lingua can define P systems within a number of different cell–like models: active membrane P systems with membrane division rules or membrane creation rules, transition P systems, symport/antiport P systems, stochastic P systems and probabilistic P systems. Each model follows semantics restrictions, which define several constraints for the rules (number of objects on each side, whether membrane creation and/or membrane division are allowed, and so on), and which indicate the way rules are applied on configurations.

A Java [22] library called pLinguaCore has been developed as a software framework for cell–like P systems. It includes parsers to handle input files (either in XML or in P–Lingua format), and furthermore the parsers check possible programming errors (both lexical/syntactical and semantical).

The library includes several built–in simulators to generate P system computations for the supported models, and it can export several output file formats to represent P systems (at the current stage, XML and binary file formats) in order to get interoperability between different software environments.

The term P–Lingua 2.0 refers to the software framework under GNU GPL license [21] consisting of the above mentioned library together with a user interface called pLinguaPlugin. It is not a closed software because developers with knowledge of Java can include new components to the library: new supported models, built–in simulators for the supported models, parsers to process new input file formats and generators for new output file formats. In order to facilitate those tasks, a website for users and developers of P–Lingua 2.0 [24] has been created. It contains technical information about standard programming methods to expand the pLinguaCore library. These methods have been used on all the existent components. The website also contains a download section, tutorials, user manuals, information about projects using P–Lingua, and other useful stuff.

Furthermore, pLinguaCore is not a stand–alone product, it is created to be used inside other software applications. In order to illustrate this idea, the paper includes an application using pLinguaCore for describing and simulating ecosystems by means of P systems.

2   Models

The library pLinguaCore is able to accept input files (either in P–Lingua or XML file formats) that define P systems within the supported models. As mentioned in the Introduction, Java developers can include new models to the library by using standard programming methods, easing the task. The current supported models are enumerated below.

2.1   Transition P System Model

The basic P systems were introduced in [14] by Gh. Păun.

A *transition P system* of degree $q \geq 1$ is a tuple of the form

$$\Pi = (\Gamma, L, \mu, \mathcal{M}_1, \ldots, \mathcal{M}_q, (R_1, \rho_1), \ldots, (R_q, \rho_q), i_o)$$

where:

- Γ is an alphabet whose elements are called *objects*.
- L is a finite set of labels.
- μ is a membrane structure consisting of q membranes with the membranes (and hence the regions, the space between a membrane and the immediately inner membranes, if any) injectively labelled with elements of L; as usual, we represent the membrane structures by strings of matching labelled parentheses.
- $\mathcal{M}_i$, $1 \leq i \leq q$, are strings which represent multisets over Γ associated with the q membranes of μ.
- R_i, $1 \leq i \leq q$, are finite sets of *evolution rules* over Γ, associated with the membranes of μ. An evolution rule is of the form $u \rightarrow v$, where u is a string over Γ and $v = v'$ or $v = v'\delta$, being v' a string over $\Gamma \times (\{here, out\} \cup \{in_j : 1 \leq j \leq q\})$.
- ρ_i, $1 \leq i \leq q$, are strict partial orders over R_i.
- i_o, $1 \leq i_o \leq q$, is the label of an elementary membrane (the *output membrane*).

The objects to evolve in a step and the rules by which they evolve are chosen in a non–deterministic manner, but in such a way that in each region we have a maximally parallel application of rules. This means that we assign objects to rules, non–deterministically choosing the rules and the objects assigned to each rule, but in such a way that after this assignation no further rule can be applied to the remaining objects.

2.2 Symport/Antiport P System Model

Symport/antiport rules were incorporated in the framework of P systems in [13].

A *P system with symport/antiport rules* of degree $q \geq 1$ is a tuple of the form

$$\Pi = (\Gamma, L, \mu, \mathcal{M}_1, \ldots, \mathcal{M}_q, E, R_1, \ldots, R_q, i_o)$$

where:

- Γ is the alphabet of objects,
- L is the finite set of labels for membranes (in general, one uses natural numbers as labels), μ is the membrane structure (of degree $q \geq 1$), with the membranes labelled in a one-to-one manner with elements of L,
- $\mathcal{M}_1, \ldots, \mathcal{M}_q$ are strings over Γ representing the multisets of objects present in the q compartments of μ in the initial configuration of the system.
- $E \subseteq \Gamma$ is the set of objects supposed to appear in the environment in arbitrarily many copies.
- R_i, $1 \leq i \leq q$, are finite sets of rules associated with the q membranes of μ. The rules can be of two types (by Γ^+ we denote the set of all non-empty strings over Γ, with λ denoting the empty string):
  - *Symport rules*, of the form (x, in) or (x, out), where $x \in \Gamma^+$. When using such a rule, the objects specified by x enter or exit, respectively, the membrane with which the rule is associated. In this way, objects are sent to or imported from the surrounding region – which is the environment in the case of the skin membrane.
  - *Antiport rules*, of the form $(x, out; y, in)$, where $x, y \in \Gamma^+$. When using such a rule for a membrane i, the objects specified by x exit the membrane and those specified by y enter from the region surrounding membrane i; this is the environment in the case of the skin membrane.
- $i_o \in L$ is the label of a membrane of μ, which indicates the *output* region of the system.

The rules are used in the non-deterministic maximally parallel manner, standard in membrane computing.

2.3 Active Membranes P System Model

With membrane division rules. P systems with membrane division were introduced in [15], and in this model the number of membranes can increase exponentially in polynomial time. Next, we define P systems with active membranes using 2-division for elementary membranes, with polarizations, but without cooperation and without priorities (and without permitting the change of membrane labels by means of any rule).

A *P system with active membranes* using 2-division for elementary membranes of degree $q \geq 1$ is a tuple $\Pi = (\Gamma, L, \mu, \mathcal{M}_1, \ldots, \mathcal{M}_q, R, i_o)$, where:

- Γ is an alphabet of symbol-objects.
- L is a finite set of labels for membranes.

- μ is a membrane structure, of m membranes, labelled (not necessarily in a one-to-one manner) with elements of L.
- $\mathcal{M}_1, \ldots, \mathcal{M}_q$ are strings over Γ, describing the initial multisets of objects placed in the q regions of μ.
- R is a finite set of rules, of the following forms:
  - (a) $[a \rightarrow \omega]_h^\alpha$ for $h \in L, \alpha \in \{+, -, 0\}$, $a \in \Gamma$, $\omega \in \Gamma^*$: This is an object evolution rule, associated with a membrane labelled with h and depending on the polarization of that membrane, but not directly involving the membrane.
  - (b) $a[\ ]_h^{\alpha_1} \rightarrow [b]_h^{\alpha_2}$ for $h \in L$, $\alpha_1, \alpha_2 \in \{+, -, 0\}$, $a, b \in \Gamma$: An object from the region immediately outside a membrane labelled with h is introduced in this membrane, possibly transformed into another object, and, simultaneously, the polarization of the membrane can be changed.
  - (c) $[a]_h^{\alpha_1} \rightarrow b[\ ]_h^{\alpha_2}$ for $h \in L$, $\alpha_1, \alpha_2 \in \{+, -, 0\}$, $a, b \in \Gamma$: An object is sent out from membrane labelled with h to the region immediately outside, possibly transformed into another object, and, simultaneously, the polarity of the membrane can be changed.
  - (d) $[a]_h^\alpha \rightarrow b$ for $h \in L$, $\alpha \in \{+, -, 0\}$, $a, b \in \Gamma$: A membrane labelled with h is dissolved in reaction with an object. The skin is never dissolved.
  - (e) $[a]_h^{\alpha_1} \rightarrow [b]_h^{\alpha_2} [c]_h^{\alpha_3}$ for $h \in L$, $\alpha_1, \alpha_2, \alpha_3 \in \{+, -, 0\}$, $a, b, c \in \Gamma$: An elementary membrane can be divided into two membranes with the same label, possibly transforming some objects and the polarities.
- $i_o \in L$ is the label of a membrane of μ, which indicates the *output* region of the system.

These rules are applied according to the following principles:

- All the rules are applied in parallel and in a maximal manner. In one step, one object of a membrane can be used by only one rule (chosen in a non-deterministic way), but any object which can evolve by one rule of any form, must do it (with the restrictions below indicated).
- If a membrane is dissolved, its content (multiset and internal membranes) is left free in the surrounding region.
- If at the same time a membrane labelled by h is divided by a rule of type (e) and there are objects in this membrane which evolve by means of rules of type (a), then we suppose that the evolution rules of type (a) are used before division is produced. Of course, this process takes only one step.
- The rules associated with membranes labelled by h are used for all copies of this membrane. At one step, a membrane can be the subject of *only one* rule of types (b)-(e).

With membrane creation rules. Membrane creation rules were first considered in [9], [10].

A *P system with membrane creation* of degree $q \geq 1$ is a tuple of the form

$$\Pi = (\Gamma, L, \mu, \mathcal{M}_1, \ldots, \mathcal{M}_q, R, i_o)$$

where:

- Γ is the alphabet of objects.
- L is a finite set of labels for membranes.
- μ is a membrane structure consisting of q membranes labelled (not necessarily in a one-to-one manner) with elements of L.
- $\mathcal{M}_1, \ldots, \mathcal{M}_q$ are strings over Γ, describing the initial multisets of objects placed in the q regions of μ.
- R is a finite set of rules of the following forms:
  - (a) $[a \to v]_h$ where $h \in L$, $a \in \Gamma$, and v is a string over Γ describing a multiset of objects. These are *object evolution rules* associated with membranes and depending only on the label of the membrane.
  - (b) $a[\,]_h \to [b]_h$ where $h \in L$, $a, b \in \Gamma$. These are *send-in communication rules*. An object is introduced in the membrane possibly modified.
  - (c) $[a]_h \to [\,]_h\, b$ where $h \in L$, $a, b \in \Gamma$. These are *send-out communication rules*. An object is sent out of the membrane possibly modified.
  - (d) $[a]_h \to b$ where $h \in L$, $a, b \in \Gamma$. These are *dissolution rules*. In reaction with an object, a membrane is dissolved, while the object specified in the rule can be modified.
  - (e) $[a \to [v]_{h_2}]_{h_1}$ where $h_1, h_2 \in L$, $a \in \Gamma$, and v is a string over Γ describing a multiset of objects. These are *creation rules*. In reaction with an object, a new membrane is created. This new membrane is placed inside the membrane of the object which triggers the rule and has associated an initial multiset and a label.
- $i_o \in L$ is the label of a membrane of μ, which indicates the *output* region of the system.

Rules are applied according to the following principles:

- Rules from (a) to (d) are used as usual in the framework of membrane computing, that is, in a maximally parallel way. In one step, each object in a membrane can only be used for applying one rule (non-deterministically chosen when there are several possibilities), but any object which can evolve by a rule of any form must do it (with the restrictions below indicated).
- Rules of type (e) are used also in a maximally parallel way. Each object a in a membrane labelled with h_1 produces a new membrane with label h_2 placing in it the multiset of objects described by the string v.
- If a membrane is dissolved, its content (multiset and interior membranes) becomes part of the immediately external one. The skin membrane is never dissolved.
- All the elements which are not involved in any of the operations to be applied remain unchanged.
- The rules associated with the label h are used for all membranes with this label, independently of whether or not the membrane is an initial one or it was obtained by creation.
- Several rules can be applied to different objects in the same membrane simultaneously. The exception are the rules of type (d) since a membrane can be dissolved only once.

2.4 Probabilistic P System Model

A probabilistic approach in the framework of P systems was first considered by
A. Obtulowicz in [12].

A *probabilistic P system* of degree $q \geq 1$ is a tuple

$$\Pi = (\Gamma, \mu, \mathcal{M}_1, \ldots, \mathcal{M}_q, R, \{c_r\}_{r \in R}, i_o)$$

where:

- Γ is the alphabet (finite and nonempty) of objects (the working alphabet).
- μ is a membrane structure, consisting of q membranes, labeled $1, 2, \ldots, q$.
  The skin membrane is labeled by 0. We also associate electrical charges with
  membranes from the set $\{0, +, -\}$, neutral and positive.
- $\mathcal{M}_1, \ldots, \mathcal{M}_q$ are strings over Γ, describing the multisets of objects initially
  placed in the q regions of μ.
- R is a finite set of evolution rules. An evolution rule associated with the
  membrane labelled by i is of the form $r : u[\, v \,]_i^{\alpha} \xrightarrow{c_r} u'[\, v' \,]_i^{\beta}$, where u, v, u', v'
  are a multiset over Γ, $\alpha, \beta \in \{0, +, -\}$ and c_r is a real number between 0
  and 1 associated with the rule such that:
    - for each $u, v \in M(\Gamma)$, $h \in H$ and $\alpha \in \{0, +\}$, if $r_1, \ldots, r_t$ are the rules
      whose left–hand side is $u[\, v \,]_h^{\alpha}$, then $\sum_{j=1}^{t} c_{r_j} = 1$
- $i_o \in L$ is the label of a membrane of μ, which indicates the *output* region of
  the system.

We assume that a global clock exists, marking the time for the whole system
(for all compartments of the system); that is, all membranes and the application
of all rules are synchronized.

The q-tuple of multisets of objects present at any moment in the q *regions*
of the system constitutes the *configuration* of the system at that moment. The
tuple $(\mathcal{M}_1, \ldots, \mathcal{M}_q)$ is the initial configuration of the system.

We can pass from one configuration to another one by using the rules from R
as follows: at each transition step, the rules to be applied are selected according
to the probabilities assigned to them, all applicable rules are simultaneously
applied, and all occurrences of the left–hand side of the rules are consumed, as
usual. Rules with the same left–hand side and whose right–hand side has the
same polarization can be applied simultaneously.

2.5 Stochastic P System Model

The original motivation of P systems was not to provide a comprehensive and
accurate model of the living cell, but to imitate the computational nature of
operations that take place in cell membranes. Most P system models have been
proved to be Turing complete and computationally efficient, in the sense that
they can solve computationally hard problems in polynomial time, by trading
time for space. Most research in P systems focus on complexity classes and
computational power.

However, P systems have been used recently to model biological phenomena very successfully. Models of oscillatory systems [4], signal transduction [18], gene regulation control [16], quorum sensing [17] and metapopulations [19] have been presented.

We introduce in this section the specification of stochastic P systems, that constitute the framework for modelling biological phenomena.

A *stochastic P system* of degree $q \geq 1$ is a tuple

$$\Pi = (\Gamma, L, \mu, \mathcal{M}_1, \ldots, \mathcal{M}_q, R_{l_1}, \ldots, R_{l_m})$$

where:

- Γ is a finite alphabet of symbols representing objects.
- $L = \{l_1, \ldots, l_m\}$ is a finite alphabet of symbols representing labels for the membranes.
- μ is a membrane structure containing $q \geq 1$ membranes identified in a one to one manner with values in $\{1, \ldots, q\}$ and labelled with elements from L.
- $M_i = (l_i, w_i, s_i)$, for each $1 \leq i \leq q$, initial configuration of the membrane i, $l_i \in L$ is the label, $w_i \in \Gamma^*$ is a finite multiset of objects and s_i is a finite set of strings over Γ.
- $R_{l_t} = \{r_1^{l_t}, \ldots, r_{k_{l_t}}^{l_t}\}$, for each $1 \leq t \leq m$, is a finite set of rewriting rules associated with membranes of label $l_t \in L$. Rules are of one of the following two forms:
  - Multiset rewriting rules:

    $$r_j^{l_t} : u[w]_l \xrightarrow{c_j^{l_t}} u'[w']_l$$

    with $u, w, u', w' \in \Gamma^*$ some finite multisets of objects and l a label from L. A multiset of objects, u is represented as $u = a_1 + \cdots + a_m$, with $a_1, \ldots, a_m \in \Gamma$. The empty multiset will be denoted by λ and we will write o^n instead of $\overbrace{o + \cdots + o}^{n}$. The multiset u placed outside of the membrane labelled with l and the multiset w placed inside of that membrane are simultaneously replaced with a multiset u' and w' respectively.
  - String rewriting rules:

    $$r_j^{l_t} : [u_1 + s_1; \ldots; u_p + s_p]_l \xrightarrow{c_j^{l_t}} [u_1' + s_{1,1}' + \cdots + s_{1,i_1}'; \ldots; u_p' + s_{p,1}' + \cdots + s_{p,i_p}']$$

    A string s is represented as $s = \langle o_1.o_2.\cdots.o_j \rangle$, where $o_1, o_2, \ldots, o_j \in \Gamma$. Each multiset of objects u_j and string s_j, $1 \leq j \leq p$, are replaced by a multiset of objects u_j' and strings $s_{j,1}', \ldots, s_{j,i_j}'$.

A constant $c_j^{l_t}$ is associated with each rule and will be referred to as *stochastic constant* and is needed to calculate the propensity of the rule according to the current context of the membrane to which this rule corresponds.

Rules in stochastic P systems model biochemical reactions. The *propensity* a_j of a reaction R_j is defined so that $a_j dt$ represents the probability that R_j will occur in the infinitesimal time interval $[t, t + dt]$ [7].

Applications of the rules and the semantics of stochastic P systems can vary, depending on which algorithm is used to simulate the model. At the present stage, two algorithms have been implemented and integrated as simulators within the pLinguaCore library. They will be discussed in Section 3.1.

3   Simulators

In [3], only one simulator was implemented, since there was only one model to simulate. However, as new models have been included, new simulators have been developed inside the pLinguaCore library, providing at least one simulator for each supported model.

All the current simulators can step backwards, but this option should be set before the simulation starts.

The library also takes into account the existence of different simulation algorithms for the same model and provides means for selecting a simulator among the ones which are suitable to simulate the P system, by checking its model.

Next, simulation algorithms for Stochastic and Probabilistic P systems are explained, but pLinguaCore integrates simulators for all supported models.

3.1   Simulators for Stochastic P Systems

In the original approach to membrane computing P systems evolve in a non-deterministic and maximally parallel manner (that is, all the objects in every membrane that can evolve by a rule must do it [14]). When trying to simulate biological phenomena, like living cells, the classical non-deterministic and maximally parallel approach is not valid anymore. First, biochemical reactions, which are modeled by rules, occur at a specific rate (determined by the propensity of the rule), therefore they can not be selected in an arbitrary and non-deterministic way. Second, in the classical approach all time step are equal and this does not represent the time evolution of a real cell system.

The strategies to replace the original approach are based on Gillespie's Theory of Stochastic Kinetics [7]. As mentioned in Section 2.5, a constant $c_j^{l_t}$ is associated to each rule. This provides P systems with a stochastic extension. The constant $c_j^{l_t}$ depends on the physical properties of the molecules involved in the reaction modeled by the rule and other physical parameters of the system and it represents the probability per time unit that the reaction takes place. Also, it is used to calculate the propensity of each rule which determines the probability and time needed to apply the rule.

Two different algorithms based on the principles stated above have been currently implemented and integrated in pLinguaCore.

Multicompartimental Gillespie Algorithm. The Gillespie [7] algorithm or SSA (Stochastic Simulation Algorithm) was developed for a single, well-mixed and fixed volume/compartment. P systems generally contain several compartments or membranes. For that reason, an adaptation of this algorithm was presented in [20] and it can be applied in the different regions defined by the compartmentalised structure of a P system model. The next rule to be applied in each compartment and the waiting time for this application is computed using a *local* Gillespie algorithm. The Multicompartimental Gillespie Algorithm can be broadly summarized as follows:

Repeat until a prefixed simulation time is reached:

1. Calculate for each membrane $i, 1 \leq i \leq m$ and for each rule $r_j \in R_{l_i}$ the propensity, a_j, by multiplying the stochastic constant $c_j^{l_i}$ associated to r_j by the number of distinct possible combinations of the objects and substrings present of the left-side of the rule with respect to the current contents of membranes involved in the rule.

2. Compute the sum of all propensities

$$a_0 = \sum_{i=1}^{m} \sum_{r_j \in R_{l_i}} a_j$$

3. Generate two random numbers r_1 and r_2 from the uniform distribution in the unit interval and select τ_i and j_i according to

$$\tau_i = \frac{1}{a_0} \ln(\frac{1}{r_1})$$

$$j_i = \text{the smallest integer satisfying} \sum_{j=1}^{j_i} a_j > r_2 a_0$$

In this way, we choose τ_i according to an exponential distribution with parameter a_0.

4. The next rule to be applied is r_{j_i} and the waiting time for this rule is τ_i. As a result of the application of this rule, the state of one or two compartments may be changed and has to be updated.

Multicompartimental Next Reaction Method. The Gillespie Algorithm is an exact numerical simulation method appropiate for systems with a small number of reactions, since it takes time proportional to the number of reactions (i.e., the number of rules). An exact algorithm which is also efficient is presented in [6], the Next Reaction Method. It uses only a single random number per simulation event (instead of two) and takes time proportional to the logarithm of the number of reactions. We have adapted this algorithm to make it compartimental.

The idea of this method is to be extremely sensitive in recalculating a_j and t_i, recalculate them only if they change. In order to do that, a data structure called *dependency graph* [6] is introduced.

Let $r : u[v]_l \xrightarrow{c} u'[v']_l$ be a given rule with propensity a_r and let the parent membrane of l be labelled with l'. We define the following sets:

- DependsOn$(a_r) = \{(b,t) : b$ is an object or string whose quantity affect the value a_r and $t = l$ if $b \in v$ and $t = l'$ if $b \in u\}$

  Generally, DependsOn$(a_r) = \{(b,l) : b \in v\} \cup \{(b,l') : b \in u\}$

- Affects$(r) = \{(b, t) : b$ is an object or string whose quantity is changed when the rule r is executed and $t = l$ if $b \in v \vee b \in v'$ and $t = l'$ if $b \in u \vee b \in u'\}$. Generally, Affects$(r) = \{(b, l) : b \in v \vee b \in v'\} \cup \{(b, l') : b \in u \vee b \in u'\}$

Definition 1. *Given a set of rules $R = R_{l_1} \cup \cdots \cup R_{l_m}$, the dependency graph is a directed graph $G = (V, E)$, with vertex set $V = R$ and edge set $E = \{(v_i, v_j) : $ Affects$(v_i) \cap$ DependsOn$(a_{v_j}) \neq \emptyset\}$*

In this way, if there exists an edge $(v_i, v_j) \in E$ and v_i is executed, as some objects affected by this execution are involved in the calculation of a_{v_j}, this propensity would have to be recalculated. The dependency graph depends only on the rules of the system and is static, so it is built only once.

The times τ_i, that represent the waiting time for each rule to be applied, are stored in an *indexed priority queue*. This data structure, discussed in detail in [6], has nice properties: finding the minimum element takes constant time, the number of nodes is the number of rules $|R|$, because of the indexing scheme it is possible to find any arbitrary reaction in constant time and finally, the operation of updating a node (only when τ_i is changed, which we can detect using to the dependency graph) takes $\log |R|$ operations.

The Multicompartimental Next Reaction Method can be broadly summarized as follows:

1. Build the dependency graph, calculate the propensity a_r for every rule $r \in R$ and generate τ_i for every rule according to an exponential distribution with parameter a_r. All the values τ_r are stored in a priority queue. Set $t \leftarrow 0$ (this is the global time of the system).
2. Get the minimum τ_μ from the priority queue, $t \leftarrow t + \tau_\mu$. Execute the rule r_μ (this is the next rule scheduled to be executed, because its waiting time is least).
3. For each edge (μ, α) in the dependency graph recalculate and update the propensity a_α and
   - if $\alpha \neq \mu$, set

$$\tau_\alpha \leftarrow \frac{a_{\alpha,old}(\tau_\alpha - \tau_\mu)}{a_{\alpha,new}} + \tau_\mu$$

   - if $\alpha = \mu$, generate a random number r_1, according to an exponential distribution with parameter a_μ and set $\tau_\mu \leftarrow \tau_\mu + r_1$

   Update the node in the indexed priority queue that holds τ_α.
4. Go to 2 and repeat until a prefixed simulation time is reached.

Both Multicompartimental Gillespie Algorithm and Multicompartimental Next Reaction Method are the core of the Direct Stochastic Simulator and Efficient Stochatic Simulator, respectively. One of them, which can be chosen in runtime, will be executed when compiling and simulating a P-Lingua file that starts with @model<stochastic>. See Section 4.1 for more details about the syntax.

3.2   Simulators for Probabilistic P Systems

Two different simulation algorithms have been created in this paper and integrates within the pLinguaCore library for the Probabilistic P system model. The first one is called Uniform Random Distribution Algorithm. The second one gives a better efficiency by using the binomial distribution, and it is called Binomial Random Distribution Algorithm.

Uniform Random Distribution Algorithm. Next, we describe how this algorithm determines the applicability of the rules to a given configuration.

(a) Rules are classified into sets so that all the rules belonging to the same set have the same left–hand side.
(b) Let $\{r_1, \ldots, r_t\}$ be one of the said sets of rules. Let us suppose that the common left-hand side is $u\ [v]_i^\alpha$ and their respective probabilistic constants are $c_{r_1}, \ldots, c_{r_t}$. In order to determine how these rules are applied to a give configuration, we proceed as follows:
  - It is computed the greatest number N so that u^N appears in the father membrane of i and v^N appears in membrane i.
  - N random numbers x such that $0 \leq x < 1$ are generated.
  - For each k $(1 \leq k \leq t)$ let n_k be the amount of numbers generated belonging to interval $[\ \sum_{j=0}^{k-1} c_{r_j}\ ,\ \sum_{j=0}^{k} c_{r_j})$ (assuming that $c_{r_0} = 0$).
  - For each k $(1 \leq k \leq t)$, rule r_k is applied n_k times.

Binomial Random Distribution Algorithm. Next, we describe how this algorithm determines the applicability of the rules to a given configuration.

(a) Rules are classified into sets so that all the rules belonging to the same set have the same left–hand side.
(b) Let $\{r_1, \ldots, r_t\}$ be one of the said sets of rules. Let us suppose that the common left-hand side is $u\ [v]_i^\alpha$ and their respective probabilistic constants are $c_{r_1}, \ldots, c_{r_t}$. In order to determine how these rules are applied to a give configuration, we proceed as follows:
(c)) Let $F(N, p)$ a function that returns a discrete random number within the binomial distribution $B(N, p)$
  - It is computed the greatest number N so that u^N appears in the father membrane of i and v^N appears in membrane i.
  - let $d = 1$
  - For each k $(1 \leq k \leq t - 1)$ do
    * let c_{r_k} be $\frac{c_{r_k}}{d}$
    * let n_k be $F(N, c_{r_k})$
    * let N be $N - n_k$
    * let q be $1 - c_{r_k}$
    * let d be $d * q$
  - let n_t be N
  - For each k $(1 \leq k \leq t)$, rule r_k is applied n_k times.

4   Formats

As well as models and simulators, new file formats to define P systems have been included in P-Lingua 2.0. Although XML format and P–Lingua format were included on the first version of the software [3], those formats have been upgraded to allow representation of P systems which have cell-like structure. As P–Lingua 2.0 provides backwards compatibility, all valid actions in the first version are still valid. Furthermore, a new format has been included: the binary format (suitable for the forthcoming Nvidia CUDA simulator [11]).

Formats are classified in two sorts: **Input formats** (whose files can be read by pLinguaCore) and **Output formats** (whose files can be generated by pLinguaCore). Some formats may belong to both categories.

One format which is worth showing up is the P–Lingua format. This input format allows to specify P systems in a very intuitive, friendly and straightforward way. Another asset to bear in mind is that the parser for P–Lingua inside the pLinguaCore library is capable of locating errors on files specified on this format.

4.1   P-Lingua Format

In the version of P-Lingua presented in [3] only P systems with active membranes and division rules were considered and therefore, possible to be defined in the P-Lingua language. New models have been added and consequently the syntax has been modified and extended, in order to support them. The current syntax of the P-Lingua language is defined as follows.

Valid identifiers. We say that a sequence of characters forms a `valid identifier` if it does not begin with a numeric character and it is composed by characters from the following:

```
a b c d e f g h i j k l m n o p q r s t u v w x y z
A B C D E F G H I J K L M N O P Q R S T U V W X Y Z
0 1 2 3 4 5 6 7 8 9 _
```

Valid identifiers are widely used in the language: to define module names, parameters, indexes, membrane labels, alphabet objects and strings.

The following text strings are reserved words in the language: `def`, `call`, `@mu`, `@ms`, `@model`, `@lambda`, `@d`, `let`, `@inf`, `@debug`, `main`, `-->`, `#` and they cannot be used as valid identifiers.

Variables. Four kind of variables are permitted in P-Lingua: `Global variables`, `Local variables`, `indexes`, `Parameters`.

Variables are used to store numeric values and their names are valid identifiers. We use 64 bits (signed) in double precision.

Global variables definition

Global variables must be declared out of any program module and they can be accessed from all of the program modules (see 4.1). The name of a global variable global_variable_name must be a valid identifier. The syntax to define a global variable is the following:

```
global_variable_name = numeric_expression;
```

Local variables definition

Local variables can only be accessed from the module in which they were declared and they must only be defined inside module definitions. The name of a local variable local_variable_name must be a valid identifier. The syntax to define a local variable is the following:

```
let local_variable_name = numeric_expression;
```

Indexes and parameters can be consider local variables used in 4.1 and 4.1 respectively.

Identifiers for electrical charges. In P-Lingua, we can consider electrical charges by using the + and − symbols for positive and negative charges respectively, and no one for neutral charge. It is worth mentioning that polarizationless P systems are included.

Membrane labels. There are three ways of writing membrane labels in P-Lingua: the first one is just a natural number; the second one is to denote the label as a valid identifier and the third one is by numeric expressions that represent natural numbers between brackets.

Numeric expressions. Numeric expressions can be written by using * (multiplication), / (division), % (module), + (addition), − (subtraction) and ^ (potence) operators with integer or real numbers and/or variables, along with the use of parentheses. It is possible to write numbers by using exponential notation. For example, $3 * 10^{-5}$ is written 3e-5.

Objects. The objects of the alphabet of a P system are written using valid identifiers, and the inclusion of sub-indexes is permitted. For example, $x_{i,2n+1}$ and Yes are written as x{i,2*n+1} and Yes respectively.

The multiplicity of an object is represented by using the * operator. For example, x_i^{2n+1} is written as x{i}*(2*n+1).

Strings. Strings are enclosed between < and > and made by concatenating valid identifiers with the character ., that is <identifier1. ... .identifierN>. For example, <cap.RNAP.op>.

Substrings. Substrings are used in string rewriting rules and the syntax is similar to strings, but it is possible to use the character ? to represent any arbitrary sequence of valid identifiers concatenated by .. The empty sequence is included. For example, `<cap.?.NAP.op>` is a substring of the string `<cap.op.op.op.NAP.op>` and of the string `<cap.NAP.op>`.

Model specification. As this programming language supports more than one model, it is necessary to specify in the beggining of the file which is the model of the P system defined. Not each type of rule is allowed in every model, for example, membrane creation rules are not permitted in P systems with symport/antiport rules. The built-in compiler of P-Lingua detects such error. Models are specified by using `@model<model_name>` and at this stage, the allowed models are:

```
@model<membrane_division>

@model<membrane_creation>

@model<transition_psystem>

@model<probabilistic_psystem>

@model<stochastic_psystem>

@model<symport_antiport_psystem>
```

Modules definition. Similarities between various solutions to **NP**-complete numerical problems by using families of recognizing P systems are discussed in [8]. Also, a cellular programming language is proposed based on libraries of subroutines. Using these ideas, a P-Lingua program consists of a set of programming modules that can be used more times by the same, or other, programs.

The syntax to define a module is the following.

```
def module_name(param1,..., paramN)
{
   sentence0;
   sentence1;
   ...
   sentenceM;
}
```

The name of a module, `module_name`, must be a valid and unique identifier. The parameters must be valid identifiers and cannot appear repeated. It is possible to define a module without parameters. Parameters have a numerical value that is assigned at the module call (see below).

All programs written in P-Lingua must contain a `main` module without parameters. The compiler will look for it when generating the output file.

In P-Lingua there are sentences to define the membrane structure of a P system, to specify multisets, to define rules, to define variables and to call to other modules. Next, let us see how such sentences are written.

Module calls. In P-Lingua, modules are executed by using calls. The format of an sentence that calls a module for some specific values of its parameters is given next:

```
call module_name(value1, ..., valueN);
```

where value$i$ is a numeric expression or a variable.

Definition of the initial membrane structure of a P system. In order to define the initial membrane structure of a P system, the following sentence must be written:

```
@mu = expr;
```

where expr is a sequence of matching square brackets representing the membrane structure, including some identifiers that specify the label and the electrical charge of each membrane.
   Examples:

1. $[[\,]_2^0]_1^0 \equiv$ @mu = [[] '2] '1
2. $[[\,]_b^0[\,]_c^-]_a^+ \equiv$ @mu = +[[] 'b, -[] 'c] 'a

Definition of multisets. The next sentence defines the initial multiset associated to the membrane labelled by label.

```
@ms(label) = list_of_objects;
```

where label is a membrane label and list_of_objects is a comma-separated list of objects. The character # is used to represent an empty multiset.
   If a stochastic P system is being defined (that is, the file starts with @model<stochastic>), strings are also permitted in the initial content of a membrane:

```
@ms(label) = list_of_objects_and_strings;
```

list_of_objects_and_strings is a comma-separated list of objects and/or strings.

Union of multisets. P-Lingua allows to define the union of two multisets (recall that the input multiset is "added" to the initial multiset of the input membrane) by using a sentence with the following format.

```
@ms(label) += list_of_objects;
```

For stochastic P systems, it would be

```
@ms(label) += list_of_objects_and_strings;
```

Definition of rules. The definition of rules has been significantly extended in this version of P-Lingua. A general rule is defined as follow (most elements are optional):

$$u[v[w_1]_{h_1}^{\alpha_1} \ldots [w_n]_{h_n}^{\alpha_n}]_h^{\alpha} \xrightarrow{k} x[y[z_1]_{h_1}^{\beta_1} \ldots [z_n]_{h_n}^{\beta_n}]_h^{\beta}[s]_h^{\gamma}$$

where $u, v, w_1, \ldots, w_n, x, y, z_1, \ldots, z_n$ are multisets of objects or strings, $h, h_1, \ldots, h_n$ are labels, $\alpha, \alpha_1, \ldots, \alpha_n, \beta, \beta_1, \ldots, \beta_n, \gamma$ are electrical charges and k is a numerical value.

The P-Lingua sintax for such a rule is:

```
uα[vα₁[w1]'h1...αₙ[wN]'hN]'h -->
                    xβ[yβ₁[z1]'h1...βₙ[zN]'hN]'h γ[s]'h :: k
```

where u, v, w1...wN, x, y, z1...zN, s are comma-separated list of objects or strings (it is possible to use the character # in order to represent the empty multiset), h,h1,..., hN are labels, $\alpha, \alpha_1, \ldots, \alpha_n, \beta, \beta_1, \ldots, \beta_n, \gamma$ are identifiers for electrical charges and k is a numeric expression.

As mentioned before, not each type of rule is permitted in every model. Below we enumerate the possible types of rules, classified by the model in which they are allowed.

@model<mebrane_division>

1. The format to define evolution rules of type $[a \rightarrow v]_h^{\alpha}$ is given next:

   ```
   α[a --> v]'h
   ```

2. The format to define send-in communication rules of type $a[\,]_h^{\alpha} \rightarrow [b]_h^{\beta}$ is given next:

   ```
   aα[]'h -->β[b]
   ```

3. The format to define send-out communication rules of type $[a]_h^{\alpha} \rightarrow b[\,]_h^{\beta}$ is given next:

   ```
   α[a]'h --> β[]b
   ```

4. The format to define division rules of type $[a]_h^{\alpha} \rightarrow [b]_h^{\beta}[c]_h^{\gamma}$ is given next:

   ```
   α[a]'h -->β[b]γ[c]
   ```

5. The format to define dissolution rules of type $[a]_h^{\alpha} \rightarrow b$ is given next:

   ```
   α[a]'h --> b
   ```

@model<membrane_creation>

1. Rules 1, 2, 3 and 5 of **@model<membrane_division>** can be defined in this model, with the same format.
2. The format to define membrane creation rules of type $[a]_h^{\alpha} \rightarrow [[b]_{h_1}^{\beta}]_h^{\alpha}$ is given next:

   ```
   α[a]'h --> α[β[b]'h1]'h
   ```

@model<transition_psystem>

1. The format to define evolution rules of type $[u[u_1]_{h_1}, \ldots, [u_N]_{h_N} \rightarrow v[v_1]_{h_1}, \ldots, [v_N]_{h_N}, \lambda]_h$ is given next:

   [u [u1]'h1 ... [uN]'hN --> v [v1]'h1, ... [vN]'hN, @d]'h

   @d is a new keyword representing the containing membrane is marked to dissolved.

@model<symport_antiport_psystem>

1. The format to define symmetric communication rules of type $a[b]_h^\alpha \rightarrow b[a]_h^\alpha$ is given next:

   αa[b]'h --> βb[a]'h

@model<probabilistic_psystem>

1. The format to define rules of type $u[v]_h^\alpha \xrightarrow{p} u_1[v_1]_h^\beta$ is given next:

   uα[v]'h --> u1β[v1]'h::p

@model<stochastic_psystem>

1. The format to define multiset rewriting rules of type $u[v]_h \xrightarrow{c} u_1[v_1]_h$ is given next:

   u[v]'h --> u1 [v1]'h::c

2. The format to define string rewriting rules of type $[u + s]_h \xrightarrow{c} [v + r]_h$ is given next:

   [u,s]'h --> [v,r]'h::c

- α, β and γ are identifiers for electrical charges.
- a, b and c are objects of the alphabet.
- u, u1, v, v1, ..., vN are comma-separated lists of objects that represents a multiset.
- s and r are comma-separated lists of substrings.
- h, h1, ..., hN are labels.
- p and c are real numeric expressions. The result of evaluating p must be between 0 and 1, and the result of evaluating c must be greater or equal than 0.

Some examples:

- $[x_{i,1} \rightarrow r_{i,1}^4]_2^+ \equiv$ +[x{i,1} --> r{i,1}*4]'2
- $d_k[]_2^0 \rightarrow [d_{k+1}]_2^0 \equiv$ d{k}[]'2 --> [d{k+1}]
- $[d_k]_2^+ \rightarrow []_2^0 d_k \equiv$ +[d{k}]'2 --> []d{k}
- $[d_k]_2^0 \rightarrow [d_k]_2^+ [d_k]_2^- \equiv$ [d{k}]'2 --> +[d{k}]-[d{k}]
- $[a]_2^- \rightarrow b \equiv$ -[a]'2 --> b
- $Y_{i,j}[]_2 \xrightarrow{k_{i,8}} [B^{k_i,12}]_2 \equiv$ Y{i,j}[]'2 --> [B*k{i,12}]'2::k{i,8}
- $[RNAP+ < cap.\omega.op >]_m \xrightarrow{c} [< cap.\omega.RNAP.op >]_m \equiv$
  [RNAP,<cap.?.op>]'m --> [<cap.?.RNAP.op>]'m::c

Parametric sentences. In P-Lingua, it is possible to define parametric sentences by using the following format:

```
sentence : range1, ..., rangeN, restriction1, ...,
restrictionN;
```

where **sentence** is a sentence of the language, or a sequence of sentences in brackets, and **range1, ..., rangeN** is a comma-separated list of ranges with the format:

```
min_value <= index <= max_value
```

where **min_value** and **max_value** are numeric expressions, integer numbers or variables, and **index** is a variable that can be used in the context of the sentence. It is possible to use the operator $<$ instead of $<=$.

And **restriction1, ..., restrictionN** are optional restrictions for the indexes values which the next syntax:

```
value1 <> value2
```

where **value1** and **value2** are numeric expressions, integer numbers or variables.

The sentence will be repeated for each possible values of each **index**.

Some examples of parametric sentences:

1. $[d_k]_2^0 \rightarrow [d_k]_2^+ [d_k]_2^- : 1 \leq k \leq n \equiv$
   `[d{k}]'2 --> +[d{k}]-[d{k}] : 1<= k <= n;`

2. $[x_{i,j} \rightarrow x_{i,j-1}]_2^+ : 1 \leq i \leq m, 2 \leq j \leq n, i \neq j \equiv$
   `+[x{i,j} --> x{i,j-1}]'2 : 1<=i<=m,2<=j<=n,i<>j;`

Inclusion of comments. The programs in P-Lingua can be commented by writing phrases into the text strings /* and */.

Inclusion of debug information. Each rule sentence can optionally include a debug message which will be presented every time the rule is executed by the simulator. The syntax to write a debug message associated to a rule definition is defined as follows:

```
rule_definition @debug "debug message"
```

5   Command-Line Tools

P-Lingua 1.0 provided command-line tools for simulating P systems and compiling files which specify P systems [3]. In P-Lingua 2.0, the command-line tool general syntax has changed but, as it provides backwards compatibility, all valid actions in P-Lingua 1.0 are still valid in P-Lingua 2.0, as well.

5.1   Compilation Command-Line Tool

The command-line tool general syntax for compiling input files is defined as follows:

```
plingua [-input_format] input_file [-output_format]
output_file [-v verbosity_level] [-h]
```

The command header `plingua` reports the system to compile the P system specified on a file to a file specified on another, whereas the file `input_file` contains the program that we want to be compiled, and `output_file` is the name of the file that is generated [3]. Optional arguments are in square brackets:

- The option `-input_format` defines the format followed by `input_file`, which should be an input format.
- At this stage, valid input formats are:
  - P-Lingua
  - XML
- If no input format is set, the P-Lingua format is assumed.
- The option `-output_format` defines the format followed by `output_file`, which should be an output format.
- At this stage, valid output formats are:
  - XML
  - bin
- If no input format is set, the XML format is assumed by default.
- The option `-v` verbosity level is a number between 0 and 5 indicating the level of detail of the messages shown during the compilation process [3].
- The option `-h` displays some help information [3].

5.2   Simulation Command-Line Tool

The simulations are launched from the command line as follows:

```
plingua_sim [-input_format] input_file -o output_file [-v
verbosity level] [-h] [-to timeout] [-st steps] [-mode
simulatorID] [-a] [-b]
```

The command header `plingua_sim` reports the system to simulate the P system specified on a file, whereas `input_xml` is an XML document where a P system is formatted on, and output file is the name of the file where the report about the simulated computation will be saved [3]. Optional arguments are in brackets:

- The option `-input_format` defines the format followed by `input_file`, which should be an input format.
- The option `-v` verbosity level is a number between 0 and 5 indicating the level of detail of the messages shown during the compilation process [3]. If no value is specified, by default it is 3.

- The option -h displays some help information [3].
- The option -to sets a timeout for the simulation defined in timeout (in milliseconds), so when the time out has elapsed the simulation is halted. If the simulation has reached a halting configuration before the time out has elapsed this option has no effect.
- The option -st sets a maximum number of steps the simulation can take (defined in steps), so when the time out has elapsed the simulation comes to a halt. If the simulation has reached a halting configuration or the time out has elapsed (in case the option -to is set) before the specified number of steps have been taken this option has no effect.
- The option -mode sets the specific simulator to simulate the P system (defined in simulatorID). This option reports an error in case the simulator defined by simulatorID is not a valid simulator for the P system model.
- The option -a defines if the simulation can take alternative steps. This option reports an error if the simulator does not support alternative steps.
- The option -b defines if the simulation can step backwards. As every simulator supports stepping backwards, this option does not report errors.

6   pLinguaCore

pLinguaCore © is a JAVA library which performs all functions supported by P-Lingua 2.0, that is, models definition, simulators and formats. This library reports the rules and membrane structure read from a file where a P system is defined, detects errors in the file, reports them. And, if the P system is defined in P-Lingua language, locates the error on the file. This library performs simulations by using the simulators implemented as well as taking into account all options defined. It reports the simulation process, by displaying the current configuration as text and reporting the elapsed time. Eventually, this library translates files, which define a P system, between formats, for instance, from P-Lingua language format to binary format. For more information and library documentation, please browse the P-Lingua website [24]. This library is free software published under GNU GPL license [21], so everyone who is interested can change and distribute this library respecting the license conditions.

7   A Tool for Simulating Ecosystems Based on P-Lingua

The Bearded Vulture (Gypaetus barbatus) is an endangered species in Europe that feeds almost exclusively on bone remains of wild and domestic ungulates. In [1], it is presented a first model of an ecosystem related to the Bearded Vulture in the Pyrenees (NE Spain), by using probabilistic P systems where the inherent stochasticity and uncertainty in ecosystems are captured by using probabilistic strategies. In order to validate experimentally the designed P system (see figure 1) the authors have developed a simulator that allows them to analyze the evolution of the ecosystem under different initial conditions. That software application is focused on a particular P system, specifically, the initial model of

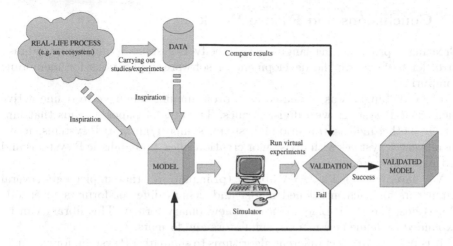

Fig. 1. Validation process

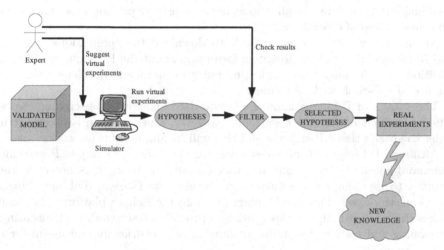

Fig. 2. Virtual experimentation

the ecosystem presented in [1]. With the aim of improving the model, the authors are adding ingredients to it, such as new species and a more complex behaviour for the animals. In this sense, a second version of the model is presented in [2]

A new GPL [21] licensed JAVA application with a friendly user-interface sitting on the pLinguaCore library has been developed. This application provides a flexible way to check, validate and improve computational models of ecosystem based on P systems instead of designing new software tools each time new ingredients are added to the models. Furthermore, it is possible to change the initial parameters of the modelled ecosystem in order to make the virtual experiments suggested by experts (see figure 2). These experiments will provide results that can be interpreted in terms of hypotheses. Finally, some of these hypotheses will be selected by the experts in order to be checked in real experiments.

8   Conclusions and Future Work

Creating a programming language to specify P systems is an important task in order to facilitate the development of software applications for membrane computing.

In [3], P-Lingua was presented as a programming language to define active membrane P systems with division rules. The present paper extends that language to other models: transition P systems, symport/antiport P systems, active membrane P systems with division or creation rules, probabilistic P systems and stochastic P systems.

We have developed a JAVA library (pLinguaCore) that implements several simulators for each mentioned model and defines different formats to encode P systems, like the P-Lingua one or a new binary format. This library can be expanded to define new models, simulators and formats.

It is possible to select different algorithms to simulate a P system, for example, there are two different algorithms for stochastic P systems. The library can be used inside other software applications, in this sense, we present a tool for virtual experimentation of ecosystems.

An internet website [24] is available to download the applications, libraries and source-code, as well as provide information about the P-Lingua project. In addition, this site aims to be a meeting point for users and developers through the use of web-tools such as forums.

The syntax of P-Lingua language is standard enough for specifying several different models of cell–like P systems. However, a new version is necessary in order to specify tissue P systems and this will be aim of a future work.

Although P-Lingua 2.0 provides a way to simulate and compile P systems, command-line tools are usually not user-friendly. It means it is not easy and intuitive to use them. For this purpose, a new user interface called pLinguaPlugin has been developed. This one is integrated into the Eclipse platform [23], so it makes the most of Eclipse's capabilities to provide a framework for translating, developing and testing P systems. It aims to be user-friendly and useful for P system researchers.

Acknowledgement

The authors acknowledge the support of the project TIN2006–13425 of the Ministerio de Educación y Ciencia of Spain, cofinanced by FEDER funds, and the support of the "Proyecto de Excelencia con Investigador de Reconocida Valía" of the Junta de Andalucía under grant TIC04200.

References

1. Cardona, M., Colomer, M.A., Pérez–Jiménez, M.J., Sanuy, D., Margalida, A.: Modeling ecosystems using P systems: The Bearded Vulture, a case study. In: Corne, D.W., Frisco, P., Paun, G., Rozenberg, G., Salomaa, A. (eds.) WMC 2008. LNCS, vol. 5391, pp. 137–156. Springer, Heidelberg (2009)

2. Cardona, M., Colomer, M.A., Margalida, A., Pérez–Hurtado, I., Pérez–Jiménez, M.J., Sanuy, D.: P System based model of an ecosystem of the scavenger birds. In: Păun, G., et al. (eds.) WMC 2009. LNCS, vol. 5957. Springer, Heidelberg (2010)

3. Díaz–Pernil, D., Pérez–Hurtado, I., Pérez–Jiménez, M.J., Riscos–Núñez, A.: A P-lingua programming environment for membrane computing. In: Proce. 9th Workshop on Membrane Computing, pp. 155–172 (2008)

4. Fontana, F., Bianco, L., Manca, V.: P systems and the modelling of biochemical oscillations. In: Freund, R., Păun, G., Rozenberg, G., Salomaa, A. (eds.) WMC 2005. LNCS, vol. 3850, pp. 199–208. Springer, Heidelberg (2006)

5. Freund, R., Verlan, S.: A formal framework for static (tissue) P systems. In: Eleftherakis, G., Kefalas, P., Păun, G., Rozenberg, G., Salomaa, A. (eds.) WMC 2007. LNCS, vol. 4860, pp. 271–284. Springer, Heidelberg (2007)

6. Gibson, M.A., Bruck, J.: Efficient exact stochastic simulation of chemical systems with many species and many channels. J. Phys. Chem. 104, 1876–1889 (2000)

7. Gillespie, D.T.: Exact stochastic simulation of coupled chemical reactions. J. Phys. Chem. 81, 2340–2361 (1977)

8. Gutiérrez–Naranjo, M.A., Pérez–Jiménez, M.J., Riscos–Núñez, A.: Towards a programming language in cellular computing. Electronic Notes in Theoretical Computer Science 123, 93–110 (2005)

9. Ito, M., Martín–Vide, C., Păun, G.: A characterization of Parikh sets of ET0L languages in terms of P systems. In: Ito, M., Păun, G., Yu, S. (eds.) Words, semigroups and transducers, pp. 239–254. Word Scientific, Singapore (2001)

10. Madhu, M., Krithivasan, K.: P systems with membrane creation: Universality and efficiency. In: Margenstern, M., Rogozhin, Y. (eds.) MCU 2001. LNCS, vol. 2055, pp. 276–287. Springer, Heidelberg (2001)

11. Martínez–del–Amor, M.A., Pérez–Hurtado, I., Pérez–Jiménez, M.J., Cecilia, J.M., Guerrero, G.D., García, J.M.: Simulation of recognizer P systems by using many-core GPUs. In: Păun, G., et al. (eds.) WMC 2009. LNCS, vol. 5957. Springer, Heidelberg (2010)

12. Obtulowicz, A.: Probabilistic P systems. In: Păun, G., Rozenberg, G., Salomaa, A., Zandron, C. (eds.) WMC 2002. LNCS, vol. 2597, pp. 377–387. Springer, Heidelberg (2003)

13. Păun, A., Păun, G.: The power of communication: P systems with symport/antiport. New Generation Computing 20(3), 295–305 (2002)

14. Păun, G.: Computing with membranes. Journal of Computer and System Sciences 61(1), 108–143 (2000)

15. Păun, G.: P systems with active membranes. Journal of Automata, Languages and Combinatorics 1, 75–90 (2001)

16. Pérez–Jiménez, M.J., Romero–Campero, F.J.: Modelling gene expression control using P systems: The Lac operon, a case study. BioSystems 91, 438–457 (2008)

17. Pérez–Jiménez, M.J., Romero–Campero, F.J.: A model of the quorum sensing system in Vibrio Fischeri using P systems. Artificial Life 14, 95–109 (2008)

18. Pérez–Jiménez, M.J., Romero–Campero, F.J.: P Systems, a new computational modelling tool for systems biology. In: Priami, C., Plotkin, G. (eds.) Transactions on Computational Systems Biology VI. LNCS (LNBI), vol. 4220, pp. 176–197. Springer, Heidelberg (2006)

19. Pescini, D., Besozzi, D., Mauri, G., Zandron, C.: Dynamical probabilistic P systems. International Journal of Foundations of Computer Science 17(1), 183–195 (2006)

20. Romero–Campero, F.J.: P Systems, a Computational Modelling Framework for Systems Biology. Doctoral Thesis, University of Seville, Department of Computer Science and Artificial Intelligence (2008)
21. The GNU General Public License, http://www.gnu.org/copyleft/gpl.html
22. Java web page, http://www.java.com/
23. The Eclipse Project, http://www.eclipse.org
24. The P-Lingua website, http://www.p-lingua.org

Characterizing Tractability by
Tissue-Like P Systems

Rosa Gutiérrez-Escudero[1], Mario J. Pérez-Jiménez[1], and Miquel Rius-Font[2]

[1] Research Group on Natural Computing
Department of Computer Science and Artificial Intelligence
University of Sevilla
Avda. Reina Mercedes s/n, 41012 Sevilla, Spain
{rgutierrez,marper}@us.es
[2] Department of Applied Mathematics IV
Universitat Politécnica de Catalunya
Edifici C3, Despatx 016, Av. del Canal Olímpic, s/n
08860 Castelldefels, Spain
mrius@ma4.upc.edu

Abstract. In the framework of recognizer cell–like membrane systems it is well known that the construction of exponential number of objects in polynomial time is not enough to efficiently solve **NP**–complete problems. Nonetheless, it may be sufficient to create an exponential number of membranes in polynomial time.

In this paper, we study the computational efficiency of recognizer tissue P systems with communication (symport/antiport) rules and division rules. Some results have been already obtained in this direction: (a) using communication rules and making no use of division rules, only tractable problems can be efficiently solved; (b) using communication rules with length three and division rules, **NP**–complete problems can be efficiently solved. In this paper, we show that the length of communication rules plays a relevant role from the efficiency point of view for this kind of P systems.

1 Introduction

Membrane Computing is a branch of Natural Computing and starts from the assumption that the processes taking place within the compartmental structure of a living cell can be interpreted as computations [9]. The computational devices in Membrane Computing are called *P systems*. Roughly speaking, a P system consists of a membrane structure, in the compartments of this structure there are multisets of objects which evolve according to given rules in a synchronous, non–deterministic, maximally parallel manner[1].

In recent years, many different models of P systems have been proposed and proved to be computationally universal. The most studied variants are characterized by a *cell-like* membrane structure, where communication happens between

---

[1] An informal overview can be found in [11] and further bibliography at [15].

G. Păun et al. (Eds.): WMC 2009, LNCS 5957, pp. 289–300, 2010.
© Springer-Verlag Berlin Heidelberg 2010

a membrane and the surrounding one. In this model, the membrane structure is hierarchical and the graph of the neighbourhood relation between compartments is a tree.

We shall focus here on another type of P systems, the so-called (because of their membrane structure) *tissue P Systems*. Instead of considering a hierarchical arrangement, membranes are modeled as nodes of an undirected graph. The biological inspiration for this variant is twofold: intercellular communication and cooperation between neurons. The common mathematical model of these two mechanisms is a net of processors dealing with symbols and communicating these symbols along channels specified in advance. Communication between cells is based on symport/antiport rules[2]. Symport rules move a number of objects across a membrane together in the same direction, whereas antiport rules move objects across a membrane in opposite directions.

Since the initial definition of tissue P systems several research lines have been developed and other variants have arisen. One of the most interesting variants of tissue P systems was presented in [12] where the definition of tissue P systems is combined with that of P systems with active membranes, yielding the model of *tissue P systems with cell division*. This model has been studied in depth in [1], where the importance of the cell division rules with respect to the computational power of the model is shown. Working with tissue P systems without division rules it is not possible to solve computationally hard problems [2] (unless **P=NP**). We focus now on the influence of the length of communication rules on the computational power of tissue P systems with cell division. In particular, when limiting this length to 1, only tractable problems can be efficiently solved. A proof of this result is presented here.

The paper is organized as follows. In Section 2, we recall some definitions related to tissue P systems (further information can be found in the literature, see [15]). Section 3 is devoted to formalizing the concept of polynomial–time solvability of decision problems by recognizer tissue P systems. In Section 4, we introduce a dependency graph for tissue P systems and use this technique to prove the main result of the paper. Finally, the last section contains some remarks and raises open questions and future work directions.

2   Recognizer Tissue P Systems

Firstly, the concept of *tissue P system of degree $q \geq 1$ with cell division* is introduced.

Definition 1. *A tissue P system of degree $q \geq 1$ with cell division is a tuple*

$$\Pi = (\Gamma, \Sigma, \Omega, \mathcal{M}_1, \ldots, \mathcal{M}_q, R, i_{in}, i_{out})$$

where:

1. *Γ is a finite alphabet (called working alphabet) whose elements are called objects;*
2. *Σ is a finite alphabet (called input alphabet) strictly contained in Γ ;*

---

[2] This method of communication for P systems was introduced in [8].

3. $\Omega \subseteq \Gamma \setminus \Sigma$ *is a finite alphabet, describing the set of objects located in the environment in arbitrarily many copies each;*
4. $\mathcal{M}_1, \ldots, \mathcal{M}_q$ *are strings over Γ, describing the multisets of objects placed in the q cells of the system;*
5. R *is a finite set of rules, of the following forms:*
   (a) Communication rules: $(i, u/v, j)$, *for* $i, j \in \{0, 1, 2, \ldots, q\}, i \neq j$, *and* $u, v \in \Gamma^*; 1, 2, \ldots, q$ *identify the cells of the system, 0 is the environment.*
   (b) Division rules: $[\, a\,]_i \rightarrow [\, b\,]_i [\, c\,]_i$, *where* $i \in \{1, 2, \ldots, q\}$ *and* $a, b, c \in \Gamma$.
6. $i_{in} \in \{1, \ldots, q\}$ *is the input cell, and* $i_{out} \in \{0, 1, \ldots, q\}$ *is the output cell.*

When applying a communication rule $(i, u/v, j)$, the objects of the multiset represented by u are sent from region i to region j and, simultaneously, the objects of multiset v are sent from region j to region i. We say that the sum of the lengths of u and v is the *length* of the rule.

When applying a division rule $[\, a\,]_i \rightarrow [\, b\,]_i [\, c\,]_i$, under the influence of object a, the cell with label i is divided in two cells with the same label; in the first copy the object a is replaced by b, in the second copy the object a is replaced by c; all other objects are replicated and copies of them are placed in the two new cells.

The rules of such a system are applied in a non-deterministic maximally parallel way as it is customary in membrane computing. In each step, all cells which can evolve must evolve in a maximally parallel way (in each step we apply a multiset of rules which is maximal, no further rule can be added), with the following important remark: if a cell divides, then the division rule is the only one which is applied for that cell at that step; its objects do not evolve by means of communication rules. In other words, before division a cell interrupts all its communication channels with the other cells and with the environment; the new cells resulting from division will interact with other cells or with the environment only at the next step – providing that they do not divide once again. The label of a cell precisely identifies the rules which can be applied to it.

A configuration of Π is described by the multisets of objects over Γ associated with all the cells present in the system and the multiset over $\Gamma \setminus \Omega$ associated with the environment (the objects in the environment which are in finitely many copies). For two configurations C_1, C_2 of Π, we write $C_1 \Rightarrow_\Pi C_2$, and we say that we have a *transition* from C_1 to C_2, if we can pass from C_1 to C_2 by applying the rules from R.

The initial configuration of the system is $(\emptyset, \mathcal{M}_1, \ldots, \mathcal{M}_q)$. For each multiset m over the input alphabet, the initial configuration of the system associated with it is $(\emptyset, \mathcal{M}_1, \ldots, \mathcal{M}_{i_{in}} \cup m, \ldots, \mathcal{M}_q)$. Then, m is an *input multiset* of every computation $\mathcal{C} = \{C_i\}_{i<r}$ such that C_0 is the initial configuration of Π associated with m.

All computations start from an initial configuration and proceed as stated above; only halting computations give a result, which is encoded by the objects in the output cell i_{out} in the last configuration. From now on, we will consider that the output is collected in the environment (that is, $i_{out} = 0$, and thus we will omit i_{out} in the definition of tissue P systems). In this way, if Π is a tissue

P system and $\mathcal{C} = \{C_i\}_{i<r}$ is a halting computation of Π, then the answer of computation $\mathcal{C}$ is

$$Output(\mathcal{C}) = \Psi_{\Gamma \setminus \Omega}(M_{r-1,0})$$

where Ψ is the Parikh function, and $M_{r-1,0}$ is the multiset over $\Gamma \setminus \Omega$ associated with the environment at the halting configuration C_{r-1}.

Let us recall that **NP**–completeness has been usually studied in the framework of *decision problems*, that is problems whose solution is either *yes* or *no*. More formally, a decision problem is a pair (I_X, θ_X) where I_X is a language over a finite alphabet whose elements are called *instances*, and θ_X is a total Boolean function over I_X.

Each decision problem $X = (I_X, \theta_X)$ has a language L_X over the alphabet of I_X associated with it, defined as follows: $L_X = \{a \in I_X : \theta_X(a) = 1\}$. Reciprocally, each language L over an alphabet Σ has a decision problem, X_L associated with it as follows: $I_{X_L} = \Sigma^*$, and $\theta_{X_L} = \{(x,1) : x \in L\} \cup \{(x,0) : x \notin L\}$.

Recognizer cell-like P systems were introduced in [14]. They are the natural framework to study and solve decision problems within Membrane Computing, since deciding whether an instance of a given problem has an affirmative or negative answer is equivalent to deciding if a string belongs or not to the language associated with the problem.

In literature, recognizer cell-like P systems are associated with P systems with *input* in a natural way. The data encoding an instance of the decision problem has to be provided to the P system in order to compute the appropriate answer. This is done by codifying each instance as a multiset placed in an *input membrane*. The output of the computation (**yes** or **no**) is sent to the environment in the last step of the computation. In this way, cell-like P systems with input and external output are devices which can be seen as black boxes, in the sense that the user provides the data before the computation starts, and then waits *outside* the P system until it sends to the environment the output in the last step of the computation.

In order to use these computational devices for solving decision problems, *recognizer tissue P systems* are introduced.

Definition 2. *A tissue P system with cell division of degree $q \geq 1$*

$$\Pi = (\Gamma, \Sigma, \Omega, \mathcal{M}_1, \ldots, \mathcal{M}_q, R, i_{in})$$

is a recognizer system if the following holds:

1. *The working alphabet Γ has two distinguished objects* **yes** *and* **no**, *present in at least one copy in some initial multisets $\mathcal{M}_1, \ldots, \mathcal{M}_q$, but not present in Ω.*
2. *All computations halt.*
3. *If $\mathcal{C} = \{C_i\}_{i<r}$ is a computation of Π, then either the object* **yes** *or the object* **no** *(but not both) must have been released into the environment, and only in the last step of the computation.*

Given a recognizer tissue P system with cell division, and a computation $\mathcal{C} = \{C_i\}_{i<r}$ of Π ($r \in \mathbf{N}$), we define the result of $\mathcal{C}$ as follows:

$$Output(\mathcal{C}) = \begin{cases} \text{yes, if } \Psi_{\{\text{yes,no}\}}(M_{r-1,0}) = (1,0) \\ \qquad \wedge \Psi_{\{\text{yes,no}\}}(M_{k,0}) \quad = (0,0) \text{ for } k = 0, \dots, r-2 \\ \text{no, } \text{ if } \Psi_{\{\text{yes,no}\}}(M_{r-1,0}) = (0,1) \\ \qquad \wedge \Psi_{\{\text{yes,no}\}}(M_{k,0}) \quad = (0,0) \text{ for } k = 0, \dots, r-2 \end{cases}$$

That is, $\mathcal{C}$ is an accepting computation (respectively, rejecting computation) if the object yes (respectively, no) appears in the environment (only) in the halting configuration of $\mathcal{C}$.

3 Polynomial–Time Solvability by Recognizer Tissue P Systems

In this section, the definition of polynomial–time (uniform) solvability of decision problems by a family of cell–like P systems is extended to solvability by a family of tissue P systems.

Definition 3. *We say that a decision problem $X = (I_X, \theta_X)$ is solvable in polynomial time by a family $\mathbf{\Pi} = \{\Pi(n) : n \in \mathbb{N}\}$ of recognizer tissue P systems with cell division if the following hold:*

- *The family $\mathbf{\Pi}$ is polynomially uniform by Turing machines, that is, there exists a deterministic Turing machine which constructs the system $\Pi(n)$ from $n \in \mathbb{N}$ in polynomial time with respect n.*
- *There exists a pair (cod, s) of polynomial-time computable functions over I_X (called a polynomial encoding of I_X in $\mathbf{\Pi}$) such that:*
  - *For each instance $u \in I_X$, $s(u)$ is a natural number and $cod(u)$ is an input multiset of the system $\Pi(s(u))$.*
  - *The family $\mathbf{\Pi}$ is polynomially bounded with regard to (X, cod, s); that is, there exists a polynomial function p, such that for each $u \in I_X$ every computation of $\Pi(s(u))$ with input $cod(u)$ is halting and, moreover, it performs at most $p(|u|)$ steps.*
  - *The family $\mathbf{\Pi}$ is sound with regard to (X, cod, s); that is, for each $u \in I_X$, if there exists an accepting computation of $\Pi(s(u))$ with input $cod(u)$, then $\theta_X(u) = 1$.*
  - *The family $\mathbf{\Pi}$ is complete with regard to (X, cod, s); that is, for each $u \in I_X$, if $\theta_X(u) = 1$, then every computation of $\Pi(s(u))$ with input $cod(u)$ is an accepting one.*

From the soundness and completeness conditions above we deduce that every P system $\Pi(n)$ is *confluent*, in the following sense: every computation of a system with the *same* input multiset must always give the *same* answer.

We denote by $\mathbf{PMC}_{TDC}$ the set of all decision problems which can be solved by means of recognizer tissue P systems with cell division in polynomial time.

This class is closed under polynomial–time reduction and under complement (see [13] for a similar result for cell-like P systems). We also denote by $\mathbf{PMC}_{TDC(k)}$ the set of all decision problems which can be solved by means of recognizer tissue P systems with cell division in polynomial time, by using communication rules whose length is, at most, k.

4   Dependency Graph Associated with Tissue P Systems

Let Π be a tissue P system with cell division and let all communication rules be of length 1. In this case, each rule of the system can be activated by a single object. Hence, there exists in a certain sense, a *dependency* between the object triggering the rule and the object or objects produced by its application. This dependency allows us to adapt the ideas developed in [5] for cell-like P systems with active membranes to tissue P systems with cell division and communication rules of length 1.

We can consider a general pattern $(a, i) \rightarrow (b_1, j) \dots (b_s, j)$ where $i, j \in \{0, 1, 2, \dots, q\}, i \neq j$, and $a, b \in \Gamma$. Communication rules correspond to the case $s = 1$ and $b_1 = a$, and division rules correspond to the case $s = 2$ and $j = i \neq 0$. The above pattern can be interpreted as follows: from the object a in the cell (or in the environment) labelled with i we can *reach* the objects $b_1, \dots, b_s$ in the cell (or in the environment) labelled with j.

Without loss of generality we can assume that all communication rules in the system obey the syntax $(i, a/\lambda, j)$, since every rule of the form $(j, \lambda/a, i)$ can be rewritten to follow the above syntax, with equivalent semantics.

Next, we formalize these ideas in the following definition.

Definition 4. *Let $\Pi = (\Gamma, \Sigma, \Omega, \mathcal{M}_1, \dots, \mathcal{M}_q, R, i_{in})$ be a tissue P system of degree $q \geq 1$ with cell division. Let $H = \{0, 1, \dots, q\}$. The dependency graph associated with Π is the directed graph $G_\Pi = (V_\Pi, E_\Pi)$ defined as follows:*

$$V_\Pi = \{(a, i) \in \Gamma \times H : \exists j \in H \; ((i, a/\lambda, j) \in R \; \vee \; (j, a/\lambda, i) \in R) \; \vee$$
$$\exists b, c \in \Gamma \; ([a]_i \rightarrow [b]_i[c]_i \in R \; \vee \; [b]_i \rightarrow [a]_i[c]_i \in R)\},$$
$$E_\Pi = \{((a, i), (b, j)) : (a = b \; \wedge \; (i, a/\lambda, j) \in R) \; \vee$$
$$\exists c \in \Gamma \; ([a]_i \rightarrow [b]_i[c]_i \in R \wedge j = i)\}.$$

In what follows, every algorithm is analyzed under the *uniform cost criterion*, that is, each basic instruction/operation takes constant time.

Proposition 1. *Let $\Pi = (\Gamma, \Sigma, \Omega, \mathcal{M}_1, \dots, \mathcal{M}_q, R, i_{in})$ be a tissue P system with cell division, in which the length of all communication rules is 1. Let $H = \{0, 1, \dots, q\}$. There exists a deterministic Turing machine that constructs the dependency graph G_Π associated with Π, in polynomial time (that is, the run–time is bounded by a polynomial function depending on the total number of rules).*

Proof. A deterministic algorithm that, given a P system Π with the set R of rules, constructs the corresponding dependency graph, is the following:

```
Input: Π (with R as its set of rules)
Vₙ ← ∅; Eₙ ← ∅
for each rule r ∈ R of Π do
    if r = (i, a/λ, j) then
        Vₙ ← Vₙ ∪ {(a,i),(a,j)}; Eₙ ← Eₙ ∪ {((a,i),(a,j))}
    if r = [a]ᵢ → [b]ᵢ[c]ᵢ then
        Vₙ ← Vₙ ∪ {(a,i),(b,i),(c,i)};
        Eₙ ← Eₙ ∪ {((a,i),(b,i)),((a,i),(c,i))}
```

The running time of this algorithm is bounded by $O(|R|) \subseteq O(|\Gamma|^3 \cdot q^2)$. $\qquad\square$

Proposition 2. *Let* $\Pi = (\Gamma, \Sigma, \Omega, \mathcal{M}_1, \ldots, \mathcal{M}_q, R, i_{in})$ *be a tissue P system with cell division, in which the length of all communication rules is 1. Let* $H = \{0, 1, \ldots, q\}$. *Let* Δ_Π *be defined as follows:*

$$\Delta_\Pi = \{(a,i) \in \Gamma \times H : \text{there exists a path (within the dependency graph)} \\ \text{from } (a,i) \text{ to } (\text{yes}, 0)\}.$$

Then, there exists a Turing machine that constructs the set Δ_Π *in polynomial time (that is, the run–time is bounded by a polynomial function depending on the total number of rules).*

Proof. We can construct the set Δ_Π from Π as follows:

- We construct the dependency graph G_Π associated with Π.
- Then we consider the following algorithm:

```
Input: Gₙ = (Vₙ, Eₙ)
    Δₙ ← ∅
    for each (a,i) ∈ Vₙ do
        if reachability (Gₙ, (a,i), (yes,0)) = yes then
            Δₙ ← Δₙ ∪ {(a,i)}
```

The running time of this algorithm is of order[3] $O(|V_\Pi| \cdot (|V_\Pi| + E_\Pi))$, hence it is of order $O(|\Gamma|^3 \cdot q^3)$. $\qquad\square$

---

[3] The Reachability Problem is the following: *given a (directed or undirected) graph, G, and two nodes a, b, determine whether or not the node b is reachable from a, that is, whether or not there exists a path in the graph from a to b.* It is easy to design an algorithm running in polynomial time solving this problem. For example, given a (directed or undirected) graph, G, and two nodes a, b, we consider a depth–first–search with source a, and we check if b is in the tree of the computation forest whose root is a. The total running time of this algorithm is $O(|V|+|E|)$, that is, in the worst case is quadratic in the number of nodes. Moreover, this algorithm needs to store a linear number of items (it can be proved that there exists another polynomial–time algorithm which uses $O(\log^2(|V|))$ space).

Notation: Let $\Pi = (\Gamma, \Sigma, \Omega, \mathcal{M}_1, \ldots, \mathcal{M}_q, R, i_{in})$ be a tissue P system with cell division. Let m be a multiset over Σ. Then we denote $\mathcal{M}_j^* = \{(a, j) : a \in \mathcal{M}_j\}$, for $1 \leq j \leq q$, and $m^* = \{(a, i_{in}) : a \in m\}$.

Below, we characterize accepting computations of a recognizer tissue P system with cell division and communication rules of length 1 by distinguished paths in the associated dependency graph.

Lemma 1. *Let $\Pi = (\Gamma, \Sigma, \Omega, \mathcal{M}_1, \ldots, \mathcal{M}_q, R, i_{in})$ be a recognizer confluent tissue P system with cell division in which the length of all communication rules is 1. The following assertions are equivalent:*

(1) There exists an accepting computation of Π.
(2) There exists $(a_0, i_0) \in \bigcup_{j=1}^q \mathcal{M}_j^*$ and a path in the dependency graph associated with Π, from (a_0, i_0) to $(\mathbf{yes}, 0)$.

Proof. (1) $\Rightarrow$ (2) First, we show that for each accepting computation $\mathcal{C}$ of Π there exists $(a_0, i_0) \in \bigcup_{j=1}^q \mathcal{M}_j^*$ and a path $\gamma_{\mathcal{C}}$ in the dependency graph associated with Π from (a_0, i_0) to $(\mathbf{yes}, 0)$. By induction on the length n of $\mathcal{C}$.

If $n = 1$, a single step is performed in $\mathcal{C}$ from C_0 to C_1. A rule of the form $(j, \mathbf{yes}/\lambda, 0)$, with $a \in \Gamma, j \neq 0$, has been applied in that step. Then, $(\mathbf{yes}, j) \in \mathcal{M}_j^*$, for some $j = 1, \ldots, q$. Hence, $((\mathbf{yes}, j), (\mathbf{yes}, 0))$ is a path in the dependency graph associated with Π.

Let us suppose that the result holds for n. Let $\mathcal{C} = (C_0, C_1, \ldots, C_n, C_{n+1})$ be an accepting computation of Π. Then $\mathcal{C}' = (C_1, \ldots, C_n, C_{n+1})$ is an accepting computation of the system $\Pi' = (\Gamma, \Sigma, \Omega, \mathcal{M}_1', \ldots, \mathcal{M}_q', R, i_{in})$, being $\mathcal{M}_j'$ the contents of cell j in configuration C_1, for $1 \leq j \leq q$. By induction hypothesis there exists an object b_0 in a cell i_0 from C_1, and a path in the dependency graph associated with Π' from (b_0, i_0) to $(\mathbf{yes}, 0)$. If (b_0, i_0) is an element of configuration C_0 (that means that in the first step a division rule has been applied to cell i_0), then the result holds. Otherwise, there is an element (a_0, j_0) in C_0 producing (b_0, i_0). So, there exists a path $\gamma_{\mathcal{C}}$ in the dependency graph associated with Π from (a_0, j_0) to $(\mathbf{yes}, 0)$.

(2) $\Rightarrow$ (1). Let us see that for each $(a_0, i_0) \in \bigcup_{j=1}^q \mathcal{M}_j^*$ and for each path in the dependency graph associated with Π from (a_0, i_0) to $(\mathbf{yes}, 0)$, there exists an accepting computation of Π. By induction on the length n of the path.

If $n = 1$, we have a path $((a_0, i_0), (\mathbf{yes}, 0))$. Then, $a_0 = \mathbf{yes}$ and the computation $\mathcal{C} = (C_0, C_1)$ where the rule $(i_0, \mathbf{yes}/\lambda, 0)$ belongs to a multiset of rules m_0 that produces configuration C_1 from C_0 is an accepting computation of Π.

Let us suppose that the result holds for n. Let

$$((a_0, i_0), (a_1, i_1), \ldots (a_n, i_n), (\mathbf{yes}, 0))$$

be a path in the dependency graph of length $n + 1$. If $(a_0, i_0) = (a_1, i_1)$, then the result holds by induction hypothesis. Otherwise, let C_1 be the configuration of Π reached from C_0 by the application of a multiset of rules containing the rule that produces (a_1, i_1) from (a_0, i_0). Then $((a_1, i_1), \ldots (a_n, i_n), (\text{yes}, 0))$ is a path of length n in the dependency graph associated with the system

$$\Pi' = (\Gamma, \Sigma, \Omega, \mathcal{M}'_1, \ldots, \mathcal{M}'_q, R, i_{in})$$

where $\mathcal{M}'_j$ is the content of cell j in configuration C_1, for $1 \leq j \leq q$. By induction hypothesis, there exists an accepting computation $\mathcal{C}' = (C_1, \ldots, C_t)$ of Π'. Hence, $\mathcal{C} = (C_0, C_1, \ldots, C_t)$ is an accepting computation of Π. $\qquad \square$

Next, given a family $\mathbf{\Pi} = (\Pi(n))_{n \in \mathbf{N}}$ of recognizer tissue P system with cell division in which the length of all communication rules is 1, solving a decision problem, we will characterize the acceptance of an instance of the problem, w, using the set $\Delta_{\Pi(s(w))}$ associated with the system $\Pi(s(w))$ that processes the given instance w. More precisely, the instance is accepted by the system if and only if there is an object in the initial configuration of the system $\Pi(s(w))$ with input $cod(w)$ such that there exists a path in the associated dependency graph starting from that object and reaching the object **yes** in the environment.

Proposition 3. *Let $X = (I_X, \theta_X)$ be a decision problem. Let $\mathbf{\Pi} = (\Pi(n))_{n \in \mathbf{N}}$ be a family of recognizer tissue P system with cell division solving X in which the length of all communication rules is 1, according to Definition 3. Let (cod, s) be the polynomial encoding associated with that solution. Then, for each instance w of the problem X the following assertions are equivalent:*

(a) $\theta_X(w) = 1$ (that is, the answer to the problem is yes for w).

(b) $\Delta_{\Pi(s(w))} \cap ((cod(w))^* \cup \bigcup_{j=1}^{q} \mathcal{M}_j^*) \neq \emptyset$, where $\mathcal{M}_1, \ldots, \mathcal{M}_q$ are the initial multisets of the system $\Pi(s(w))$.

Proof. Let $w \in I_X$. Then $\theta_X(w) = 1$ if and only if there exists an accepting computation of the system $\Pi(s(w))$ with input multiset $cod(w)$. From Lemma 1, we know this condition is equivalent to the following: in the initial configuration of $\Pi(s(w))$ with input multiset $cod(w)$ there exists at least one object $a \in \Gamma$ in a cell labelled with i such that in the dependency graph the node $(\text{yes}, 0)$ is reachable from (a, i).

Hence, $\theta_X(w) = 1$ if and only if $\Delta_{\Pi(s(w))} \cap \mathcal{M}_j^* \neq \emptyset$ for some $j \in \{1, \ldots, q\}$, or $\Delta_{\Pi(s(w))} \cap (cod(w))^* \neq \emptyset$. $\qquad \square$

Theorem 1. $\mathbf{P} = \boldsymbol{PMC}_{TDC(1)}$

Proof. We have $\mathbf{P} \subseteq \mathbf{PMC}_{TDC(1)}$ because $\mathbf{PMC}_{TDC(1)}$ is a nonempty class closed under polynomial–time reduction. Next, we show that $\mathbf{PMC}_{TDC(1)} \subseteq \mathbf{P}$.

Let $X \in \mathbf{PMC}_{TDC(1)}$ and let $\mathbf{\Pi} = (\Pi(n))_{n \in \mathbf{N}}$ be a family of recognizer tissue P systems with cell division solving X, according to Definition 3. Let (cod, s) be the polynomial encoding associated with that solution.

We consider the following deterministic algorithm:

```
Input: An instance w of X
  - Construct the system Π(s(w)) with input multiset cod(w).
    q(w) ← degree of Π(s(w))
  - Construct the dependency graph G_{Π(s(w))} associated with Π(s(w)).
  - Construct the set Δ_{Π(s(w))} as indicated in Proposition 2
    answer ← no; j ← 1
    while j ≤ q(w) ∧ answer = no do
        if Δ_{Π(s(w))} ∩ M*_j ≠ ∅ then
          answer ← yes
        j ← j + 1
    endwhile
    if Δ_{Π(s(w))} ∩ (cod(w))* ≠ ∅ then
        answer ← yes
```

On the one hand, the answer of this algorithm is **yes** if and only if there exists a pair (a, i) belonging to $\Delta_{\Pi(s(w))}$ such that the symbol a appears in the cell labelled with i in the initial configuration (with input the multiset $cod(w)$).

On the other hand, a pair (a, i) belongs to $\Delta_{\Pi(s(w))}$ if and only if there exists a path from (a, i) to $(\text{yes}, 0)$, that is, if and only if we can obtain an accepting computation of $\Pi(s(w))$ with input $cod(w)$. Hence, the algorithm above described solves the problem X.

The cost to determine whether or not $\Delta_{\Pi(s(w))} \cap \mathcal{M}^*_j \neq \emptyset$ (or $\Delta_{\Pi(s(w))} \cap (cod(w))^* \neq \emptyset$) is of order $O(|\Gamma(w)|^2 \cdot q(w)^2)$, being $\Gamma(w)$ the working alphabet of $\Pi(s(w))$.

Hence, the running time of this algorithm can be bounded by $f(|w|) + O(|R(w)|) + O(|\Gamma(w)|^3 \cdot q(w)^3) + O(q(w) \cdot |\Gamma(w)|^2 \cdot q(w)^2) + O(|\Gamma(w)|^2 \cdot q(w)^2) \subseteq f(|w|) + O(|\Gamma(w)|^3 \cdot q(w)^3)$, where f is the (total) cost of the (considered) polynomial encoding from X to $\mathbf{\Pi}$ and $R(w)$ (respectively, $\Gamma(w)$ and $q(w)$) is the set of rules (resp. the working alphabet and the degree) of the system $\Pi(s(w))$. But from Definition 3, we deduce that all involved parameters are polynomials in $|w|$. That is, the algorithm is polynomial in the size $|w|$ of the input. □

In [3], a polynomial time solution of the **Vertex Cover** problem was given by using a family of recognizer tissue P systems with cell division and communication rules of length at most 3. Hence, $\mathbf{NP} \cup \mathbf{co\text{-}NP} \subseteq \mathbf{PMC}_{TDC(3)}$.

Hence, in the framework of recognizer tissue P systems with cell division, the length of the communication rules provides a borderline between efficiency and non-efficiency. Specifically, a frontier is obtained when we pass from length 1 to length 3.

5   Final Remarks and Future Work

It is known [2] that tissue P systems with communication rules and without division rules can efficiently solve only tractable problems. It is also well known that by adding division rules we can efficiently solve **NP**–complete problems in linear time by using communication rules with length at most 3 [3].

In order to obtain new borderlines between tractability and intractability of problems, we study the possibility to restrict the length of communication rules to 1, allowing division rules. By using the dependency graph technique of cell–like P systems, we have shown that only tractable problems can be efficiently solved in that scenario.

Several questions regarding the role of the length remain open, for example:

- What happens if we consider tissue P systems using communication rules of length at most 2?
- In the solution provided in [3], antiport rules of length at most 3 were used. Would it be possible to provide another solution in which all rules of length 3 were symport?

Other open issues related to tissue P systems that may be interesting are:

- Analyzing a new role for the environment. More specifically, consider in the initial configuration only objects with finite multiplicity in the environment. It seems that this new scenario would be equivalent to tissue P systems without environment, with a new distinct cell with no division rules associated. Is it still possible to solve **NP**–complete problems in polynomial time in this new framework, permitting division rules?
- Considering variations in the semantics of division rules, for example, dispensing with replication or with evolution. Division rules without replication would obey the syntax $[\,a\,]_i \rightarrow [\;]_i[\,u\,]_i$, where $i \in \{1, 2, \ldots, q\}$, $a \in \Gamma$ and $u \in \Gamma^*$, meaning that under the influence of object a, the cell with label i is divided in two cells with the same label. The first copy contains all objects of the original cell except for a and in the second copy the content of the original cell is replaced by the multiset u. Division rules without evolution would be either of the form $[\,a\,]_i \rightarrow [\;]_i[\;]_i$ or $[\,a\,]_i \rightarrow [\,a\,]_i[\,a\,]_i$, where $i \in \{1, 2, \ldots, q\}$ and $a \in \Gamma$. In both cases, under the influence of object a, the cell with label i is divided in two cells. All objects are replicated and copies of them are placed in the two new cells, except for a in the first case.

Acknowledgement

The authors acknowledge the support of the project TIN2006–13425 of the Ministerio de Educación y Ciencia of Spain, cofinanced by FEDER funds, and the support of the Project of Excellence with *Investigador de Reconocida Valía* of the Junta de Andalucía, grant P08-TIC-04200.

References

1. Díaz–Pernil, D.: Sistemas celulares de tejidos: Formalización y eficiencia computacional. Ph D. Thesis, University of Sevilla (2008)
2. Díaz–Pernil, D., Pérez–Jiménez, M.J., Romero–Jiménez, A.: Efficient simulation of tissue-like P systems by transition cell-like P systems. Natural Computing, http://dx.doi.org/10.1007/s11047-008-9102-z
3. Díaz–Pernil, D., Pérez–Jiménez, M.J., Riscos–Núñez, A., Romero–Jiménez, A.: Computational efficiency of cellular division in tissue-like membrane systems. Romanian Journal of Information Science and Technology 11(3), 229–241 (2008)
4. Frisco, P., Hoogeboom, H.J.: Simulating counter automata by P systems with symport/antiport. In: Păun, G., Rozenberg, G., Salomaa, A., Zandron, C. (eds.) WMC 2002. LNCS, vol. 2597, pp. 288–301. Springer, Heidelberg (2003)
5. Gutiérrez–Naranjo, M.A., Pérez–Jiménez, M.J., Riscos–Núñez, A., Romero–Campero, F.J.: On the power of dissolution in P systems with active membranes. In: Freund, R., Păun, G., Rozenberg, G., Salomaa, A. (eds.) WMC 2005. LNCS, vol. 3850, pp. 224–240. Springer, Heidelberg (2006)
6. Gutiérrez–Naranjo, M.A., Pérez–Jiménez, M.J., Riscos–Núñez, A., Romero–Campero, F.J., Romero–Jiménez, A.: Characterizing tractability by cell-like membrane systems. In: Subramanian, K.G., Rangarajan, K., Mukund, M. (eds.) Formal models, languages and applications, ch. 9, pp. 137–154. World Scientific, Singapore (2006)
7. Martín–Vide, C., Pazos, J., Păun, Gh., Rodríguez–Patón, A.: Tissue P systems. Theoretical Computer Science 296, 295–326 (2003)
8. Păun, A., Păun, Gh.: The power of communication: P systems with symport/antiport. New Generation Computing 20(3), 295–305 (2002)
9. Păun, Gh.: Computing with membranes. Journal of Computer and System Sciences 61(1), 108–143 (2000)
10. Păun, Gh.: Membrane Computing. An Introduction. Springer, Berlin (2002)
11. Păun, Gh., Pérez–Jiménez, M.J.: Recent computing models inspired from biology: DNA and membrane computing. Theoria 18(46), 72–84 (2003)
12. Păun, Gh., Pérez–Jiménez, M.J., Riscos–Núñez, A.: Tissue P Systems with cell division. International Journal of Computers, Communications & Control III(3), 295–303 (2008); A preliminary version in Păun, Gh., et al. (eds.) Second Brainstorming Week on Membrane Computing, Sevilla, Report RGNC 01/2004, pp. 380–386 (2004)
13. Pérez–Jiménez, M.J., Romero–Jiménez, A., Sancho–Caparrini, F.: Complexity classes in cellular computing with membranes. Natural Computing 2(3), 265–285 (2003)
14. Pérez–Jiménez, M.J., Romero–Jiménez, A., Sancho–Caparrini, F.: A polynomial complexity class in P systems using membrane division. Journal of Automata, Languages and Combinatorics 11(4), 423–434 (2003); A preliminary version in Csuhaj-Varjú, E., et al. (eds.) Proc. Fifth Intern. Workshop on Descriptional Complexity of Formal Systems, DCFS 2003, Budapest, Hungary, July 12-14, 284–294 (2003)
15. P systems web page, http://ppage.psystems.eu/

Searching Previous Configurations
in Membrane Computing

Miguel A. Gutiérrez-Naranjo and Mario J. Pérez-Jiménez

Research Group on Natural Computing
Department of Computer Science and Artificial Intelligence
University of Sevilla
Avda. Reina Mercedes s/n, 41012, Sevilla, Spain
{magutier,marper}@us.es

Abstract. Searching all the configurations C' which produce a given configuration C is an extremely hard task. The current approximations are based on heavy hand-made calculus by considering the specific features of the given configuration. In this paper we present a general method for characterizing all the configurations C' which produce a given configuration C in the framework of transition P systems without cooperation and without dissolution.

1 Introduction

Given a computational model with a universal clock, where the time is considered in a discrete way and the transition from a state to the next one is produced by a set of rules, it is usual to wonder about the previous state of a given one. Note that the determinism of the model does not make the solution easier, since the determinism of the computation does not lead to the determinism of the reverse computation. One can pass deterministically from S to S_0 and from S' to S_0, but given S_0, the reversed computation is not deterministic. A special situation is considered when the rules are *reversible*, i.e., rules for which one can change the left hand side and right hand side of the rule and the new rule suits to the syntactic constraints of the considered P system model. In this case, it suffices to apply the reversed rules to S_1 according to the computational model to obtain the desired states (it was studied for P systems in [1]).

In this paper we study the problem of characterizing the set of configurations of a P system that produce a given configuration in one transition step. We study the case in which the P system is not necessarily deterministic and the rules are not reversible in general. In our study, we modify the representation for rules and configurations used in [2,4] by introducing the notion of order between pairs as in [3]. We use Linear Algebra as a tool for computing and consider a restricted version of transition P systems without cooperation where the membrane structure does not change along the computation.

The paper is organized as follows: first we expose an example that shows the necessity of finding a method for computing backwards, avoiding the heavy

G. Păun et al. (Eds.): WMC 2009, LNCS 5957, pp. 301–315, 2010.
© Springer-Verlag Berlin Heidelberg 2010

calculus based on specific features of the given configuration. Next, our P system model is briefly introduced and a representation for configurations and rules in such a P system is presented. In Section 6 we prove our main result: Computing the set of all the configurations C' which produce a given configuration C can be reduced to find solutions of a system of linear equations with values in $\mathbb{N}$. In Section 7 we provide a general method of calculus based on our theorem. Finally, some conclusions and new open research lines are presented.

2   Motivation

Let us start with a P system Π with working alphabet $\Gamma = \{a, b, c\}$, set of labels $H = \{e, s\}$, membrane structure $\mu = [\,[\,]_e\,]_s$ and the following set of rules R:

$$\begin{array}{ll}
\textbf{Rule 1: } [\,a \to b^2 c\,]_e & \textbf{Rule 4: } [\,b \to a\,]_s \\
\textbf{Rule 2: } [\,a\,]_e \to a\,[\,]_e & \textbf{Rule 5: } a\,[\,]_e \to [\,c\,]_e \\
\textbf{Rule 3: } [\,b \to c^2\,]_s & \textbf{Rule 6: } [\,c \to a\,]_e
\end{array}$$

In Section 3, we will give a detailed description of the P system model studied in this paper, but by now it is enough to know that all the rules are applied in a non-deterministic maximal parallel way as usual in the general framework of Membrane Computing (see [5] for details).

Let us consider now the configuration $C' = [\,[a^2 b\,]_e\, a^2 c\,]_s$, i.e., the configuration in which the multiset placed in the membrane labelled by e is $a^2 b$ and the multiset in the membrane s is $a^2 c$. Our problem is to find the configuration (or configurations) C such that we can pass from C to C' in *one* transition step. In other words, we want to characterize *all* the configurations C such that produce C' in one transition step.

We can reason in the following way:

- We find two objects a in the membrane labelled by e in the configuration C'. Since rules 1 and 2 consume all the objects in the membrane e from the previous configuration C, we conclude that such pair of objects a must be produced by the application of rule(s) of Π. It is easy to check that only rule 6 produces objects a in membrane e, then the number of objects c in configuration C must be at least 2. If we look at the set of rules again, we observe that object c in membrane e only triggers rule 6. Hence, if the number of objects c in e is higher than 2 we conclude that the number of objects a in the membrane e in the configuration C must be greater than 2. Therefore, we conclude that the number of objects c in the membrane e in configuration C is exactly equal to 2.
- We find one object b in the membrane labelled by e in configuration C'. The unique rule that can produce it is rule 1, but the application of the rule produces at least two objects b in membrane e. Then we conclude that rule 1 is not applied. The occurrence of such object b can only be explained by considering its occurrence in configuration C. As one can check, no rule is triggered by object b in the membrane e, then the number of objects b in membrane e in the configuration C equals to 1.

– No object c are placed in the membrane e in C'. All such objects from
the previous configuration C are consumed by rule 6, so no object c in the
membrane e imply that rules 1 and 5 have not been triggered. From the
previous paragraph, it is known that rule 5 has not been applied. Since all
the objects a in membrane s send objects e into membrane c by means of
rule 5 and the numbers of objects c in such membrane in configuration C'
is zero, we conclude that in configuration C no objects a are placed in the
membrane s.

– We find one object c in the membrane labelled by s in configuration C'.
The unique rule that can produce it is rule 3, but the application of the rule
produces at least two objects c in membrane s. Then we conclude that rule
3 is not applied. The occurrence of such object b can only be explained by
considering its occurrence in configuration C. As one can check, no rule is
triggered by the object c in the membrane s, then the number of objects c
in membrane s in the configuration C equals 1.

– Finally, we find two objects a in the membrane labelled by s in the configuration C'. Since rule 5 consumes all the objects in the membrane e from
the previous configuration C, we conclude that such objects a must be produced by the application of rule(s) of Π. Rules 2 and 4 produce objects a in
membrane s. Rule 2 is triggered by an object a in the membrane e and rule
4 is triggered by an object b in membrane s. We can also check that all the
objects b in s produce objects a. Nonetheless, an object a in the membrane
e can trigger rules 1 and 2. Fortunately, we have seen that rule 1 is not
triggered, so can conclude that all the objects a in membrane e trigger rule
2. We conclude that the number of objects a in membrane e in the configuration C and the number of objects b in the membrane s must be less than
or equal to 2 and the sum of both numbers must be exactly equal to 2.

Bearing in mind these considerations, there are exactly three configurations
C such that produce C' in one transition step:

– $C_1 = [\,[\,bc^2\,]_e\, b^2 c\,]_s$, i.e., $w_e = bc^2$ and $w_s = b^2 c$. It is easy to check that by
applying the rules 4 and 6 we obtain the configuration $C' = [\,[\,a^2 b\,]_e\, a^2 c\,]_s$.

– $C_2 = [\,[\,abc^2\,]_e\, bc\,]_s$, i.e., $w_e = abc^2$ and $w_s = bc$. In this case, C' is obtained
by applying the rules 2, 4 and 6.

– $C_3 = [\,[\,a^2 bc^2\,]_e\, c\,]_s$, i.e., $w_e = a^2 bc^2$ and $w_s = c$. In this case, C' is obtained
by applying the rules 2 and 6.

A question arises in a natural way: Could this reasoning be automated? In
other words, given a P system and a configuration C', is there an algorithm such
that outputs the set $\mathcal{C}$ of configurations C and produce C' in one transition step?

We can even go beyond. We wonder if there exists an algorithm such that
it takes a P system Π as input and it outputs a mapping $\mathcal{R}_\Pi$ which, for *every*
configuration C' of Π, $\mathcal{R}_\Pi(C')$ is the set of all computations C such that C' is
reachable from C in one computational step. In this paper, we will give a positive
answer to both questions. Before, we need to stress the relationship between P
systems and Linear Algebra.

3   The P System Model

Throughout this paper, we will consider a restricted form of transition P systems without dissolution and without output membrane. Considering an output membrane is irrelevant for our study, since we are not interested in the objects placed in a particular membrane, but in the computation process itself. We also restrict the type of rules. Cooperation is not allowed and then rules are triggered by only one object.

Namely, along this paper a P system of degree m is a tuple

$$\Pi = (\Gamma, H, \mu, w_1, \ldots, w_m, R), \text{ where:}$$

- Γ is the working alphabet whose elements are called *objects*;
- $H = \{1, \ldots, m\}$ is the set of labels;
- μ is the membrane structure of the P system and membranes are bijectively labelled with the elements of H;
- $w_1, \ldots, w_m$ are strings that represent multisets over Γ associated with each membrane of μ;
- $R = \{R_1, \ldots, R_m\}$ is the set of sets of rules, where R_i with $i \in \{1, \ldots, m\}$ are finite sets of *evolution rules* over Γ. The type of evolution rules of R_i depends on the membrane structure μ. Let $j_1, \ldots, j_r$ be the labels of membranes immediately inside the membrane i. An evolution rule of R_i is of the form $a \to v$, where $a \in \Gamma$ and v is an string over Γ^i_{tar}, where $\Gamma^i_{tar} = \Gamma \times TAR_i$, for $TAR_i = \{here, out\} \cup \{in_{j_k} \mid k \in \{1, \ldots, r\}\}$.

The symbols $here$, out and in_{j_k} are called *target commands*. The rules are applied in a non-deterministic maximally parallel way. Given a rule $a \to v$, the effect of applying this rule in a compartment i is to remove the object a and to insert the objects specified by v in the regions designated by the target commands associated with the objects from v. In particular,

- if v contains $(a, here)$, the object a will be placed in the same region where the rule is applied;
- if v contains (a, out), the object a will be placed in the compartment that surrounds the region where the rule is applied;
- if v contains (a, in_j), the object a will be placed in compartment j, provided that j is immediately inside i.

In one step, each object in a membrane can only be used for one rule (non deterministically chosen when there are several possibilities), but any object which can evolve by a rule of any form must do it. All the elements which are not involved in any of the rules to be applied remain unchanged. Several rules can be applied to different objects in the same cell simultaneously.

Along the computation, the multisets associated with the membranes can change, but the alphabet Γ, the set of labels H, the membrane structure μ and the set of rules R are constant. We call the 4-uple (Γ, H, μ, R) the *skeleton* of the P system.

Notice that the P system presented in Section 2 is a particular case of this P system model with a slight change of notation in the rules:

1. Notation $[a \rightarrow v]_h$ where $h \in H$, $a \in \Gamma$ and v is a string over Γ is a short notation to indicate that the rule $a \rightarrow (v_1, here) \ldots (v_n, here)$ belongs to the set of rules R_h, with $v = v_1 \ldots v_n$.
2. Notation $a[\,]_h \rightarrow [v]_h$ where $h \in H$, $a \in \Gamma$ and v is a string over Γ is a short notation to indicate that the rule $a \rightarrow (v_1, in_h) \ldots (v_n, in_h)$ belongs to the set of rules R_{h^*}, with h^* the label of the membrane surrounding the membrane h and $v = v_1 \ldots v_n$.
3. Notation $[a]_h \rightarrow v[\,]_h$ where $h \in H$, $a \in \Gamma$ and v is a string over Γ is a short notation to indicate that the rule $a \rightarrow (v_1, out) \ldots (v_n, out)$ belongs to the set of rules R_h, with $v = v_1 \ldots v_n$.

4   Changing the Point of View

The key idea of the present paper is to consider an algebraic representation for the configurations and the rules of a P system. The starting point is the representation used in [2], but we introduce several changes.

First, our elementary objects are pairs of type $(a, h) \in \Gamma \times H$ meaning that object $a \in \Gamma$ is placed in the membrane (labelled by) $h \in H$. Roughly speaking, transitions in P systems are performed by rules in which the occurrence of an element a_0 in a membrane h_0 produces the occurrence of β_1 copies of element a_1 in membrane h_1, β_2 copies of element a_2 in membrane h_2, etc.

More formally, the rules in the P system model presented above can be reformulated as follows:

$$(a_0, h_0) \rightarrow (a_1, h_1)^{\beta_1} (a_2, h_2)^{\beta_2} \ldots (a_n, h_n)^{\beta_n}$$

Note that, for all $i \in \{1, \ldots, n\}$, if $h_0 = h_i$ then, (a_i, h_i) is equivalent to the pair $(a_i, here)$. Otherwise, if $h_0 \neq h_i$ both membranes must be adjacent (one membrane is the father of the other one). If h_0 is the father of h_i, then the pair (a_i, h_i) is $equivalent$, in some sense, to (a_i, in_{h_i}). Finally, if h_i is the father of h_0, then the pair (a_i, h_i) is $equivalent$ to (a_i, out). For each $i \in \{1, \ldots, n\}$, β_i represents the multiplicity of (a_i, h_i) in the right-hand side (RHS) of the rule.

The second basic idea in the representation appears in [3] as well. It consists on settling a total order in the set $\Gamma \times H$. Along the paper, in order to simplify the notation, given an alphabet Γ and a set of labels H, d will denote the cardinal $\Gamma \times H$. Let us consider a total order $\mathcal{O}$ on the set $\Gamma \times H$, $\mathcal{O} : \{1, \ldots, d\} \rightarrow \Gamma \times H$. By using this order, we represent $\Gamma \times H$ as the finite sequence $\langle \gamma_1, \ldots, \gamma_d \rangle$, where γ_i is the i-th pair of $\Gamma \times H$ in the order $\mathcal{O}$.

By using this order, each rule

$$(a_0, h_0) \rightarrow (a_1, h_1)^{\beta_1} (a_2, h_2)^{\beta_2} \ldots (a_n, h_n)^{\beta_n}$$

can be represented as

$$\gamma \rightarrow \gamma_1^{\alpha_1} \gamma_2^{\alpha_2} \ldots \gamma_d^{\alpha_d}$$

where $(a_0, h_0) = \gamma$ and for all $i \in \{1, \ldots, d\}$:

- If there exists $j \in \{1, \ldots, n\}$ such that $\gamma_i = (a_j, h_j)$ then $\alpha_i = \beta_j$.
- Otherwise $\alpha_i = 0$.

We say that $\gamma \to \gamma_1^{\alpha_1} \gamma_2^{\alpha_2} \ldots \gamma_d^{\alpha_d}$ is the *pairwise* representation of the rule.

The use of an order on $\Gamma \times H$ leads us to a more homogeneous representation of rule $\gamma \to \gamma_1^{\alpha_1} \gamma_2^{\alpha_2} \ldots \gamma_d^{\alpha_d}$. It can be represented by a pair $\langle \gamma, v \rangle$ where γ (the LHS of the rule) belongs to $\Gamma \times H$, and v is a vector of dimension d whose components are in $\mathbb{N}$. Formally, we have the following definition:

Definition 1. *Let us consider a P system Π with Γ the alphabet and H the set of labels. Let $\Gamma \times H$ be the ordered set $\langle \gamma_1, \ldots, \gamma_d \rangle$. The algebraic representation of the rule*

$$\gamma \to \gamma_1^{\alpha_1} \gamma_2^{\alpha_2} \ldots \gamma_d^{\alpha_d}$$

is the pair (γ, v) where $v = (\alpha_1, \ldots, \alpha_d)$. We say that v represents the right-hand side of the rule r_i.

Remark 1: Given an order $\langle \gamma_1, \ldots, \gamma_d \rangle$ on $\Gamma \times H$, a pair $\langle \gamma, v \rangle$ where $\gamma \in \Gamma \times H$ and v is a vector of dimension d (with values in $\mathbb{N}$) defines a unique rule and vice-versa, each rule having a unique algebraic representation.

Remark 2: If the P system is not deterministic, then there exists at least one $\gamma \in \Gamma \times H$ such that there exists two different vectors v_1 and v_2 such that pairs $\langle \gamma, v_1 \rangle$ and $\langle \gamma, v_2 \rangle$ represent two different rules.

Let us see an example of this algebraic representation.

Example 1. Let us consider the skeleton of the P system considered in Section 2 with $\Gamma = \{a, b, c\}$, $H = \{e, s\}$, $\mu = [\,[\,]_e\,]_s$ and R the set of rules

Rule 1: $[a \to b^2 c]_e$	**Rule 4:** $[b \to a]_s$
Rule 2: $[a]_e \to a[\,]_e$	**Rule 5:** $a[\,]_e \to [c]_e$
Rule 3: $[b \to c^2]_s$	**Rule 6:** $[c \to a]_e$

The set of objects is $\Gamma = \{a, b, c\}$ and the set of labels is $H = \{e, s\}$. Let us consider the following total order in $\Gamma \times H$

$$\langle (a, e), (b, e), (c, e), (a, s), (b, s), (c, s) \rangle$$

The six rules of the P system can be settled as

r_1: $(a, e) \to (b, e)^2 (c, e)$	r_4: $(b, s) \to (a, s)$
r_2: $(a, e) \to (a, s)$	r_5: $(a, s) \to (c, e)$
r_3: $(b, s) \to (c, s)^2$	r_6: $(c, e) \to (a, e)$

By using the previous total order in $\Gamma \times H$, these rules have the following algebraic representation

Rule 1: $\langle (a, e), (0, 2, 1, 0, 0, 0) \rangle$	**Rule 4:** $\langle (b, s), (0, 0, 0, 1, 0, 0) \rangle$
Rule 2: $\langle (a, e), (0, 0, 0, 1, 0, 0) \rangle$	**Rule 5:** $\langle (a, s), (0, 0, 1, 0, 0, 0) \rangle$
Rule 3: $\langle (b, s), (0, 0, 0, 0, 0, 2) \rangle$	**Rule 6:** $\langle (c, e), (1, 0, 0, 0, 0, 0) \rangle$

4.1   Configurations

A *configuration* of such a P system is the description of the multiset placed in the membranes of the P system in a given instant. Formally, given a P system with working alphabet Γ and set of labels H, a configuration C is a multiset over $\Gamma \times H$, $C : \Gamma \times H \rightarrow \mathbb{N}$, and we denote by $C(a, m)$ the multiplicity of object a in the membrane labelled by m of that configuration. The support of C, $supp(C)$, is defined as $supp(C) = \{(a, m) \in \Gamma \times H \mid C(a, m) \neq 0\}$ and, as usual in multisets theory, C will be represented as $\{(a, m)^{C(a,m)} \mid (a, m) \in supp(C)\}$. For example, the configuration of our example $[\,[\,b\,]_e \, c^3\,]_s$ can be represented as $\{(b, e), (c, s)^3\}$.

From the idea of setting an order on $\Gamma \times H$, the representation of a configuration via a vector is quite natural.

Definition 2. *Let us consider a P system Π with Γ the alphabet, H the set of labels and order $\langle \gamma_1, \dots, \gamma_d \rangle$ on $\Gamma \times H$. An algebraic representation of a configuration $C : \Gamma \times H \rightarrow \mathbb{N}$ is a vector*

$$\boldsymbol{C} = (C(\gamma_1), \dots, C(\gamma_d))$$

that is, the j-th component in $\boldsymbol{C}$ is a number representing the multiplicity of the j-th element of $\Gamma \times H$.

Let us remark that, if the order on $\Gamma \times H$ is set, then there exists a bijective correspondence between a configuration C and its algebraic representation $\boldsymbol{C}$.

Example 2. As we saw before, the initial configuration $[\,[\,b\,]_e \, c^3\,]_s$ can be expressed as the multiset $C = \{(b, e), (c, s)^3\}$. If we consider order

$$\langle (a, e), (b, e), (c, e), (a, s), (b, s), (c, s) \rangle$$

then the algebraic representation of the configuration is $\boldsymbol{C} = (0, 1, 0, 0, 0, 3)$.

In order to formalize the concept of computation with this new representation, we fix some notations. We denote by RHS_r the right-hand side of rule r and for all $\sigma \in \Gamma \times H$, $|RHS_r(\sigma)|$ denotes the multiplicity of σ in the multiset RHS_r.

Example 3. Let us consider the pairwise representation of the rule $r_1 : (a, e) \rightarrow (b, e)^2(c, e)$, then $RHS_{r_1} = (b, e)^2(c, e)$ and $|RHS_{r_1}(b, e)| = 2$.

Definition 3. *Let us consider an alphabet Γ, a set of labels H and the set of rules R of a P system. We denote by $\mathcal{LHS}(R)$ the set of all the pairs from $\Gamma \times H$ that are the left-hand side of a rule from R. Formally*

$$\mathcal{LHS}(R) = \{\gamma \in \Gamma \times H \mid \exists r \in R \, (\gamma = LHS(r))\}$$

Example 4. Let us consider $\Gamma = \{a, b, c\}$, $H = \{e, s\}$ and R the set of rules

$$r_1 \colon (a, e) \rightarrow (c, e)^2 \qquad r_2 \colon (a, e) \rightarrow (a, s) \qquad r_3 \colon (b, e) \rightarrow (c, e)$$
$$r_4 \colon (a, s) \rightarrow (b, s) \qquad r_5 \colon (a, s) \rightarrow (b, s)(c, s)^2$$

In this case $\mathcal{LHS}(R) = \{(a, e), (b, e), (a, s)\}$.

Definition 4. *Let us consider an alphabet Γ and a set of labels H of a P system Π and let $R = \langle r_1, \ldots, r_p \rangle$ be an enumeration of its set of rules with $r_j = (LHS(r_j), v_j)$. Let $C : \Gamma \times H \to \mathbb{N}$ be a configuration of Π.*

A partition of C with respect to R is a p-tuple

$$\mathcal{P} = \langle (r_1, k_1), \ldots, (r_p, k_p) \rangle$$

such that for all $j \in \{1, \ldots, p\}$, $k_j \geq 0$ and for all $\gamma \in \mathcal{LHS}(\mathcal{R})$

$$\sum_{LHS(r_j)=\gamma} k_j = C(\gamma)$$

Example 5. Let us consider an alphabet $\Gamma = \{a, b, c\}$ a set of labels $H = \{e, s\}$, $\mu = [[\,]_e\,]_s$ and R the set of rules from example 4

$$r_1 \colon (a, e) \to (c, e)^2 \qquad r_2 \colon (a, e) \to (a, s) \qquad r_3 \colon (b, e) \to (c, e)$$
$$r_4 \colon (a, s) \to (b, s) \qquad r_5 \colon (a, s) \to (b, s)(c, s)^2$$

Let us consider a configuration with algebraic representation $C = \langle 3, 0, 1, 7, 4, 1 \rangle$ associated with order $\langle (a, e), (b, e), (c, e), (a, s), (b, s), (c, s) \rangle$ of $\Gamma \times H$. In this case, one possible partition of C with respect to R is

$$\mathcal{P} = \langle (r_1, 2), (r_2, 1), (r_3, 0), (r_4, 2), (r_5, 5) \rangle$$

the number associated to each rule is a natural number and $\mathcal{LHS}(R) = \{(a, e), (b, e), (a, s)\}$, so in order to check that $\mathcal{P}$ is a partition it suffices to check

$$\sum_{LHS(r_j)=(a,e)} k_j = k_1 + k_2 = 2 + 1 = 3 = C(a, e)$$
$$\sum_{LHS(r_j)=(b,e)} k_j = k_3 = 0 = C(b, e)$$
$$\sum_{LHS(r_j)=(a,s)} k_j = k_4 + k_5 = 2 + 5 = 7 = C(a, s)$$

The different possible partitions capture the idea of different choice of rules in the case of non-deterministic P system. Notice that in the case of a deterministic P system, there exists only one partition

$$\mathcal{P} = \langle (r_1, C(LHS(r_1))), (r_2, C(LHS(r_2))), \ldots, (r_p, C(LHS(r_p))) \rangle$$

In order to obtain a new configuration C' from a given configuration C and from the set of rules $\{r_1, \ldots, r_p\}$, we need to describe the multiplicity of any $\sigma \in \Gamma \times H$ in C'. For the calculus of such multiplicity we need

- A partition $\mathcal{P} = \langle (r_1, k_1), \ldots, (r_p, k_p) \rangle$ of C with respect to R.
- The set $\mathcal{LHS}(R)$

In such multiplicity, each rule $r_i : \gamma_i \to RHS_{r_i}$ adds the multiplicity of σ in the right hand side of the rule multiplied by the value k_i in the partition $\mathcal{P}$. If the object is not consumed by any rule, we also add the multiplicity in the original configuration.

Formally, for every $\sigma \in \Gamma \times H$ we have:

$$C'(\sigma) = \begin{cases} \sum_{i=1}^{i=p} k_i \cdot |RHS_{r_i}(\sigma)| & \text{if } \sigma \in \mathcal{LHS}(R) \\ \sum_{i=1}^{i=p} k_i \cdot |RHS_{r_i}(\sigma)| + C(\sigma) & \text{if } \sigma \notin \mathcal{LHS}(R) \end{cases}$$

Example 6. Let us come back again to our P system Π with alphabet $\Gamma = \{a, b, c\}$, set of labels $H = \{e, s\}$, membrane structure $\mu = [\,[\,]_e\,]_s$ and the set of rules R

Rule 1: $[\,a \to b^2c\,]_e$ **Rule 4:** $[\,b \to a\,]_s$
Rule 2: $[\,a\,]_e \to a\,[\,]_e$ **Rule 5:** $a\,[\,]_e \to [\,c\,]_e$
Rule 3: $[\,b \to c^2\,]_s$ **Rule 6:** $[\,c \to a\,]_e$

Let us consider configuration $C_1 = [\,[\,bc^2\,]_e\,b^2c\,]_s$, i.e., $w_e = bc^2$ and $w_s = b^2c$. It is easy to check that by applying rules 4 and 6 we obtain configuration $C' = [\,[\,a^2b\,]_e\,a^2c\,]_s$. Such configuration can also be obtained by considering the multiplicity of each pair in $\Gamma \times H$ and using the previous formula. First we consider the partition $\mathcal{P} = \langle (r_1, 0), (r_2, 0), (r_3, 0), (r_4, 2), (r_5, 0), (r_6, 2) \rangle$ and $\mathcal{LHS}(R) = \{(a, e), (b, s), (a, s), (c, e)\}$. Then, for example,

$$C'(a, s) = k_1 \cdot 0 + k_2 \cdot 1 + k_3 \cdot 0 + k_4 \cdot 1 + k_5 \cdot 0 + k_6 \cdot 0 = 2 \cdot 1 = 2$$
$$C'(b, e) = k_1 \cdot 2 + k_2 \cdot 0 + k_3 \cdot 0 + k_4 \cdot 0 + k_5 \cdot 0 + k_6 \cdot 0 + C(b, e) = 0 \cdot 2 + 1 = 1$$

and the remaining multiplicities in configuration C' can be obtained in a similar way.

5 Matrix Associated with the Skeleton

After defining the algebraic representation of rules and configurations, we define a numerical matrix associated with the skeleton of a P system. The next definition of *extended* set of rules will be used in the definition of the matrix.

Definition 5. *Let Γ be the alphabet, H the set of labels and R the set of rules of a P system where R is a set of rules in its pairwise form. The extended set of rules of R in this skeleton, R^* is the set of rules R together with the identity rule $\gamma \to \gamma$ for all the $\gamma \in \Gamma \times H$ such that there is no rule in R with γ in its left-hand side.*

Considering identity rules, we obtain P systems whose computations never stop. In this paper, we are interested only in the evolution of computation in time and not in halting conditions. Let us remark two important considerations related with the extended set of rules:

- If R^* is the extended set of rules of R, then $\mathcal{LHS}(R^*) = \Gamma \times H$.
- Consequently, if C is a configuration of a P system Π with $\langle \gamma_1, \ldots, \gamma_d \rangle$ an order on $\Gamma \times H$ and $\mathcal{P}^* = \langle (r_1, k_1), \ldots, (r_p, k_p) \rangle$ is a partition of a configuration C of a P system with respect to its extended set of rules, then configuration C' that can be obtained from C in one computation step following such partition is $C'(\gamma_j) = \sum_{i=1}^{i=p} k_i \cdot |RHS_{r_i}(\gamma_j)|$ for all $j \in \{1, \ldots, d\}$.

Example 7. Let us consider again the skeleton of example 1, and its set of rules,

$$r_1: (a, e) \to (b, e)^2(c, e) \qquad r_4: (b, s) \to (a, s)$$
$$r_2: (a, e) \to (a, s) \qquad\qquad r_5: (a, s) \to (c, e)$$
$$r_3: (b, s) \to (c, s)^2 \qquad\qquad r_6: (c, e) \to (a, e)$$

Note that the pairs γ from $\Gamma \times H$ such that there is no rule in R with γ as its left-hand side are (b, e) and (c, s), therefore to obtain R^* we have to add to R the rules

$$r_7: (b, e) \to (b, e) \qquad r_8: (c, s) \to (c, s)$$

Obviously, the set of rules R^* has also an algebraic representation

Rule 1: $\langle (a, e), (0, 2, 1, 0, 0, 0) \rangle$ **Rule 5:** $\langle (a, s), (0, 0, 1, 0, 0, 0) \rangle$
Rule 2: $\langle (a, e), (0, 0, 0, 1, 0, 0) \rangle$ **Rule 6:** $\langle (c, e), (1, 0, 0, 0, 0, 0) \rangle$
Rule 3: $\langle (b, s), (0, 0, 0, 0, 0, 2) \rangle$ **Rule 7:** $\langle (b, e), (0, 1, 0, 0, 0, 0) \rangle$
Rule 4: $\langle (b, s), (0, 0, 0, 1, 0, 0) \rangle$ **Rule 8:** $\langle (c, s), (0, 0, 0, 0, 0, 1) \rangle$

With the help of the concept of extended set of rules, we define the matrix associated with a skeleton.

Definition 6. *Let us consider skeleton $Sk = (\Gamma, H, \mu, R)$ of a P system and let $\langle r_1, \ldots, r_p \rangle$ be an enumeration of the extended set of rules R^* of R in its algebraic form. The matrix associated with skeleton Sk, M_{Sk} is the matrix whose rows are vectors $\boldsymbol{v_1}, \ldots, \boldsymbol{v_p}$, where for each i with $1 \leq i \leq p$, $\boldsymbol{v_i}$ is the vector which represents the right-hand side of rule r_i.*

Before showing an example, some remarks are necessary.

– The matrix associated with a skeleton depends on the skeleton, as well as on the enumeration of the rules of the extended set and the order on $\Gamma \times H$. A different enumeration produces a different order in the rows of the matrix.
– In case of deterministic P systems, the number of rules in the extended set, p, and the number of pairs in $\Gamma \times H$, d are the same and we have a square matrix[1]. In general, M_{Sk} is a $d \times p$ matrix with $d \leq p$.

Example 8. If we consider the skeleton of example 7 and the enumeration of the eight rules of the extended set R^* and the usual order on $\Gamma \times H$, $\langle (a, e), (b, e), (c, e), (a, s), (b, s), (c, s) \rangle$

Rule 1: $\langle (a, e), (0, 2, 1, 0, 0, 0) \rangle$ **Rule 5:** $\langle (a, s), (0, 0, 1, 0, 0, 0) \rangle$
Rule 2: $\langle (a, e), (0, 0, 0, 1, 0, 0) \rangle$ **Rule 6:** $\langle (c, e), (1, 0, 0, 0, 0, 0) \rangle$
Rule 3: $\langle (b, s), (0, 0, 0, 0, 0, 2) \rangle$ **Rule 7:** $\langle (b, e), (0, 1, 0, 0, 0, 0) \rangle$
Rule 4: $\langle (b, s), (0, 0, 0, 1, 0, 0) \rangle$ **Rule 8:** $\langle (c, s), (0, 0, 0, 0, 0, 1) \rangle$

---

[1] This kind of matrices were studied in [3].

we have the following matrix

$$M_{Sk} = \begin{pmatrix} 0 & 2 & 1 & 0 & 0 & 0 \\ 0 & 0 & 0 & 1 & 0 & 0 \\ 0 & 0 & 0 & 0 & 0 & 2 \\ 0 & 0 & 0 & 1 & 0 & 0 \\ 0 & 0 & 1 & 0 & 0 & 0 \\ 1 & 0 & 0 & 0 & 0 & 0 \\ 0 & 1 & 0 & 0 & 0 & 0 \\ 0 & 0 & 0 & 0 & 0 & 1 \end{pmatrix}$$

6 Computing Backwards

The definition of these algebraic objects allows us to define an algebraic method to characterize the set of configurations $\mathcal{C}$ which can produce a given configuration C_0 in one computation step. First, we need to find the solutions of a system of linear equations.

Definition 7. *Let Π be a P system, $\langle r_1, \ldots, r_p \rangle$ an enumeration of its set of extended rules, M_{Sk} the matrix associated with the skeleton of Π based on that enumeration of R^* and let C_0 be the vectorial representation of a configuration C_0. We define the solution set of M_{Sk} and C_0 and we will denote it by $SOL(M_{Sk}, C_0)$ the set of real-valued vectors x with dimension p such that $C_0 = x \cdot M_{Sk}$.*

Notice that according to the definition, $SOL(M_{Sk}, C_0)$ can be the empty set. It is well known in Linear Algebra that if the range of the matrix M_{Sk} and the range of the matrix M_{Sk} augmented with the vector of coefficients C_0 is not the same, then the system of equations has no solution.

$SOL(M_{Sk}, C_0)$ is a manifold of dimension p minus the range of the matrix M_{Sk} embedded in a vectorial space of dimension p, but the study of the algebraic properties of such manifold is out of the scope of this paper.

Example 9. Let us come back to our main example. If we take the matrix M_{Sk} from example 8, configuration $C' = [[a^2 b]_e a^2 c]_s$ from Section 2 and algebraic representation $C' = (2, 1, 0, 2, 0, 1)$, then in order to get $SOL(M_{Sk}, C')$ we need to solve the system

$$(2, 1, 0, 2, 0, 1) = (x_1, x_2, x_3, x_4, x_5, x_6, x_7, x_8) \begin{pmatrix} 0 & 2 & 1 & 0 & 0 & 0 \\ 0 & 0 & 0 & 1 & 0 & 0 \\ 0 & 0 & 0 & 0 & 0 & 2 \\ 0 & 0 & 0 & 1 & 0 & 0 \\ 0 & 0 & 1 & 0 & 0 & 0 \\ 1 & 0 & 0 & 0 & 0 & 0 \\ 0 & 1 & 0 & 0 & 0 & 0 \\ 0 & 0 & 0 & 0 & 0 & 1 \end{pmatrix}$$

or equivalently,

$$x_6 = 2x_2 + x_4 = 2, 2x_1 + x_7 = 1, 2x_3 + x_8 = 1, x_1 + x_5 = 0.$$

Then, $SOL(M_{Sk}, \mathbf{C}')$ is the following 3-dimensional manifold embedded in an 8-dimensional vectorial space

$$SOL(M_{Sk}, \mathbf{C}') = \{(\alpha, \beta, \gamma, 2 - \beta, -\alpha, 2, 1 - 2\alpha, 1 - 2\gamma) \mid \alpha, \beta, \gamma \in \mathbb{R}\}$$

Definition 8. *Let Π be a P system and an order $\langle \gamma_1, \ldots, \gamma_d \rangle$ on $\Gamma \times H$, $\langle r_1, \ldots, r_p \rangle$ an enumeration of its set of extended rules, M_{Sk} the matrix associated with the skeleton of Π based on that enumeration of R^* and let $\mathbf{C}$ be the vectorial representation of a configuration C. We define the* constructor mapping *as*

$$\psi_\Pi : SOL(M_{Sk}, \mathbf{C}) \to \mathbb{R}^d$$

such that for all $(x_1, \ldots, x_p) \in SOL(M_{Sk}, \mathbf{C}')$, $\psi_\Pi((x_1, \ldots, x_p)) = (y_1, \ldots, y_d)$ verifying for all $i \in \{1, \ldots, d\}$,

$$y_i = \sum_{\gamma_i = LHS(r_k)} x_k$$

Notice that the set $SOL(M_{Sk}, \mathbf{C})$ depends on the way in which the set of extended rules is enumerated, but $\psi_\Pi(SOL(M_{Sk}, \mathbf{C}))$ is independent of such enumeration. Obviously, if all the coordinates of $\mathbf{x} \in SOL(M_{Sk}, \mathbf{C}')$ are natural numbers, then all the coordinates of $\psi(\mathbf{x})$ are also natural numbers.

Example 10. Following with the set $SOL(M_{Sk}, \mathbf{C}')$ from Example 9 and order $\langle ((a, e), (b, e), (c, e), (a, s), (b, s), (c, s) \rangle$ on $\Gamma \times H$, we have

$$y_1 = \sum_{(a,e)=LHS(r_k)} x_k = x_1 + x_2 = \alpha + \beta$$
$$y_2 = \sum_{(b,e)=LHS(r_k)} x_k = x_7 = 1 - 2\alpha$$
$$y_3 = \sum_{(c,e)=LHS(r_k)} x_k = x_6 = 2$$
$$y_4 = \sum_{(a,s)=LHS(r_k)} x_k = x_5 = -\alpha$$
$$y_5 = \sum_{(b,s)=LHS(r_k)} x_k = x_3 + x_4 = 2 + \gamma - \beta$$
$$y_6 = \sum_{(c,s)=LHS(r_k)} x_k = x_8 = 1 - 2\gamma$$

Therefore $\psi_\Pi(SOL(M_{Sk}, \mathbf{C}))$ is a 3-dimensional manifold embedded in an 6-dimensional vectorial space

$$\psi_\Pi(SOL(M_{Sk}, \mathbf{C})) = \{(\alpha + \beta, 1 - 2\alpha, 2, -\alpha, 2 + \gamma - \beta, 1 - 2\gamma) \mid \alpha, \beta, \gamma \in \mathbb{R}\}$$

Finally, we only consider the elements of $SOL(M_{Sk}, \mathbf{C})$ such that all its coordinates are natural numbers. We prove below that the image of such vectors by means of the constructor mapping represent the searched configurations.

Definition 9. *Let Π be a P system, $\langle r_1, \ldots, r_p \rangle$ an enumeration of its set of extended rules, M_{Sk} the matrix associated with the skeleton of Π based on that enumeration of R^* and let $\mathbf{C}$ be the vectorial representation of a configuration C. We define*

- $NSOL(M_{Sk}, C)) = \{(x_1, \ldots, x_p) \in SOL(M_{Sk}, C)) \mid x_i \in \mathbb{N}, 1 \le i \le n\}.$
- A constructed configurations C_1 of Π is a configuration such that $C_1 \in \psi_\Pi(NSOL(M_{Sk}, C)).$

Example 11. If we take $\psi_\Pi(SOL(M_{Sk}, C))$ from example 10

$$\psi_\Pi(NSOL(M_{Sk}, C)) = \left\{ \begin{array}{c} (\alpha + \beta, 1 - 2\alpha, 2, -\alpha, 2 + \gamma - \beta, 1 - 2\gamma) \mid \\ \alpha, \beta, \gamma \in \mathbb{R}, \ \alpha + \beta \in \mathbb{N}, \ 1 - 2\alpha \in \mathbb{N}, \\ -\alpha \in \mathbb{N}, \ 2 + \gamma - \beta \in \mathbb{N}, \ 1 - 2\gamma \in \mathbb{N} \end{array} \right\}$$

The set $\psi_\Pi(NSOL(M_{Sk}, C))$ has only three elements

$$C_1 = (0, 1, 2, 0, 2, 1) \quad C_2 = (1, 1, 2, 0, 1, 1) \quad C_3 = (2, 1, 2, 0, 0, 1)$$

which correspond to the three configurations obtained in Section 2. Next we prove that the result holds in the general case.

Theorem 1. *Let Π be a P system with skeleton $Sk = (\Gamma, H, \mu, R)$ and let C be a configuration of Π. Let $\langle \gamma_1, \ldots, \gamma_d \rangle$ be an order on $\Gamma \times H$ and $\langle r_1, \ldots, r_p \rangle$ an enumeration of the extended set of rules R^* of R. Let M_{Sk} be the matrix associated with the skeleton Sk following such order and enumeration. Then, the configuration C_1 produces C in one computation step if and only if $C_1 \in \psi_\Pi(NSOL(M_{Sk}, C)).$*

Proof. Let us consider a configuration C_1 such that $C_1 \in \psi_\Pi(NSOL(M_{Sk}, C))$. Such configuration is a multiset C_1 on the set $\Gamma \times H$ such that for all $i \in \{1, \ldots, n\}$, $C_1(\gamma_i) \in \mathbb{N}$.

$C_1 \in \psi_\Pi(NSOL(M_{Sk}, C))$ if and only if there exist $(x_1, \ldots, x_p) \in SOL(M_{Sk}, C)$ with $x_i \in \mathbb{N}$ for all $i \in \{1, \ldots, p\}$ such that $\psi_\Pi(x_1, \ldots, x_n) = (C_1(\gamma_1), \ldots, C_1(\gamma_d))$. By definition of the constructor mapping $\psi_\Pi : SOL(M_{Sk}, C) \to \mathbb{R}^d$ we have for all $i \in \{1, \ldots, d\}$,

$$C_1(\gamma_i) = \sum_{\gamma_i = LHS(r_k)} x_k$$

On the other hand, we also know that $(x_1, \ldots, x_p) \in SOL(M_{Sk}, C)$, i.e.,

$$(C(\gamma_1), \ldots, C(\gamma_d)) = (x_1, \ldots, x_d) \cdot M_{Sk}$$

By construction of the matrix M_{Sk}, the previous equality means that for all $i \in \{1, \ldots, n\}$,

$$C(\gamma_i) = \sum_{j=1}^{p} x_j \cdot |RHS_{r_j}(\gamma_i)|$$

To sum up, $C_1 \in \psi_\Pi(NSOL(M_{Sk}, C))$ if and only if there exist $(x_1, \ldots, x_p)$ such that for all $i \in \{1, \ldots, p\}$

(a) $x_i \in \mathbb{N}$
(b) $C_1(\gamma_i) = \sum_{\gamma_i = LHS(r_k)} x_k$
(c) $C(\gamma_i) = \sum_{j=1}^{p} x_j \cdot |RHS_{r_j}(\gamma_i)|$

Since R^* is a set of extended rules, $\mathcal{LHS}(R^*)$ is the set $\Gamma \times H$. Bearing this equality in mind, properties (a) and (b) claim that $\mathcal{P}^* = \langle (r_1, x_1), \ldots, (r_p, x_p) \rangle$ is a partition of C_1 with respect to R^* and property (c) claims that the configuration C can be obtained from C_1 by using the partition $\mathcal{P}^*$.

On the other hand, if C_1 produces C in one computation step, then there exist a vector $(x_1, \ldots, x_n)$ such that $\langle (r_1, x_1), \ldots, (r_p, x_p) \rangle$ is a partition of C_1 with respect to R^* verifying properties (a), (b) and (c) and therefore $C_1 \in \psi_\Pi(NSOL(M_{Sk}, C))$.

7   A General Method

After the proof of Theorem 1, we come back to the questions asked at the end of Section 2. We wondered if there exists an algorithm such that it takes a P system Π as input and it outputs a mapping $\mathcal{R}_\Pi$ which, for *every* configuration C' of Π, $\mathcal{R}_\Pi(C')$ is the set of all computations C such that C' is obtained from C in one computational step. A method for computing such algorithm is the following:

Given a P system Π with skeleton $Sk = (\Gamma, H, \mu, R)$,

1. Fix an order $\langle \gamma_1, \ldots, \gamma_d \rangle$ for $\Gamma \times H$.
2. Consider the pairwise representation of the rules in R according to such order.
3. Consider the extended set of rules R^* from R and fix an enumeration $\langle r_1, \ldots, r_p \rangle$ of the rules from R^* in its algebraic representation.
4. Define matrix M_{Sk} following the orders $\langle \gamma_1, \ldots, \gamma_d \rangle$ and $\langle r_1, \ldots, r_p \rangle$.

Matrix M_{Sk} is the same for all configurations. Next we provide a method for finding all the configurations C' such that C' produce a given configuration C in one computation step.

Given a configuration C of Π

1. Obtain the algebraic representation C of C according to the order $\langle \gamma_1, \ldots, \gamma_d \rangle$.
2. Find all the vectors x with natural coordinates such that $C = x \cdot M_{Sk}$. The set of all these vectors is called $NSOL(M_{Sk}, C)$.
3. For each $x \in NSOL(M_{Sk}, C)$, we consider $C_x = (y_1 \ldots, y_d)$ where, for all $i \in \{1, \ldots, n\}$

$$y_i = \sum_{\gamma_i = LHS(r_k)} x_k$$

4. The set $\{C_x \mid x \in NSOL(M_{Sk}, C)\}$ is the set of the algebraic representations of all the configurations such that produce C in one computation step.

8   Conclusions and Future Work

In this paper, we provide a general method for finding all the configurations that produce a given one in one computational step. For that purpose, we have used an algebraic representation of rules and configurations and a matrix associated with the skeleton of the P systems.

The key step of the algorithm is to find all the vectors of natural numbers that are solutions of a system of linear equations. In such a system, the number of equations is the number of objects in the alphabet multiplied by the number of labels. The number of variables in the system is the cardinal of the set of extended rules which is at least the same as the number of equations and has no upper bound.

The problem of finding the solutions with natural values of a system of linear equations is a problem involving heavy tasks, specially if we consider a high number of variables and equations (which is the usual case for P systems). Nonetheless, currently there exist some powerful software tools able to deal with large numerical matrices and solve the corresponding systems under the restriction of finding natural-valued vectors.

In this way, we hope that this method can be useful for researchers interested in computing backwards in Membrane Computing, since it can consider the problem of finding the previous configurations as a computationally hard problem of Integer Programming.

Finally, this work can be extended in several ways. Not only by going deeper in the concept of computing backwards along a computation (and not only in one step) but exploring if these ideas can be extended to other P system models.

Acknowledgment. The authors acknowledge the support of project TIN2006-13425 of the Ministerio de Educación y Ciencia of Spain, cofinanced by FEDER funds, and the support of the Project of Excellence with *Investigador de Reconocida Valía* of the Junta de Andalucía, grant P08-TIC-04200.

References

1. Agrigoroaiei, O., Ciobanu, G.: Dual P systems. In: Corne, D.W., Frisco, P., Paun, G., Rozenberg, G., Salomaa, A. (eds.) WMC 2008. LNCS, vol. 5391, pp. 95–107. Springer, Heidelberg (2009)
2. Cordón-Franco, A., Gutiérrez-Naranjo, M.A., Pérez-Jiménez, M.J., Riscos-Núñez, A.: Exploring computation trees associated with P systems. In: Mauri, G., Păun, G., Jesús Pérez-Jímenez, M., Rozenberg, G., Salomaa, A. (eds.) WMC 2004. LNCS, vol. 3365, pp. 278–286. Springer, Heidelberg (2005)
3. Gutiérrez-Naranjo, M.A., Pérez-Jiménez, M.J.: Efficient computation in rational-valued P systems. Mathematical Structures in Computer Science (in press)
4. Gutiérrez–Naranjo, M.A., Pérez–Jiménez, M.J., Riscos–Núñez, A., Romero–Campero, F.J.: On the power of dissolution in P systems with active membranes. In: Freund, R., Păun, G., Rozenberg, G., Salomaa, A. (eds.) WMC 2005. LNCS, vol. 3850, pp. 224–240. Springer, Heidelberg (2006)
5. Păun, Gh.: Membrane Computing. An Introduction. Springer, Berlin (2002)

Modelling Signalling Networks with Incomplete Information about Protein Activation States: A P System Framework of the KaiABC Oscillator

Thomas Hinze[1], Thorsten Lenser[2], Gabi Escuela[2], Ines Heiland[1],
and Stefan Schuster[1]

Friedrich-Schiller University Jena
[1]School of Biology and Pharmacy, Department of Bioinformatics
[2]Department of Computer Science, Bio Systems Analysis Group
Ernst-Abbe-Platz 1–4, D-07743 Jena, Germany
{thomas.hinze,thorsten.lenser,gabi.escuela}@uni-jena.de,
{heiland.ines,stefan.schu}@uni-jena.de

Abstract. Reconstruction of signal transduction network models based on incomplete information about network structure and dynamical behaviour is a major challenge in current systems biology. In particular, interactions within signalling networks are frequently characterised by partially unknown protein phosphorylation and dephosphorylation cascades at a submolecular description level. For prediction of promising network candidates, reverse engineering techniques typically enumerate the reaction search space. Considering an underlying amount of phosphorylation sites, this implies a potentially exponential number of individual reactions in conjunction with corresponding protein activation states. To manage the computational complexity, we extend P systems with string-objects by a subclass for protein representation able to process wild-carded together with specific information about protein binding domains and their ligands. This variety of reactants works together with assigned term-rewriting mechanisms derived from discretised reaction kinetics. We exemplify the descriptional capability and flexibility of the framework by discussing model candidates for the circadian clock formed by the KaiABC oscillator found in the cyanobacterium *Synechococcus elongatus*. A simulation study of its dynamical behaviour demonstrates effects of superpositioned protein abundance courses based on regular expressions corresponding to dedicated protein activation states.

1 Introduction

Biological signalling networks have been identified to exhibit a universal capability to process information [14,17]. They can be viewed as complex computational devices of the cell, triggering and directing responses to external stimuli. It turns out that successive formation or decomposition of protein complexes in conjunction with domain-specific protein binding (as during phosphorylation by kinases) plays a central role in biological signal transduction based on submolecular assembly [1]. In this context, resulting biomolecules act as information carriers of

G. Păun et al. (Eds.): WMC 2009, LNCS 5957, pp. 316–334, 2010.

astonishing storage capacity and structural plasticity. For example, the tumor suppressor protein p53 is equipped with 27 phosphorylation sites [3]. It could theoretically assume up to $2^{27} = 34,217,728$ different activation states. Having in mind that each of these states is able to form an individual constituent of a reaction network incorporating all distinguishable states of up to several hundred interacting proteins, the potential dimension of those protein signalling networks is obvious.

In a typical scenario of exploring coupled intracellular *modules* – functional network units – the present knowledge on involved constituents and topology lacks some detailed information with regard to comprising the entirety of individual molecular interactions. Hence, an integrative setup, prediction, and reconstruction of network model candidates based on incomplete data is a challenging task in systems biology since it requires unconventional techniques to cope with the combinatorial complexity of exhaustive search within the underlying reaction space [15]. A variety of reverse engineering approaches emerged to tackle enumerative reaction network reconstruction at different levels of abstraction (cf. [10,16]).

While the steady-state behaviour might be sufficient to characterise a metabolic network (cf. [12]), the function of a protein signalling network depends heavily on its temporal evolution [26]. Oscillators based on phosphorylation/dephosphorylation cycles represent significant examples [20,22,27]. Thus, the aspect of *dynamical behaviour* should be reflected in the choice of the preferred modelling approach. For that purpose, ordinary differential equations (ODEs) derived from appropriate kinetics are commonly employed. Since this method usually assumes each individual protein activation state to act as a separate species, it easily leads to an exponential growth of the number of distinct ODEs (addressed amongst others in [7]). An opportunity to temporarily unify several activation states by one dedicated species could be a keystone to overcome this insufficiency.

Inspired by this initial idea, we propose a P systems framework able to specify proteins together with relevant properties by string-objects. In contrast to species names in ODEs, phenotypic information about a protein is represented by a character string. Each individual protein property is allowed to be marked as present, absent, or arbitrary. In the latter case, placeholders known from regular expressions denote unassigned protein properties. Consequently, reaction rules may also contain placeholders processed by a matching relation for association of available particles to reactants given within rules. Furthermore, our P systems framework combines the ability to manage specific string-objects with discretised reaction kinetics. Incomplete information about protein activation states can be handled by setting placeholders if required. While they enable a unification of several activation states when specifying a protein on the one hand, placeholders contribute to trace the variety of potential effects by embedding wild-cards into reaction rules on the other hand. Thus, a bottom-up strategy for the modelling of signalling networks by successive knowledge integration can benefit from the proposed framework. Along with intermediate results coming from simulation of a partially wild-carded system, synergies between wetlab experimental setup and model refinement considering structural dynamics might emerge. Inclusion of reaction kinetics into the formalism of P systems was explained in [18] exemplified by metabolic networks, supplemented by signalling and gene regulatory networks [13]. A previous formulation of periodic and quasi-periodic processes based on symbol objects without inner structure is given in [5]. The BioNetGen framework [6] allows handling of string pattern to constitute species. However, its expressive capability of reaction kinetics excludes stoichiometry.

The paper is organised in two main sections: Firstly, we define the P systems framework Π_{CSM} (Cell Signalling Module) with emphasis on the combination of reaction kinetics and wild-carded representation of proteins as string-objects. Matching strategies accomplish the handling of incomplete information. In order to provide formalisms to select reactants for rule-based rewriting, we adopt the strategy of loose matching [13]. It is expressed by a relation between strings forming objects and strings acting as patterns in rewriting rules. The loose matching checks whether there is at least one common wild-card free representation for both strings. So, it is intended to generate a maximal variety of potential effects. A more general matching approach able to find patterns common to a set of strings has been specified by the Angluin pattern language [2]. In order to enable detailed studies on the temporal evolution of the system, we replace the maximally parallel rewriting from the original framework [23] with a mechanism that is based on reaction kinetics. For each rewriting rule, the number of applications per turn is given by a kinetic function, depending on the current configuration of the system. This way, a deterministic system evolution is obtained. The formal system definition is followed by a comprehensive application scenario: Section 3 demonstrates the suitability of the framework for discussing model candidates of the circadian clock formed by the KaiABC oscillator found in cyanobacterium *Synechococcus elongatus*. Since the detailed mechanism of this biochemical oscillation is partially unknown, various models have been developed recently e.g. [8,19,29]. We show their integration into the P systems framework Π_{CSM} in terms of an intersecting superposition of consistent elements flanked by wild-carded completion. A simulation study of the dynamical system's behaviour discloses effects of superpositioned protein abundance courses based on regular expressions corresponding to dedicated protein activation states.

2 System Description

Multiset Prerequisites

Let A be an arbitrary set and $\mathbb{N}$ the set of natural numbers including zero. $\mathcal{P}(A)$ denotes the power set of A. A multiset over A is a mapping $F : A \longrightarrow \mathbb{N} \cup \{\infty\}$. $F(a)$, also denoted as $[a]_F$, specifies the multiplicity of $a \in A$ in F. Multisets can be written as an elementwise enumeration of the form $\{(a_1, F(a_1)), (a_2, F(a_2)), \ldots\}$ since $\forall (a, b_1), (a, b_2) \in F : b_1 = b_2$. The support $\mathrm{supp}(F) \subseteq A$ of F is defined by $\mathrm{supp}(F) = \{a \in A \mid F(a) > 0\}$. A multiset F over A is said to be empty iff $\forall a \in A \; : \; F(a) = 0$. The cardinality $|F|$ of F over A is $|F| = \sum_{a \in A} F(a)$. Let F_1 and F_2 be multisets over A. F_1 is a subset of F_2, denoted as $F_1 \subseteq F_2$, iff $\forall a \in A \; : \; (F_1(a) \leq F_2(a))$. Multisets F_1 and F_2 are equal iff $F_1 \subseteq F_2 \wedge F_2 \subseteq F_1$. The intersection $F_1 \cap F_2 = \{(a, F(a)) \mid a \in A \wedge F(a) = \min(F_1(a), F_2(a))\}$, the multiset sum $F_1 \uplus F_2 = \{(a, F(a)) \mid a \in A \wedge F(a) = F_1(a) + F_2(a)\}$, and the multiset difference $F_1 \ominus F_2 = \{(a, F(a)) \mid a \in A \wedge F(a) = \max(F_1(a) - F_2(a), 0)\}$ form multiset operations. Multiplication of a multiset $F = \{(a, F(a)) \mid a \in A\}$ with a scalar c, denoted $c \cdot F$, is defined by $\{(a, c \cdot F(a)) \mid a \in A\}$. The term $\langle A \rangle = \{F : A \longrightarrow \mathbb{N} \cup \{\infty\}\}$ describes the set of all multisets over A.

Definition of System Components

A P system for a cell signalling module (CSM) is a construct

$$\Pi_{\mathrm{CSM}} = (V, V', R_1, \dots, R_r, f_1, \dots, f_r, A, C, \Delta\tau)$$

where V and V' are two alphabets (not necessarily disjoint); without loss of generality $\#, \neg, * \notin V \cup V'$. The regular set

$$S = V^+ \cdot \left(\{\#\} \cdot \left((V')^+ \cup \{\neg\} \cdot (V')^+ \cup \{*\}\right)\right)^*$$

describes the syntax for string-objects. The leftmost substring from V^+ holds the protein identifier, followed by a finite number of protein property substrings from $(V')^+$ which are separated by $\#$. For example, consider the string-object C:D#p#*#¬q identifying protein (complex) C:D with specified property p, a second arbitrary property ($*$), and without property q. Each protein property substring expresses a specific additional information about the protein, for instance whether it is activated by carrying a ligand at a certain binding site. Two kinds of meta symbols are allowed. The symbol $\neg$ excludes the subsequent property but permits all other properties at this substring position. The placeholder $*$ stands for an arbitrary (also unknown or unspecified) protein property substring. This way, uncertainty about the properties of proteins can be explicitly expressed. String-objects can be dynamically processed by reaction rules:

$R_i \in \langle S \rangle \times \langle S \rangle$ is a reaction rule composed of two finite multisets

$f_i : \langle S \rangle \longrightarrow \mathbb{N}$ is a function corresponding to kinetics of reaction R_i

$A \in \langle S \rangle$ is a multiset of axioms representing the initial molec. configuration

$C \in \mathbb{R}_+$ spatial capacity of the module (vessel or compartment)

$\Delta\tau \in \mathbb{R}_+$ time discretisation interval

We explain the system evolution of Π_{CSM} within three consecutive subsections. Based on the specification of the system configuration, we define an iteration scheme that updates this configuration from time t to time $t + 1$. The update includes processing of reactions given by the rules R_i ($i = 1, \dots, r$). For this purpose, an appropriate matching between wild-carded strings representing reactants and those stated in the current configuration is required. Then, a reaction is executed by removing the multiset of matching reactants from the current configuration followed by adding the corresponding products. In order to consider kinetic issues, each reaction can be multiply processed. Therefore, the number of turns is provided by the function f_i.

Dynamical System Behaviour

A P system of the form Π_{CSM} evolves by successive progression of its *configuration* $L_t \in \langle S \rangle$ at discrete points in time $t \in \mathbb{N}$ for what we assume a global clock. Two consecutive dates t and $t+1$ specify a time span $\Delta\tau$ (discretisation interval). A system step at time t consists of two modification stages per reaction $1, \ldots, r$. Firstly, the multiset of reactants is determined and removed from L_t. Afterwards, the corresponding multiset of products is added. To cope with conflicts that can occur if the available amount of reactants cannot satisfy all matching reactions, we prioritise the reaction rules by their index: $R_1 > R_2 > \ldots > R_r$. Thus, we keep determinism of the system evolution and enable mass conservation.

$$L_0 = L_{0,0} = A$$

$$L_{t,1} = \begin{cases} L_{t,0} \ominus Reactants_{t,1} \uplus Products_{t,1} & \text{if } Reactants_{t,1} \subseteq L_{t,0} \\ L_{t,0} & \text{otherwise} \end{cases}$$

$$L_{t,2} = \begin{cases} L_{t,1} \ominus Reactants_{t,2} \uplus Products_{t,2} & \text{if } Reactants_{t,2} \subseteq L_{t,1} \\ L_{t,1} & \text{otherwise} \end{cases}$$

$$\vdots$$

$$L_{t+1} = L_{t,r} = \begin{cases} L_{t,r-1} \ominus Reactants_{t,r} \uplus Products_{t,r} & \text{if } Reactants_{t,r} \subseteq L_{t,r-1} \\ L_{t,r-1} & \text{otherwise} \end{cases}$$

Let $R_j = (A_j, B_j) \in \langle S \rangle \times \langle S \rangle$ be a reaction rule with $\mathrm{supp}(A_j) = \{a_1, \ldots, a_p\}$ and $\mathrm{supp}(B_j) = \{b_1, \ldots, b_q\}$. In terms of a chemical denotation, it can be written as

$$A_j(a_1)\, a_1 + \ldots + A_j(a_p)\, a_p \longrightarrow B_j(b_1)\, b_1 + \ldots + B_j(b_q)\, b_q$$

where $A_j(a_1), \ldots, A_j(a_p)$ represent stoichiometric factors of reactants $a_1, \ldots, a_p$, and $B_j(b_1), \ldots, B_j(b_q)$ stoichiometric factors of products $b_1, \ldots, b_q$, respectively. All reactant strings that match to the pattern a_k are provided by a dedicated relation $Match(a_k)$ (see next subsection for definition). A combination of reactant strings from L_t matching the left hand side of R_j forms a multiset of string-objects used to apply the reaction once. Since the kinetic law, described by the corresponding scalar function f_j, returns the number of applications of reaction rule R_j within one step, the multiset of string-objects extracted from L_t to act as reactants for R_j can be written as $Reactants_{t,j}$:

$$Reactants_{t,j} = \biguplus_{e_1 \in Match(a_1)} \cdots \biguplus_{e_p \in Match(a_p)} f_j\left(\{(e_1, \infty), \ldots, (e_p, \infty)\} \cap L_{t,j-1}\right) \cdot$$

$$\{(e_1, A_j(a_1)), \ldots, (e_p, A_j(a_p))\}$$

Accordingly, the multiset of products resulting from reaction rule R_j is determined by the multiset $Products_j(t)$:

$$Products_{t,j} = \biguplus_{e_1 \in Match(a_1)} \cdots \biguplus_{e_p \in Match(a_p)} f_j\left(\{(e_1, \infty), \ldots, (e_p, \infty)\} \cap L_{t,j-1}\right) \cdot$$

$$\{(b_1, B_j(b_1)), \ldots, (b_q, B_j(b_q))\}$$

Matching

Let the regular set S be a syntax description for string-objects. In the symmetric relation $Match$, two string-objects match iff there is at least one common representation without wild-cards. This loose strategy requires a minimum degree of similarity between objects with incomplete information. Uncertainty is interpreted as arbitrary replacements within the search space given by S.

$$Match \subseteq S \times S$$
$$Match = \bigcup_{m \in \mathbb{N}} \{(p\#p_1\#p_2 \ldots \#p_m, \ s\#s_1\#s_2 \ldots \#s_m) \mid (p = s) \wedge$$

$$\forall j \in \{1, \ldots, m\} : [(p_j = s_j) \vee (p_j = *) \vee (s_j = *) \vee$$
$$((p_j = \neg q) \wedge (s_j \neq q)) \vee ((s_j = \neg q) \wedge (p_j \neq q))]\}$$

Matching of a single string-object $w \in S$ to the entire set S is defined by

$$Match(w) = \{s \in S \mid (w, s) \in Match\}$$

Consequently, we define the matching of a language $L \subseteq S$ by the function $Match : \mathcal{P}(S) \longrightarrow \mathcal{P}(S)$ with

$$Match(L) = \bigcup_{w \in L} Match(w).$$

Discrete Reaction Kinetics

Within the P systems framework Π_{CSM}, we formulate reaction kinetics by specification of scalar functions f_j attached to corresponding reactions R_j ($j = 1, \ldots, r$). Each scalar function converts the current configuration L_t, a multiset of string-objects, into the number of turns for application of rewriting rule R_j:

$$f_j(L_t) = \left\lfloor k_j \prod_{\forall \alpha \in Match(A_j) \cap Match(L_t) : (R_j = (A_j, B_j))} \hat{f}(L_t(\alpha))^{|Match(A_j) \cap \{(\alpha, \infty)\}|} \right\rfloor \tag{1}$$

whereas the auxiliary term α passes through all string-objects present in L_t which also form reactants in R_j. The multiplicity $L_t(\alpha)$ of occurrences of α acts as argument for a kinetic law $\hat{f}(L_t(\alpha))$. Examples adopted from mass-action, Michaelis-Menten, and Hill kinetics are shown in Figure 1.

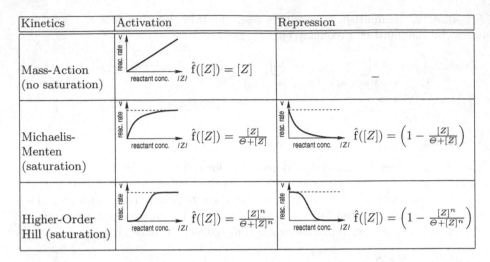

Kinetics	Activation	Repression
Mass-Action (no saturation)	$\hat{f}([Z]) = [Z]$	–
Michaelis-Menten (saturation)	$\hat{f}([Z]) = \frac{[Z]}{\Theta+[Z]}$	$\hat{f}([Z]) = \left(1 - \frac{[Z]}{\Theta+[Z]}\right)$
Higher-Order Hill (saturation)	$\hat{f}([Z]) = \frac{[Z]^n}{\Theta+[Z]^n}$	$\hat{f}([Z]) = \left(1 - \frac{[Z]^n}{\Theta+[Z]^n}\right)$

Fig. 1. Overview of several widely used kinetic laws $\hat{f}([Z])$ dependent on reactant concentration $[Z]$. Parameters: threshold $\Theta \in \mathbb{R}_+$, Hill coefficient $n \in \mathbb{N}_+$.

Relations to ODE-Based Reaction Kinetics

For a reaction system with a total number of n species $(i = 1, \ldots, n)$ and r reactions $(j = 1, \ldots, r)$

$$a_{1,j}Z_1 + a_{2,j}Z_2 + \ldots + a_{n,j}Z_n \xrightarrow{\hat{k}_j} b_{1,j}Z_1 + b_{2,j}Z_2 + \ldots + b_{n,j}Z_n$$

the corresponding ODEs

$$\frac{\mathrm{d}\,[Z_i]}{\mathrm{d}\,t} = \sum_{j=1}^{r}\left(\hat{k}_j \cdot (b_{i,j} - a_{i,j}) \cdot \prod_{l=1}^{n}\hat{f}_j([Z_l])^{a_{l,j}}\right) \quad \text{with} \quad i = 1, \ldots, n. \quad (2)$$

describe the temporal systems behaviour by consideration of stoichiometric coefficients $a_{i,j} \in \mathbb{N}$ (reactants) and $b_{i,j} \in \mathbb{N}$ (products) as well as a kinetic law $\hat{f}_j([Z_i]) : \mathbb{R}_+ \to \mathbb{R}_+$ that maps a species concentration $[Z_i]$ into an effective reaction rate [9]. All initial concentrations $[Z_i](0) \in \mathbb{R}_+$, $i = 1, \ldots, n$ are allowed to be set according to the needs of the reaction system.

A species concentration $[Z_i] := \frac{z_i}{C}$ is defined as fraction of its molecular amount $z_i = \text{supp}(\{(Z_i, z_i)\})$ with respect to the spatial system capacity $C \in \mathbb{R}_+$.

A correspondence between the reaction rate k_j (employed in Π_{CSM} by function f_j attached to reaction R_j) and the kinetic constant $\hat{k}_j$ utilised in ODE (2) can be obtained by the Euler method of integrating differential equations. Discretisation of (2) with respect to time and concentration value results in:

$$\frac{\frac{z_{i,t+1}-z_{i,t}}{C}}{\Delta\tau} = \sum_{j=1}^{r}\left(\hat{k}_j \cdot (b_{i,j} - a_{i,j}) \cdot \prod_{l=1}^{n}\hat{f}_j([Z_l])^{a_{l,j}}\right)$$

$$z_{i,t+1} - z_{i,t} = C \cdot \Delta\tau \cdot \sum_{j=1}^{r} \left(\hat{k}_j \cdot (b_{i,j} - a_{i,j}) \cdot \prod_{l=1}^{n} \hat{f}_j([Z_l])^{a_{l,j}} \right)$$

By setting $k_j = \hat{k}_j \cdot C \cdot \Delta\tau$, we obtain:

$$z_{i,t+1} - z_{i,t} = k_1(b_{i,1} - a_{i,1}) \prod_{l=1}^{n} \hat{f}_1([Z_l])^{a_{l,1}} + \ldots + k_r(b_{i,r} - a_{i,r}) \prod_{l=1}^{n} \hat{f}_r([Z_l])^{a_{l,r}}$$

Replacing $k_j \cdot \hat{f}_j([Z_l])^{a_{l,j}}$ by the discretised (and hence approximated) scalar function $f_j(L_t)$ from Equation (1) leads to:

$$z_{i,t+1} - z_{i,t} \approx (b_{i,1} - a_{i,1}) \cdot f_1(L_t) + \ldots + (b_{i,r} - a_{i,r}) \cdot f_r(L_t)$$

Since the stoichiometric coefficients $a_{i,j}$ and $b_{i,j}$ of each reaction $R_j = (A_j, B_j)$ in Π_{CSM} are expressed by multisets A_j (reactants) and B_j (products), we write:

$$z_{i,t+1} - z_{i,t} = (B_1(b_i) - A_1(a_i)) \cdot f_1(L_t) + \ldots + (B_r(b_i) - A_r(a_i)) \cdot f_r(L_t)$$

From that, we achieve the update scheme for species Z_i present in L_t with $z_{i,t}$ copies at time t by processing reaction R_j:

$$z_{i,t+1} = z_{i,t} - A_j(Z_i) \cdot f_j(L_t) + B_j(Z_i) \cdot f_j(L_t)$$

By extension from a single species to the entire configuration along with inclusion of matching, we finally yield

$$L_{t+1,j} = L_{t,j} \ominus Reactants_{t,j} \uplus Products_{t,j}$$

in accordance to the iteration scheme for Π_{CSM} evolution. The conversion of thresholds Θ occurring in Michaelis-Menten or Hill terms from the ODE approach into the Π_{CSM} framework can be done by parameter fitting or regression that maps the concentration-based gradient into an amount-based counterpart.

3   The KaiABC Oscillator – A Circadian Clock

Biological Background

Circadian rhythms embody an interesting biological phenomenon that can be seen as a widespread property of life. The coordination of biological activities into daily cycles provides an important advantage for the fitness of diverse organisms [4,25]. Based on self-sustained biochemical oscillations, circadian clocks are characterised by a period of approximately 24h that persists under constant conditions (like constant darkness or constant light). Their ability for compensation of temperature in the physiological range enables then to maintain the period in case of environmental changes. Furthermore, circadian clocks can be entrained. This property allows a gradual reset of the underlying oscillatory system for adjustment by exposure to external stimuli like light/dark or temperature cycles. A variety of metabolic, cell signalling, and gene regulatory processes is

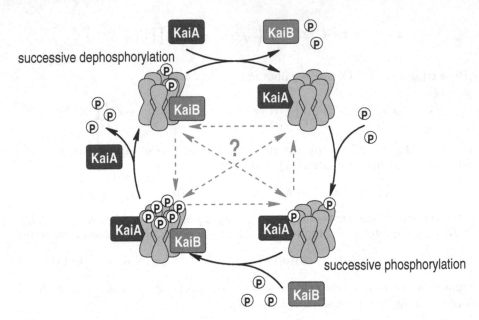

Fig. 2. Reaction cycle of the KaiABC oscillator characterised by four phases and incomplete information about interphase feedback loops, arranged from descriptions of the oscillatory mechanism given in [11,20]. A corresponding minimal model of the four-phase cycle has been proposed in [4].

synchronised or controlled by circadian clocks. Chemically, they utilise an individual cycling reaction scheme including one or more feedback loops. Most of the circadian clocks comprise gene transcription and translation feedback loops [24].

Surprisingly, the prokaryotic cyanobacterium *Synechococcus elongatus* was discovered to carry a post-translational circadian clock even functioning *in vitro* [27]. Three key clock proteins KaiA, KaiB, and KaiC with known atomic structure could be identified [21]. KaiC as the focal protein rhythmically oscillates between hypophosphorylated and hyperphosphorylated forms [22]. The spatial structure of KaiC represents a homohexamer shaped as a "double doughnut" with 6 phosphorylation twin sites at the interfaces between monomeric subunits. Presence of the supplementary protein KaiA specifically enhances KaiC phosphorylation while KaiBC complex formation activates KaiC dephosphorylation [20]. The KaiABC circadian oscillator appears as a reaction cycle consisting of four consecutive phases [11], see Figure 2: KaiAC complex formation releasing KaiB, successive KaiAC phosphorylation, KaiABC complex formation, and successive KaiABC dephosphorylation in conjunction with KaiA dissociation. Each of these phases takes approximately 6h. There is some evidence for further interactions between the aforementioned protein complexes and intermediate products in terms of negative feedback loops stabilising the oscillation. However, the detailed mechanism is still unclear and gives room for hypotheses reflected in a couple of model candidates [4]. A current study raises the question whether clock-protein expression could still be involved in its general function [28].

Review of Modelling Approaches

In this section, we briefly compare three current model candidates [8,19,29] beyond a minimal model [4] able to capture the dynamical behaviour of the KaiABC oscillator in accordance with wetlab experimental data. Assumptions on unknown parts of the oscillator mechanism result from empirical studies. Here, an underlying reaction network topology is hypothesised and afterwards filled with appropriate parameter values obtained by fitting using an exhaustive search.

KaiA sequestration has been suggested in [8]. The resulting model identifies a total number of 15 interacting species where $C^0, \ldots, C^6$ correspond to the amount of phosphorylated monomeric subunits within KaiC. Accordingly, $BC^0, \ldots, BC^6$ are species names for complex KaiBC. B indicates KaiB. KaiA is assumed to be sequestered by the KaiC/KaiBC complexes and hence not modelled explicitly. Instead, it is interpreted as an inhibiting factor causing negative feedback loops. See Figure 3 **A** for the reaction network topology.

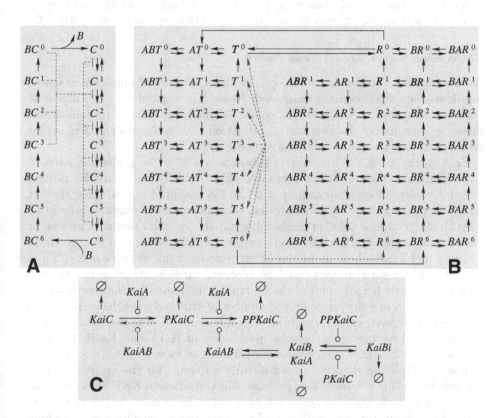

Fig. 3. Comparison of KaiABC oscillator network topologies adapted from [8] (**A**), [29] (**B**), and [19] (**C**). Dashed lines indicate relevant feedback loops for sustained oscillation.

Following the idea of a quick KaiC monomer shuffle, in [29] a network topology containing 54 dedicated species is proposed. There are two categories of species marked as "tense" (T) for those employed in the phosphorylation phase and "relaxed" (R) for the dephosphorylation phase. Indexes attached to T and R ranging from 0 to 6 comprise the number of currently phosphorylated monomeric subunits while association of KaiA and/or KaiB complexes is denoted by concatenation of A or B to the species names. Figure 3 **B** illustrates the network topology by usage of dashed arrows for monomer shuffle.

A different description has been introduced in [19] managing on 7 species (by neglecting intermediate products of protein degradation). Inspired by the insight that distinction of two states is sufficient to obtain robust oscillations of KaiC phosphorylation, a cascade of elementary cell signalling motifs is proposed. In this two-stage scenario, three phosphates from species $KaiC$ can be added and removed per stage by catalysts $KaiA$ and $KaiAB$, respectively. Additionally, the model formulates the complex formation of $KaiAB$ which is catalysed by the three-fold phosphorylated protein $PKaiC$. Vice versa, its decomposition is supported by the six-fold phosphorylated protein $PPKaiC$. Decay reactions for each protein complete the model candidate's network topology, see Figure 3 **C**.

Conversion to the Π_{CSM} Framework

We demonstrate a conversion of the core oscillator extracted from different model candidates into the P systems framework Π_{CSM}. The capability of this algebraic approach is to cope with a potential combinatorial complexity of protein states, shown by formulating reaction and transduction rules using placeholders ($*$) for arbitrary or unknown molecular constituents.

Each of the six KaiC monomeric subunits is said to be phosphorylated iff both phosphorylation sites are saturated. Theoretically, the KaiABC protein complex could induce a maximum of $2^8 = 256$ potential states. This amount results from the general assumption that each monomeric subunit is able to be individually phosphorylated or dephosphorylated in combination with present or absent association of KaiA and KaiB, respectively. In terms of a distinction of 8 binary digits from these molecular configurations, a full network of $2 \cdot \binom{256}{2} =$ $65,280$ bi-molecular reactions could be spanned. Since KaiC turns out to be a highly symmetric homohexamer, the individual monomeric subunits cannot be distinguished in practice. Instead, the number of attached phosphates is utilised that varies in a seven-stage range from 0 up to 6. In addition to the combinatorial variety caused by present or absent association of KaiA and KaiB, KaiABC possess $7 \cdot 4 = 28$ states from a biochemical point of view.

For the P systems description, we identify a module for the cycling reaction scheme sketched in Figure 2. Key proteins KaiA, KaiB, and KaiC resulting from expression of corresponding genes are assumed to be present in the module *ab initio*. Considering the core oscillator, 17 reaction rules along with loose matching correspond to the four-phase reaction cycle. Successive KaiC phosphorylation in the presence of KaiA is expressed by rules R_1 to R_6 followed by successive dephosphorylation in the presence of KaiB within rules R_7 to R_{12}. Finally, R_{13}

and R_{14} formulate inhibiting KaiA/KaiB exchange acting as negative feedback loops, and R_{15} up to R_{17} reflect protein degradation. A kinetic function f is attached to each reaction rule that follows from discretised Michaelis-Menten kinetic laws in concert with linear mass-action kinetics for protein degradation.

$$\Pi_{KaiABC} = (V, V', R_1, \ldots, R_{17}, f_1, \ldots, f_{17}, A, C, \Delta\tau)$$

$$V = \{C\} \cup \ldots\ldots\ldots\ldots\ldots\ldots\text{identifier of the focal protein KaiC}$$
$$\{A, B\}\ldots\ldots\ldots\ldots\ldots\ldots\text{identifiers of proteins KaiA and KaiB}$$
$$V' = \{A, B\} \cup \ldots\ldots\ldots\ldots\ldots\text{KaiA, KaiB within a complex associated to KaiC}$$
$$\{0, 1, 2, 3, 4, 5, 6\}\ldots\ldots\text{number of attached phosphates}$$

$$R_1 = C\#\neg A\#B\#0 + A \longrightarrow C\#A\#\neg B\#1 + B$$
$$R_2 = C\#A\# * \#1 + A \longrightarrow C\#A\# * \#2 + A$$
$$R_3 = C\#A\# * \#2 + A \longrightarrow C\#A\# * \#3 + A$$
$$R_4 = C\#A\# * \#3 + A \longrightarrow C\#A\# * \#4 + A$$
$$R_5 = C\#A\# * \#4 + A \longrightarrow C\#A\# * \#5 + A$$
$$R_6 = C\#A\#\neg B\#5 + B \longrightarrow C\#\neg A\#B\#6 + A$$
$$R_7 = C\# * \#B\#6 + B \longrightarrow C\# * \#B\#5 + B$$
$$R_8 = C\# * \#B\#5 + B \longrightarrow C\# * \#B\#4 + B$$
$$R_9 = C\# * \#B\#4 + B \longrightarrow C\# * \#B\#3 + B$$
$$R_{10} = C\# * \#B\#3 + B \longrightarrow C\# * \#B\#2 + B$$
$$R_{11} = C\# * \#B\#2 + B \longrightarrow C\# * \#B\#1 + B$$
$$R_{12} = C\# * \#B\#1 + B \longrightarrow C\# * \#B\#0 + B$$
$$R_{13} = C\#\neg A\#B\#* + A \longrightarrow C\#A\#\neg B\#* + B$$
$$R_{14} = C\#A\#\neg B\#* + B \longrightarrow C\#\neg A\#B\#* + A$$
$$R_{15} = A \longrightarrow \emptyset$$
$$R_{16} = B \longrightarrow \emptyset$$
$$R_{17} = C\# * \# * \#* \longrightarrow \emptyset$$

$$f_1(L_t) = \left\lfloor k_1 \cdot \frac{L_t(C\#\neg A\#B\#0)}{\Theta_{1,1} + L_t(C\#\neg A\#B\#0)} \cdot \frac{L_t(A)}{\Theta_{1,2} + L_t(A)} \right\rfloor$$

$$f_2(L_t) = \left\lfloor k_2 \cdot \frac{L_t(C\#A\# * \#1)}{\Theta_{2,1} + L_t(C\#A\# * \#1)} \cdot \frac{L_t(A)}{\Theta_{2,2} + L_t(A)} \right\rfloor$$

$$f_3(L_t) = \left\lfloor k_3 \cdot \frac{L_t(C\#A\# * \#2)}{\Theta_{3,1} + L_t(C\#A\# * \#2)} \cdot \frac{L_t(A)}{\Theta_{3,2} + L_t(A)} \right\rfloor$$

$$f_4(L_t) = \left\lfloor k_4 \cdot \frac{L_t(C\#A\# * \#3)}{\Theta_{4,1} + L_t(C\#A\# * \#3)} \cdot \frac{L_t(A)}{\Theta_{4,2} + L_t(A)} \right\rfloor$$

$$f_5(L_t) = \left\lfloor k_5 \cdot \frac{L_t(C\#A\# * \#4)}{\Theta_{5,1} + L_t(C\#A\# * \#4)} \cdot \frac{L_t(A)}{\Theta_{5,2} + L_t(A)} \right\rfloor$$

$$f_6(L_t) = \left\lfloor k_6 \cdot \frac{L_t(C\#A\#\neg B\#5)}{\Theta_{6,1} + L_t(C\#A\#\neg B\#5)} \cdot \frac{L_t(B)}{\Theta_{6,2} + L_t(B)} \right\rfloor$$

$$f_7(L_t) = \left\lfloor k_7 \cdot \frac{L_t(C\# * \#B\#6)}{\Theta_{7,1} + L_t(C\# * \#B\#6)} \cdot \frac{L_t(B)}{\Theta_{7,2} + L_t(B)} \right\rfloor$$

$$f_8(L_t) = \left\lfloor k_8 \cdot \frac{L_t(C\# * \#B\#5)}{\Theta_{8,1} + L_t(C\# * \#B\#5)} \cdot \frac{L_t(B)}{\Theta_{8,2} + L_t(B)} \right\rfloor$$

$$f_9(L_t) = \left\lfloor k_9 \cdot \frac{L_t(C\# * \#B\#4)}{\Theta_{9,1} + L_t(C\# * \#B\#4)} \cdot \frac{L_t(B)}{\Theta_{9,2} + L_t(B)} \right\rfloor$$

$$f_{10}(L_t) = \left\lfloor k_{10} \cdot \frac{L_t(C\# * \#B\#3)}{\Theta_{10,1} + L_t(C\# * \#B\#3)} \cdot \frac{L_t(B)}{\Theta_{10,2} + L_t(B)} \right\rfloor$$

$$f_{11}(L_t) = \left\lfloor k_{11} \cdot \frac{L_t(C\# * \#B\#2)}{\Theta_{11,1} + L_t(C\# * \#B\#2)} \cdot \frac{L_t(B)}{\Theta_{11,2} + L_t(B)} \right\rfloor$$

$$f_{12}(L_t) = \left\lfloor k_{12} \cdot \frac{L_t(C\# * \#B\#1)}{\Theta_{12,1} + L_t(C\# * \#B\#1)} \cdot \frac{L_t(B)}{\Theta_{12,2} + L_t(B)} \right\rfloor$$

$$f_{13}(L_t) = \left\lfloor k_{13} \cdot \left(1 - \frac{L_t(C\#\neg A\#B\#*)}{\Theta_{13,1} + L_t(C\#\neg A\#B\#*)}\right) \cdot \left(1 - \frac{L_t(A)}{\Theta_{13,2} + L_t(A)}\right) \right\rfloor$$

$$f_{14}(L_t) = \left\lfloor k_{14} \cdot \left(1 - \frac{L_t(C\#A\#\neg B\#*)}{\Theta_{14,1} + L_t(C\#A\#\neg B\#*)}\right) \cdot \left(1 - \frac{L_t(B)}{\Theta_{14,2} + L_t(B)}\right) \right\rfloor$$

$$f_{15}(L_t) = k_{15} \cdot L_t(A)$$

$$f_{16}(L_t) = k_{16} \cdot L_t(B)$$

$$f_{17}(L_t) = k_{17} \cdot L_t(C\# * \# * \#*)$$

$$A \in \langle \{C\# * \# * \#*\} \rangle$$

Simulation Case Study

Using the KaiABC circadian oscillator we conducted a simulation case study to demonstrate the practicability of the modelling approach addressed before. The reaction scheme formulated by the P system Π_{KaiABC} exhibits a high degree of symmetry among its constituents. The main reaction cycle is composed of 12 consecutive feedforward reactions flanked by widespread negative feedback loops. They affect each intermediate product within the reaction cycle following the intention of an inhibiting KaiA/KaiB exchange independent of the phosphorylation state.

For simulation of the dynamical behaviour of Π_{KaiABC}, we empirically parameterise and initialise the system in a symmetric way to obtain phase-shifted protein abundance courses which stably oscillate with a period of approximately 24 hours. To avoid a transient oscillation phase, the initial amounts of protein constituents were set directly at the discrete limit cycle. This constraint is reflected in the following multiset of axioms:

$$
\begin{aligned}
A = \{ & (C\#\neg A\#B\#0, 470), (C\#A\#\neg B\#1, 351), (C\#A\#\neg B\#2, 198), \\
& (C\#A\#\neg B\#3, 135), (C\#A\#\neg B\#4, 148), (C\#A\#\neg B\#5, 210), \\
& (C\#\neg A\#B\#6, 282), (C\#\neg A\#B\#5, 364), (C\#\neg A\#B\#4, 463), \\
& (C\#\neg A\#B\#3, 541), (C\#\neg A\#B\#2, 586), (C\#\neg A\#B\#1, 571), \\
& (A, 2520), (B, 2520) \}
\end{aligned}
$$

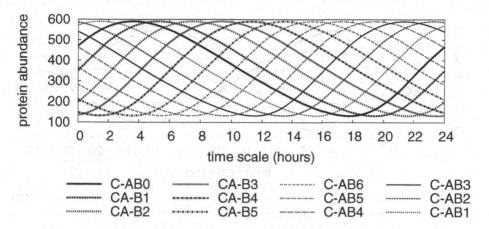

Fig. 4. Temporal courses of 12 specific KaiABC subproducts representing the process status of the reaction cycle. Kinetic parameters and initial amounts adjusted in a way to obtain a period of ≈ 24 hours and symmetry among individual oscillations.

Each KaiC protein within the pattern C# * # * #* keeps an average amount of 360 copies (arbitrarily chosen).

Figure 4 shows the corresponding individual protein abundance courses resulting from following parameter setting for the discrete iteration scheme: $\Theta_{i,1} = 79.2, \Theta_{i,2} = 554.4, \hat{k}_i = 360.0$ for $i \in \{1, \ldots, 12\}$; $\Theta_{i,1} = 64.8, \Theta_{i,2} = 453.6, \hat{k}_i = 412.8$ for $i \in \{13, 14\}$, and $\hat{k}_{15} = \hat{k}_{16} = 508.1, \hat{k}_{17} = 254.6$; C = 1.2, $\Delta\tau = 0.05$. The iteration scheme for system evolution was implemented in the programming language C to obtain the course data.

Based on the individual protein abundance courses depicted in Figure 4, Figure 5 illustrates the effect of subsuming KaiABC subproducts according to their number of attached phosphates ranging from 0 to 6. Association of KaiA and KaiB is neglected here resulting in consideration of regular expressions C# * # * #i for $i = 0, \ldots, 6$. The simulation shows that medium phosphorylation levels possess smaller amplitudes than minor or major phosphorylation levels. Due to symmetry reasons, KaiABC subproducts carrying three phosphates double the frequency of oscillation. Hence, the reaction system is able to act as a scaler. This feature could be useful to control downstream processes at a subcircadian granularity.

Classification of KaiABC subproducts with regard to association of KaiA and KaiB leads to simulation results depicted in Figure 6. As expected, both courses proceed in opposite direction emphasising the mutually exclusive association of KaiA and KaiB to KaiC.

Further simulation studies could explore the effects of different temperatures to the network behaviour. To this end, modified forms of Arrhenius terms based on the Boltzmann constant instead of the universal gas constant might be utilised to replace each reaction parameter k_j. In this way, a possible capability of

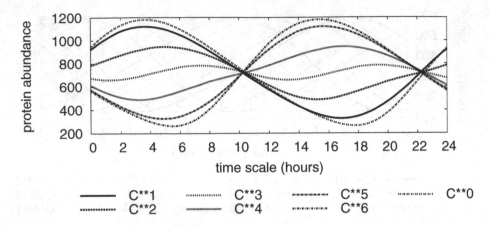

Fig. 5. Temporal courses of KaiABC subproducts subsumed by their level of phosphorylation ranging from 0 to 6. Kinetic parameters and initial amounts adjusted in a way to obtain a period of ≈ 24 hours and symmetry among individual oscillations.

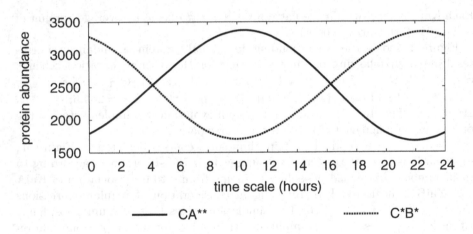

Fig. 6. Temporal courses of KaiABC subproducts separated into two groups by association of KaiA resp. KaiB to KaiC. Kinetic parameters and initial amounts adjusted in a way to obtain a period of ≈ 24 hours and symmetry among individual oscillations.

temperature compensation or entrainment is investigable and can be applied to fine-tuning of the model.

Extensions of the System

In this section, we address specialties of the different modelling approaches [8,19,29] in the context of their conversion into the P systems framework by additional wild-carded reactions. Each of these reactions subsumes a variety of individually interacting components that form feedback loops capable of stabilising or destabilising the oscillating behaviour of the whole system. Kinetic laws within system extensions also employ discretised Michaelis-Menten kinetics for enzymatic processes and linear mass-action kinetics for protein degradation.

Premature dissociation or association of KaiA or KaiB can destabilise the oscillatory behaviour by damping effects. In contrast, spontaneous dephosphorylation and monomer shuffle amplify the influence of feedbacks within the reaction system. This makes the network behaviour more sensitive to slight parameter changes. Toggling KaiB between an active and an inactive form as well as inhibition of KaiC phosphorylation catalysed by KaiB is able to break the symmetry among the reaction cycle.

Premature KaiA association [29]:

$$A + C\#\neg A\# * \#* \longrightarrow C\#A\# * \#*$$

Premature KaiA dissociation [29]:

$$C\#A\# * \#* \longrightarrow A + C\#\neg A\# * \#*$$

Premature KaiB association [29]:

$$B + C\# * \#\neg B\#* \longrightarrow C\# * \#B\#*$$

Premature KaiB dissociation [29]:

$$C\# * \#B\#* \longrightarrow B + C\# * \#\neg B\#*$$

Spontaneous dephosphorylation [8,29]:

$$C\# * \# * \#6 \longrightarrow C\# * \# * \#5$$
$$C\# * \# * \#5 \longrightarrow C\# * \# * \#4$$
$$C\# * \# * \#4 \longrightarrow C\# * \# * \#3$$
$$C\# * \# * \#3 \longrightarrow C\# * \# * \#2$$
$$C\# * \# * \#2 \longrightarrow C\# * \# * \#1$$
$$C\# * \# * \#1 \longrightarrow C\# * \# * \#0$$

Monomer shuffle in absence of KaiA and KaiB [29]:

$$C\#\neg A\#\neg B\#* \longrightarrow C\#\neg A\#\neg B\#*$$

Toggling KaiB between active and inactive form [19]: A new species Bi is introduced that denotes KaiB in its inactive form. KaiC in its partial or complete phosphorylated state then catalyses the toggling reactions.

$$B + C\# * \# * \#3 \longrightarrow Bi + C\# * \# * \#3$$
$$Bi + C\# * \# * \#6 \longrightarrow B + C\# * \# * \#6$$

Inhibition of KaiC phosphorylation [8]: Here, the additional string-object $C\# * \#B\#i$, $i \in \{0, \ldots, 3\}$ acts as an inhibiting factor for phosphorylating reactions $R_1, \ldots, R_6$.

4   Conclusions

Coping with incomplete information about protein activation states can be seen as a challenging task in systems biology. Particularly, the number of individual protein interactions that can potentially occur grows exponentially with regard to the number of binding sites for activation. In order to conduct exhaustive studies about the variety of potential behavioural scenarios of an entire network that includes unknown parts, all corresponding subnetworks covering these unknown parts have to be considered. Incorporation of regular expressions for representation of proteins and their activation states enables usage of placeholder symbols to express arbitrariness or uncertainty about components within those states. In this way, a wild-carded representation may subsume a combinatorial variety of individual activation states.

Accordingly, the proposed P systems framework Π_{CSM} intends to combine advantages of processing regular expressions that represent molecular entities with the corresponding dynamical behaviour of an entire reaction network resulting from superpositioning of individual molecular abundance courses. To this end, we have integrated string-objects into a deterministic framework able to emulate discretised forms of reaction kinetics in concert with dedicated matching strategies in order to identify reactants from the current system configuration. A simulation study of the KaiABC oscillator demonstrates the practicability of this approach.

From an algebraic point of view, oscillations that occur in structural or configural dynamics of P systems can be detected using a backtracking mechanism along with the temporal system evolution: By monitoring the overall configurations over time, a derivation tree is obtained. Stable oscillations appear as recurring, but nonadjacent overall configurations along a path through the derivation tree. Equipping P systems analysis tools with such a backtracking mechanism is a promising idea for futural work.

Acknowledgements

We gratefully acknowledge funding from the German Federal Ministry of Education and Research (BMBF, project no. 0315260A) within the Research Initiative in Systems Biology (FORSYS).

References

1. Alon, U.: An Introduction to Systems Biology: Design Principles of Biological Circuits. Chapman & Hall, Boca Raton (2006)
2. Angluin, D.: Finding patterns common to a set of strings. Journal of Computer and System Sciences 21, 46–62 (1980)
3. Arkin, A.P.: Synthetic cell biology. Current Opinion in Biotechnology 12(6), 638–644 (2001)
4. Axmann, I.M., Legewie, S., Herzel, H.: A minimal circadian clock model. Genome Inform. 18, 54–64 (2007)

5. Bernardini, F., Manca, V.: Dynamical aspects of P systems. BioSystems 70, 85–93 (2003)
6. Blinov, M.L., Faeder, J.R., Goldstein, B., Hlavacek, W.S.: BioNetGen: Software for Rule-Based Modeling of Signal Transduction Based on the Interactions of Molecular Domains. Bioinformatics 20, 3289–3292 (2004)
7. Blinov, M.L., Faeder, J.R., Goldstein, B., Hlavacek, W.S.: A network model of early events in epidermal growth factor receptor signaling that accounts for combinatorial complexity. BioSystems 83, 136–151 (2006)
8. Clodong, S., Dühring, U., Kronk, L., Wilde, A., Axmann, I.M., Herzel, H., Kollmann, M.: Functioning and robustness of a bacterial circadian clock. Molecular Systems Biology 90(3), 1–9 (2007)
9. Connors, K.A.: Chemical Kinetics. VCH Publishers, Weinheim (1990)
10. Eils, R., Kriebe, A. (eds.): Computational Systems Biology. Academic Press, London (2005)
11. Golden, S.S., Cassone, V.M., LiWang, A.: Shifting nanoscopic clock gears. Nature Structural and Molecular Biology 14, 362–363 (2007)
12. Heinrich, R., Schuster, S.: The Regulation of Cellular Systems. Springer, Heidelberg (2006)
13. Hinze, T., Lenser, T., Dittrich, P.: A protein substructure based P system for description and analysis of cell signalling networks. In: Hoogeboom, H.J., Păun, G., Rozenberg, G., Salomaa, A. (eds.) WMC 2006. LNCS, vol. 4361, pp. 409–423. Springer, Heidelberg (2006)
14. Hinze, T., Fassler, R., Lenser, T., Dittrich, P.: Register machine computations on binary numbers by oscillating and catalytic chemical reactions modelled using mass-action kinetics. International Journal of Foundations of Computer Science 20(3), 411–426 (2009)
15. Klipp, E., Herwig, R., Kowald, A., Wierling, C., Lehrach, H.: Systems Biology in Practice: Concepts, Implementation, and Application. Wiley-VCH, Chichester (2006)
16. Lenser, T., Hinze, T., Ibrahim, B., Dittrich, P.: Towards evolutionary network reconstruction tools for systems biology. In: Marchiori, E., Moore, J.H., Rajapakse, J.C. (eds.) EvoBIO 2007. LNCS, vol. 4447, pp. 132–142. Springer, Heidelberg (2007)
17. Magnasco, M.O.: Chemical kinetics is Turing universal. Physical Review Letters 78(6), 1190–1193 (1997)
18. Manca, V., Bianco, L., Fontana, F.: Evolution and oscillation in P systems: Applications to biological phenomena. In: Mauri, G., Păun, G., Jesús Pérez-Jímenez, M., Rozenberg, G., Salomaa, A. (eds.) WMC 2004. LNCS, vol. 3365, pp. 63–84. Springer, Heidelberg (2005)
19. Miyoshi, F., Nakayama, Y., Kaizu, K., Iwasaki, H., Tomita, M.: A mathematical model for the Kai-protein-based chemical oscillator and clock gene expression rhythms in cyanobacteria. Journal of Biological Rhythms 22(1), 69–80 (2007)
20. Mori, T., Williams, D.R., Byrne, M.O., Qin, X., Egli, M., Mchaourab, H.S., Stewart, P.L., Johnson, C.H.: Elucidating the ticking of an in vitro circadian clockwork. PLoS Biology 5(4), 841–853 (2007)
21. Nakajima, M., Imai, K., Ito, H., Nishiwaki, T., Murayama, Y.: Reconstitution of circadian oscillation of cyanobacterial KaiC phosphorylation in vitro. Science 308, 414–415 (2005)
22. Paranjpe, D.A., Sharma, V.K.: Evolution of temporal order in living organisms. Journal of Circadian Rhythms 3, 7 (2005)

23. Păun, G.: Computing with membranes. Journal of Computer and System Sciences 61(1), 108–143 (2000)
24. Rosato, E.: Circadian Rhythms: Methods and Protocols. Springer, Heidelberg (2007)
25. Roussel, M.R., Gonze, D., Goldbeter, A.: Modeling the differential fitness of cyanobacterial strains whose circadian oscillators have different free-running periods. J. Theor. Biol. 205(2), 321–340 (2000)
26. Schuster, S., Zevedei-Oancea, I.: A theoretical framework for detecting signal transfer routes in signalling networks. Comput. Chem. Eng. 29, 597–617 (2005)
27. Tomita, J., Nakajima, M., Kondo, T., Iwasaki, H.: No transcription-translation feedback in circadian rhythm of KaiC phosphorylation. Science 307, 251–254 (2005)
28. Xu, Y., Mori, T., Johnson, C.H.: Circadian clock-protein expression in cyanobacteria: rhythms and phase-setting. EMBO Journal 19, 3349–3357 (2007)
29. Yoda, M., Eguchi, K., Terada, T.P., Sasai, M.: Monomer-shuffling and allosteric transition in KaiC circadian oscillation. PLoS ONE 5, 1–7 (2007)

Solving NP-Complete Problems by Spiking Neural P Systems with Budding Rules

Tseren-Onolt Ishdorj[1], Alberto Leporati[2], Linqiang Pan[3,4], and Jun Wang[3]

[1] Computational Biomodelling Laboratory
Åbo Akademi University
Department of Information Technologies
20520 Turku, Finland
tishdorj@abo.fi
[2] Università degli Studi di Milano – Bicocca
Dipartimento di Informatica, Sistemistica e Comunicazione
Viale Sarca 336/14, 20126 Milano, Italy
alberto.leporati@unimib.it
[3] Key Laboratory of Image Processing and Intelligent Control
Department of Control Science and Engineering
Huazhong University of Science and Technology
Wuhan 430074, Hubei, People's Republic of China
junwangjf@gmail.com, lqpan@mail.hust.edu.cn
[4] Research Group on Natural Computing
Department of CS and AI, University of Sevilla
Avda Reina Mercedes s/n, 41012 Sevilla, Spain

Abstract. Inspired by the growth of dendritic trees in biological neurons, we introduce spiking neural P systems with budding rules. By applying these rules in a maximally parallel way, a spiking neural P system can exponentially increase the size of its synapse graph in a polynomial number of computation steps. Such a possibility can be exploited to efficiently solve computationally difficult problems in deterministic polynomial time, as it is shown in this paper for the **NP**-complete decision problem SAT.

1 Introduction

Spiking neural P systems (SN P systems, for short) have been introduced in [5] as a new class of distributed and parallel computing devices, inspired by the neurophysiological behavior of neurons sending electrical impulses (*spikes*) along axons to other neurons. SN P systems can also be viewed as an evolution of P systems [19,16] corresponding to a shift from *cell-like* to *neural-like* architectures. We recall that this biological background has already led to several models in the area of neural computation, e.g., see [13,14,4].

In SN P systems the cells (also called *neurons*) are placed in the nodes of a directed graph, called the *synapse graph*. The contents of each neuron consist of a number of copies of a single object type, called the *spike*. Every cell may also contain a number of *firing* and *forgetting* rules. Firing rules allow a neuron to send

G. Păun et al. (Eds.): WMC 2009, LNCS 5957, pp. 335–353, 2010.

information to other neurons in the form of electrical impulses (also called spikes) which are accumulated at the target cells. The applicability of each rule is determined by checking the contents of the neuron against a regular set associated with the rule. In each time unit, if a neuron can use some of its rules then one of such rules must be used. The rule to be applied is nondeterministically chosen. Thus, the rules are used in a sequential manner in each neuron, but neurons function in parallel with each other. Observe that, as usually happens in membrane computing, a global clock is assumed, marking the time for the whole system, hence the functioning of the system is synchronized. When a cell sends out spikes it becomes "closed" (inactive) for a specified period of time, that reflects the refractory period of biological neurons. During this period, the neuron does not accept new inputs and cannot "fire" (that is, emit spikes). Another important feature of biological neurons is that the length of the axon may cause a time delay before a spike reaches its target. In SN P systems this delay is modeled by associating a delay parameter to each rule which occurs in the system. If no firing rule can be applied in a neuron, there may be the possibility to apply a *forgetting rule*, that removes from the neuron a predefined number of spikes.

The computational efficiency of SN P systems has been recently investigated in a series of works [2,6,9,11,10]. In [12] it has been proved that a deterministic SN P system of polynomial size cannot solve an **NP**-complete problem in a polynomial time, unless **P=NP**. Hence, under the assumption that $\mathbf{P} \neq \mathbf{NP}$, efficient solutions to **NP**-complete problems cannot be obtained without introducing features which enhance the efficiency, such as pre-computed resources, ways to exponentially grow the workspace during the computation, nondeterminism, and so on. Indeed, in the framework of SN P systems, most of the solutions to computationally hard problems exploit the power of nondeterminism [11,10,12] or use pre-computed resources of exponential size [2,6,9,7].

The possibility of using SN P systems to solve computationally hard problems by using some (possibly exponentially large) pre-computed resources has been first presented in [6], that contains a description of a uniform family of SN P systems with pre-computed resources of exponential size that solves all the instances of the **NP**-complete decision problem SAT in a polynomial time. In the present paper we complement the study exposed in [6], by describing an SN P system that first builds the necessary resources (by exponentially increasing its workspace in a polynomial time), and then uses such resources to solve the SAT problem. To this purpose, we extend the SN P systems given in [6] by introducing *neuron budding rules*. We show that SN P systems with budding rules can grow an exponential size synapse graph in a time which is polynomial with respect to the size of the instances of the problem we are going to solve. Then, the systems themselves can be used to solve such instances. All the systems we will propose work in a *deterministic* way.

The biological motivation for the mechanism that we use to expand the synapse graph of SN P systems comes from the growth of dendritic trees in biological neurons [20]. It is known that the human brain is made up of about 100 billion cells. Almost all brain cells are formed before birth. Dendrites (from the

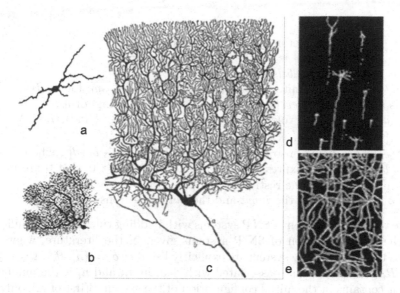

Fig. 1. A growing neuron: a. dendrites begin to emerge from a single neuron, b. dendrites developed into a cluster of touch points; c. Ramon y Cajal, Santiago. Classical drawing: Purkinje cell; d. newborn neuron dendrites, e. three months later. Photos from Tag Toys [20].

Greek, "tree") are the branched projections of a neuron. The point at which the dendrites of a cell come into contact with the dendrites of another cell is where information transfer (communication) occurs. Brain cells can grow as many as one billion of dendrite connections; the greater the number of dendrites, the more information can be processed. Dendrites grow as a result of stimulation from and interaction with their environment. With limited stimulation there is limited growth; with no stimulation, dendrites actually retract and disappear. The microscope photographs illustrated in Figure 1 show actual dendrite development. Dendrites begin to emerge from a single neuron (brain cell) and develop into a cluster of touch points seeking to connect with dendrites from other cells.

In the framework of SN P systems, the dendrite connection points are modelled as abstract neurons, while the branches of dendrite trees are modelled as abstract synapses. A new connection between dendrites coming from two different neuron cells is understood as a newly created synapse. In this way, new neurons and new synapses can be produced during the growth of a dendrite tree. The formal definition of neuron budding rule and its semantics will be given in Section 2.

2 SN P Systems with Budding Rules

A *spiking neural P system with budding rules*, of initial degree $m \geq 1$, is a construct of the form

$$\Pi = (O, \Sigma, H, syn, R, in, out),$$

where:

1. $O = \{a\}$ is the singleton alphabet (a is called *spike*);
2. $\Sigma = \{\sigma_1, \sigma_2, \ldots, \sigma_m\}$ is a finite set of initial neurons;
3. H is a finite set of *labels* for neurons;
4. $syn \subseteq H \times H$ is a finite set of *synapses*, with $(i, i) \notin syn$ for $i \in H$;
5. R is a finite set of *developmental rules*, of the following forms:
   (1) *neuron budding* rule $x[\ ]_i \rightarrow y[\ ]_j$, where $x \in \{(k, i), (i, k), \lambda\}$, $y \in \{(i, j), (j, i), \lambda\}$, $i, j, k \in H$, $i \neq k$, $i \neq j$;
   (2) *extended firing* (also called *spiking*) rule $[E/a^c \rightarrow a^p; d]_i$, where $i \in H$, E is a regular expression over a, and $c \geq 1$, $p \geq 0$, $d \geq 0$ are integer numbers, with the restriction $c \geq p$;
6. $in, out \in H$ indicate the *input* and the *output* neurons of Π.

Note that the definition of SN P systems with budding rules is slightly different from the usual definition of SN P systems given in the literature, where the neurons that occur in the system are explicitly listed as $\sigma_i = (n_i, R_i)$, $1 \leq i \leq m$, where R_i is the set of rules associated with neuron σ_i, and n_i is the number of spikes it contains in the initial configuration of the system. First of all, only the *structure* of the system is given in our definition; the presence of spikes (if any) in the initial configuration is specified at the beginning of each computation. Further, i is considered as the label of neuron σ_i. In SN P systems with budding rules it is possible to create new neurons in the course of a computation; hence the system may contain, in a given configuration, several neurons that are labelled with the same element of H. With a slight abuse of notation, in what follows we will refer to any neuron having the label $i \in H$ by calling it σ_i.

Considering the budding rule $x[\ ]_i \rightarrow y[\ ]_j$, its left hand side describes the neuron σ_i with a synapse x connected with one of its neighbouring neurons, to which the rule is supposed to be applied. The right hand side describes the result of the rule application, that is, the newly created neuron σ_j and synapse y. Note that for the sake of simplicity, in the rule notation we omit to repeat the contents of the left hand side of the rule in the right hand side. We say that the rule is *restricted* because only one neighbouring neuron is considered in each side of the rule.

A budding rule can be applied only if the neighbourhood of the associated neuron is exactly as described in the left hand side of the rule, in other words, $x = X$ where X is the current set of synapses of neuron σ_i. As a result of the rule application, a new neuron σ_j and a synapse y are established, provided that they do not already exist; if a neuron with label j already exists in the system but no synapse of type y exists, then only the synaptic connection y between the neurons σ_i and σ_j is established; no new neuron with label j is budded. We stress here that the application of budding rules does not depend on the spikes contained into the neuron. Budding rules are applied in a maximally parallel way: if the neighbourhood of neuron σ_i enables several budding rules, then all these rules are applied in parallel; as a result, several new neurons and synapses are produced (which corresponds to have several branches at a touch point in the dendrite tree). Note that the way of using neuron budding rules is different with respect to the usual way in which P systems with active membranes use cell division or cell creation rules, where at most one of these rules can be applied inside each membrane during a computation step.

Extended firing rules are defined as usually done in SN P systems. If an extended firing rule $[E/a^c \rightarrow a^p; d]_i$ has $E = a^c$, then we will write it in the simplified form $[a^c \rightarrow a^p; d]_i$; similarly, if a rule $[E/a^c \rightarrow a^p; d]_i$ has $d = 0$, then we can simply write it as $[E/a^c \rightarrow a^p]_i$; hence, if a rule $[E/a^c \rightarrow a^p; d]_i$ has $E = a^c$ and $d = 0$, then we can write $[a^c \rightarrow a^p]_i$. A rule $[E/a^c \rightarrow a^p]_i$ with $p = 0$ is written in the form $[E/a^c \rightarrow \lambda]_i$ and is called an *extended forgetting* rule. Rules of the types $[E/a^c \rightarrow a; d]_i$ and $[a^c \rightarrow \lambda]_i$ are said to be *standard*. However, even in this case we do not require that if a forgetting rule is enabled then no firing rules are also enabled at the same time in the same neuron, as it happens in standard SN P systems.

If a neuron σ_i contains k spikes and $a^k \in L(E), k \geq c$, then the rule $[E/a^c \rightarrow a^p; d]_i$ is enabled and can be applied. This means consuming (removing) c spikes (thus only $k - c$ spikes remain in neuron σ_i); the neuron is fired, and it produces p spikes after d time units. If $d = 0$, then the spikes are emitted immediately; if $d = 1$, then the spikes are emitted in the next step, etc. If the rule is used in step t and $d \geq 1$, then in steps $t, t + 1, t + 2, \ldots, t + d - 1$ the neuron is closed (this corresponds to the refractory period from neurobiology), so that it cannot receive new spikes (if a neuron has a synapse to a closed neuron and tries to send a spike along it, then that particular spike is lost). In the step $t + d$, the neuron spikes and becomes open again, so that it can receive spikes (which can be used starting with the step $t + d + 1$, when the neuron can again apply rules). Once emitted from neuron σ_i, the p spikes reach immediately all neurons σ_j such that there is a synapse going from σ_i to σ_j and which are open, that is, the p spikes are replicated and each target neuron receives p spikes; as stated above, spikes sent to a closed neuron are "lost", that is, they are removed from the system. In the case of the output neuron, p spikes are also sent to the environment. Of course, if neuron σ_i has no synapse leaving from it, then the produced spikes are lost. If the rule is a forgetting one of the form $[E/a^c \rightarrow \lambda]_i$, then, when it is applied, $c \geq 1$ spikes are removed. When a neuron is closed, none of its rules can be used until it becomes open again.

In each time unit, if a neuron σ_i can use one of its rules, then a rule from R *must* be used. If the neighbourhood of neuron σ_i enables several budding rules, then all these rules are applied in parallel. If several spiking rules are enabled in neuron σ_i, then only one of them is nondeterministically chosen. If both spiking rules and budding rules are enabled in the same computation step, then one type of rules is nondeterministically chosen. When a neuron budding rule is applied, at this step the associated neuron is closed, and thus it cannot receive spikes. In the next step, the neurons obtained by budding will be open.

The *configuration* of the system is described by its topology structure, the number of spikes associated with each neuron, and the *state* of each neuron (open or closed). We emphasize that the system introduced here contains no spikes in the initial configuration. Using the rules as described above, one can define *transitions* among configurations. Any sequence of transitions that starts in the initial configuration is called a *computation*. A computation *halts* if it reaches a configuration where all the neurons are open and no rule can be used.

In what follows, we give an example to make the application of budding rules transparent. Neither spiking nor forgetting rules are used.

An example. Let Π_1 be an SN P system with budding rules, whose initial topological structure (composed by a single neuron σ_1) is shown in the left hand side of Figure 2. Let Π_1 contain the following six budding rules:

$a.\ \lambda[\ ]_1 \to (1,2)[\ ]_2,$
$b.\ (1,2)[\ ]_2 \to (3,2)[\ ]_3,$
$c.\ (1,2)[\ ]_2 \to (2,4)[\ ]_4,$
$d.\ (2,3)[\ ]_3 \to (3,5)[\ ]_5,$
$e.\ (2,4)[\ ]_4 \to (4,6)[\ ]_6,$
$f.\ (4,6)[\ ]_6 \to (6,3)[\ ]_3.$

In the initial configuration, neuron σ_1 has no neighbourhood and only rule a. is enabled. The application of rule a. produces a new neuron σ_2 with a synapse (1,2) connecting it with σ_1. Now both neurons σ_1 and σ_2 have a neighbourhood (each one being the neighbourhood of the other), since a synaptic connection exists between them. In this circumstance, rule a. is disabled while rules b. and c. are enabled and may be applied in parallel to neuron σ_2. When these two rules are applied two new neurons σ_3 and σ_4 are created, with the associated synapses (3,2) and (2,4). In the resulting configuration, rules b. and c. are disabled since now neuron σ_2 has three neighbours. At this step only rule e. can be applied to neuron σ_4, producing a new neuron σ_6 with a synaptic connection (4,6). Note that at this step rule d. was not enabled as the synapse of neuron σ_3 is (3,2), instead of (2,3) as required by the rule. Now only rule f. is enabled, which creates only the synapse (6,3) because neuron σ_3 already exists. From now on no rule is enabled, and thus the computation halts.

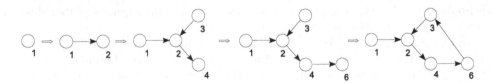

Fig. 2. Evolution of the structure of the SN P system Π_1, as the effect of the application of budding rules

3   SN P Systems Solving SAT

Let us now consider the **NP**-complete decision problem SAT [8, p. 39]. The instances of SAT depend upon two parameters: the number n of variables, and the number m of clauses. We recall that a *clause* is a disjunction of literals, occurrences of x_i or $\neg x_i$, built on a given set $X = \{x_1, x_2, \ldots, x_n\}$ of Boolean variables. Without loss of generality, we can avoid the clauses in which the same literal is repeated or both the literals x_i and $\neg x_i$, for any $1 \leq i \leq n$, occur. So doing, a clause can be seen as a *set* of at most n literals. An *assignment* to the variables $x_1, x_2, \ldots, x_n$ is a mapping $a : X \to \{0, 1\}$ that associates to each variable a truth value. The number of all possible assignments to the variables of X is 2^n. We say that an assignment *satisfies* the clause C if, assigned the truth values to all the variables which occur in C, the evaluation of C (considered as a Boolean formula) gives 1 (*true*) as a result.

We can now formally state the SAT problem as follows.

Problem 1. NAME: SAT.

- INSTANCE: a set $C = \{C_1, C_2, \ldots, C_m\}$ of clauses, built on a finite set $\{x_1, x_2, \ldots, x_n\}$ of Boolean variables.
- QUESTION: is there an assignment to the variables $x_1, x_2, \ldots, x_n$ that satisfies all the clauses in C?

Equivalently, we can say that an instance of SAT is a propositional formula $\gamma_{n,m} = C_1 \wedge C_2 \wedge \cdots \wedge C_m$, expressed in the conjunctive normal form as a conjunction of m clauses, where each clause is a disjunction of literals built using the Boolean variables $x_1, x_2, \ldots, x_n$. With a little abuse of notation, from now on we will denote by $\text{SAT}(n, m)$ the set of instances of SAT which have n variables and m clauses.

In [6], a uniform family $\{\Pi_{SAT}(\langle n, m \rangle)\}_{n,m \in \mathbb{N}}$ of SN P systems was built such that for all $n, m \in \mathbb{N}$ the system $\Pi_{SAT}(\langle n, m \rangle)$ solves all the instances of $\text{SAT}(n, m)$ in a number of steps which is quadratic in n and linear in m. Here $\langle n, m \rangle$ denotes the natural number obtained by applying the Cantor bijection to the pair (n, m) of natural numbers; so doing, the family of P systems depends upon one parameter instead of two. We assume that the reader is familiar with the construction given in [6]; for his convenience, here we summarize the structure and functioning of the system $\Pi_{SAT}(\langle n, m \rangle)$. In the next section, we are going to build such a system by means of budding rules.

Because the construction is uniform, we need a way to encode any given instance $\gamma_{n,m}$ of $\text{SAT}(n, m)$. As stated above, each clause C_i of $\gamma_{n,m}$ can be seen as a disjunction of at most n literals, and thus for each $j \in \{1, 2, \ldots, m\}$ either x_j occurs in C_i, or $\neg x_j$ occurs, or none of them occurs. In order to distinguish these three situations we define the *spike variables* α_{ij}, for $1 \leq i \leq m$ and $1 \leq j \leq n$, as variables whose values are amounts of spikes, and we assign to them the following values:

$$\alpha_{ij} = \begin{cases} a & \text{if } x_j \text{ occurs in } C_i \\ a^2 & \text{if } \neg x_j \text{ occurs in } C_i \\ \lambda & \text{otherwise} \end{cases} \tag{1}$$

So doing, clause C_i will be represented by the sequence $\alpha_{i1}\alpha_{i2}\cdots\alpha_{in}$ of spike variables; in order to represent the entire formula $\gamma_{n,m}$ we just concatenate the

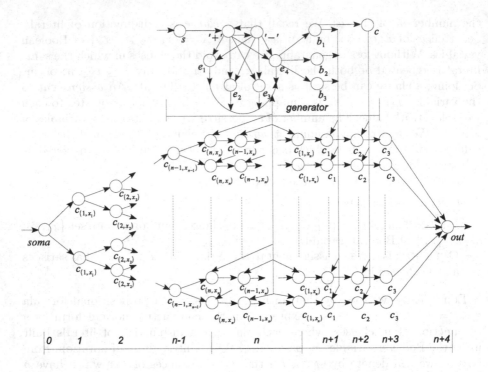

Fig. 3. A SN P system structure devoted to solve all the instances of SAT(n, m)

representations of the single clauses, thus obtaining the sequence $\alpha_{11}\alpha_{12}\cdots\alpha_{1n}$ $\alpha_{21}\alpha_{22}\cdots\alpha_{2n}\cdots\alpha_{m1}\alpha_{m2}\cdots\alpha_{mn}$. As an example, the representation of $\gamma_{3,2} = (x_1 \vee \neg x_2) \wedge (x_1 \vee x_3)$ is the sequence $aa^2\lambda a\lambda a$.

The system structure is composed of $n+5$ layers, as illustrated in Figure 3. The first layer (numbered by 0) is used to insert into the system the representation of the instance of SAT(n, m) to be solved, encoded as stated above. Note that each layer from 1 to n contains two times the neurons contained in the previous layer. In this way we obtain in the n-th layer 2^n copies of a subsystem which is a sequence of n neurons; each subsystem is bijectively associated to one of the possible assignments to the variables $x_1, x_2, \ldots, x_n$. The neurons that occur in each subsystem are of two types: f and t. The type of a neuron indicates that the corresponding Boolean variable is assigned with the Boolean value $t(rue)$ or $f(alse)$, respectively. These subsystems, together with the so called *generator*, have a very specific function in the overall SN P system: to test (in parallel) all possible assignments against a given clause.

The assignment is performed by sending 3 spikes to all the neurons labelled with t, and 4 spikes to all the neurons labelled with f. This means that neurons e in the generator will have three synapses going to neurons t and four synapses towards neurons f. All these spikes arrive every n computation steps, when the spikes indicated by the spike variables α_{ij} that correspond to a clause of $\gamma_{n,m}$

Table 1. Number of spikes resulting from the assignment in the neurons of layer n, and its effect on the truth value of the clause

	Assign. to x_j	Literal	N. of spikes	Truth value of C_i
Neuron t	true	$x_j \notin C_i$	$3+0=3$	?
	true	$x_j \in C_i$	$3+1=4$	true
	true	$\neg x_j \in C_i$	$3+2=5$	?
Neuron f	false	$x_j \notin C_i$	$4+0=4$	?
	false	$x_j \in C_i$	$4+1=5$	?
	false	$\neg x_j \in C_i$	$4+2=6$	true

are contained into the subsystems of layer n. This process is started by putting one spike in neuron s at the beginning of the computation. The delay associated with the rule contained in neuron s allows to send the first spikes from neurons e to neurons t and f exactly when the first clause is contained in layer n.

Recall our encoding of literals in the clauses (1): we have 0 spikes if the variable does not occur in the clause, 1 spike if it occurs non negated, and 2 spikes if it occurs negated. These spikes are added with those representing the assignments, and the possible results are illustrated in Table 1. From this table we can see that if a neuron labelled with t receives a total number of 4 spikes then the corresponding variable occurs non negated in the clause and is assigned the truth value *true*; we can immediately conclude that the clause is satisfied, and thus the neuron sends one spike towards the next layer. Similarly, if a neuron labelled with f receives 6 spikes then the corresponding variable occurs negated in the clause and is assigned the truth value *false*; also in this case we can immediately conclude that the clause is satisfied, and the neuron signals this event by sending one spike towards the next layer. In all the other cases we cannot conclude anything on the truth value of the clause, and thus no spike is emitted.

All the spikes which are emitted by neurons t and f are propagated through the neurons that compose layer n, until they reach the corresponding neuron σ_1 in layer $n+1$. Such a neuron is designed to make neuron σ_2 (in layer $n+2$) retain only one spike from those received by layer n. Hence, those assignments that satisfy the clause produce a single spike in the corresponding neuron σ_2; such a spike is accumulated in the associated neuron σ_3 (in layer $n+3$), that operates like a counter. When the first clause of $\gamma_{n,m}$ has been processed, the second enters into the system (in n steps) and takes place in the subsystems; then all possible assignments are tested against this clause, and so on for all the clauses. When all the m clauses of $\gamma_{n,m}$ have been processed, neurons σ_3 in layer $n+3$ contain each the number of clauses which are satisfied by the corresponding assignment. The neurons that contain m spikes fire, sending one spike to neuron σ_{out}, thus signalling that their corresponding assignment satisfies all the clauses of the instance. Neuron σ_{out} operates like an OR gate: it fires if and only if it contains at least one spike, that is, if and only if at least one of the assignments satisfies all the clauses of $\gamma_{n,m}$. Further technical details will be presented in the last part of the next section.

4   A Uniform Solution to SAT by SN P Systems with Budding Rules

In this section we show that the pre-computed structures which are used in [6] to solve the instances of SAT(n, m) can be built in a polynomial time (with respect to n, independent of m) by SN P systems with budding rules. The SN P system with budding rules that we are going to define is composed of two subsystems: a first subsystem builds the structure of a second subsystem, that solves the instances of SAT(n, m) as described in the previous section. For the sake of simplicity, we avoid to use the neuron budding and the spiking rules at the same time in each subsystem.

Formally, the SN P system with budding rules is defined as

$$\Pi = (O, \Sigma, H, syn, R, soma, out)$$

where:

1. $O = \{a\}$ is the singleton alphabet;
2. $\Sigma = \{\sigma_i \mid i \in H_0\}$ is the set of initial neurons;
3. H is a finite set of labels for neurons, and $H \supseteq H_0 = \{soma, out, e_0, e_1, e_2, e_3, b_1, b_2, b_3, c, s, +, -\}$ is the set of labels for the neurons initially given;
4. $syn \subseteq H \times H$ is a finite set of synapses, with $(i, i) \notin syn$ for $i \in H$), and $syn \supseteq syn_0 = \{(e, e_i) \mid 0 \leq i \leq 3, e \in \{+, -\}\} \cup \{(e_0, b_i) \mid 1 \leq i \leq 3\} \cup \{(b_3, c), (s, +), (+, -), (-, +), \lambda\}$ is the set of synapses initially in use;
5. $soma$ and out are the labels for the *input* and *output* neuron, respectively;
6. R is a set of *neuron budding* and *extended spiking* rules defined as follows.

Building the system structure. The system initially contains an input neuron σ_{soma}, an output neuron σ_{out}, and a sub-structure G (named the *generator*) which is composed of the set of neurons specified in Σ and the set of synapses from syn_0, arranged as illustrated in Figure 4.

The generator is governed only by neuron budding rules, and is controlled by the labels of budding neurons and by the synapses created during the computation. The system construction algorithm consists of two phases:

A. Generation of the *dendritic-tree* sub-structure (the layers from 0 to n in Figure 3) and assignment of the truth values to the n Boolean variables. The process starts from the initial neuron σ_{soma} (the root node) and produces 2^n neurons in n steps. The label of each neuron in layer n encodes an associated truth assignment.

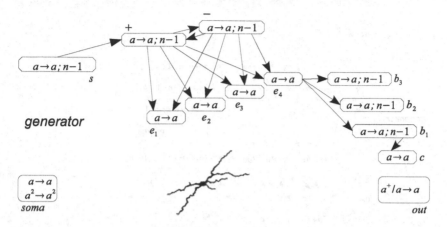

Fig. 4. The initial topological structure (newly born dendrite) of the SN P system Π: the input (*soma*) and the output (*out*) neurons, and the *generator*

B. Completion of the network structure. The neurons in the n-th layer of the system establish connections with the *generator*, according to the truth assignments represented in those neurons. The structure is then further expanded by three layers, and finally all the neurons in the last layer are connected with the output neuron σ_{out}.

Let us now describe in depth each of these phases.

Phase A. In this phase of computation, the dendritic tree (which is a complete binary tree) is generated in n steps by applying budding rules of type $\mathbf{a_1}$), described below, starting from an initial neuron σ_{soma}. The dendritic tree generation process is controlled by the labels of the neurons as well as by the synapses generated so far. It is worth to note that, since the truth assignments associated with the neurons in n-th layer are encoded in the labels of those neurons, also the truth assignments to the variables $x_1, x_2, \ldots, x_n$ are generated during the construction of the dendritic tree.

The label of a neuron σ_c in layer i is a sequence of the form

$$c = (i, x_i^{(p)}) = (i, x_i(1) = p) = (i, p, x_{k2}, \ldots, x_{ii}),$$

with $p \in \{t, f\}$, where the first entry (i) indicates the number of layer, while $x_i^{(p)}$ is a subsequence of length i formed by the Boolean values t and f that have been generated up to now, that represents a truth assignment to the variables $x_1, x_2, \ldots, x_i$. The component p in $x_i^{(p)}$ indicates that the first entry of the subsequence is exactly p.

An almost complete structure of the SN P system that solves the instances of $\text{SAT}(2, m)$ is illustrated in Figure 5. It is worth to follow its construction.

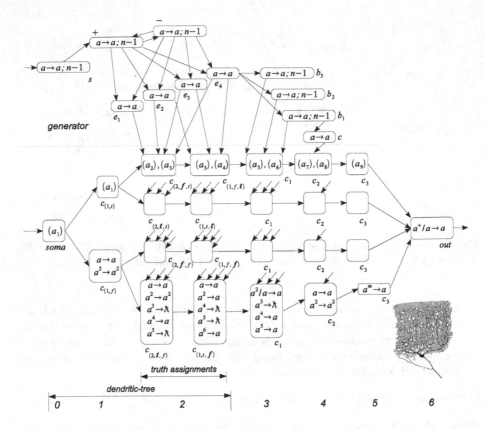

Fig. 5. An almost complete structure (maturated dendrite tree) of the P system for solving the instances of SAT$(2, m)$. The neuron budding rules used in each computation step are indicated by their labels in the corresponding neurons. Some of the spiking rules are also indicated.

a₁) $(c_{(i,x_{i-1})}, c_{(i,x_i)})[\ ]_{c_{(i,x_i)}} \to (c_{(i,x_i)}, c_{(i+1,p,x_i)})[\ ]_{c_{(i+1,p,x_i)}},$

$0 \le i \le n-1$, $p \in \{t, f\}$, $x_i \in \{t, f\}^i$, $(c_{(-1,x_{-1})}, c_{(0,x_0)}) = \lambda$, $c_{(0,\lambda)} = soma$. The computation starts by applying two rules of type **a₁)**, for $i = 0$, to the input neuron $\sigma_{c_{soma}}$. These two rules are:

$$[\ ]_{c_{soma}} \to (c_{soma}, c_{(1,t)})[\ ]_{c_{(1,t)}}, \quad \text{and} \quad [\ ]_{c_{soma}} \to (c_{soma}, c_{(1,f)})[\ ]_{c_{(1,f)}},$$

where $(c_{soma}, c_{(1,t)}), (c_{soma}, c_{(1,f)}) \in syn$.

The left hand side of each rule (where $\lambda \in syn_0$ is omitted) requires that its interaction environment be empty, i.e., no synapse exists connected to neuron $\sigma_{c_{soma}}$. As the left hand sides of both these rules are the same, and satisfy the constraints posed on the interaction environment of neuron $\sigma_{c_{soma}}$, they are applied simultaneously. As a result, two new neurons are budded: $\sigma_{c_{(1,t)}}$, with a synapse $(c_{soma}, c_{(1,t)})$ coming from the father neuron, and $\sigma_{c_{(1,f)}}$, connected with the father neuron by a synapse $(c_{soma}, c_{(1,f)})$. The symbols t and f in the neuron labels indicate the truth values *true* and *false*, respectively, and can be regarded as the two truth assignments (t) and (f) of length 1 for a single Boolean variable x_1. The first layer of the dendritic tree is thus established, and rules of type $\mathbf{a_1})$ cannot be applied anymore, since the interaction environment of neuron $\sigma_{c_{soma}}$ has changed. At the second computation step $(i = 1)$, the following two rules are enabled and can be applied to each of the newly created neurons:

$$(c_{soma}, c_{(1,t)})[\ ]_{c_{(1,t)}} \rightarrow (c_{(1,t)}, c_{(2,f,t)})[\ ]_{c_{(2,f,t)}},$$
$$(c_{soma}, c_{(1,t)})[\ ]_{c_{(1,t)}} \rightarrow (c_{(1,t)}, c_{(2,t,t)})[\ ]_{c_{(2,t,t)}}$$

for $\sigma_{c_{(1,t)}}$, and

$$(c_{soma}, c_{(1,f)})[\ ]_{c_{(1,f)}} \rightarrow (c_{(1,f)}, c_{(2,t,f)})[\ ]_{c_{(2,t,f)}},$$
$$(c_{soma}, c_{(1,f)})[\ ]_{c_{(1,f)}} \rightarrow (c_{(1,f)}, c_{(2,f,f)})[\ ]_{c_{(2,f,f)}}$$

for $\sigma_{c_{(1,f)}}$. The former pair of rules yields to two new neurons having label $c_{(2,f,t)}$ and $c_{(2,t,t)}$, respectively; the synapses specified in these rules are budded from the neuron labelled with $c_{(1,t)}$. The latter pair of rules generates two neurons with labels $c_{(2,f,f)}$ and $c_{(2,t,f)}$, respectively; the synapses mentioned in these rules go from the neuron labelled with $c_{(1,f)}$ to the newly created neurons. In the meanwhile the truth assignments (f, t), (t, t), (f, f), (t, f), for the Boolean variables x_1 and x_2, are generated at each leaf node, as illustrated in Figure 5. Since the interaction environment of neurons $\sigma_{c_{(1,t)}}$ and $\sigma_{c_{(1,f)}}$ has changed, the rules applied in this step cannot be applied anymore to these neurons.

By continuing in this way, by applying the budding rules of type $\mathbf{a_1})$ in the maximally parallel way for n computation steps, a complete binary tree of depth n having 2^n leaves (hence an exponentially large workspace) is built. The label of each leaf node encodes a truth assignment of length n, hence all possible truth assignments for the Boolean variables $x_1, x_2, \ldots, x_n$ are generated.

Phase B. The pre-computation to construct the SN P system structure continues until it converges to the output neuron in a further few steps. The main goal of this part of the construction algorithm is to design the substructure which is devoted to test the satisfiability of the clauses of the instance $\gamma_{n,m}$ of SAT(n, m) given as input against all possible truth assignments, and to determine whether there exist some assignments that satisfy all the clauses of $\gamma_{n,m}$.

The substructure is composed of 2^n subsystems, each being a sequence of n neurons $\sigma_{c_{(j,x_n)}}, 1 \leq j \leq n$, including the leaf nodes of the dendritic tree. A subsequence $x_n = (x_{n1}, x_{n2}, \ldots, x_{nn}) \in \{t, f\}^n$ in a neuron label $c_{(j,x_n)}$ represents a truth assignment, and we can abstractly assign a pair $(j, x_n(j))$ to a neuron $\sigma_{c_{(j,x_n)}}$ as its identity. Thus each subsystem represents a truth assignment formed by its neurons' identities. As stated above, a neuron with identification $(j_1, x(j_1) = t)$ has 3 synapses coming from the generator module, whereas a neuron with identity $(j_2, x(j_2) = f)$ is connected with the generator by means of 4 synapses. As we will see, these connections are used to perform assignments to the Boolean variables $x_1, x_2, \ldots, x_n$ that compose $\gamma_{n,m}$, and to check which assignments satisfy the clause of $\gamma_{n,m}$ currently under consideration.

For instance, the case in which $n = 2$ is described in Figure 5, where $2^2 = 4$ different truth assignments of length 2 have been generated for the two Boolean variables x_1, x_2. The first subsystem is composed of two neurons having labels $c_{(2,f,t)}$ and $c_{(1,f,t)}$, respectively. The former is associated with the Boolean value *false*, as $x_2 = (f, t)$ and $x_2(2) = f$, while the latter is associated with *true*, as $x_2(1) = t$; altogether they form the truth assignment (f, t). The other subsystems are similar, and are associated with the truth assignments (t, t), (f, f) and (t, f). One can see that the four truth assignments are well distinguished from each other by the layer structure of the four subsystems.

To build the substructure of n layers mentioned above, from now on two rules of types $\mathbf{a_2}$) and $\mathbf{a_3}$) are applied simultaneously to a same neuron for $n-1$ steps. The first rule creates a new neuron with an associated synapse, while the second rule creates 3 or 4 synapses to the generator block. The same process occurs during the n-th step, by means of the rules of types $\mathbf{a_3}$) and $\mathbf{a_4}$); note that in this step the rules of type $\mathbf{a_2}$) cannot be applied anymore.

$\mathbf{a_2}$) $(c_{(n+1-j,x_n)}, c_{(n-j,x_n)})[\ ]_{c_{(n-j,x_n)}} \to (c_{(n-j,x_n)}, c_{(n-1-j,x_n)})[\ ]_{c_{(n-1-j,x_n)}}$,
$p \in \{t, f\}$, $0 \le j \le n - 1$, $1 \le k \le n$, $c_{(k,0,x_k^{(p)})} = c_{(k-1,n,x_{k-1})}$,
$x_k^{(p)} = (p, x_{k-1}) \in \{t, f\}^k$.

$\mathbf{a_3}$) $(c_{(n+1-j,x_n)}, c_{(n-j,x_n)})[\ ]_{c_{(n-j,x_n(j+1)=p)}} \to (c_{(n-j,x_n(j+1)=p)}, e_i)[\ ]_{e_i}$,
$0 \le j \le n$, $p \in \{t, f\}$ and $s \le i \le 3$, where $s = 1$ if $p = t$, and $s = 0$ if $p = f$, $c_{(n,0,x_n)} = c_{(n-1,n,x_n)}$.

We are now in the $(n + 1)$-th step of the computation. When $j = 0$, both rules of types $\mathbf{a_2}$) and $\mathbf{a_3}$) are applicable to each neuron $\sigma_{c_{(n,x_n)}}$ of layer n. The former rules generate neurons $\sigma_{c_{(n-1,x_n)}}$ with a synapse $(c_{(n,x_n)}, c_{(n-1,x_n)})$. The latter type of rules creates three synapses to all neurons of type $\sigma_{c_{(n,x_n(1)=t)}}$ coming from the neurons $\sigma_{c_{e_i}}$, $1 \le i \le 3$, and four synapses to the neurons $\sigma_{c_{(n,x_n(1)=f)}}$ coming from the four neurons $\sigma_{c_{e_i}}$, $0 \le i \le 3$, of the generator block. The neuron budding rules of type $\mathbf{a_2}$) and the synapse creation rules of type $\mathbf{a_3}$) are applied simultaneously to the same neurons (leaf nodes) in layer n in the following $n - 1$ steps, since their interaction environments coincide. The effect of the application of these rules is the production of neurons having connections with the generator block.

So doing, 2^n subsystems, each one being a sequence of n neurons, are generated starting from layer n. In each subsystem, every neuron corresponding to the Boolean value *true* $(x_n(j) = t)$ is connected with the generator block by means of three synapses, while the neurons that correspond to the Boolean value *false* $(x_n(j) = f)$ are connected with the generator block by four synapses.

From the $(2n+1)$-th step of the computation on, no interaction environment of any neuron in the system allows to activate the rules of type $\mathbf{a_2}$). Hence these rules cannot be applied, but the computation continues with the next types of rules.

$\mathbf{a_4}$) $(c_{(2,x_n)}, c_{(1,x_n)})[\ ]_{c_{(1,x_n)}} \to (c_{(1,x_n)}, c_1)[\ ]_{c_1}$.

The rules of type $\mathbf{a_4}$) can be applied in parallel to the leaf nodes (neurons) of layer n; they produce the neurons σ_{c_1} forming the $(n+1)$-th layer and, meanwhile, the rules of type $\mathbf{a_3}$) create synapses from these neurons to the generator block.

$\mathbf{a_5}$) $(c_{(1,x_n)}, c_1)[\ ]_{c_1} \to (c_1, c_2)[\ ]_{c_2}$,

$\mathbf{a_6}$) $(c_{(1,x_n)}, c_1)[\ ]_{c_1} \to (b_i, c_1)[\ ]_{b_i}$, $1 \le i \le 3$.

While the rules of type $\mathbf{a_5}$) are applied to the neurons σ_{c_1} and bud neurons σ_{c_2}, the rules of type $\mathbf{a_6}$) are also applied and create three synapses coming from the neurons σ_{b_i}, $1 \le i \le 3$, to each neuron σ_{c_1}. In this way, layer $n+2$ is formed.

$\mathbf{a_7}$) $(c_1, c_2)[\ ]_{c_2} \to (c_2, c_3)[\ ]_{c_3}$,

$\mathbf{a_8}$) $(c_1, c_2)[\ ]_{c_2} \to (c, c_2)[\ ]_{c}$.

The rules of types $\mathbf{a_7}$) and $\mathbf{a_8}$) apply simultaneously to every neuron σ_2 having a synapse (c_1, c_2). As a result, a new neuron σ_{c_3} is budded with a connection (c, c_2) coming from neuron σ_c. All the neurons σ_{c_2} in the same layer are subject to the same effect, since the rules are applied in the maximally parallel way.

$\mathbf{a_9}$) $(c_2, c_3)[\ ]_{c_3} \to (c_3, out)[\ ]_{out}$.

The pre-computation of the SN P system structure is completed by forming the connections from the neurons σ_{c_3} to the output neuron σ_{out}, by means of the rules of type $\mathbf{a_9}$). These rule are applied in the maximally parallel way to all the neurons in layer $n+3$.

Summarizing, phases A and B build an empty (that is, containing no spikes) structure of an SN P system, that can be used to solve all the instances of SAT(n, m) in a linear (with respect to n) number of computation steps. The size of the structure is exponential with respect to n.

Solving SAT (Phase C). Given an instance $\gamma_{n,m}$ of SAT(n, m), we first encode it as a sequence of spike variables, as explained in Section 3, equation (1). Then, the computation of the system may start. The sequence of spikes encoding $\gamma_{n,m}$ is introduced in the system, using neuron σ_{soma}. Let us see how spiking rules are used to compute the solution, with a brief description for each.

$\mathbf{c_1}$) $[a \to a]_{c_{(i,x_i)}}$; $[a^2 \to a^2]_{c_{(i,x_i)}}$; $0 \le i \le n$, $x_i \in \{t, f\}^i$, $c_{(0,x_0)} = soma$.

$\mathbf{c_2}$) $[a \to a; n-1]_s$.

We insert 0, 1 or 2 spikes into the system by rule $\mathbf{c_1}$) using the input neuron σ_{soma}, according to the value of the spike variable α_{ij} we are considering in the representation of $\gamma_{n,m}$. In the meanwhile we insert a single spike a into neuron σ_s, to fire once the rule $\mathbf{c_2}$), thus activating the generator block.

Each spike, encoding a spike variable inserted into the input neuron, is duplicated and transmitted to the next layer of neurons. This duplication is performed n times, until 2^n replicated copies of the spike are placed in the leaf nodes (in layer n) of the dendritic tree.

c_3) $[a \rightarrow a]_{e_i}$; $0 \leq i \leq 3$,
$[a \rightarrow a; n-1]_+$; $[a \rightarrow a; n-1]_-$.

These are the spiking rules of the generator block. Each n steps, the generator provides 3 and 4 spikes, respectively, to the neurons of layer n associated with the truth values t and f. This is made in order to test the satisfiability of a clause which has propagated through the layers of the dendritic tree, by checking it against all possible truth assignments to the variables $x_1, x_2, \ldots, x_n$.

In another n steps, the 2^n copies of the clause of $\gamma_{n,m}$ take place in the corresponding subsystems located in layers from $n+1$ to $2n$, where the satisfiability of the clause against all possible truth assignments is tested. For this purpose, the spike-truth values a^4 and a^3 are assigned from the generator to the spike-variables of the clause, according to the truth assignments represented by the neurons that compose the subsystems. In fact, recall that in each subsystem every neuron corresponding to the Boolean value *true* ($x_n(j) = t$) is connected with the generator block by means of three synapses, while the neurons that correspond to the Boolean value *false* ($x_n(j) = f$) are connected with the generator by means of four synapses. The satisfiability is then checked by means of the rules of types c_4) and c_5) residing in the neurons.

c_4) $[a \rightarrow a]_{t_t}$; $[a^3 \rightarrow \lambda]_{t_t}$; $[a^2 \rightarrow a^2]_{t_1}$;
$[a^4 \rightarrow a]_{t_t}$; $[a^5 \rightarrow \lambda]_{t_t}$; $[a^2 \rightarrow a]_{t_0}$;
$t_t = c_{(j, x_n(j) = t)}, 1 \leq j \leq n$,
$t_1 = c_{(j, x_n(j) = t)}, 2 \leq j \leq n$,
$t_0 = c_{(1, x_n(n) = t)}, x_n \in \{t, f\}^n$.

These are the spiking rules that reside in the neurons of layer n, associated with the Boolean value *true* (in Figure 5 $n = 2$, $\sigma_{c_{(2,t,f)}}$ stands for *false* while $\sigma_{c_{(1,t,f)}}$ stands for *true*). The rules $a^2 \rightarrow a^2$, $a^2 \rightarrow a$, and $a \rightarrow a$ are used to transmit the spike variables a, a^2 along the subsystems. Once a clause C_i is ready to be tested for satisfiability, each neuron associated with *true* contains either one spike (a), two spikes (a^2) or is empty (λ). As a spike variable a represents the occurrence of a Boolean variable x_j in C_i, to which a *true* value (a^3) sent by the generator is assigned, resulting in a *yes* answer (a^4), then it passes to the neuron σ_{c_1} along the subsystem as an indication that C_i is satisfied by a truth assignment in which the Boolean variable x_j is true. On the other hand, if the Boolean value *true* (a^3) is assigned to a spike variable that represents the occurrence of $\neg x_j$ in C_i (a^2) or the fact that x_j does not occur in C_i (λ), then in these cases the answer is *no*, which is computed by the rules $a^3 \rightarrow \lambda$ and $a^5 \rightarrow \lambda$.

c_5) $[a \rightarrow a]_{f_f}$; $[a^4 \rightarrow \lambda]_{f_f}$; $[a^2 \rightarrow a^2]_{f_1}$;
$[a^5 \rightarrow \lambda]_{f_f}$; $[a^6 \rightarrow a]_{f_f}$; $[a^2 \rightarrow a]_{f_0}$;

$f_f = c_{(j,x_n(j)=f)}, 1 \leq j \leq n,$

$f_1 = c_{(j,x_n(j)=f)}, 2 \leq j \leq n,$

$f_0 = c_{(1,x_n(n)=f)}, x_n \in \{t,f\}^n.$

These are the spiking rules that reside in the neurons of layer n, associated with the Boolean value *false*. The functioning of these rules is similar to that of rules c_4).

c_6) $[a \rightarrow a; n-1]_{b_i}; \ 1 \leq i \leq 3,$

$[a^2/a \rightarrow a]_{c_1}; \ [a^3 \rightarrow \lambda]_{c_1}; \ [a^4 \rightarrow a]_{c_1}; \ [a^5 \rightarrow a]_{c_1}.$

Whether an assignment satisfies or not the clause under consideration, is checked by a combined functioning of the neurons with label 1 in layer $n+1$ and the neurons with label $b_i, 1 \leq i \leq 3$, in the generator.

c_7) $[a \rightarrow \lambda]_{c_2}; \ [a^2 \rightarrow a]_{c_2}; \ [a \rightarrow a]_c.$

With a combined action of neuron σ_c, neuron σ_{c_2} sends a spike to neuron σ_{c_3} if and only if the corresponding assignment satisfies the clause under consideration.

c_8) $[a^m \rightarrow a]_{c_3}; \ [a^+/a \rightarrow a]_{out}.$

Neurons with label c_3 count how many clauses of the instance $\gamma_{n,m}$ are satisfied by the corresponding truth assignments. If one of these neurons get m spikes, then it fires. Hence the number of spikes that reach neuron *out* is the number of assignments that satisfy all the clauses of $\gamma_{n,m}$. The output neuron fires if it contains at least one spike, thus signalling that the problem has a positive solution; otherwise, there is no assignment that satisfies the instance $\gamma_{n,m}$.

This stage of computation ends at the $(nm + n + 4)$-th step. The entire computation of the system thus halts in at most $nm + n + 5$ computation steps, hence in a polynomial time with respect to n and m.

In conclusion, we obtained a deterministic, polynomial time and uniform solution to SAT(n, m) in the framework of SN P systems.

5   Conclusions and Directions for Future Research

In the present paper we proposed a way to solve the **NP**-complete decision problem SAT in a polynomial time with respect to the number n of Boolean variables and the number m of clauses that compose the instances of SAT being solved. Specifically, we introduced SN P systems with *neuron budding rules*, a new feature that enhances the efficiency of SN P systems by allowing them to generate an exponential size synapse graph (regarded as the workspace of the system) in a polynomial time with respect to n.

Neuron budding rules drive the mechanism of neuron production and synapse creation, according to the interaction of neurons with their neighbourhoods (described by the synapses that connect them to other neurons). We have shown that a very restricted type of neuron budding rules, involving one or two synapses (actually, when two synapses are involved, they appear one in each side of the rule) is sufficient to solve the SAT problem. The solution is computed in two stages: the first phase builds an exponential size SN P system that contains no spikes; then, this SN P system is fed with the instance of SAT to be solved (encoded in an appropriate way) and the answer is computed. The system is deterministic, and the rules are applied in the maximally parallel way.

The idea of producing new neurons in SN P systems is not new: already in [15] neurons are generated by *division*. However, both biological motivation and mathematical formal definition are different: neuron budding in this paper depends on the connections (structure) with other neurons, while neuron division depends on the number of spikes occurring inside the neurons (that is, the contents); hence they are two different ways to increase the workspace of SN P systems.

An open question is whether SN P systems with budding rules can be used to efficiently solve other computationally difficult problems, such as *numerical* **NP**-complete problems and **PSPACE**-complete problems.

SN P systems with neuron budding rules can be extended by introducing more general rules, which in some sense capture the dynamic interaction of neurons with their neighbourhood. One possible form of such general rules is as follows: $A_i[\ ]_i B_i \to C_j[\ ]_j D_j$, where A_i, B_i and C_j, D_j are the sets of synapses coming to and going out from, respectively, the specified neurons σ_i and σ_j. Clearly, in such general rules, more than one synapse can be involved in the neighbourhood of the considered neuron.

Acknowledgments

The work of Tseren-Onolt Ishdorj was supported by BIOTARGET, a joint project between the University of Turku and Åbo Akademi University, funded by the Academy of Finland. The work of L. Pan was supported by the National Natural Science Foundation of China (Grant Nos. 60674106, 30870826, 60703047, and 60533010), Program for New Century Excellent Talents in University (NCET-05-0612), Ph.D. Programs Foundation of Ministry of Education of China (20060487014), Chenguang Program of Wuhan (200750731262), HUST-SRF (2007Z015A), and by the Natural Science Foundation of Hubei Province (2008CDB113 and 2008CDB180). The work of Alberto Leporati was partially supported by MIUR project "Mathematical aspects and emerging applications of automata and formal languages" (2007).

References

1. Chen, H., Freund, R., Ionescu, M., Păun, G., Pérez-Jiménez, M.J.: On string languages generated by spiking neural P systems. Fundamenta Informaticae 75, 141–162 (2007)
2. Chen, H., Ionescu, M., Ishdorj, T.-O.: On the efficiency of spiking neural P systems. In: Proceedings of the 8th International Conference on Electronics, Information, and Communication, Ulanbator, Mongolia, June 2006, pp. 49–52 (2006)

3. Chen, H., Ionescu, M., Ishdorj, T.-O., Păun, A., Păun, G., Pérez-Jiménez, M.J.: Spiking neural P systems with extended rules. In: Gutiérrez-Naranjo, M.A., et al. (eds.) Fourth Brainstorming Week on Membrane Computing, RGNC Report 02/2006, Research Group on Natural Computing, Sevilla University, Fénix Editora, vol. I, pp. 241–266 (2006)
4. Gerstner, W., Kistler, W.: Spiking Neuron Models. Single Neurons, Populations, Plasticity. Cambridge University Press, Cambridge (2002)
5. Ionescu, M., Păun, G., Yokomori, T.: Spiking neural P systems. Fundamenta Informaticae 71(2-3), 279–308 (2006)
6. Ishdorj, T.-O., Leporati, A.: Uniform solutions to SAT and 3-SAT by spiking neural P systems with pre-computed resources. Natural Computing 7(4), 519–534 (2008)
7. Ishdorj, T.-O., Leporati, A., Pan, L., Zeng, X., Zhang, X.: Deterministic solutions to QSAT and Q3SAT by spiking neural P systems with pre-computed resources (submitted for publication)
8. Garey, M.R., Johnson, D.S.: Computers and Intractability. A Guide to the Theory on NP–Completeness. W.H. Freeman and Company, New York (1979)
9. Leporati, A., Gutiérrez-Naranjo, M.A.: Solving SUBSET SUM by spiking neural P systems with pre-computed resources. Fundamenta Informaticae 87(1), 61–77 (2008)
10. Leporati, A., Mauri, G., Zandron, C., Păun, G., Pérez-Jiménez, M.J.: Uniform solutions to SAT and SUBSET SUM by spiking neural P systems. Natural Computing (2008), doi:10.1007/s11047-008-9091-y
11. Leporati, A., Zandron, C., Ferretti, C., Mauri, G.: Solving numerical NP-complete problems with spiking neural P systems. In: Eleftherakis, G., Kefalas, P., Păun, G., Rozenberg, G., Salomaa, A. (eds.) WMC 2007. LNCS, vol. 4860, pp. 336–352. Springer, Heidelberg (2007)
12. Leporati, A., Zandron, C., Ferretti, C., Mauri, G.: On the computational power of spiking neural P systems. International Journal of Unconventional Computing 5(5), 459–473 (2009)
13. Maass, W.: Computing with spikes. Special Issue on Foundations of Information Processing of TELEMATIK 8(1), 32–36 (2002)
14. Maass, W., Bishop, C. (eds.): Pulsed Neural Networks. MIT Press, Cambridge (1999)
15. Pan, L., Păun, G., Pérez-Jiménez, M.J.: Spiking neural P systems with neuron division and budding. In: Gutiérrez-Escudero, R., et al. (eds.) Seventh Brainstorming Week on Membrane Computing, RGNC Report 01/2009, Research Group on Natural Computing, Sevilla University, Fénix Editora, vol. II, pp. 151–167 (2009)
16. Păun, G.: Membrane Computing – An Introduction. Springer, Berlin (2002)
17. Păun, G.: Twenty six research topics about spiking neural P systems. In: Gutiérrez-Naranjo, M.A., et al. (eds.) Fifth Brainstorming Week on Membrane Computing, RGNC Report 01/2007, Research Group on Natural Computing, Sevilla University, Fénix Editora, pp. 263–280 (2007)
18. Sipser, M.: Introduction to the Theory of Computation. PWS Publishing Company, Boston (1997)
19. The P systems Web page, http://ppage.psystems.eu/
20. Think and Grow Toys, http://www.tagtoys.com/dendrites.php

Tuning P Systems for Solving the Broadcasting Problem

Raluca Lefticaru[1], Florentin Ipate[1], Marian Gheorghe[1,2], and Gexiang Zhang[3]

[1] Department of Computer Science, University of Piteşti
Str. Târgu din Vale 1, 110040 Piteşti, Romania
raluca.lefticaru@gmail.com, florentin.ipate@ifsoft.ro
[2] Department of Computer Science, University of Sheffield
Regent Court, Portobello Street, Sheffield S1 4DP, UK
m.gheorghe@dcs.shef.ac.uk
[3] School of Electrical Engineering, Southwest Jiaotong University
Chengdu, 610031, P.R. China
zhgxdylan@126.com

Abstract. P systems are employed in various contexts to specify or model different problems. In certain cases the system is not entirely known. In this paper we will consider the broadcasting algorithm and present a method to determine the format of the rules of the P system used to specify the algorithm.

1 Introduction

P systems (also called membrane systems) represent a class of parallel and distributed computing devices which are inspired by the structure and the functioning of the living cells [10]. The model has been used for theoretical investigations as well as a vehicle to represent different problems from various domains.

With very few exceptions, [13], [5], [3], [4], in all previous studies the systems considered have been fully specified. There are situations when some components of a model are not known or maybe available in certain contexts and circumstances.

In the vast majority of cases, the P system rules act either within compartments or between those that share the same neighbourhood. There are only few situations (for instance, P systems with gemmation [1]) when rules of a compartment transfer objects from their current position to a destination that might be far away from their place.

In this paper we study the broadcasting algorithm defined in a P system framework [6], by considering a number of variants of P systems. We will study the dependencies between the format of the rules in each compartment and the number of its neighbours, as well as a method to automatically generate the rules in each compartment depending on the number of neighbours. This problem is also important in the context of P systems where compartments are added to or removed from them. The structure of a system can be changed either by operations belonging to the system, like in the case of P systems with active membranes, or by external means, but this aspect is not considered in this paper.

G. Păun et al. (Eds.): WMC 2009, LNCS 5957, pp. 354–370, 2010.
© Springer-Verlag Berlin Heidelberg 2010

2   Basic Concepts

A P system is a computational model, inspired by the functioning and structure of the living cell. The cell-like P systems [12] consist of: (i) a hierarchical arrangement of *membranes*, embedded in the skin membrane, the one which separates the system from its environment; (ii) *objects* occurring inside the regions delimited by membranes, coding complex chemical molecules or compounds; and (iii) *rules* assigned to the regions of the membrane structure, acting upon the objects inside and the regions themselves. A membrane without any membrane inside is called an elementary one. Each membrane defines a region. Each region contains, apart from zero or many membranes, a multiset of objects and a set, in this paper, of transformation and communication rules.

A configuration of a P system is represented by the current membrane structure and the multisets of objects occurring in each region. The system will go from one configuration to a new one by applying the rules in a non-deterministic and maximally parallel manner, i.e., at each step, in each membrane it is applied a maximal multiset of rules. The system will halt when no more rules are available to be applied. Usually, the result of the computation is obtained in a specified component of the system, called the output region.

In what follows a basic P system using transformation and communication rules is formally defined. For more details look at [12], [11].

Definition 1. *A P system is a construct*

$$\Pi = (V, \mu, M_1, \ldots, M_m, R_1, \ldots, R_m, i_0),$$

where

- *V is an alphabet; its elements are called objects;*
- *μ is a membrane structure consisting of m membranes, with the membranes and the regions labelled in a one-to-one manner with elements of a given set Λ, usually, the set $\{1, \cdots, m\}$; m is called the degree of Π;*
- *M_i, $1 \le i \le m$, are strings which represent multisets over V associated with the regions of μ;*
- *R_i, $1 \le i \le m$, are transformation-communication rules associated with the regions of μ; each rule of R_i has the form $x \to y$, where x is a non-empty multiset over V, and y defines a multiset over $\{a_j | a \in V, j \in \{here, out, 1, \cdots, m\}\}$ (a_{here} means a remains in the current region, i; subsequently here will be ignored; a_{out} indicates that a has to go out of i to the outer region; a_j, $1 \le j \le m$, shows that a goes to the region j that must be directly contained by the current membrane); applying a rule means replacing x by y and following the target indications;*
- *i_0 is a number between 1 and m which specifies the output membrane of Π.*

When a target indication, t, occurs more than once in a sequence, i.e., $a_t^1 \cdots a_t^h$ then the following shortcut notation $(a^1 \cdots a^h)_t$ is used. A P system provides a suitable framework for distributed parallel computation that develops in steps.

Indeed, any computation starts by processing the initial multisets, w_i, and then in each step the rules associated to each region are applied in a non-deterministic and maximally parallel manner. The result of a computation, a multiset of simple objects, is obtained in region i_0. We notice that the rules presented above combine both transformation and communication, being responsible for evolving the objects and transferring them to regions according to specified targets. We will consider specific contexts for applying some of these rules, namely promoters and inhibitors [2]. *Promoters* are used to show the reaction enhancing (*inhibitors* have reaction prohibiting) roles for various substances or molecules present in cells [2].

3   Broadcasting through a P System

Broadcasting messages to the nodes of a network occurs in various communications and is well-studied for different network topologies, message lengths, transmission constraints. The problem is also formulated in the context of a basic P system and its complexity has been studied [6]. A basic broadcasting problem consists in sending a message from a node of a network to all the other nodes without revisiting them. In a P system environment it involves sending the message through the tree structure of the P system. The broadcasting algorithm for P systems [6] does not discuss the format of the rules that may lead to various types of P systems and, more important, specific complexity aspects of the communication processes involved.

We will first present various variants of P systems and analyse complexity aspects related to the communication processes that occur and the dependencies between the format of the rules in a compartment and the number of its neighbours.

The broadcasting problem is presented through the P system having the membrane structure given by the tree structure in Fig. 1(a) where the message will start from membrane 9. According to the broadcasting principle, illustrated in [6], from each membrane, or node of the tree, the message is sent one level up, to its parent membrane, and to all its directly contained membranes. Initially the message from membrane 9 is sent to $6, 11, 12$. In the following step from these compartments the messages are sent to $3, 10, 15, 16$, respectively. Please note that from the membranes $15, 16, 12$ the message does no longer travel away from them. We can better illustrate how the message travels up and down the structure by representing the tree with root 9 (see Fig. 1(b)) as the associated tree structure where the message travels only downwards.

We will consider a generic node j surrounded by neighbours p, i, k; one of these may be a parent and the others children, or all of them children. The message, denoted by O, might come from any of them and travel then to the others. The message will come with other symbols that help the system implementing the algorithm. We will conceive various rules allowing the message received from one of its neighbours to travel through j towards its other neighbours. We will consider four distinct cases illustrated by different types of P systems.

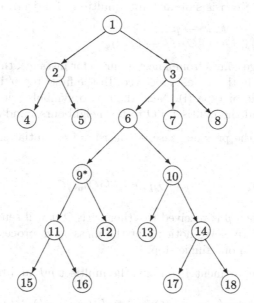

(a) Tree describing a membrane structure; the start node for broadcasting is 9

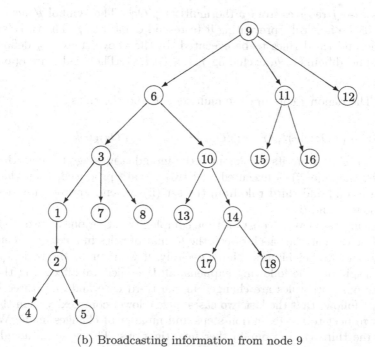

(b) Broadcasting information from node 9

Fig. 1. Trees illustrating the membrane structure of a P system and the broadcasting principle

Case 1. Initially j consists of an empty multiset of objects and the rules are

- (i) $p' \to ik$, $i' \to pk$, $k' \to pi$;
- (ii) $p \to (j'O)_p$, $i \to (j'O)_i$, $k \to (j'O)_k$.

When a message comes from a neighbour, p for instance, then the corresponding multiset, $p'O$ in this case, is received. In the first step p' is transformed into ik by using a rule of type (i). Next these two symbols trigger rules from (ii) which in turn send the multiset $j'O$ to the neighbours i and k, respectively.

Case 2. Like in the previous case, j consists of an initial empty multiset; the rules are

$$p \to (jO)_i(jO)_k, \ i \to (jO)_p(jO)_k, \ k \to (jO)_p(jO)_i.$$

In this case when p is received together with O it will send jO to i and to k by using the first rule. We notice that the message is processed and passed on to its neighbours in one single step.

Case 3. The compartment j contains the multiset pik and the rules

$$pc \to (j'Oc^{n_{j,p}})_p|\neg p', \ ic \to (j'Oc^{n_{j,i}})_i|\neg i', \ kc \to (j'Oc^{n_{j,k}})_k|\neg k'.$$

In this case j receives from p the multiset $p'Occ$. The symbol p' acts as an inhibitor of the first rule, preventing it to resend O back to p. The two c's allow the second and third rules to be executed. In the above rules $n_{j,h}$ defines the number of neighbours of h, excluding j, $h \in \{p, i, k\}$. These rules are applied in one step.

Case 4. The region j contains the multiset pik and the rules

- (i) $c \to x^2$,
- (ii) $px \to (j'Oc)_p|\neg p'$, $ix \to (j'Oc)_i|\neg i'$, $kx \to (j'Oc)_k|\neg k'$.

Once j receives from its neighbour p the intended message through the multiset $p'Oc$, the rule (i) is executed and two x's are produced; then they will allow the second and third rule from the set (ii) to send appropriate messages to neighbours i and k.

These four cases have a constant time complexity, either one or two steps. We now analyse the correlations between the format of rules in a compartment and the number of its neighbours. More precisely, if we refer to the region j then for each neighbour the following happens: all the rules are affected in the first two cases; only two rules are changed in the third case and only three in the last one. It follows that the last two cases have a lower complexity than the first two with respect to the execution steps and number of changes made. We will consider the third case in our further investigations. This case, although very attractive due to its low complexity, with respect to number of steps, and relative robustness to changes, requires to assess in advance the number of neighbours for each compartment. We will consider this case for the example described in Fig. 1(a).

Example 1. Let us consider a more general situation whereby a membrane j is included in p and contains k membranes $i_1, \ldots, i_k$. The region j consists of a multiset composed of the identifiers of the outer membrane, p, and inner membranes $i_1, \ldots, i_k$, i.e., its **close neighbours**. Formally this is given by $M_j = \{p, i_1, \cdots, i_k\}$. We will adopt this notation for multisets, instead of string based, due to numbers used as symbols in the notation below. Given the membrane structure defined by the tree in Fig. 1(a), the membrane 9 is part of membrane 6 and contains 11 and 12. The membrane structure is provided by

$$\mu = [[[\;]_4[\;]_5]_2[[[[\;]_{15}[\;]_{16}]_{11}[\;]_{12}]_9[[\;]_{13}[[\;]_{17}[\;]_{18}]_{14}]_{10}]_6[]_7[\;]_8]_3]_1$$

The initial multisets are:

$$
\begin{array}{llll}
M_1 = \{2, 3\} & M_2 = \{1, 4, 5\} & M_3 = \{1, 6, 7, 8\} & M_4 = \{2\} \\
M_5 = \{2\} & M_6 = \{3, 9, 10\} & M_7 = \{3\} & M_8 = \{3\} \\
M_9 = \{6, 11, 12\} & M_{10} = \{6, 13, 14\} & M_{11} = \{9, 15, 16\} & M_{12} = \{9\} \\
M_{13} = \{10\} & M_{14} = \{10, 17, 18\} & M_{15} = \{11\} & M_{16} = \{11\} \\
M_{17} = \{14\} & M_{18} = \{14\}
\end{array}
$$

The rules of j are:

$$pc \to (j'O)_p(c_p)^{n_{j,p}} |\neg p';$$
$$i_sc \to (j'O)_{i_s}(c_{i_s})^{n_{j,i_s}} |\neg i_s', \quad s = 1, \cdots, k;$$

where:

- like in Case 3 presented above, p', i_s' are inhibitors (a rule above is applied when there is no p' or i_s', respectively, in membrane j), O is the message that will be sent, c is an object which is associated with a communication between two membranes;
- $n_{j,p}$, n_{j,i_s} are integer values defining the number of non-visited neighbours of p, i_s, respectively; it is easy to work out the relationship between the format of a rule and the number of non-visited descendants of the neighbour associated with the rule.

We briefly describe the first two steps of the broadcasting algorithm in this case.

Step 1. In the membrane that initiates the broadcasting are injected an object O and a number of objects c, one for every neighbour.

For example, if the starting membrane is $j = 9$, like in Fig. 1(a), then we have the initial multiset M_9 and the additional symbols mentioned above leading to the multiset $\{6, 11, 12, O, c, c, c\}$; the rules are

$$R_9 = \{r_{9,6} : 6c \to (9'O)_6(c_6)^{n_{9,6}} |\neg 6',$$
$$r_{9,11} : 11c \to (9'O)_{11}(c_{11})^{n_{9,11}} |\neg 11',$$
$$r_{9,12} : 12c \to (9'O)_{12}(c_{12})^{n_{9,12}} |\neg 12'\}.$$

After these rules are applied in membrane 9, the objects $6, 11, 12, c, c, c$ are consumed and only an O remains in this membrane showing that the message has been received.

Step 2. Since this step onwards it is easy to follow the route of messages travelling through the system by representing it as a tree with root 9 as in Fig. 1(b). If in Step 1 we consider $n_{9,6} = 2$, $n_{9,11} = 2$ and $n_{9,12} = 0$, then in the membranes $6, 11, 12$ which are neighbours of 9, the multisets will be: $\{3, 9, 10, 9', O, c, c\}$, $\{9, 15, 16, 9', O, c, c\}$, $\{9, 9', O\}$, respectively; the rules will be:

$$R_6 = \{r_{6,3} : 3c \rightarrow (6'O)_3(c_3)^{n_{6,3}} | \neg 3',$$
$$r_{6,9} : 9c \rightarrow (6'O)_9(c_9)^{n_{6,9}} | \neg 9',$$
$$r_{6,10} : 10c \rightarrow (6'O)_{10}(c_{10})^{n_{6,10}} | \neg 10'\}$$

$$R_{11} = \{r_{11,9} : 9c \rightarrow (11'O)_9(c_9)^{n_{11,9}} | \neg 9',$$
$$r_{11,15} : 15c \rightarrow (11'O)_{15}(c_{15})^{n_{11,15}} | \neg 15',$$
$$r_{11,16} : 16c \rightarrow (11'O)_{16}(c_{16})^{n_{11,16}} | \neg 16'\}$$

$$R_{12} = \{r_{12,9} : 9c \rightarrow (12'O)_9(c_9)^{n_{12,9}} | \neg 9'\}.$$

The rules $r_{6,3}$, $r_{6,10}$, $r_{11,15}$, $r_{11,16}$ are applied and the following multisets are obtained $\{O\}$, $\{9, 9', O\}$, $\{9, 9', O\}$, $\{9, 9', O\}$, in regions $9, 6, 11, 12$, respectively.

If in Step 1 we consider $n_{9,6} = 0$ or $n_{9,6} = 1$, then at least one of the rules $r_{6,3}$ or $r_{6,10}$ cannot be applied as a c is missing and then in the corresponding hierarchy of compartments the message O is not received[1]. The multiset associated with region 6 becomes $\{3, 9, 9', O\}$, where 3 is the non-visited compartment together with its neighbours.

If in Step 1 it is considered $n_{9,6} > 2$ then the multiset is $\{3, 9, 10, 9', O, c^{n_{9,6}}\}$, and by applying the two existing rules, it becomes $\{9, 9', O, c^{n_{9,6}-2}\}$.

The process restarts from the compartments that have been affected by the communication rules in Step 2.

From this example we observe the following regarding the values $n_{j,i}$ involved.

- If the values $n_{j,i}$ are appropriately chosen then in each membrane we will eventually get an O and no c.
- If $n_{j,i}$ is less than the expected value then for at least one hierarchy of compartments the message O does not travel to it.
- If $n_{j,i}$ has a bigger value then in some compartments we will have some more c's.
- Some $n_{j,i}$ do not count, i.e., those where the inhibitors i' are present. For instance: $n_{6,9}$, $n_{11,9}$, $n_{15,11}$ etc.
- For the membrane structure given in Fig. 1(a), the solution is: $n_{9,6} = 2$, $n_{9,11} = 2$, $n_{9,12} = 0$, $n_{6,3} = 3$, $n_{6,10} = 2$, $n_{11,15} = 0$, $n_{11,16} = 0$, $n_{3,1} = 1$, $n_{3,7} = 0$, $n_{3,8} = 0$, $n_{10,13} = 0$, $n_{10,14} = 2$, $n_{1,2} = 2$, $n_{14,17} = 0$, $n_{14,18} = 0$, $n_{2,4} = 0$, $n_{2,5} = 0$; the other $n_{i,j}$ do not count.

---

[1] $r_{6,9}$ can not be applied due to the inhibitor $9'$.

- The number of $n_{i,j}$ values that are relevant is the same as the number of pairs parent-child in the membrane structure and is equal to the number of compartments minus 1.
- By using the above values $n_{i,j}$, the P system will end up with the multisets below, where M_j is this multiset for the compartment j:

$$M_1 = \{3, 3', O\} \quad M_2 = \{1, 1', O\} \quad M_3 = \{6, 6', O\}$$
$$M_4 = \{2, 2', O\} \quad M_5 = \{2, 2', O\} \quad M_6 = \{9, 9', O\}$$
$$M_7 = \{3, 3', O\} \quad M_8 = \{3, 3', O\} \quad M_9 = \{O\}$$
$$M_{10} = \{6, 6', O\} \quad M_{11} = \{9, 9', O\} \quad M_{12} = \{9, 9', O\}$$
$$M_{13} = \{10, 10', O\} \ M_{14} = \{10, 10', O\} \ M_{15} = \{11, 11', O\}$$
$$M_{16} = \{11, 11', O\} \ M_{17} = \{14, 14', O\} \ M_{18} = \{14, 14', O\}$$

- Given the non-determinism of the P system, for the same values of some parameters we can have different number of messages sent. For instance if $n_{9,6} = 1$, then $M_6 = \{3, 9, 10, 9', O, c\}$. If $r_{6,3}$ is applied then the hierarchy of compartments starting with 10 remains without messages (5 compartments without O). Similarly, if $r_{6,10}$ is applied then the 7 compartments occurring in the subtree rooted in 3 remained non-visited – see Fig. 1(b).

4 Tuning the P System

In order to tune the system the values $n_{i,j}$ have to be identified. In the following a further transformation of the system is provided together with a more abstract representation.

The X-machine associated to the P system. According to the broadcasting problem defined above the values $n_{i,j}$ have to be found and we will apply an evolutionary approach using genetic algorithms to find these values. In order to apply it we will transform the cell-like structure of the system into a tree based structure. For a membrane structure μ we will consider as tree root the node from which the broadcast starts. For the P system presented in Example 1, node 9 will be the tree root - see Fig. 1(b). We can further abstract the problem and define each communication between two nodes i, j as a function $f_{i,j}$ with $n_{i,j}$ as its parameter describing the number of non-visited neighbours. It is easy to observe that the functions emerging from the same node will be executed in parallel, maybe together with other functions emerging from other nodes, they are independent of each other and an interleaving strategy can be adopted. In this case sequences of functions can be considered. A state machine or an X-machine can be defined by considering all possible interleavings of the arcs coming out of the nodes of a subtree. In the case presented in Example 1 the initial node is 9 and we distinguish three cases; when a c will be in 9 then we have three non-deterministic choices from 9 to each of the neighbours, the arcs being $f_{9,x}$ where $x \in \{6, 11, 12\}$; when two c's are in 9 then there are 6 non-deterministic choices: for each state defined by a pair $\{x, y\}$, $x, y \in \{6, 11, 12\}$, $x \neq y$, two non-deterministic sequences $f_{9,x}, f_{9,y}$ and $f_{9,y}, f_{9,x}$

can be conceived; for three or more c's there are again six non-deterministic choices from 9 to the state $\{6, 11, 12\}$, given by all the possible combinations of sequences of three functions $f_{9,x}$, $x \in \{6, 11, 12\}$. From each of the above seven states, $\{6\}, \{11\}, \{12\}, \{6, 11\} \{6, 12\}, \{11, 12\}, \{6, 11, 12\}$ the construction of the machine follows the following steps: the arcs of the subtrees of roots specified by these states are shuffled. All shuffled routes starting in a given state are equivalent as the order of executing these functions does not matter.

5   Experiments and Results

The experiments performed aimed to determine the unknown elements of a P system, more precisely the values $n_{i,j}$, using genetic algorithms. Considering that the structure of the P system contains m compartments, the number of parameter values that should be discovered is $m - 1$. In order to determine these values $n_{i,j}$, only the tree structure of the P system was used. Each candidate solution was encoded by an integer vector with $m - 1$ components, ranging from 0 to 10 and, consequently, the search space size was 11^{m-1}. The JGAP package (Java Genetic Algorithms and Genetic Programming Package) [9] was used for an elitist genetic algorithm implementation. The crossover operator has a great impact on the success of the genetic algorithm and the one chosen for this problem was the uniform crossover [7] (it is not part of the current JGAP version, but the package can be quickly extended with other operators). For selection we used a *BestChromosomesSelector* with the rate 0.8, which takes the

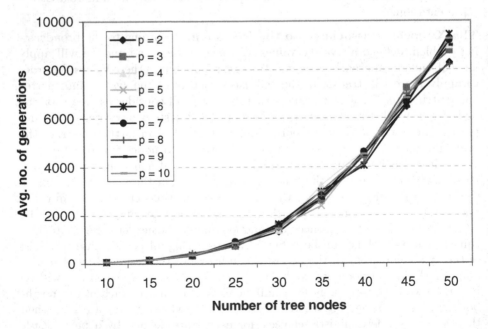

Fig. 2. Average number of generations for trees with fixed number of sons p

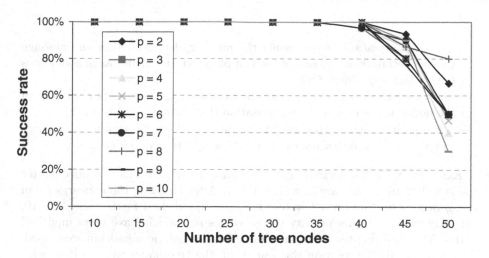

Fig. 3. Success rate for trees with a fixed number of sons p

top 80% individuals into the next generation, according to their fitness. The mutation operator employed had a 1/12 mutation rate.

The experiments performed considered trees having different number of nodes: 10, 15, 20, 25, 30, 35, 40, 45, 50. Obviously, it is more difficult to find a solution for a tree with 50 nodes (49 unknown variables) than for a tree with 10 nodes (and 9 unknown parameters). Due to the fact that the tree structure might have (or not) an influence on the problem considered, the following types of trees were considered:

1. *Trees with fixed number of sons:* each node has exactly p sons, excepting the leafs and eventually the last non-leaf node. For example, if the tree has $m = 10$ nodes and we consider $p = 3$, the root and its direct descendants will have exactly three sons. If $m = 10$ and $p = 4$, then the tree will have four direct descendants from the root, four for another node and only one descendant for another node.
2. *Trees with a random number of sons:* each non-leaf node can have a different number of sons, randomly chosen, with an equal probability, from the set $\{1,\dots,p\}$.

In both cases, for each number of nodes $m \in \{10, 15, 20, 25, 30, 35, 40, 45, 50\}$ (corresponding to compartments in the P system) we considered all the values $p \in \{2, 3, 4, 5, 6, 7, 8, 9, 10\}$. A tree was generated according to the structural criterion 1 or 2 and the unknown parameters values $n_{i,j}$ were searched using a genetic algorithm. The fitness function simulated a broadcasting (transmission) in the tree, starting from the root and using the parameters $n_{i,j}$. At the end of the transmission, each candidate solution was evaluated by counting the unvisited nodes and the extra messages sent to the nodes. For this we used the formula

$$fitness = \lambda \cdot no\_of\_unvisited\_nodes + no\_of\_extra\_messages,$$

where:

- *no_of_unvisited_nodes* represents the number of nodes where the message was not received; at the end of the computation, the membrane (node) does not contain any object O;
- *no_of_extra_messages* represents the number of extra objects c, present in the nodes at the end of the computation that cannot be consumed;
- λ is a positive penalty (or weight) parameter which gives more importance to the *no_of_unvisited_nodes* or to the *no_of_extra_messages*.

Experimentally, we noticed that a function for which $\lambda > 1$ guided better the search than in the case in which $\lambda = 1$. After checking the convergence of the genetic algorithm on a few test trees, we decided to further use $\lambda = 10$, this way giving a higher penalty to the values $n_{i,j}$ which leave more unvisited nodes. The following termination criteria for the genetic algorithm were used: A) *fitness* $= 0$ (the solution was found: all the tree nodes were visited, with no extra messages sent) and B) the maximum allowed number of generations (10000) was reached. The population size used in these experiments was in all cases of 20 individuals.

For each combination, given by the structural criterion 1 or 2, the number of nodes in the tree $m \in \{10, 15, 20, \ldots, 50\}$ and the number of sons for each node $p \in \{2, 3, \ldots, 10\}$, the genetic algorithm was run 30 times. After each run, the best solution obtained, its fitness and the current generation were retained. The Tables 1, 2, 3, 4 present, for each set of 30 runs the following information: $m =$ number of nodes in the tree, $p =$ number of sons, the search space dimension for each case and the success rate for the 30 runs. Also, the mean and the standard deviation are shown for the best fitness function values (MF, SF) and for the

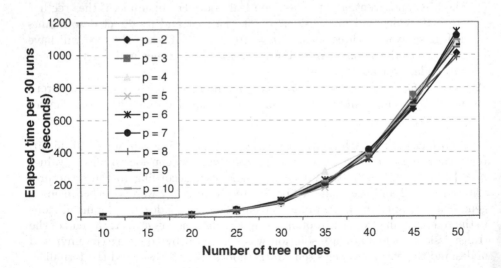

Fig. 4. Elapsed time for trees with a fixed number of sons p

Table 1. Statistics for trees with m nodes and fixed number of sons p

m	p	Space size	Succ.	MF	SF	MG	SG	Dur.
10	2	2.36E+09	100.0 %	0.00	0.00	65.50	29.60	1
10	3	2.36E+09	100.0%	0.00	0.00	62.77	25.38	1
10	4	2.36E+09	100.0%	0.00	0.00	66.13	33.93	1
10	5	2.36E+09	100.0%	0.00	0.00	63.13	21.65	1
10	6	2.36E+09	100.0%	0.00	0.00	68.43	28.07	1
10	7	2.36E+09	100.0%	0.00	0.00	59.07	22.45	1
10	8	2.36E+09	100.0%	0.00	0.00	63.77	26.55	1
10	9	2.36E+09	100.0%	0.00	0.00	63.50	20.34	1
10	10	2.36E+09	100.0%	0.00	0.00	67.10	29.93	1
15	2	3.80E+14	100.0%	0.00	0.00	161.07	76.80	5
15	3	3.80E+14	100.0%	0.00	0.00	153.53	63.31	5
15	4	3.80E+14	100.0%	0.00	0.00	161.53	61.55	5
15	5	3.80E+14	100.0%	0.00	0.00	140.13	51.84	5
15	6	3.80E+14	100.0%	0.00	0.00	154.70	76.02	5
15	7	3.80E+14	100.0%	0.00	0.00	154.03	71.19	5
15	8	3.80E+14	100.0%	0.00	0.00	156.83	43.10	5
15	9	3.80E+14	100.0%	0.00	0.00	168.27	69.02	5
15	10	3.80E+14	100.0%	0.00	0.00	153.80	63.36	5
20	2	6.12E+19	100.0%	0.00	0.00	363.10	188.67	14
20	3	6.12E+19	100.0%	0.00	0.00	327.23	124.68	13
20	4	6.12E+19	100.0%	0.00	0.00	388.43	158.48	15
20	5	6.12E+19	100.0%	0.00	0.00	370.63	213.60	15
20	6	6.12E+19	100.0%	0.00	0.00	364.13	135.54	14
20	7	6.12E+19	100.0%	0.00	0.00	348.33	152.45	14
20	8	6.12E+19	100.0%	0.00	0.00	442.20	178.49	17
20	9	6.12E+19	100.0%	0.00	0.00	354.73	154.01	14
20	10	6.12E+19	100.0%	0.00	0.00	357.30	142.46	14
25	2	9.85E+24	100.0%	0.00	0.00	782.77	250.66	39
25	3	9.85E+24	100.0%	0.00	0.00	730.87	228.65	37
25	4	9.85E+24	100.0%	0.00	0.00	834.83	316.04	42
25	5	9.85E+24	100.0%	0.00	0.00	855.97	358.82	43
25	6	9.85E+24	100.0%	0.00	0.00	808.17	369.63	41
25	7	9.85E+24	100.0%	0.00	0.00	915.60	306.14	46
25	8	9.85E+24	100.0%	0.00	0.00	721.67	327.35	36
25	9	9.85E+24	100.0%	0.00	0.00	847.90	336.24	42
25	10	9.85E+24	100.0%	0.00	0.00	859.47	313.92	43
30	2	1.59E+30	100.0%	0.00	0.00	1329.13	488.55	83
30	3	1.59E+30	100.0%	0.00	0.00	1616.13	850.51	100
30	4	1.59E+30	100.0%	0.00	0.00	1377.40	516.99	86
30	5	1.59E+30	100.0%	0.00	0.00	1522.27	642.85	94

Table 2. Statistics for trees with m nodes and fixed number of sons p

m	p	Space size	Succ.	MF	SF	MG	SG	Dur.
30	6	1.59E+30	100.0%	0.00	0.00	1653.27	549.80	102
30	7	1.59E+30	100.0%	0.00	0.00	1523.37	585.38	94
30	8	1.59E+30	100.0%	0.00	0.00	1493.30	512.72	92
30	9	1.59E+30	100.0%	0.00	0.00	1643.13	659.68	101
30	10	1.59E+30	100.0%	0.00	0.00	1452.80	490.25	89
35	2	2.55E+35	100.0%	0.00	0.00	2678.50	972.70	200
35	3	2.55E+35	100.0%	0.00	0.00	2611.27	932.03	199
35	4	2.55E+35	100.0%	0.00	0.00	3237.53	1537.68	284
35	5	2.55E+35	100.0%	0.00	0.00	2398.23	609.77	181
35	6	2.55E+35	100.0%	0.00	0.00	2984.70	844.46	228
35	7	2.55E+35	100.0%	0.00	0.00	2808.17	869.71	209
35	8	2.55E+35	100.0%	0.00	0.00	2810.83	929.17	211
35	9	2.55E+35	100.0%	0.00	0.00	2673.77	857.52	199
35	10	2.55E+35	100.0%	0.00	0.00	3004.70	1077.03	220
40	2	4.11E+40	100.0%	0.00	0.00	4476.57	1587.89	392
40	3	4.11E+40	96.7%	0.03	0.18	4464.57	1544.74	397
40	4	4.11E+40	96.7%	0.03	0.18	4643.00	1670.95	412
40	5	4.11E+40	100.0%	0.00	0.00	4592.83	1640.95	407
40	6	4.11E+40	100.0%	0.00	0.00	4046.40	1184.09	358
40	7	4.11E+40	96.7%	0.03	0.18	4607.63	1851.40	415
40	8	4.11E+40	100.0%	0.00	0.00	4175.73	1212.75	366
40	9	4.11E+40	100.0%	0.00	0.00	4471.87	1566.23	394
40	10	4.11E+40	100.0%	0.00	0.00	4513.00	1465.79	390
45	2	6.63E+45	93.3%	0.07	0.25	6464.13	2055.62	667
45	3	6.63E+45	80.0%	0.27	0.58	7239.77	1869.96	753
45	4	6.63E+45	86.7%	0.17	0.46	6700.20	2103.96	695
45	5	6.63E+45	90.0%	0.10	0.31	6964.23	2095.90	725
45	6	6.63E+45	80.0%	0.27	0.58	6541.43	2342.57	682
45	7	6.63E+45	90.0%	0.13	0.43	6620.33	1898.96	690
45	8	6.63E+45	86.7%	0.17	0.46	6975.67	1719.86	723
45	9	6.63E+45	76.7%	0.27	0.52	6864.73	2304.19	714
45	10	6.63E+45	90.0%	0.10	0.31	6811.03	1914.10	689
50	2	1.07E+51	66.7%	0.47	0.78	8275.27	1624.30	1011
50	3	1.07E+51	50.0%	0.80	0.89	8766.20	1434.48	1063
50	4	1.07E+51	40.0%	0.80	0.89	8981.07	1523.70	1073
50	5	1.07E+51	46.7%	0.67	0.76	9189.17	1309.14	1100
50	6	1.07E+51	50.0%	0.57	0.63	9443.93	991.21	1145
50	7	1.07E+51	50.0%	0.70	0.84	9189.77	1131.65	1117
50	8	1.07E+51	80.0%	0.33	0.76	8198.33	1419.55	985
50	9	1.07E+51	50.0%	0.73	0.87	8985.03	1310.33	1061
50	10	1.07E+51	30.0%	1.03	0.89	9262.57	1306.81	1092

Table 3. Statistics for trees with m nodes and variable number of sons between $\{1, \ldots, p\}$

m	p	Space size	Succ.	MF	SF	MG	SG	Dur.
10	2	2.36E+09	100.0%	0.00	0.00	63.27	32.43	1
10	3	2.36E+09	100.0%	0.00	0.00	58.27	23.10	1
10	4	2.36E+09	100.0%	0.00	0.00	58.80	17.81	1
10	5	2.36E+09	100.0%	0.00	0.00	72.73	25.61	1
10	6	2.36E+09	100.0%	0.00	0.00	66.50	24.56	1
10	7	2.36E+09	100.0%	0.00	0.00	64.47	29.34	1
10	8	2.36E+09	100.0%	0.00	0.00	57.10	20.66	1
10	9	2.36E+09	100.0%	0.00	0.00	57.13	17.19	1
10	10	2.36E+09	100.0%	0.00	0.00	66.30	25.95	1
15	2	3.80E+14	100.0%	0.00	0.00	155.43	67.94	5
15	3	3.80E+14	100.0%	0.00	0.00	148.43	49.25	5
15	4	3.80E+14	100.0%	0.00	0.00	158.93	73.09	5
15	5	3.80E+14	100.0%	0.00	0.00	163.67	62.21	5
15	6	3.80E+14	100.0%	0.00	0.00	159.40	60.32	5
15	7	3.80E+14	100.0%	0.00	0.00	152.60	55.89	5
15	8	3.80E+14	100.0%	0.00	0.00	147.73	49.97	5
15	9	3.80E+14	100.0%	0.00	0.00	152.03	63.03	5
15	10	3.80E+14	100.0%	0.00	0.00	153.90	58.05	5
20	2	6.12E+19	100.0%	0.00	0.00	381.27	155.23	14
20	3	6.12E+19	100.0%	0.00	0.00	379.93	149.30	15
20	4	6.12E+19	100.0%	0.00	0.00	356.53	94.02	15
20	5	6.12E+19	100.0%	0.00	0.00	358.53	161.98	14
20	6	6.12E+19	100.0%	0.00	0.00	353.97	151.56	14
20	7	6.12E+19	100.0%	0.00	0.00	328.90	110.80	13
20	8	6.12E+19	100.0%	0.00	0.00	382.50	149.76	15
20	9	6.12E+19	100.0%	0.00	0.00	407.43	179.24	16
20	10	6.12E+19	100.0%	0.00	0.00	376.80	196.99	15
25	2	9.85E+24	100.0%	0.00	0.00	882.93	475.38	44
25	3	9.85E+24	100.0%	0.00	0.00	772.13	249.86	39
25	4	9.85E+24	100.0%	0.00	0.00	823.43	364.62	41
25	5	9.85E+24	100.0%	0.00	0.00	863.57	363.86	43
25	6	9.85E+24	100.0%	0.00	0.00	833.20	495.31	42
25	7	9.85E+24	100.0%	0.00	0.00	842.67	318.03	42
25	8	9.85E+24	100.0%	0.00	0.00	834.87	291.64	42
25	9	9.85E+24	100.0%	0.00	0.00	822.13	423.69	40
25	10	9.85E+24	100.0%	0.00	0.00	834.23	323.54	42
30	2	1.59E+30	100.0%	0.00	0.00	1549.30	680.43	94
30	3	1.59E+30	100.0%	0.00	0.00	1519.50	622.27	93
30	4	1.59E+30	100.0%	0.00	0.00	1785.00	589.93	109
30	5	1.59E+30	100.0%	0.00	0.00	1475.80	761.75	94

Table 4. Statistics for trees with m nodes and variable number of sons between $\{1, \ldots, p\}$

m	p	Space size	Succ.	MF	SF	MG	SG	Dur.
30	6	1.59E+30	100.0%	0.00	0.00	1657.53	621.15	105
30	7	1.59E+30	100.0%	0.00	0.00	1691.43	555.99	108
30	8	1.59E+30	100.0%	0.00	0.00	1423.43	517.50	91
30	9	1.59E+30	100.0%	0.00	0.00	1579.10	497.21	99
30	10	1.59E+30	100.0%	0.00	0.00	1451.13	502.28	92
35	2	2.55E+35	100.0%	0.00	0.00	2864.53	1163.11	218
35	3	2.55E+35	100.0%	0.00	0.00	2700.27	832.91	205
35	4	2.55E+35	100.0%	0.00	0.00	2822.23	1200.13	216
35	5	2.55E+35	100.0%	0.00	0.00	3073.17	1217.87	236
35	6	2.55E+35	100.0%	0.00	0.00	2491.67	781.10	192
35	7	2.55E+35	100.0%	0.00	0.00	2426.27	785.50	186
35	8	2.55E+35	100.0%	0.00	0.00	2681.93	859.48	203
35	9	2.55E+35	100.0%	0.00	0.00	2952.83	1151.27	226
35	10	2.55E+35	100.0%	0.00	0.00	2610.67	1133.39	193
40	2	4.11E+40	100.0%	0.00	0.00	4112.87	1169.85	364
40	3	4.11E+40	100.0%	0.00	0.00	4785.10	1656.32	420
40	4	4.11E+40	100.0%	0.00	0.00	4557.37	1544.00	401
40	5	4.11E+40	100.0%	0.00	0.00	4120.03	1417.94	363
40	6	4.11E+40	100.0%	0.00	0.00	4534.57	1634.49	404
40	7	4.11E+40	100.0%	0.00	0.00	4819.10	1268.60	422
40	8	4.11E+40	100.0%	0.00	0.00	4281.47	1744.27	375
40	9	4.11E+40	96.7%	0.03	0.18	4317.23	1719.93	378
40	10	4.11E+40	100.0%	0.00	0.00	4343.13	1500.82	384
45	2	6.63E+45	83.3%	0.17	0.38	6733.47	2096.95	690
45	3	6.63E+45	90.0%	0.10	0.31	7226.00	1993.56	745
45	4	6.63E+45	83.3%	0.17	0.38	6764.33	2299.93	693
45	5	6.63E+45	80.0%	0.20	0.41	7079.87	1975.24	725
45	6	6.63E+45	93.3%	0.07	0.25	6686.43	1886.45	686
45	7	6.63E+45	93.3%	0.07	0.25	6328.57	1980.41	649
45	8	6.63E+45	86.7%	0.13	0.35	6766.33	2225.37	690
45	9	6.63E+45	83.3%	0.17	0.38	6997.10	1954.87	717
45	10	6.63E+45	93.3%	0.07	0.25	6519.57	1973.89	671
50	2	1.07E+51	40.0%	0.80	0.89	9223.73	1262.43	1088
50	3	1.07E+51	40.0%	0.80	0.81	9554.10	699.02	1145
50	4	1.07E+51	33.3%	0.87	0.82	9311.43	1200.56	1097
50	5	1.07E+51	50.0%	0.67	0.76	8869.97	1615.60	1040
50	6	1.07E+51	63.3%	0.37	0.49	8775.83	1498.18	1034
50	7	1.07E+51	53.3%	0.60	0.77	8571.30	1755.25	1014
50	8	1.07E+51	56.7%	0.67	0.99	8782.73	1450.53	1051
50	9	1.07E+51	40.0%	0.73	0.69	8849.47	1684.96	1042
50	10	1.07E+51	50.0%	0.70	0.88	8994.07	1635.41	1063

number of generations (MG, SG), after 30 runs. The last column from the table shows the cumulated duration of the 30 runs, expressed in seconds.

We will refer only to results obtained for trees with fixed number of sons as for trees with random number of descendants the results are very similar. The average number of generations (Fig. 2) and the time elapsed to get the solution (Fig. 4) grow proportional to the number of nodes in the tree. The maximum allowed number of generations for the GA was set to 10000. Consequently, the success rates were very high for trees with less than 45 nodes (for which the solution was found in less generations) and then almost halves for trees with 50 nodes (Fig. 3).

6   Conclusions

In this paper a method to determine the rules of a P system that models the broadcasting algorithm is introduced. Naturally, the number of unknown parameter values $n_{i,j}$ increases with the compartments number and consequently the search space size grows also. The search space size is obviously $c^{no\text{-}par}$, where c is the number of possible values for one parameter $n_{i,j}$ and no_par is the number of unknown parameters. The average number of generations and the elapsed time needed to find a solution increase when the search space is very large. If the maximum allowed number of generations is not high enough, the GA might end unsuccessful. One possible solution to overcome this is to increase the maximum allowed number of generations for the GA. Other solutions can rely on using hybrid approaches, i.e. combining GAs with local search techniques (like hill climbing) and developing new GAs operators, suited for this problem (the crossover operator has in particular a great impact on the GA).

The method is described in a more general context of an abstract X-machine that captures some specific aspects of the P system, namely the size of the rules. Given that similar approaches to map P systems into X-machines prove to be very effective in testing these systems [8], we can conclude that such testing strategies developed for associated X-machines can be applied in the case of the broadcasting problem as well. Hence, we can provide a powerful method to estimate the P system that models the broadcasting problem and then test the implementation based on this model.

Further studies will aim to improve the precision and efficiency of the method discussed in this paper and to extend it to other classes of P systems.

Acknowledgements

The research of RL, FI and MG is supported by CNCSIS grant IDEI no. 496/2009, *An integrated evolutionary approach to formal modelling and testing* (EvoMT). The research of GZ is supported by the National Natural Science Foundation of China (60702026), the Scientific and Technological Funds for Young Scientists of Sichuan and the Open Foundation of Engineering Research Centre of Safety Transportation of the Ministry of Education of China. The authors would like to thank all the referees for their helpful comments.

References

1. Besozzi, D., Zandron, C., Mauri, G., Sabadini, N.: P systems with gemmation of mobile membranes. In: Restivo, A., Ronchi Della Rocca, S., Roversi, L. (eds.) ICTCS 2001. LNCS, vol. 2202, pp. 136–153. Springer, Heidelberg (2001)
2. Bottoni, P., Martín-Vide, C., Păun, Gh., Rozenberg, G.: Membrane systems with promoters/inhibitors. Acta Informatica 38(10), 695–720 (2002)
3. Castellini, A., Manca, V.: Learning regulation functions of metabolic systems by artificial neural networks. In: Rothlauf, F. (ed.) GECCO 2009, pp. 193–200. ACM, New York (2009)
4. Castellini, A., Manca, V., Suzuki, Y.: Metabolic P system flux regulations by artificial neural networks. In: Păun, G., Pérez-Jiménez, M.J., Riscos-Núñez, A., Rozenberg, G., Salomaa, A. (eds.) WMC 2009. LNCS, vol. 5957. Springer, Heidelberg (2009)
5. Cavaliere, M., Mardare, R.: Partial knowledge in membrane systems: A logical approach. In: Hoogeboom, H.J., Păun, G., Rozenberg, G., Salomaa, A. (eds.) WMC 2006. LNCS, vol. 4361, pp. 279–297. Springer, Heidelberg (2006)
6. Ciobanu, G.: Distributed algorithms over communicating membrane systems. Biosystems 70(2), 123–133 (2003)
7. Drake, S.: Uniform crossover revisited: Maximum disruption in real-coded GAs. In: Cantú-Paz, E., Foster, J.A., Deb, K., Davis, L., Roy, R., O'Reilly, U.-M., Beyer, H.-G., Kendall, G., Wilson, S.W., Harman, M., Wegener, J., Dasgupta, D., Potter, M.A., Schultz, A., Dowsland, K.A., Jonoska, N., Miller, J., Standish, R.K. (eds.) GECCO 2003. LNCS, vol. 2724, pp. 1576–1577. Springer, Heidelberg (2003)
8. Ipate, F., Gheorghe, M.: Testing non-deterministic stream X-machine models and P systems. Electr. Notes Theor. Comput. Sci. 227, 113–126 (2009)
9. Meffert, K., et al.: JGAP - Java Genetic Algorithms and Genetic Programming Package, http://jgap.sf.net
10. Păun, Gh.: Computing with membranes. Journal of Computer and System Sciences 61(1), 108–143 (2000)
11. Păun, Gh.: Membrane computing. An introduction. Springer, Berlin (2002)
12. Păun, Gh., Rozenberg, G.: A guide to membrane computing. Theoretical Computer Science 287(1), 73–100 (2002)
13. Romero-Campero, F.J., Cao, H., Cámara, M., Krasnogor, N.: Structure and parameter estimation for cell systems biology models. In: Ryan, C., Keijzer, M. (eds.) GECCO 2008, pp. 331–338. ACM, New York (2008)

An Improved Membrane Algorithm for Solving Time-Frequency Atom Decomposition

Chunxiu Liu[1], Gexiang Zhang[1], Hongwen Liu[1],
Marian Gheorghe[2,3], and Florentin Ipate[3]

[1] School of Electrical Engineering, Southwest Jiaotong University
Chengdu, 610031, P.R. China
liucx2007@163.com, zhgxdylan@126.com, hongwenliu@163.com
[2] Department of Computer Science, The University of Sheffield
Regent Court, Portobello Street, Sheffield, S1 4DP, UK
M.Gheorghe@dcs.shef.ac.uk
[3] Department of Computer Science and Mathematics
University of Piteşti, Romania
florentin.ipate@ifsoft.ro

Abstract. To decrease the computational complexity and improve the search capability of quantum-inspired evolutionary algorithm based on P systems (QEPS), a real-observation QEPS (RQEPS) was proposed. RQEPS is a hybrid algorithm combining the framework and evolution rules of P systems with active membranes and real-observation quantum-inspired evolutionary algorithm (QEA). The RQEPS involves a dynamic structure including membrane fusion and division. The membrane fusion is helpful to enhance the information communication among individuals and the membrane division is beneficial to reduce the computational complexity. An **NP**-complete problem, the time-frequency atom decomposition of noised radar emitter signals, is employed to test the effectiveness and practical capabilities of the RQEPS. The experimental results show that RQEPS is superior to QEPS, the greedy algorithm and binary-observation QEA in terms of search capability and computational complexity.

1 Introduction

In 1998, Gheorghe Păun proposed membrane computing (P systems) [15][16]. A P system, employing various features to specify the structure and functionality of the living cells, is a membrane structure with objects in its membranes, with specified evolution rules like transformation/communication, merging and dividing membranes [15]. Until now, using the advantages of the new distributed parallel computing model and evolutionary algorithms (EAs), the combination technique of them, membrane algorithm, is applied to solve various complex problems. In [13] and [14], a membrane algorithm with a nested membrane structure was introduced to solve the travelling salesman problem as well as the min storage problem [10]. In [7]-[9], a hybrid algorithm combining a P system with a conventional genetic algorithm (CGA) was proposed to solve single-objective and

G. Păun et al. (Eds.): WMC 2009, LNCS 5957, pp. 371–384, 2010.

multi-objective numerical optimization problems. In [20], a hybrid distributed EA with membrane systems was presented to solve some continuous optimization problems. In [22], a membrane algorithm combining one level membrane structure with binary-observation quantum-inspired evolutionary algorithms (bQEA), called a QEA based on P systems (QEPS), was proposed to solve knapsack problems, and the experimental results show that QEPS performs better than its counterpart bQEA. But there are some drawbacks such as discretization error and Hamming cliff [6][24], when bQEA is used to solve numerical optimization problems. In [24], a real-observation QEA (RQEA) was proposed for numerical optimization problems to overcome the disadvantages of bQEA.

By combining RQEA with P systems having active membranes, this paper proposes an improved membrane algorithm, called a real-observation QEPS (RQEPS), to reduce the computational complexity and improve the search capability of QEPS [22][11]. In RQEPS, the real-observation rules are employed to connect quantum-inspired bit (Q-bit) representation and real-valued variables in each elementary membrane. Then all the elementary membranes are merged into one and all individuals in elementary membranes enter the merged membrane, where a copy of the best individual is sent out to the skin membrane. The recombination is operated on all individuals in the merged membrane to exchange the information among individuals. To demonstrate the effectiveness and applicability of the introduced method, experiments are carried out on the time-frequency atom decomposition (TFAD) of noised radar emitter signals to extend the application of the membrane algorithm. The experimental results show that RQEPS performs better than the greedy algorithm (GrA) [12], bQEA [6] and QEPS [22][11].

The TFAD is an approach that decomposes any signal into a linear combination of waveforms selected from a redundant dictionary of time-frequency atoms, which localized well both in time and frequency [12]. Differing from Fourier and Wavelet transforms, the information in TFAD is not diluted across the whole basis. Unlike Wigner and Cohen class distributions, the energy distribution obtained by TFAD does not include interference terms [12]. Hence, TFAD has become an important analysis technique in signal processing and harmonic analysis [12][17] [5]. One of the most successful methods for signal representations in overcomplete dictionaries to solve this problem is the greedy algorithm (GrA) [12], but the extremely high computational load greatly blocks its practical applications. In [18][3][19][2], conventional genetic algorithms (CGAs) were introduced into TFAD to reduce the computational cost. However, due to slow convergence and premature convergence, it is difficult for CGAs to guide individuals toward better solutions in the search space. This paper uses a novel algorithm combining the framework of P systems with RQEA to reduce the computational load and improve the signal representation in the TFAD.

The remainder of this paper is organized as follows. Section 2 describes the TFAD and the pseudocode algorithm of EAs-based TFAD. Section 3 presents the detailed algorithm for RQEPS. Section 4 discusses the number of elementary membranes, and conducts extensively comparative experiments on noised radar emitter signals. Finally, conclusions are drawn in Section 5.

2 Time-Frequency Atom Decomposition

The TFAD is an approach to select satisfactory time-frequency atoms $g_\gamma(t)_{\gamma \in \Gamma}$ from a redundant time-frequency atom dictionary $D = (g_\gamma(t))$ to decompose a signal into a linear combination of waveforms [12]. Let f be the original signal, $f \in H$, where H is a Hilbert space. When the signal f is decomposed up to the order $item$, f_{item} can be represented as

$$f_{item} = \sum_{n=0}^{item} \langle R^n f, g_{\gamma_n} \rangle g_{\gamma_n} + R^{item+1} f, \tag{1}$$

where g_{γ_n} satisfies

$$|\langle R^n f, g_{\gamma_n} \rangle| = \sup_{\gamma \in \Gamma} |\langle R^n f, g_\gamma \rangle|, \tag{2}$$

where $\Gamma = R^+ \times R^2$ is a set of indexes γ, and $R^{n+1} f$ is the residual signal

$$R^{n+1} f = R^n f - \langle R^n f, g_{\gamma_n} \rangle g_{\gamma_n}. \tag{3}$$

According to the conclusion [12]: $\lim_{item \to \infty} ||R^{item+1} f|| = 0$, the signal f_{item} can be represented as

$$f_{item} = \sum_{n=0}^{item} \langle R^n f, g_{\gamma_n} \rangle g_{\gamma_n}. \tag{4}$$

The problem of selecting a series of atoms to optimally approximate a signal in a redundant time-frequency atom dictionary is NP-hard [1]. One of the most successful methods to solve this problem is the greedy algorithm (GrA) [12]. GrA used a greedy strategy, in which the time-frequency atoms were selected one by one from an over-complete dictionary to best match the structure of

---

Begin
   Initialization of TFAD; % Initial iteration $item$=1;
   While (not termination condition) **do**
      Set parameters of time-frequency atom ;
      Search the suboptimal time-frequency atom in D
      using EAs (RQEPS);
      Compute $|\langle R^{item} f, g_{\gamma_{item}} \rangle g_{\gamma_{item}}|$;
      $R^{item} f \leftarrow (R^{item} f - \langle R^{item} f, g_{\gamma_{Item}} \rangle g_{\gamma_{Item}})$;
      $item = item + 1$;
   End while
End begin

Fig. 1. Pseudocode algorithm for EAs-based TFAD

signals [12][21]. However, as usual, the time-frequency dictionary is very large, so it is almost impossible for GrA to conduct the full search and represent the signals within a finite time, which seriously limits the practical application of TFAD. By the way, TFAD is a NP-hard problem. To decrease the computational efforts of TFAD, EAs were introduced into TFAD to search the suboptimal time-frequency atom from redundant time-frequency atom dictionaries [21]. The pseudocode algorithm for EAs-based TFAD is shown in Fig. 1. In this paper, an improved membrane algorithm, RQEPS is introduced into TFAD to decrease the computational complexity and improve the search capability, which will be presented in the next section.

3   An Improved Membrane Algorithm

The structure of an improved membrane algorithm, RQEPS is shown in Fig. 2, where the elementary membranes $1, 2, \cdots, m$, embedded in the skin membrane 0, contain multisets of objects and evolution rules. In the computing process, all elementary membranes may be merged into one m_{in} for information communication and the merged membrane m_{in} may be divided into the same number of elementary membranes $1, 2, \cdots, m$. The pseudocode algorithm of RQEPS is presented in Fig. 3 and the detailed description is as follows.

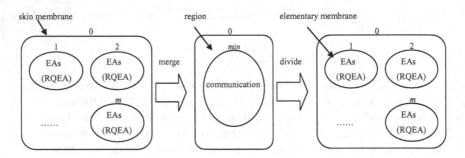

Fig. 2. The structure of RQEPS

(i) The membrane structure $[_0[_1]_1, [_2]_2, \cdots, [_m]_m]_0$ is considered, in which the skin membrane S_0 contains m elementary membranes. The initial multisets:

$$S_0 = \lambda,$$
$$S_1 = p_1 p_2 \cdots p_{n_1}, \; n_1 \leq pop,$$
$$S_2 = p_{n_1+1} p_{n_1+2} \cdots p_{n_2}, \; n_1 + n_2 \leq pop,$$
$$\cdots$$
$$S_m = p_{n_{(m-1)}+1} p_{n_{(m-1)}+2} \cdots p_{n_m}, \; n_1 + n_2 + \cdots + n_m \leq pop,$$

where pop is the dimension of the population, and p_i, $1 \leq i \leq pop$, is a Q-bit individual of length n, which is represented as

$$p_i^t = \begin{bmatrix} \alpha_{i1} | \alpha_{i2} | \cdots | \alpha_{in} \\ \beta_{i1} | \beta_{i2} | \cdots | \beta_{in} \end{bmatrix}, \tag{5}$$

```
Begin
(i)     Initializing the membrane structure; % gen=0;
        While (not termination condition) do
(ii)        Performing RQEA in all elementary membranes;
(iii)       Merging all elementary membranes into one and
                performing communication rules;
(iv)        Dividing the merged membrane;
            gen=gen+1;
        End while
End begin
```

Fig. 3. Pseudocode algorithm for RQEPS

```
a) Set the iterations for each elementary membranes;
   For i=1: m do
        t=0;
b)      Generate R(t) by observing P(t);
c)      Evaluate R(t) and store the best solution among R(t);
   While (not termination condition) do
        t=t+1;
d)          Update P(t) using Q-gates;
e)          Make R(t) by observing the states of P(t);
f)          Evaluate R(t) and store the best solution among R(t);
   End while
End for
```

Fig. 4. Pseudocode algorithm for RQEA

where α_{ij}, β_{ij} are random numbers ranged from 0 to 1, and $|\alpha_{ij}|^2 + |\beta_{ij}|^2 = 1$, $(i = 1, 2, \cdots, pop, j = 1, 2, \cdots, n)$.

(ii) The RQEA is performed in all elementary membranes. The pseudocode algorithm for RQEA is shown in Fig. 4, and the detailed description is as follows.

a) The evolutionary generation t_i for RQEA in the ith elementary membrane is set to a uniformly random integer.

b) The states $R(t)$ in $P(t)$ are observed, where $R(t) = \{a_1^t, a_2^t, \cdots, a_n^t\}$, and a_i^t $(i = 1, 2, \cdots, n)$ is an observed state of an individual p_i^t $(i = 1, 2, \cdots, n)$. a_i^t is a real number of length n, that is $a_i^t = b_1 b_2 \cdots b_n$, where b_j^t $(j = 1, 2, \cdots, n)$ is a real number between 0 and 1. The observed states $R(t)$ are generated in probabilistic way. For instance, as for the probability amplitude $[\alpha, \beta]$ of a Q-bit, a random number r in the range $[0, 1]$ is generated. If $r < 0.5$, the corresponding observed value is set to $|\alpha|^2$, otherwise, the value is set to $|\beta|^2$.

c) Each individual is evaluated to give a measure of its fitness, and the best individual is stored. The fitness is evaluated to adapt the specific problem. In this paper, the fitness function is chosen as $|\langle R^{item}, g_{\gamma_{item}} \rangle g_{\gamma_{item}}|$, shown in Fig. 1.

d) In this step, the Q-bit individuals in $P(t)$ are updated by using quantum-inspired gates (Q-gates). A Q-gate is given by

$$G = \begin{bmatrix} \cos\theta & -\sin\theta \\ \sin\theta & \cos\theta \end{bmatrix}, \tag{6}$$

where θ is the Q-gate rotation angle, and is defined as $\theta = k \cdot f(\alpha, \beta)$, where the value of k is chosen as [23]

$$k = 0.1\pi e^{-t/t_i}, \tag{7}$$

and $f(\alpha, \beta)$ are shown in Table 1.

The steps e) and f) are similar to steps b) and c), respectively.

Table 1. Look-up table of function $f(\alpha, \beta)$[24], where $sign$ is a symbolic function

		$f(\alpha, \beta)$	
$\xi_1 > 0$	$\xi_2 > 0$	$|\xi_1| \geq |\xi_2|$	$|\xi_1| < |\xi_2|$
True	Ture	$+1$	-1
True	False	$sign(\alpha_1, \alpha_2)$	
False	True	$-sign(\alpha_1, \alpha_2)$	
False	False	$sign(\alpha_1, \alpha_2)$	$-sign(\alpha_1, \alpha_2)$
$\xi_1, \xi_2 = 0$ or $\pi/2$		± 1	

(iii) Except for the skin membrane, all elementary membranes are merged into one m_{in}, and consequently the objects of all elementary membranes enter the membrane m_{in}. Subsequently, the communication rules are performed in the membrane m_{in}, that is, a copy of the best element P_{best}, selected in merged membrane, is sent out to the skin membrane. The recombination operation conducted in the merged membrane is used to exchange the information among individuals, which is shown in Fig. 5, where p_i and p_j are any arbitrary two individuals in m_{in} and p'_i and p'_j are the recombined individuals.

$$\begin{cases} p_i \begin{bmatrix} \alpha_{i1} | \alpha_{i2} | ... | \alpha_{ih} | ... | \alpha_{in} \\ \beta_{i1} | \beta_{i2} | ... | \beta_{ih} | ... | \beta_{in} \end{bmatrix} \\ p_j \begin{bmatrix} \alpha_{j1} | \alpha_{j2} | ... | \alpha_{jh} | ... | \alpha_{jn} \\ \beta_{j1} | \beta_{j2} | ... | \beta_{jh} | ... | \beta_{jn} \end{bmatrix} \end{cases} \Rightarrow \begin{cases} p'_i \begin{bmatrix} \alpha_{i1} | \alpha_{i2} | ... | \beta_{jh} | ... | \alpha_{in} \\ \beta_{i1} | \beta_{i2} | ... | \alpha_{jh} | ... | \beta_{in} \end{bmatrix} \\ p'_j \begin{bmatrix} \alpha_{j1} | \alpha_{j2} | ... | \beta_{ih} | ... | \alpha_{jn} \\ \beta_{j1} | \beta_{j2} | ... | \alpha_{ih} | ... | \beta_{jn} \end{bmatrix} \end{cases}$$

Fig. 5. The recombination operation

(iv) The membrane m_{in} is divided into the same structure with the m elementary membranes. In the process of division, the copies of objects $p_1 p_2 \cdots p_{n_1}$ are sent into the membrane S_1; the copies of objects $p_{n_1+1} p_{n_1+2} \cdots p_{n_2}$ are sent into the membrane S_2 and the rest may be deduced by analogy. Finally, the copy of P_{best} is sent from the skin membrane to each compartment to determine the Q-gate rotation angle at the next generation.

RQEPS is an improved algorithm of the QEPS [22]. The differences between these two approaches are as follows.

(a) They use different observation rules: binary-observation rules in QEPS [22] vs. real-observation rules in RQEPS. In RQEPS, a quantum-inspired state, corresponding to an optimization variable, observed by a real-observation rule is a real-valued number. But an optimization variable in QEPS needs several quantum-inspired states, which correspond with a string of binary bits in the binary-observation process. Without encoding and decoding processes, the real-observation rule is more suitable for solving numerical optimization problems.

(b) Preliminary use of membrane fusion and division is considered in RQEPS.

(c) Recombination operations are employed in merged membrane to exchange the information among individuals.

4   Experimental Results

In this section, how to choose the number m of elementary membranes will be first discussed by using a linear frequency-modulated radar emitter signal with

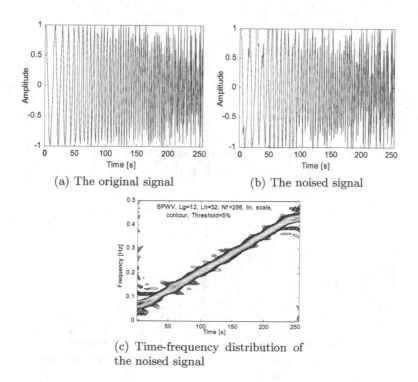

(a) The original signal   (b) The noised signal

(c) Time-frequency distribution of
the noised signal

Fig. 6. A radar emitter signal

10 dB signal-to-noise rate (SNR), shown in Fig. 6. And then the comparative experiments are carried out on the signal to demonstrate the effectiveness and applicability of the introduced method.

4.1 Parameter Setting

In this subsection, experiments on the noised signal are carried out to investigate the effects of the number m of elementary membranes on the performance of RQEPS for TFAD. Experimental environment is chosen as: the maximal number of iterations $item$ is set to 30 as the termination condition of TFAD. The time-frequency atom uses Gabor function

$$g_\gamma(t) = \frac{1}{\sqrt{s}} g(\frac{t-u}{s}) \cos(vt + w),\tag{8}$$

where the index $\gamma = (s, u, v, w)$ is a set of parameters and s, u, v, w are scale, translation, frequency and phase, respectively. They are discretized as follows: $\gamma = (a^j, pa^j \Delta u, ka^{-j} \Delta \xi, i \Delta w)$, $a = 2, \Delta u = 1/2$, $\Delta \xi = \pi$, $\Delta w = \pi/6$, $0 < j < \log_2 N$, $0 \leq p \leq N2^{-j+1}$, $0 \leq k < 2^{j+1}$, $0 \leq i \leq 12$, where N is the length of the signal f [12].

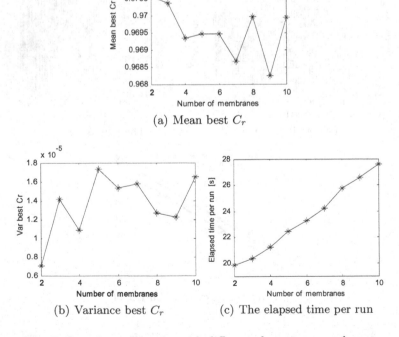

(a) Mean best C_r

(b) Variance best C_r (c) The elapsed time per run

Fig. 7. Experimental results with different elementary membranes

In RQEPS, the population size *pop* is set to 10. The parameter m varies from 2 to 10. According to the investigation of the effect of the parameter t_i $(i = 1, 2, \cdots, m)$ on the QEPS performances in [22], the RQEA's iteration t_i is set to a uniformly random integer ranged from 1 to 9. The number n of a Q-bit individual and the maximal evolutionary generation *gen* are set to 4 and 40, respectively. These experiments are carried out on the computer with 1.5 GHz CPU, 768 MB EMS memory and 80GB hard disk using the software MATLAB 7.1. The experimental results over 30 runs as the number of elementary membranes are shown in Fig. 7, which illustrates that the elapsed time, the mean best and the variance best of the correlation ratio C_r between the original signal f and the restored signal f_{res}. The correlation ratio C_r of f and f_{res} is defined as [25]

$$C_r = \frac{\langle f, f_{res} \rangle}{\|f\| \cdot \|f_{res}\|}, \tag{9}$$

The experimental results in Fig. 7(a) and 7(b) show that the mean and the variance of the best correlation ratio C_r show a broad range of variability with respect to the number of different elementary membranes, but the best results are obtained in two cases including 2 elementary membranes. As shown in Fig.7(c), the elapsed time has a steady increase with the number of the elementary membranes. Thus, to obtain the balance between the elapsed time and the correlation ratio, the number of elementary membranes could be assigned as 2.

4.2   Comparative Experiments

To verify the validity of RQEPS, the noised signal above is used to conduct the experiments with the same computer, in which bQEA [6], GrA [12] and QEPS [22][11] are brought into comparisons with RQEPS.

In bQEA, population size *pop*, the number n of binary bits and the maximal evolutionary generation g are set to 10, 40 and 200, respectively. In QEPS, according to [11], the number m of elementary membranes is set to 9; the number

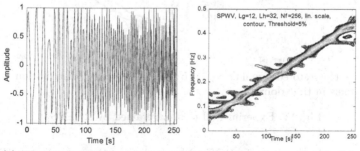

(a) The restored signal using 30 atoms in time-domain

(b) Time-frequency distribution of 30 atoms

Fig. 8. Experimental results obtained by bQEA

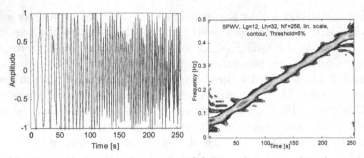

(a) The restored signal using 30 (b) Time-frequency distribution
atoms in time-domain   of 30 atoms

Fig. 9. Experimental results obtained by GrA

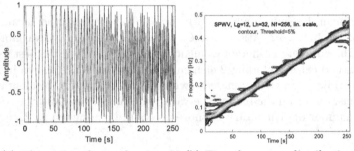

(a) The restored signal using 30 (b) Time-frequency distribution
atoms in time-domain   of 30 atoms

Fig. 10. Experimental results obtained by QEPS

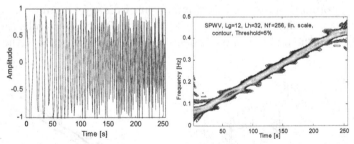

(a) The restored signal using 30 (b) Time-frequency distribution
atoms in time-domain   of 30 atoms

Fig. 11. Experimental results obtained by RQEPS

n of binary bits is set to 40. In RQEPS, according to the experiments discussed
in the above subsection, the number m of elementary membranes is set to 2;
the number n of a Q-bit individual is set to 4. In both RQEPS and QEPS, the
parameter t_i ($i = 1, 2, \cdots, m$) is set to a uniformly random integer ranged from

Table 2. Parameters of 30 atoms of a noised LFM radar emitter signal

	1	2	3	4	5	6	7	8	9	10
s	19.63	22.71	26.83	43.94	29.27	28.97	28.41	33.70	33.25	12.32
u	99.43	136.29	209.41	55.67	177.86	22.28	237.06	76.69	157.50	6.46
v	1.31	1.63	3.89	0.85	2.05	0.57	2.68	1.00	1.85	5.67
w	3.71	3.51	4.64	5.08	4.15	3.06	4.37	2.17	2.45	1.56
	11	12	13	14	15	16	17	18	19	20
s	33.87	34.51	90.38	1.73	31.82	10.29	10.29	31.73	22.65	12.32
u	120.19	193.40	45.05	0.19	49.52	80.16	224.93	251.23	100.67	196.40
v	4.85	2.20	5.36	4.40	0.68	1.14	3.77	3.42	1.50	3.95
w	3.91	2.87	4.09	3.99	0.75	3.46	3.21	0.05	3.63	4.24
	21	22	23	24	25	26	27	28	29	30
s	13.42	25.77	8.12	15.26	25.39	12.92	9.10	17.11	6.33	28.53
u	181.09	37.99	5.28	100.34	122.04	45.84	81.22	152.62	132.19	243.11
v	2.07	0.59	0.02	4.23	1.84	2.41	5.78	1.38	5.85	0.52
w	1.61	3.21	0.58	1.45	2.27	2.91	1.87	4.08	2.72	3.92

Table 3. Performance comparisons of bQEA, GrA, QEPS and RQEPS

	Correlation ratio C_r		Computing time per
	Mean	Var	run (Second)
bQEA	0.9505	7.2387e-5	43.25
GrA	0.9668	1.1476e-31	723.39
QEPS	0.9670	1.2400e-5	44.59
RQEPS	0.9706	7.0583e-6	19.87

Table 4. Results of parametric statistical test t-test

Control Algorithm	bQEA	GrA	QEPS
RQEPS	8.0113e-18	1.1684e-10	4.7336e-05

1 to 9; the population size *pop* and the maximal evolutionary generation *gen* are set to 10 and 40, respectively. In all algorithms, the maximal number of iterations item is set to 30 as the termination condition of TFAD. Experimental results are shown in Fig. 8 to Fig. 11, Table 2, Table 3 and Table 4.

Table 2 lists the parameters of the 30 Gabor atoms. Fig. 8 to Fig. 11 show the restored signals using the 30 decomposed time-frequency atoms and their time-frequency distributions of the 30 time-frequency atoms which are obtained by bQEA, GrA, QEPS and RQEPS, respectively. As shown in Fig.6 and Fig. 8 to Fig.11, it can be seen that the time-frequency distribution obtained by RQEPS is nearly identical with that of the original radar emitter signals, and the correlation ratio is the highest which reaches 0.9801, while the correlation ratio obtained by GrA is only 0.9668, which illustrates that RQEPS is more suitable for decomposing a signal into time-frequency atoms than bQEA, GrA and QEPS, in terms of search capability.

The experimental results over 30 runs are shown in Table 3 and Table 4. From Table 3, it can be seen that RQEPS gains the mean of the best correlation ratio C_r 0.9706, which is better than 0.9670, 0.9668 and 0.9505 obtained by QEPS, GrA and bQEA, respectively. Moreover, the computing time of RQEPS is 36.4061, 2.2441, and 2.1766 times as small as that of GrA, QEPS and bQEA. If the experiments are conducted in a parallel-distributed way on several machines, the computing time could be greatly reduced.

In table 4, a parametric statistical analysis t-test is applied to analyse whether there is a significant difference over one optimization problem between two algorithms [4]. We employ a 95% confidence Student t-test. The t-test results in Table 4 are far smaller than the level of significance 0.05, which implies that RQEPS really outperforms the QEPS, GrA and bQEA by introducing the active membranes with mergence and division operations, real-observation and recombination operations.

5 Conclusions

This paper proposes an improved membrane algorithm (RQEPS), by combining the framework and evolution rules of P systems with RQEA. RQEPS is characterized by active membranes with membrane fusion and division to strengthen the information communication among individuals and decrease the computational complexity, respectively, the evolutionary rules in RQEA and transformation/communication like-rules in P systems to evolve the system. The TFAD of noised radar emitter signals is considered as an application example to test the effectiveness and practicality of the introduced method. Experimental results show that RQEPS performs better than QEPS, GrA and bQEA, in terms of search capability and convergent speed.

The possible interplay between evolutionary algorithms and membrane computing represents a challenging and promising research topic. This paper introduces RQEA into P systems to solve time-frequency atom decomposition. However, how to select evolutionary algorithms within elementary membranes and communication rules in the merged membrane to solve different complex problems, in order to obtain more efficient methods, is an ongoing and challenging issue.

Acknowledgements

The authors would like to thank the anonymous referees for their valuable comments. The research of GZ is supported by the National Natural Science Foundation of China (60702026), the Scientific and Technological Funds for Young Scientists of Sichuan (09ZQ026-040) and the Open Foundation of Engineering

Research Centre of Safety Transportation of the Ministry of Education of China. The research of MG and FI is supported by CNCSIS grant no.643/2009, *An integrated evolutionary approach to formal modelling and testing*.

References

1. Davis, G., Mallat, S., Avellaneda, M.: Adaptive greedy approximation. Journal of Constructive Approximation 13(1), 57–98 (1997)
2. Ferreira da Silva, A.R.: Evolutionary-based methods for adaptive signal representation. Signal Processing 81, 927–944 (2001)
3. Figueras i Ventura, R.M., Vandergheynst, P.: Matching pursuit through genetic algorithms. LTS-EPFL Tech. Rep. (2001)
4. Garcia, S., Molina, D., Lozano, M., Herrera, F.: A study on the use of non-parametric tests for analyzing the evolutionary algorithms' behaviour: a case study on the CEC 2005 Special Session on Real Parameter Optimization. Journal of Heuristics (2005), doi:10.1007/s10732-008-9080
5. Gribonval, R., Bacry, E.: Harmonic decomposition of audio signals with matching pursuit. IEEE Transactions on Signal Processing 51, 101–111 (2003)
6. Han, K.H., Kim, J.H.: Quantum-inspired evolutionary algorithm for a class of combinatorial optimization. IEEE Transactions on Evolutionary Computation 6, 580–593 (2002)
7. Huang, L., He, X.X., Wang, N., Xie, Y.: P Systems based multi-objective optimization algorithm. Progress in Natural Science 17, 458–465 (2007)
8. Huang, L., Wang, N.: An optimization algorithm inspired by membrane computing. In: Jiao, L., Wang, L., Gao, X.-b., Liu, J., Wu, F. (eds.) ICNC 2006. LNCS, vol. 4222, pp. 49–52. Springer, Heidelberg (2006)
9. Huang, L., Wang, N., Zhao, J.H.: Multiobjective optimization for controllers. Acta Automatica Sinica 34, 472–477 (2008)
10. Leporati, A., Pagani, D.: A membrane algorithm for the min storage problem. In: Hoogeboom, H.J., Păun, G., Rozenberg, G., Salomaa, A. (eds.) WMC 2006. LNCS, vol. 4361, pp. 443–462. Springer, Heidelberg (2006)
11. Liu, C.X., Zhang, G.X., Zhu, Y.H., Fang, C., Liu, H.W.: A quantum-inspired evolutionary algorithm based on P systems for radar emitter signals. In: Proc. Fourth International Conference on Bio-Inspired Computing: Theories and Applications, pp. 24–28 (2009)
12. Mallat, S.G., Zhang, Z.F.: Matching pursuits with time-frequency dictionaries. IEEE Transactions on Signal Processing 41, 3397–3415 (1993)
13. Nishida, T.Y.: An approximate algorithm for NP-complete optimization problems exploiting P systems. In: Proc. Brainstorming Workshop on Uncertainty in Membrane Computing, pp. 185–192 (2004)
14. Nishida, T.Y.: Membrane algorithms. In: Freund, R., Păun, G., Rozenberg, G., Salomaa, A. (eds.) WMC 2005. LNCS, vol. 3850, pp. 55–66. Springer, Heidelberg (2006)
15. Păun, Gh.: Computing with membranes. Journal of Computer and System Sciences 61, 108–143 (2000)
16. Păun, Gh., Rozenberg, G.: A guide to membrane computing. Theoretical Computer Science 287, 73–100 (2002)
17. Qian, S., Chen, D.: Signal representation using adaptive normalized Gaussian functions. Signal Processing 36, 1–11 (1994)

18. Stefanoiu, D., Ionescu, F.L.: A genetic matching pursuit algorithm. In: Proc. 7th International Symposium on Signal Processing and Its Applications, pp. 577–580 (2003)

19. Vesin, J.: Efficient implementation of matching pursuit using a genetic algorithm in the continuous space. In: Proc. 10th European Signal Processing Conference, pp. 2–5 (2000)

20. Zaharie, D., Ciobanu, G.: Distributed evolutionary algorithms inspired by membranes in solving continuous optimization problems. In: Hoogeboom, H.J., Păun, G., Rozenberg, G., Salomaa, A. (eds.) WMC 2006. LNCS, vol. 4361, pp. 536–553. Springer, Heidelberg (2006)

21. Zhang, G.X.: Time-frequency atom decomposition with quantum-inspired evolutionary algorithm. In: Circuits, Systems and Signal Processing (accepted, 2009)

22. Zhang, G.X., Gheorghe, M., Wu, C.Z.: A quantum-inspired evolutionary algorithm based on P systems for a class of combinatorial optimization. Fundamenta Informaticae 87, 93–116 (2008)

23. Zhang, G.X., Li, N., Jin, W.D.: Novel quantum genetic algorithm and its applications. Frontiers of Electrical and Electronic Engineering in China 1(1), 31–36 (2006)

24. Zhang, G.X., Rong, H.N.: Real-observation quantum-inspired evolutionary algorithm for a class of numerical optimization problems. In: Shi, Y., van Albada, G.D., Dongarra, J., Sloot, P.M.A. (eds.) ICCS 2007. LNCS, vol. 4490, pp. 989–996. Springer, Heidelberg (2007)

25. Zhang, G.X., Rong, H.N., Jin, W.D., Hu, L.Z.: Radar emitter signal recognition based on resemblance coefficient features. In: Tsumoto, S., Słowiński, R., Komorowski, J., Grzymała-Busse, J.W. (eds.) RSCTC 2004. LNCS (LNAI), vol. 3066, pp. 665–670. Springer, Heidelberg (2004)

A Region-Oriented Hardware Implementation for Membrane Computing Applications

Van Nguyen, David Kearney, and Gianpaolo Gioiosa

School of Computer and Information Science
University of South Australia
{Van.Nguyen,David.Kearney,Gianpaolo.Gioiosa}@unisa.edu.au

Abstract. We have recently developed a prototype hardware implementation of membrane computing based on reconfigurable computing technology called Reconfig-P. The existing hardware design treats reaction rules as the primary computational entities and represents regions only implicitly. In this paper, we describe and evaluate an alternative hardware design that more directly reflects the intuitive conceptual understanding of a P system and therefore promotes the extensibility of Reconfig-P. A key feature of the design is the fact that regions, rather than reaction rules, are the primary computational entities. More specifically, in the design, regions are represented as loosely coupled processing units which communicate objects by message passing. Experimental results show that for many P systems the region-oriented and rule-oriented designs exhibit similar performance and hardware resource consumption.

1 Introduction

We have recently developed a prototype hardware implementation of membrane computing based on reconfigurable computing technology called Reconfig-P. The existing hardware design treats reaction rules as the primary computational entities and represents regions only implicitly. Consequently there is not always a direct mapping between the components of the intuitive conceptual understanding of a P system and the hardware components. Such indirectness is a byproduct of our attempt to simplify the hardware circuit and thereby promote the performance and efficiency of Reconfig-P. Nevertheless, a more faithful rendering of the intuitive conceptual understanding of a P system in hardware would have benefits for the extensibility of Reconfig-P. In particular, it would facilitate the process of augmenting Reconfig-P to support additional types of P systems. In this paper, we describe and evaluate an alternative hardware design that more directly reflects the intuitive conceptual understanding of a P system and therefore promotes the extensibility of Reconfig-P. A key feature of the design is the fact that regions, rather than reaction rules, are the primary computational entities. More specifically, in the design, regions are represented as loosely coupled processing units which communicate objects by message passing.

The contents of the paper are as follows. In Section 2, we discuss the background to the research described in the paper. In Section 3, we describe the

G. Păun et al. (Eds.): WMC 2009, LNCS 5957, pp. 385–409, 2010.

region-oriented hardware design. In Section 4, we explain some aspects of our implementation of regions in hardware. In Section 5, we present the results of an empirical analysis of the hardware resource consumption and performance of hardware circuits using the region-oriented design. Finally, in Section 6, we draw some conclusions regarding the significance of our contributions.

2   Background

2.1   The Intuitive Conceptual Understanding of a P System

Although in one sense a P system is a pure mathematical construct, in another sense a P system is seen as having non-mathematical properties. For example, in an informal discussion of P systems one might speak of membranes 'dissolving', of regions being 'inside' other regions, or of objects being 'consumed' by reaction rules. The very frequent use of such physicalistic metaphors in describing the operation of a P system is, of course, a result of the fact that P systems have since their introduction been modelled after biological cells. The biological interpretation of a P system, far from being dispensable, provides one with a means of intuitively grasping the computational characteristics of P systems.

According to what we call the *intuitive conceptual understanding of a P system*, a P system comprises a hierarchy of membranes, each of which defines a region that contains a collection of objects and is associated with a set of reaction rules. The P system evolves in a series of stages. At each stage, the reaction rules in every region are applied. The application of the reaction rules in a region results in the occurrence of an object transformation process within the region. The object transformation processes in the different regions occur independently. Sometimes an object transformation process results in the movement of objects between regions. Therefore, although the processes in the different regions occur independently, they may influence each other indirectly by influencing their respective inputs for the next stage of the evolution of the P system.

Given the intuitive conceptual understanding of a P system, in the context of implementing P systems on a computing platform, it is natural to regard a P system as a collection of distributed processing units (the object transformation processes occurring in the different regions) that interact only by means of message passing (the transfer of objects).

2.2   Current Status of Reconfig-P

Reconfig-P [6] [7] is an implementation of membrane computing based on reconfigurable hardware (specifically, a field-programmable gate array[1]) that is able

---

[1] A standard field-programmable gate array (FPGA) consists of a matrix of configurable logic blocks (CLBs). The CLBs, which are connected by means of a network of wires, can be used to implement logic or memory. The functionality of the logic blocks and the connections between them can be modified by loading configuration data from a host computer. In this way, any custom digital circuit can be mapped onto the FPGA, thereby enabling it to execute a variety of applications.

to execute P systems at high performance. It exploits the reconfigurability of the hardware by constructing and synthesising a customised hardware circuit for the specific P system to be executed. The hardware circuit is constructed using the hardware specification language Handel-C [2].

To maximise performance and minimise hardware resource consumption, the current version of Reconfig-P takes a minimalistic approach to the implementation of the features of a P system in hardware. According to this approach, only those features of the intuitive conceptual understanding of a P system absolutely necessary to the computational operation of a P system are implemented explicitly as processing units or data structures. As a consequence, some features that are of primary importance in the conceptual understanding of a P system are not explicitly represented as components of the hardware circuits generated by the current version of Reconfig-P. Most significantly, membranes and the regions defined by membranes are not explicitly represented. Instead, the existing implementation represents these features implicitly as logical constructions arising from the connections that exist between processing units corresponding to the reaction rules and arrays corresponding to the multisets of objects available in the regions of the P system. In other words, the conceptual model of a P system underlying the design of the current version of Reconfig-P includes only reaction rules and multisets of objects as primary features; membranes and regions are not directly represented in the model, but must be inferred on the basis of the connections that exist between the reaction rules and multisets of objects.

2.3 Motivation for the Alternative Hardware Design

Although it promotes performance and efficiency, the hardware design used in the existing version of Reconfig-P has some disadvantages. These disadvantages diminish the elegance, understandability (and therefore maintainability) and extensibility of Reconfig-P. First, by deviating from the intuitive conceptual understanding of a P system, the design is not as elegant and understandable as it could be. Second, the design does not facilitate the implementation of P systems that represent membranes as active entities or include membrane-mediated rules (such as antiport rules). Third, driven by the goal of high performance, this design puts an emphasis on P systems as models of parallel computation at the expense of not fully representing P systems as models of distributed computation. These three disadvantages have motivated us to develop an alternative hardware design.

The alternative hardware design proposed in this paper, which we call the *region-oriented design*, is intended to

- promote the elegance and understandability of Reconfig-P by more closely reflecting the intuitive conceptual understanding of a P system,
- promote the extensibility of Reconfig-P by providing a framework within which the future implementation of additional types of P systems — especially P systems that include cell-to-cell connections (e.g., tissue-like P

systems [5] and spiking neural P systems [4]), represent membranes as active entities, or include membrane-mediated rules (e.g., [1], [8], [3], [9] and [10]) — can more easily be achieved, and
- facilitate an elegant region-oriented approach to the distribution and composition of the computational activities occurring in a P system.

A region-oriented approach to the distribution of the computational activities occurring in a P system is desirable because, not only does it match the intuitive conceptual understanding of how these activities are distributed in a P system, it also allows a very natural means of scaling the amount of available hardware resources to suit the size of the P system to be executed. For example, one can envision implementing a P system using multiple hardware circuits, where each hardware circuit implements the processing associated with a particular region (or subhierarchy of regions) of the P system. Indeed, the techniques developed in implementing a region-oriented approach could be adapted to allow the composition of whole P systems into larger systems. That is, these techniques could be adapted to allow hardware circuits implementing distinct P systems or parts of P systems to communicate and therefore form a larger system. For example, one could combine existing hardware circuits implementing different parts of a P system by implementing interfaces between the relevant Handel-C programs. These interfaces would effectively represent the membranes separating the different parts of the P system. This approach would enable the implementation of very large P systems.

3   The Region-Oriented Hardware Design

In this section, we provide an overview of the region-oriented hardware design. For the sake of simplicity, in this overview we do not treat aspects of the design related to nondeterministic object distribution.

3.1   Basic Characteristics of the Design

In the region-oriented hardware design, instead of being represented only implicitly, regions are implemented explicitly as hardware components. More specifically, the design has the following three key attributes:

1. Regions are implemented as core processing units.
2. Region processing units operate independently. That is, each region processing unit coordinates all the activities occurring in one particular region of the P system and is not aware of activities occurring in other regions.
3. The movement of objects between regions is implemented as message passing between region processing units.

A key aspect of the region-oriented implementation is the use of the chan (channel) construct of Handel-C to accomplish inter-region communication. The chan construct supports the implementation of synchronous communication between parallel processing units. The example Handel-C code in Figure 1 shows a channel C being used to transfer the value 8 to the register Reg.

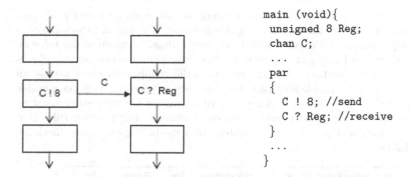

```
main (void){
    unsigned 8 Reg;
    chan C;
    ...
    par
    {
        C ! 8; //send
        C ? Reg; //receive
    }
    ...
}
```

Fig. 1. Example of a Handel-C **chan** (channel) construct being used to implement communication between two parallel branches.

3.2   Region Processing Units

Similar to the rule processing units in the rule-oriented design, the region processing units in the region-oriented design complete the execution of a transition in two phases: an *object assignment phase* and an *object production phase*[2]. In the object assignment phase, a region processing unit determines the maximum number of instances, and hence the applicability status, for each reaction rule in the region in the current transition. In the object production phase, a region processing unit carries out the consumption, production and communication of objects for the reaction rules in the region based on their maximum number of instances. In the case of P systems that contain reaction rules with relative priorities, the region processing unit must calculate the maximum numbers of instances for those reaction rules with higher priorities before doing so for those reaction rules with lower priorities. To save clock cycles, the region processing unit carries out the object consumption for the reaction rules with higher priorities in the object assignment phase rather than in the object production phase.

Object assignment phase. An important aspect of the hardware design for the object assignment phase is the way in which the region processing unit respects the relative priorities of the reaction rules (if indeed such priorities are defined) while minimising the number of clock cycles required to complete the phase by avoiding the processing of inapplicable reaction rules.

It is an assumption of the design that reaction rules in a region that consume common object types are assigned relative priorities (using the relation >, which is to be interpreted as 'has higher priority than'). Given this assumption, the set of reaction rules in a region may be partitioned into (a) a collection of singleton sets, where for each reaction rule not related by priority to any other reaction rule, there is exactly one singleton set containing that reaction rule in the collection, and there are no other singleton sets in the collection, and (b) a collection

---

[2] The object assignment phase and object production phase roughly correspond to the preparation phase and updating phase in the rule-oriented design, respectively (see [6] or [7] for details).

Table 1. An illustration of how a region processing unit determines the order in which to process reaction rules in the object assignment phase. The region processing unit begins with a preliminary order determined at compile-time, as shown in the table on the left. At the start of the object assignment phase, the region processing unit checks the applicability of the reaction rules. The results of the applicability check are shown in the table on the right (applicable reaction rules are labelled 'a', and inapplicable reaction rules 'na'). The region processing unit then updates the processing order by removing the inapplicable reaction rules from consideration. The reaction rules that the region processing unit processes immediately after the first applicability check are shown in boldface.

Execution order	Reaction rules		
1	R_{11}	R_{21}	R_{31}
2	R_{12}	R_{22}	R_{32}
3	R_{13}	R_{23}	R_{33}
4	R_{14}	R_{24}	R_{34}
5	R_{15}	R_{25}	

Execution order	Reaction rules		
1	R_{11}:a	R_{21}:na	R_{31}:na
2	R_{12}:na	R_{22}:na	**R_{32}:a**
3	R_{13}:a	R_{23}:na	R_{33}:na
4	R_{14}:na	R_{24}:na	R_{34}:a
5	R_{15}:a	**R_{25}:a**	

of totally $>$-ordered sets, where each reaction rule related by priority to another reaction rule is in exactly one totally $>$-ordered set, and if two reaction rules have relative priorities then they belong to the same totally $>$-ordered set. In the example illustrated in Table 1, the columns correspond to totally $>$-ordered sets of reaction rules. The totally $>$-ordered sets are: $T_1 = \{R_{11}, R_{12}, R_{13}, R_{14}, R_{15}\}$, $T_2 = \{R_{21}, R_{22}, R_{23}, R_{24}, R_{25}\}$ and $T_3 = \{R_{31}, R_{32}, R_{33}, R_{34}\}$. From the sets in the partition formed in this way, one or more partially time-ordered sets of reaction rules can be constructed that may be interpreted as indicating the possible temporal orders in which the region processing unit can process the reaction rules in the object assignment phase. The constraints on the possible temporal orders are that (a) reaction rules with the same priority should be processed at the same time, and (b) reaction rules with relative priorities should be processed one after the other according to their priorities. To maximise the performance of the implementation, we use a technique that avoids the processing of inapplicable reaction rules. Naturally, such a technique must be applied at run-time. The technique involves checking the applicability status of reaction rules both at the beginning of the phase and whenever any objects have been assigned to a reaction rule, and using this applicability information to determine the temporal order in which the currently applicable reaction rules should be processed in order to minimise the total number of clock cycles used in the remainder of the phase. For the example shown in Table 1, after checking the applicability of the reaction rules at the beginning of the object assignment phase, the region processing unit determines that only reaction rules R_{11}, R_{13}, R_{15}, R_{25}, R_{32} and R_{34} are applicable. Based on this information and the totally $>$-ordered sets T_1, T_2 and T_3, it determines that the currently most time-efficient way of processing the reaction rules is to first process R_{11}, R_{25} and R_{32} in parallel, then process R_{13} and R_{34} in parallel, and finally process R_{15}. It then proceeds to process R_{11}, R_{25} and R_{32} in parallel. After doing this, it again checks the applicability of the reaction rules, and based on the applicability information obtained re-evaluates the temporal order in which the currently applicable reaction rules should be processed. The region processing unit continues in this way until no reaction rules are applicable.

It is desirable to implement the dynamic determination of the partially time-ordered set of executable reaction rules in as few clock cycles as possible. In our current implementation, the number of clock cycles required to perform this task is minimal (zero if no logic depth reduction is applied). See Section 4 for details about the implementation.

Object production phase. In the object production phase, a region processing unit (a) updates the multiplicities of the object types in its region and attempts to send objects to and receive objects from the other regions, and then (b) updates the multiplicities of the object types in its region based on the objects it has received from other regions. All of the updating and communication tasks are accomplished in a massively parallel manner.

To resolve resource conflicts that may occur in the object production phase (i.e., situations in which the multiplicity of an object type is to be updated by more than one parallel process), the region-oriented design includes two resource conflict resolution strategies: the *space-oriented strategy* and the *time-oriented strategy*. These strategies are similar to those adopted in the rule-oriented hardware design (see [6] or [7]). In the space-oriented strategy, copy registers are created for those object types whose multiplicities are to be updated by more than one parallel process, and the relevant parallel processing units store the updated multiplicity values in their assigned copy registers. The time-oriented strategy involves interleaving the operations of distinct parallel processes so that update operations which would conflict if executed in the same clock cycle are executed in different clock cycles.

The space-oriented strategy is implemented in basically the same way in both the rule-oriented and region-oriented hardware designs, with a couple of differences. The first difference is that, whereas in the rule-oriented design a special *multiset replication coordinator* processing unit needs to be introduced to coordinate the values stored in the copy registers, in the region-oriented design this coordination task can be performed by the already introduced region processing unit. The second difference is that in the region-oriented design copy registers do not need to be introduced for processing units sending objects to the region from other regions, because the already introduced register dedicated to the storage of the data received over the relevant communication channel can be used as a copy register. As in the rule-oriented design, in the region-oriented design, when the space-oriented strategy is used, the object production phase takes two clock cycles to complete. In the first clock cycle, register updates implementing the production of objects by reaction rules local to the region are performed. In the second clock cycle, the values stored in the various registers representing multiplicity values of object types (including registers associated with channels) are coordinated, and the new multiplicity values for the object types in the region are stored in the original registers that store such values.

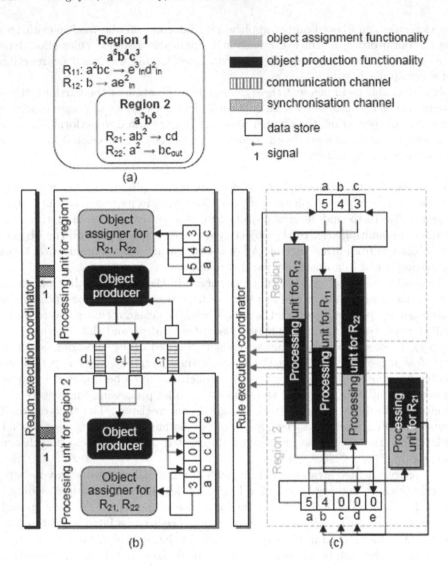

Fig. 2. An illustration of the region-oriented hardware design in comparison to the rule-oriented hardware design for a sample P system. (a) A sample P system. (b) The region-oriented design for the sample P system in which regions are implemented as processing units that communicate via channels. (c) The rule-oriented design for the sample P system in which reaction rules are implemented as processing units.

The time-oriented strategy is implemented differently in the two designs. In the rule-oriented design, the way in which updates are interleaved over time can be completely determined at compile-time, and so can be hard-coded into the source code defining the hardware circuit (see [6] or [7]). In the region-oriented design, a distinction is drawn between *internal objects* and *external objects* for a region in a particular transition. The internal objects of a region in a transition are those objects produced in the transition by one of the reaction rules associated with the region. Objects sent to the region during the transition from other regions are external to the region. While it is possible to determine at compile-time the appropriate interleaving for updating operations occasioned solely by the production of internal objects, the interleaving for updating operations occasioned wholly or partly by the receipt of external objects must be determined at run-time. This is because, to preserve the independence of region processing units, information about when it might receive external objects is unavailable to the relevant region processing unit. To accomplish the run-time determination of the interleaving, an approach based on the use of semaphores is used.

3.3 Synchronisation

It is necessary to synchronise the execution of the object assignment phases of distinct region processing units. Without such synchronisation, it would be possible for objects produced in the object production phase for one region to be sent to another region still in its object assignment phase, thereby improperly interfering with the results of the object assignment phase in that region.

Unlike in the rule-oriented design, where synchronisation of the object assignment phases of reaction rules across regions is implemented explicitly using signals and flags, the synchronisation of the execution of the object assignment phases of distinct region processing units in the region-oriented design is implicitly achieved by having region processing units communicate over channels at the beginning of the object production phase.

Channels are also used to perform explicit synchronisation at the end of each transition. The region-oriented design includes a *region execution coordinator* which is responsible for coordinating the execution of the region processing units so that the transition-by-transition evolution of the P system can be realised. The region execution coordinator is connected to each of the region processing units via dedicated synchronisation channels. Once a region processing unit has completed all of its tasks for a particular transition, it sends a signal down its synchronisation channel. Once the region execution coordinator has received a signal from every region processing unit, it triggers a new transition.

One potential problem associated with the use of channels to implement the movement of objects between regions is the occurrence of deadlock. Handel-C channels operate in a synchronous manner. That is, once a pair of processing units have started engaging in a communication, neither the sending nor the receiving processing unit can move on to perform other tasks until the communication has been accomplished. Consequently, unless the operations of sending and receiving objects among region processing units are conducted in an appropriate order, deadlock can occur. To prevent deadlock from occurring we ensure that the channel communications for different regions are carried out in distinct parallel branches of execution.

3.4 Extensibility of the Design

The region-oriented hardware design makes it possible to implement P systems with features that require the explicit presence of membranes in an intuitive way. In particular, each membrane can be implemented as a processing unit associated with two region processing units (corresponding to the inner and outer regions of the membrane) (see Figure 3). Such a membrane processing unit could, for example, mediate the exchange of objects between regions effected by antiport rules. For an antiport rule to be applicable, enough objects of the right types need to be available in both regions. As each of the two region processing units for the two regions do not know the multiset of objects available in the other region, it is not possible for the region processing units to implement the antiport rule on their own — a membrane processing unit is also required. Nevertheless, it is still possible for the membrane processing unit to remain quite independent from the two region processing units. One way in which the region-oriented hardware design could be augmented to implement antiport rules is as follows. The region processing units for the inner and outer regions send objects to a membrane processing unit. The membrane processing unit attempts to couple objects in the way specified by the antiport rule, and sends coupled objects to their destination regions and returns uncoupled objects to their regions of origin. In this way, not only do the region processing units not know about each other's multiset of objects, but the membrane processing unit does not need to know this information either. It is sufficient that both region processing units know about the existence of the membrane processing unit.

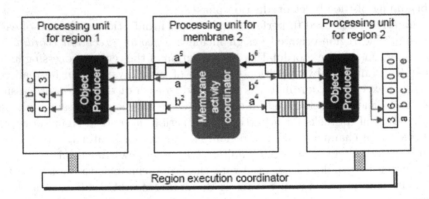

Fig. 3. An illustration of the implementation of a P system with an antiport rule (aa, in; b, out) using the region-oriented design. In this example, two instances of the antiport rule are executed.

4   Implementing Regions in Hardware

In this section, we describe how the regions of a P system are implemented using Handel-C when the region-oriented hardware design is adopted.

4.1 Atomic Operations Associated with the Application of Rules

From one perspective, the overall behaviour of a P system emerges from the application of reaction rules. At the implementation level, the execution of a single application of a reaction rule involves the execution of a certain number of instances of each of a set of logically atomic operations:

$$\text{Rule execution} = (p\text{DIV}, q\text{MIN}, r\text{MUL}, s\text{SUB}, t\text{COM}, u\text{ADD}), \text{ where}$$

$p = 0$ or 1, q, r, s, t, $u \geq 0$, DIV denotes the operation of dividing the multiplicity of the objects of a given type available in the region by the number of objects of that type required for the application of one instance of the reaction rule, MIN denotes the operation of computing the maximum number of instances of the reaction rule that can be applied in the current transition, MUL denotes the operation of computing the number of objects of a particular object type to be consumed/produced by the reaction rule in the current transition, SUB denotes the operation of reducing the multiplicity of a particular object type available in the region (by a certain amount), COM denotes the operation of sending (or attempting to send) a certain number of objects of a particular type to a particular region, and ADD denotes the operation of increasing the multiplicity of a particular object type available in the region (by a certain amount).

In Reconfig-P, each of the above operations is realised as an atomic operation. These atomic operations are the building blocks for the construction of any particular hardware circuit. The names of the operations reflect the main computational operations involved in their implementation. Mapping the atomic operations onto a hardware circuit requires making decisions about their temporal granularity. At fine granularity, an operation is performed over multiple clock cycles and therefore needs to be decomposed into suboperations. At coarse granularity, multiple operations are combined and performed in one clock cycle. Although assigning a logically atomic operation a fine granularity at implementation results in a greater number of clock cycles, it often reduces logic depth, and therefore can lead to an increased system clock rate.

To determine the appropriate degree of granularity for a given logically atomic operation, it is necessary to examine the implementation characteristics of the operation in terms of hardware resource consumption and logic depth. Multiplication and division can generate complicated combinatorial circuits and therefore in general are expensive to implement in one clock cycle. However, in the specific case of the execution of a P system, in both multiplication and division operations one of the operands is a constant. This significantly reduces the logic depth of the combinatorial circuits that implement the operations[3]. Addition and subtraction are relatively inexpensive operations and, based on the

---

[3] The Xilinx Virtex-II FPGA used in the implementation contains hardware multipliers that allow efficient and high-performance implementation of multiplication operations [11]. However, where one of the operands is a constant, multiplications can be more efficiently implemented on slices using either bitshifts or constant coefficient multipliers.

performance results for the current version of Reconfig-P (reported in [6]), do not compromise the performance of the hardware circuit. Given these considerations, in the hardware implementation the default scenario is that each of the logically atomic operations is performed in one clock cycle. However, to accommodate situations in which a large number of processing units is required and therefore the system clock rate would otherwise be compromised significantly, P Builder has the ability to generate the hardware circuit in such a way that the logically atomic operations are performed over several clock cycles.

4.2 Implementations of the Logically Atomic Operations

We now describe how we have implemented the logically atomic operations identified in the previous section in hardware.

DIV and MUL. The DIV and MUL operations are implemented in a similar way. The obvious implementation approach is to devote a separate piece of hardware to the execution of each of the operations for each reaction rule. However, this approach would lead to unnecessary duplication of hardware resources because it is often the case that different reaction rules consume/produce the same number of objects for an object type (i.e., the multiplicity of the consumed/produced object type is the same in the definitions for the reaction rules). Duplication of hardware resources can be particularly problematic when Handel-C is used as the specification language, since the Handel-C compiler generates distinct pieces of hardware for the same division or multiplication operation if this operation occurs in different places in the source code. Our solution to the problem of unnecessary hardware duplication is to have distinct DIV/MUL operations which share the same operands implemented as a single processing unit, and for the collection of all such processing units to be implemented as a pool of servers. The DIV and MUL servers continuously perform their respective division/multiplication operations. They execute their operation in one clock cycle, and then store the result in an output register. Each of the servers has direct access to the data for both operands for its operation, and so operates totally

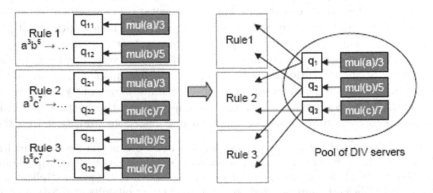

Fig. 4. An illustration of the implementation of a pool of DIV servers

independently from its clients. A client processing unit that needs to evaluate one of the relevant divisions or multiplications reads the appropriate output register at the appropriate clock cycle to obtain the result. As there is one division pool and one multiplication pool per region (rather than for the P system as a whole), our implementation approach does not cause routing problems. Figure 4 shows an example of a pool of DIV servers.

MIN. For reasons similar to those described above, a pool of processing units is used to perform MIN operations. By default, each MIN operation executes in one clock cycle. However, as the logic depth of a MIN operation is linearly proportional to the numbers of object types consumed by the reaction rules in the region, a MIN operation can easily be subjected to logic depth reduction.

SUB. As reaction rules with relative priorities are not processed simultaneously, there are two implementations of the SUB operation: SUB_M and SUB_F. SUB_M is used for those reaction rules that are unrelated by priority to any other reaction rule. It is implemented in Handel-C as a macro expression, which corresponds to a single unshareable piece of hardware with both operands hard-coded. SUB_F is used for reaction rules with relative priorities. It is implemented as a Handel-C function, which corresponds to a single shareable piece of hardware. Since in general the subtractions performed by different reaction rules have different operands, to make SUB_F processing units shareable among the processes implementing the application of reaction rules, the implementation of a SUB_F processing unit operates at the level of object types rather than at the level of reaction rules. More specifically, there is a SUB_F processing unit for each object type.

ADD. All ADD operations (which are used in the implementation of the production of objects by reaction rules, a process which is not subject to any temporal constraints) can in principle be executed simultaneously. However, unless appropriate precautions are taken, the parallel execution of ADD operations can result in parallel processes attempting to update the same register at the same time. There are two main strategies for the avoidance of such update conflicts: the time-oriented strategy and the space-oriented strategy (see Section 3.2).

When the space-oriented strategy is used, for each copy register there is one ADD_M processing unit responsible for updating that register. Each ADD_M processing unit is implemented as a Handel-C macro. This allows all updating operations in the object production phase to be completed in one clock cycle.

When the time-oriented strategy is used, there are three types of processing units implemented. The first type, called ADD_M, is used to update the multiplicity value for an object type with no conflicts. The second type, called ADD_F, is used to update the multiplicity value for a local object type with conflicts. The third type, called ADD_S, is used to update the multiplicity value for an external object type. ADD_S implements semaphore-based interleaving using the `trysema` and `releasesema` constructs provided by Handel-C.

COM. COM operations, which apply only to the region-oriented design, are implemented using channels (see Section 3). Whenever it is possible for objects to move from one region to another region, the implementation includes a channel connecting the region processing unit for the source region to the region process-ing unit for the destination region. There are various ways in which one could implement inter-region communication using channels. The approach one takes influences the number of channels required, as well as the amount of processing needed to complete sending and receiving operations. We will now briefly discuss three possible implementation methods.

In the first method, for every reaction rule r in a region x that sends objects (of any type) to a region y, there is exactly one channel connecting the region processing units for x and y. This channel is used only for the distribution of objects produced by r. Therefore, the data sent over the channel must allow the region processing unit for y to determine which object types are being sent and how many of each type are being sent. To avoid making the definition of r available to the region processing unit for y (and thereby compromising the independence of this region processing unit), this could be achieved by having the region processing unit for x send an n-tuple over the channel, where n is the

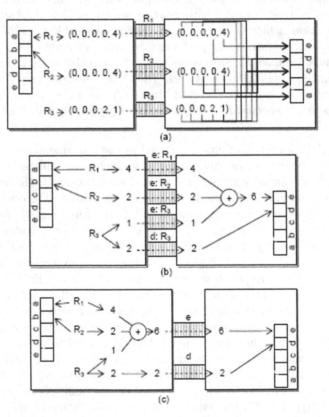

Fig. 5. An illustration of different strategies of implementing inter-region communica-tion using channels. Diagram (a) illustrates the first method mentioned in the text, di-agram (b) illustrates the second method, and diagram (c) illustrates the third method.

number of object types found in the whole P system (not only those produced by r destined for y), which contains for each object type found in the whole P system the multiplicity of that object type being sent to y. Upon receiving the n-tuple, the region processing unit for y would proceed to update the multiset array for y. Obviously, if there are multiple reaction rules in x that produce objects destined for y, there will be multiple channels between the region processing units for x and y. The region processing unit for y would need to coordinate the data received over these channels, as it would receive data relating to the same object type on different channels.

As it is possible to determine at compile-time which types of objects might be produced by which reaction rules and sent to which regions, it is possible to hard-code the relevant pieces of this information in the implementation of the receiving region processing unit. The second method of implementing inter-region communication illustrates this possibility. To implement this method, we would need to relax (albeit to a minimal extent) our requirement that region processing units be independent of each other. In the method, for each reaction rule r in a region x and for each object type o produced by r to be sent to a region y, there is exactly one channel connecting the region processing units for x and y. This channel is used only for the distribution of objects of type o produced by r destined for y. Assume that the region processing unit for y has access to information about which channel is associated with which object type. Then the region processing unit for x needs to send only the multiplicity value for o (i.e., the number of objects of type o that are to be sent in the current transition) down the channel. As in the first method, because the region processing unit for y might receive objects of the same type on different channels, it needs to coordinate the data received over the different channels before proceeding to update the multiset array for y.

In the third method, for every object type that might be produced in a region x and sent to a region y, there is exactly one channel between the region processing units for x and y. Again assume that the region processing unit for y knows which channel is associated with which object type. In this scenario, the region processing unit for x needs to evaluate for each object type the total number of objects of that type to send before engaging in the relevant channel communication. Once it has done this, it sends a single value down the channel. The region processing unit for y simply stores this value in the appropriate register of the multiset array for y.

Figure 5 illustrates the three methods of implementing inter-region communication described above.

We now discuss the relative merits of the three methods of implementing inter-region communication. As regards faithfulness to the biological inspiration of P systems, we rank the third method highest. The first method is perhaps the least in keeping with the biological inspiration of P systems. If we regard the channels in the implementation as representations of cellular transport mechanisms (such as ion channels or osmosis), and reaction rules as representations of chemical reactions, then according to the first method each cellular transport mechanism facilitates the transportation of only the products of a single chemical reaction. In the general case, this is biologically unrealistic. The second method is also quite removed from the biological inspiration of P systems in that cellular transport mechanisms would again be regarded as facilitating the transportation of only the products of a specific chemical reaction. The third method is the most biologically realistic because, in this method, cellular transport mechanisms would be regarded as facilitating the transportation of single types of chemicals (such as potassium ions), as is commonly found in biological cells. As regards the extent to which the independence of region processing units is preserved, the first method ranks highest, with the second and third methods being roughly equivalent. Even so, neither the adoption of the second method nor the adoption of the third method would result in a significant reduction in the independence of region processing units. This is because in these methods the information a region processing unit possesses about other region processing units is available only in an implicit sense. The information is embedded into the very structure of the region processing unit, and so the region processing unit does not explicitly refer to this information when carrying out its operations. As regards efficiency, the third method ranks highest, both in terms of the number of channels used and the amount of processing required. Based on the considerations just outlined, we decided to adopt the third method when implementing the region-oriented design.

4.3   Linking and Synchronisation

In the previous section, we described the hardware components that implement the logically atomic operations. To realise operations occurring at the level of reaction rules, at the level of regions, or at the level of the entire P system, it is necessary to link and synchronise the execution of these basic components. In this section, we describe how the components are linked and synchronised to accomplish some of the processing performed by a region processing unit. We have chosen to focus on this particular case because it is fundamentally important to the region-oriented design.

Figure 6 shows a high-level UML activity diagram for the object assignment phase of the execution of a region processing unit (see Section 3.2 for a description of this phase). Hardware components for logically atomic operations (described in Section 4.2) are represented as shaded boxes in the diagram. This section contains a description of how the other aspects of the diagram — the control flow, linking and synchronisation represented by arrows, solid bars and unshaded boxes — are implemented in hardware.

Linking and synchronisation within a region processing unit. To implement the simple internal control flow within a region processing unit, we use the basic control constructs provided by Handel-C. For example, the arrows in the activity diagram shown in Figure 6 are implemented using the seq construct, the solid bars are implemented as par constructs, and diamonds are implemented using conditional constructs such as if.

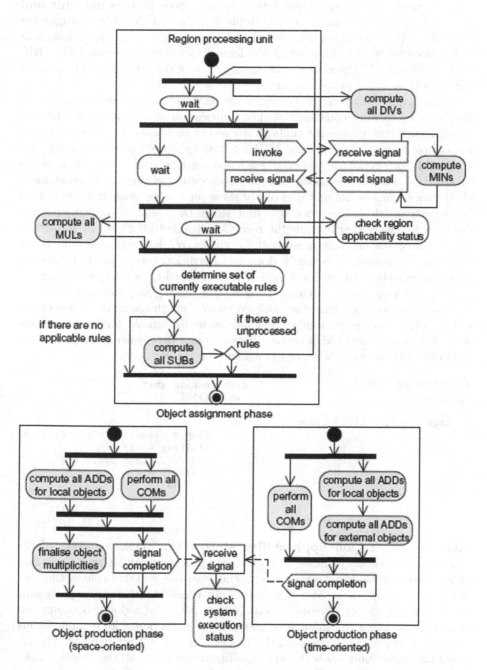

Fig. 6. A UML activity diagram presenting high-level views of the implementations of the object assignment and object production phases of a region processing unit

Linking and synchronisation between a region processing unit and external processing units. In the implementation of the object assignment phase of a region processing unit, it is necessary to link the region processing unit with processing units implementing the logically atomic operations DIV, MIN, MUL, SUB and ADD, and to synchronise the execution of the region processing unit with these other processing units.

In our implementation, processing units may be categorised according to whether they execute constantly without invocation or execute only when invoked. Among the processing units that execute constantly are the processing units implementing the DIV and MUL operations as well as a processing unit responsible for checking whether at least one reaction rule in the region is applicable (see below). Due to the continuous execution of these processing units, when a region processing unit uses one of these processing units, it needs to read the register in which the processing unit stores the result of its computation. However, to ensure that it reads the result applicable to the current transition, the region processing unit must wait for the currently applicable data to be stored in the register. This can be done by inserting the appropriate number of **delay** statements in the relevant section of the Handel-C code implementing the region processing unit or, preferably, by having the region processing unit perform other processing during the clock cycles over which the external processing unit is performing the currently applicable computation. As for the processing units that must be invoked, a region processing unit can invoke these processing units efficiently by using a set of signals and flags as follows:

```
//Processing unit 1                //Processing unit 2
  while(1) {                         while(1) {

    signal = 1; //clock cycle x        par {
    ...                                  flag = signal; //clock cycle x
}                                        if (flag == 1) {
                                           ... //clock cycle x+1
                                         } else
                                           delay;
                                       }
                                     }
```

Checking the region applicability status. Our implementation of the region-oriented design includes for each region processing unit a processing unit which is responsible for checking whether at least one reaction rule in the relevant region is applicable. This processing unit is used by the region processing unit for the purpose of preemptive termination. After the region processing unit calculates the maximum number of instances of a reaction rule, it immediately records the applicability status of the reaction rule in a 1-bit register. The external processing unit reads the applicability registers for all the reaction rules in the region, computes whether there is at least one applicable reaction rule, and then writes the result to a 1-bit output register. Therefore there is a single **delay** statement in the Handel-C code implementing the region processing unit just before the code implementing the reading of the output register.

Reporting completion. As mentioned in Section 3.3, our implementation of the region-oriented design includes a region execution coordinator processing unit, which is responsible for checking whether all region processing units have completed their executions for the current transition, and triggering a new system transition when this condition is satisfied. In the implementation, once it has completed its operations for the current transition, a region processing unit signals this fact to the region execution coordinator via a synchronisation channel (see Figure 2). The following Handel-C code shows how this is achieved in the case where there are two region processing units.

```
//Region processing unit 1
 while(1) {

    ...

    synChan1 !1; //report completion
}
//Region processing unit 2
 while(1) {

    ...

    synChan2 !1; //report completion
}
//Region execution coordinator
 while(1) {

    par {
        synChan1 ? temp1; //receive completion signal on first channel
        synChan2 ? temp2; //receive completion signal on second channel
    }

    //trigger new transition
}
```

Determining which reaction rules are applicable. As discussed in Section 3.2, when processing reaction rules in the object assignment phase, it is advantageous for a region processing unit to check the applicability of the reaction rules. If a reaction rule is inapplicable, it need not be processed further. The implementation approach for the applicability checking operation that most readily comes to mind is the use of if and else constructs. However, because the Handel-C compiler enforces an else implementation with every if implementation, unless one is willing to spread the operation over multiple clock cycles, this approach will in general result in a deeply nested if-else construction. This problem does not arise if goto statements are used instead of else constructs. Consequently, in our implementation we use goto statements. Such statements are inserted just before the code implementing the processing of a reaction rule, and allow this code to be skipped. If it is found that a reaction rule is inapplicable, the control will jump to the part of the code for the reaction rule with the highest priority out of all the remaining reaction rules. Because if and goto statements take zero clock cycles to execute, no clock cycles are wasted in determining which reaction rule should be processed next. Taking this implementation approach allows the various reaction rules to be processed in a consistent manner, and therefore greatly simplifies the control flow required for the processing of the reaction rules.

5   Evaluation of the Region-Oriented Design

In this section, we evaluate our new region-oriented hardware design. More specifically, we report on the hardware resource consumption and clock rates exhibited by hardware circuits implementing P systems using the region-oriented design, and compare the results obtained with those obtained for hardware circuits implementing P systems using the rule-oriented design. We conclude the section with comments about the performance and scalability of the region-oriented hardware design.

5.1   Details of the Experiments

In the experiments, hardware circuits were synthesised for a set of input P systems, according to different implementation strategies. The target hardware platform was a Virtex-II RC2000.

Table 2 describes the characteristics of the input P systems used in the experiments, including the number of regions and reaction rules in the P system, the number of objects (i.e., the product of the number of object types and the number of regions), the number of inter-region communications of object types in the definitions of reaction rules, the number of communication channels used in the implementation of the P system, and the total number of resource conflicts. In the table, P systems P1 through to P5 are used to test the effect of increasing the size of the input P system, and P6 and P7 are used to investigate the effect of using channels for the communication of objects and the effect of using semaphores for the dynamic updating of multisets of objects, respectively. Unlike P systems P1 through to P5, which have region hierarchies, P systems P6 and P7 contains regions connected in a tissue-like fashion. P7 was included in order to facilitate the testing of the effect of having large numbers of communications and channels.

Table 2. Details of the input P systems used in the experiments

P system	Rules	Regions	Total objects	Inter-region communications	Channels	Total conflicts
P1	10	1	3	80	0	21
P2	20	2	6	16	5	32
P3	30	3	9	36	9	32
P4	40	4	12	44	12	40
P5	50	5	15	49	15	42
P6	50	25	75	74	64	42
P7	50	25	200	319	315	45

5.2   Experimental Results

Table 3 shows the experimental results for the hardware resource consumption, which is defined in terms of the percentage of the lookup tables (LUTs) consumed, and clock rates exhibited by circuits implementing the input P systems. Figure 9 illustrates the experimental results in graphical form.

Table 3. Experimental results for the hardware resource consumption and clock rates exhibited by circuits implementing the P systems listed in Table 2 according to various implementation strategies.

P system	Resource consumption (%LUT)				P system	Clock rate (MHz)			
	Region-oriented		Rule-oriented			Region-oriented		Rule-oriented	
	Space-oriented	Time-oriented	Space-oriented	Time-oriented		Space-oriented	Time-oriented	Space-oriented	Time-oriented
P1	2.03	1.82	2.25	2.08	P1	63.84	65.78	81.77	71.94
P2	3.82	3.59	4.24	3.75	P2	60.07	62.79	75.53	77.70
P3	5.72	5.65	6.49	5.79	P3	58.15	52.63	74.35	66.67
P4	7.34	7.39	8.29	7.69	P4	63.69	58.93	80.90	81.17
P5	9.20	9.15	10.43	9.33	P5	58.90	60.10	67.70	70.44
P6	12.28	12.03	12.00	11.81	P6	65.74	66.20	67.95	70.95
P7	14.20	16.79	13.00	13.32	P7	58.58	58.61	62.11	58.64

Efficiency of hardware circuits using the region-oriented design. Figures 7 and 8 illustrate the hardware circuits generated for the input P systems P5 (which contains 5 regions in a hierarchical structure) and P6 (which contains 25 regions connected in an arbitrary graph), respectively. In keeping with the desired logical independence of regions in the region-oriented design, the regions are faithfully realised as separate, decoupled processing units on the hardware circuits.

Hardware resource consumption. The results of the experiments demonstrate that for P systems P1 through to P5 region-oriented circuits tend to be slightly more efficient in terms of hardware resource consumption than rule-oriented circuits. This is because (a) there are fewer core processing units to realise (since the number of regions is usually smaller than the number of reaction rules) in region-oriented circuits than in rule-oriented circuits, and (b) the number of channels used to implement inter-region communication, the main extra resource used in the region-oriented design, is minimised in our design and is therefore relatively small in general. As expected, for P systems P6 and P7, rule-oriented circuits are more efficient than region-oriented circuits. Among region-oriented circuits, those circuits using the time-oriented resource conflict strategy ('region-oriented time-oriented circuits') generally consume fewer hardware resources than those using the space-oriented resource conflict strategy ('region-oriented space-oriented circuits').

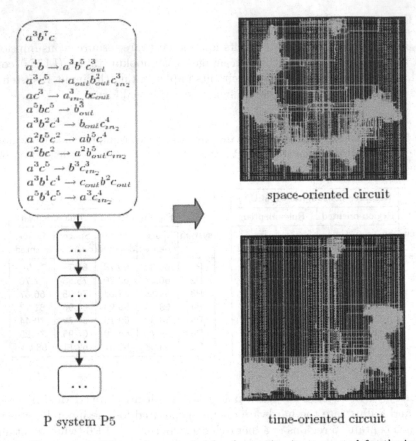

Fig. 7. Illustration of the region-oriented hardware circuits generated for the input P system P5 in space-oriented mode and time-oriented mode

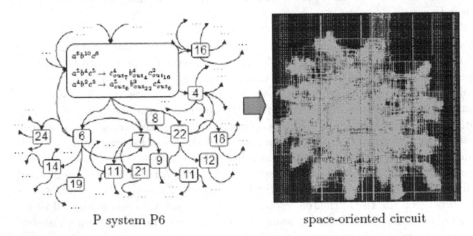

Fig. 8. Illustration of the region-oriented hardware circuit generated for the tissue-like P system P6 in space-oriented mode

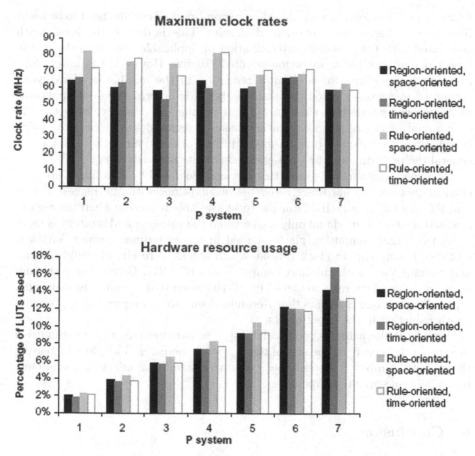

Fig. 9. Graphs of the experimental results presented in Table 3

However, when the number of regions becomes large (for example, in P system P6), the hardware resource consumption exhibited by region-oriented circuits is similar to that exhibited by rule-oriented circuits. If the number of communications is also large, and the number of channels used is large due to the specific characteristics of these communications (as is the case in, for example, P system P7), the hardware resource consumption exhibited by region-oriented circuits is greater than that exhibited by rule-oriented circuits. It is notable that in this case the region-oriented time-oriented circuits consume more hardware resources than the region-oriented space-oriented circuits. This is because our time-oriented conflict resolution strategy performs static interleaving for updating operations only for local objects (which account for only 2% of all updating operations in the case of P system P7) and therefore has to rely on Handel-C semaphores for the updating of external objects (a method which is less efficient in terms of hardware resource consumption).

Clock rates. The clock rates achieved by region-oriented circuits tend to be lower than those achieved by rule-oriented circuits. This is due to the logic depth associated with the dynamic determination of applicable reaction rules in the object assignment phase in region-oriented circuits. However, the lower clock rates are compensated by a possible reduction in the number of clock cycles consumed in region-oriented circuits: the dynamic determination of applicable reaction rules guarantees that an optimal number of clock cycles is used in each round of the object assignment phase in region-oriented circuits, which is something that cannot be guaranteed in rule-oriented circuits. Therefore, in general the performance of region-oriented circuits is satisfactory.

The experimental results show that circuits generated by Reconfig-P are very efficient in terms of hardware resource consumption, with the biggest P system P7 consuming only 16.8% of the total available resources when the region-oriented strategy is used and only 13.3% when the rule-oriented strategy is used.

As the target computing platform used in the experiments was a Virtex-II RC2000, the maximum clock rate at which a hardware circuit could execute and communicate with the host computer was 65 MHz. Given this maximum clock rate, the clock rates achieved by all the generated circuits are satisfying, especially when one considers that Reconfig-P was not configured to apply logic-depth reduction in the experiments.

Reconfig-P also achieves good scalability. The hardware resource consumption increases sub-linearly as the size of the P system increases. Therefore increasing the size of the input P system does not have a significant effect on the circuits (especially rule-oriented circuits).

6   Conclusion

In this paper, we have presented two main research contributions:

1. an elegant (region-oriented) hardware design that more directly reflects the intuitive conceptual understanding of a P system and therefore promotes the extensibility of Reconfig-P, and
2. an implementation of the region-oriented design which exhibits good performance and scalability, and facilitates the future implementation of additional types of P systems (especially those P systems with active membranes or membrane-mediated rules).

Since the existing rule-oriented hardware design and the region-oriented hardware design proposed in this paper have their own advantages and disadvantages, they should be regarded as complementary approaches to the design of a hardware implementation for membrane computing applications. To enhance the versality of Reconfig-P, it is desirable to include both designs in Reconfig-P. In the next phase of our research, we intend to develop a strategy for the efficient and seamless integration of these two alternative hardware designs and other possible implementation strategies into Reconfig-P.

References

1. Bernardini, F., Manca, V.: P systems with boundary rules. In: Păun, G., Rozenberg, G., Salomaa, A., Zandron, C. (eds.) WMC 2002. LNCS, vol. 2597, pp. 107–118. Springer, Heidelberg (2003)
2. Celoxica Ltd.: Handel-C Language Reference Manual (2005), http://babbage.cs.qc.edu/courses/cs345/Manuals/HandelC.pdf
3. Freund, R., Oswald, M.: P systems with activated/prohibited membrane channels. In: Păun, G., Rozenberg, G., Salomaa, A., Zandron, C. (eds.) WMC 2002. LNCS, vol. 2597, pp. 261–269. Springer, Heidelberg (2003)
4. Ionescu, M., Păun, G., Yokomori, T.: Spiking neural P systems with an exhaustive use of rules. Intern. J. Unconventional Computing 3, 135–154 (2007)
5. Martín-Vide, C., Păun, G., Pazos, J., Rodríguez-Patón, A.: Tissue P systems. Theoretical Computer Science 296, 295–326 (2003)
6. Nguyen, V., Kearney, D., Gioiosa, G.: Balancing performance, flexibility and scalability in a parallel computing platform for membrane computing applications. In: Eleftherakis, G., Kefalas, P., Păun, G., Rozenberg, G., Salomaa, A. (eds.) WMC 2007. LNCS, vol. 4860, pp. 385–413. Springer, Heidelberg (2007)
7. Nguyen, V., Kearney, D., Gioiosa, G.: An implementation of membrane computing using reconfigurable hardware. Computing and Informatics 27, 551–569 (2008)
8. Păun, A., Păun, G.: The power of communication: P systems with symport/antiport. New Generation Computing 20, 295–305 (2002)
9. Păun, G.: Computing with membranes – A variant: P systems with polarized membranes. Intern. J. Foundations of Computer Science 11, 167–182 (2000)
10. Păun, G.: Membrane Computing: An Introduction. Springer, Heidelberg (2002)
11. Xilinx. Virtex-II Complete Data Sheet (2007), http://www.xilinx.com/support/documentation/data_sheets/ds031.pdf

Discovering the Membrane Topology of Hyperdag P Systems

Radu Nicolescu, Michael J. Dinneen, and Yun-Bum Kim

Department of Computer Science, University of Auckland
Private Bag 92019, Auckland, New Zealand
{radu,mjd}@cs.auckland.ac.nz, tkim021@aucklanduni.ac.nz

Abstract. In an earlier paper, we presented an extension to the families of P systems, called hyperdag P systems (hP systems), by proposing a new underlying topological structure based on the hierarchical dag structure (instead of trees or digraphs). In this paper, we develop building-block membrane algorithms for discovery of the global topological structure from the local cell point of view. In doing so, we propose more convenient operational modes and transfer modes, that depend only on each of the individual cell rules.

Finally, by extending our initial work on the visualization of hP system membranes with interconnections based on dag structures without transitive arcs, we propose several ways to represent structural relationships, that may include transitive arcs, by simple-closed planar regions, which are folded (and possibly twisted) in three dimensional space.

1 Introduction

In this paper we continue our study [8]. Specifically, we are interested to validate the adequacy of our hyperdag P system (hP system) model for describing several fundamental distributed algorithms that present relevance to networking.

For Algorithms 1 and 5 below, we extend to dags the approach pioneered by Ciobanu *et al.* in [4,3]. We also provide explicit rewriting and transfer rules, as a replacement for pseudo-code. In this process, we identify areas where our initial model was not versatile enough and we propose corresponding adjustments, that can also be retrofitted to other models of the P family, such as the refinement of the rewriting and transfer modes. We also advocate the weak policy for priority rules [10], which we believe is closer to the actual task scheduling in operating systems.

This paper focuses on basic building blocks that are relevant for network discovery (see also [7]): broadcast, convergecast, flooding, and a simple synchronization solution, that highlights the versatility of the dag structure underlying hP systems.

We have earlier proposed an algorithm to visually represent hP systems, where the underlying cell structure was restricted to a canonical dag (i.e., without transitive arcs) [8]. Nodes were represented as simple closed regions on the plane

G. Păun et al. (Eds.): WMC 2009, LNCS 5957, pp. 410–435, 2010.

(with possible nesting or overlaps) and channels by direct containment relationships of the regions. In this paper, we extend this planar representation by
presenting several plausible solutions that enable us to visualize any hP system,
modelled as an arbitrary dag, in the plane. Additionally, for these solutions,
we discuss their advantages and limitations. Finally, in Section 6, we describe
a new algorithm for representing general hP systems, where transitive arcs are
not excluded.

2 Preliminaries

We assume that the reader is familiar with the basic terminology and notations
[8]: relations, graphs, nodes (vertices), arcs, directed graphs, directed acyclic
graphs (dags), canonical dags (dags without transitive arcs), trees, node height
(number of arcs on the longest path to a descendant), topological order, set or
multiset based hypergraphs, simple closed curves (Jordan curves), alphabets,
strings and multisets over an alphabet.

We also assume familiarity with transition P systems and their planar representation [10] and with hP systems [8].

Without giving all functional details, we recall here the basic notations and
the definition of hP systems. Given a set of objects O, we define the following sets
of tagged objects: $O_\uparrow = \{o_\uparrow \mid o \in O\}$, $O_\downarrow = \{o_\downarrow \mid o \in O\}$, $O_\leftrightarrow = \{o_\leftrightarrow \mid o \in O\}$,
$O_{go} = \{o_{go} \mid o \in O\}$, $O_{out} = \{o_{out} \mid o \in O\}$. Intuitively, the $_\uparrow$, $_\downarrow$, $_\leftrightarrow$ tags indicate
objects that will be transferred to parents, children, siblings, respectively; the
$_{go}$ tags indicate transfer to all neighbors (parents, children and siblings); the $_{out}$
tags indicate transfer to the environment.

Definition 1 (Hyperdag P systems). *An hP system of order m is a system*
$\Pi = (O, \sigma_1, \ldots, \sigma_m, \delta, I_{out})$, *where:*

1. *O is an ordered finite non-empty alphabet of objects;*
2. *$\sigma_1, \ldots, \sigma_m$ are cells, of the form $\sigma_i = (Q_i, s_{i,0}, w_{i,0}, P_i)$, $1 \le i \le m$, where:*
   - *Q_i is a finite set of states;*
   - *$s_{i,0} \in Q_i$ is the initial state;*
   - *$w_{i,0} \in O^*$ is the initial multiset of objects;*
   - *P_i is a finite set of multiset rewriting rules of the form $sx \rightarrow s'x'u_\uparrow v_\downarrow w_\leftrightarrow$*
     $y_{go} z_{out}$, where $s, s' \in Q_i$, $x, x' \in O^*$, $u_\uparrow \in O_\uparrow^*$, $v_\downarrow \in O_\downarrow^*$, $w_\leftrightarrow \in O_\leftrightarrow^*$,
     $y_{go} \in O_{go}^*$ and $z_{out} \in O_{out}^*$, with the restriction that $z_{out} = \lambda$ for all
     $i \in \{1, \ldots, m\} \setminus I_{out}$;
3. *δ is a set of dag parent-child arcs on $\{1, \ldots, m\}$, i.e., $\delta \subseteq \{1, \ldots, m\} \times$*
   $\{1, \ldots, m\}$, representing duplex channels between cells;
4. *$I_{out} \subseteq \{1, \ldots, m\}$ indicates the output cells, the only cells allowed to send*
   objects to the "environment".

The dynamic operations of hP systems, i.e., the configuration changes via object
rewriting and object transfer, are a natural extension of similar operations used
by transition and neural P systems. Our earlier paper, [8], describes the dynamic
behavior of hP systems, in more detail.

We measure the *runtime complexity* of a P system in terms of *P-steps*, where a P-step corresponds to a transition on a parallel P machine. If no more transitions are possible, the hP system halts. For halted hP systems, the *computational result* is the multiset of objects emitted *out* (to the "environment"), over all the time steps, from the output cells I_{out}. The *numerical result* is the set of vectors consisting of the object multiplicities in the multiset result. Within the family of P systems, two systems are *functionally equivalent* if they yield the same computational result.

Example 2. Figure 1 shows the structure of an hP system that models a computer network. Four computers are connected to "Ethernet Bus 1", the other four computers are connected to "Ethernet Bus 2", while two of the first group and two of the second group are at the same time connected to a wireless cell. In this figure we also suggest that "Ethernet Bus 1" and "Ethernet Bus 2" are themselves connected to a higher level communication hub, in a generalized hypergraph.

We have already shown, [8], that our hP systems can simulate all transition P systems [10] and all symmetric neural P systems [9], with the same number of steps and object transfers. To keep the arguments simple, we have only considered systems without additional features, such as dissolving membranes, priorities or

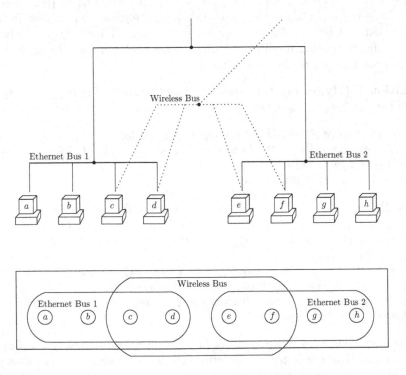

Fig. 1. A computer network and its corresponding hypergraph representation

polarities. However, our definition of hP systems can also be extended, as needed, with additional features, in a straightforward manner, and we do so in this paper.

Model Refinements

- As initially defined [8], the rules are applied according to the current cell state s, in the rewriting mode $\alpha(s) \in \{min, par, max\}$, and the objects are sent out in the transfer mode $\beta(s) \in \{one, spread, repl\}$. In this paper, we propose a refinement to these modes and allow that *the rewriting and transfer modes to depend on the rule used* (instead of the state), as long as there are no conflicting requirements. We will highlight the cases where this mode extension is essential.
- We also consider rules with *priorities*, in their *weak* interpretation [10]. In the current paper, *lower numbers* (i.e., first enumerated) indicate *higher priorities*. In the *weak* interpretation of the priority, rules are applied in decreasing order of their priorities—where a lower priority rule can only be applied after all higher priority rules have been applied (as required by the rewriting modes). In contrast, in the *strong* interpretation, a lower priority rule cannot be applied at all, if a higher priority rule was applied. We will highlight the cases where the weak interpretation is required.

3   Basic Algorithms for Network Discovery–Without IDs

In this section and the following, we study several basic distributed algorithms for network discovery, adapted to hP systems. Essentially, all cells start in the same state and with the same or similar (set of) rules, but there are several different scenarios:

1. Initially, cells know nothing about the structure in which they are linked, and must even discover their local neighborhood (i.e., their parents, children, siblings), as well as some global model topology characteristics (such as various dag measures or shortest paths).
2. As above, but each cell has its own ID (identifier) and is allowed to have custom rules for this ID.
3. As above, each cell has its own ID and also knows the details of its immediate neighbors (parents, children and, optionally, siblings).

Algorithm 1: Broadcast to all descendants.

Precondition: Cells do not need any inbuilt knowledge about the network topology. All cells start in state s_0, with the same rules. The initiating cell has an additional object a, that is not present in any other cell.

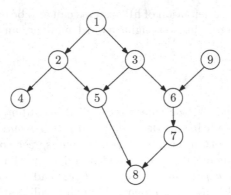

Fig. 2. Sample dag for illustrating our algorithms

Postcondition: All descendant cells are eventually visited and enter state s_1.

Rules:

1. $s_0 a \rightarrow s_1 a_{\downarrow}$, with $\alpha = min$, $\beta = repl$.
2. $s_1 a \rightarrow s_1$, with $\alpha = par$.

Proof. This is a *deterministic* algorithm. Rule 1 is applied exactly once, when a cell is in state s_0 and it contains an a. This a is consumed, the cell enters state s_1 and another a is sent to all the children, replicated as necessary. Additional a's may appear in a cell, because, in a dag structure, a cell may have more than one parent. Rule 2 is applicable in state s_1 and silently discards any additional a's, without changing the state and without interacting with other cells. All a's will eventually disappear from the system—however, cells themselves may never know that the algorithm has completed and no other a's will come from their parents. By induction, all descendants will receive an a and enter state s_1. $\square$

Remarks 3.

- This broadcast algorithm can be initiated anywhere in the dag. However, it is probably most useful when initiated on a dag source, or on all sources at the same time (using the same object a or a different object for each source).
- This algorithm completes after $h + 1$ P-steps, where h is the *height* of the initiating node.
- State s_1 may be reached before the algorithm completes and cannot be used as a termination indicator.
- Several other broadcasting algorithms can be built in a similar manner, such as *broadcast to all ancestors* or *broadcast to all reachable cells* (ancestors and descendants).
- This algorithm family follows the approach used by Ciobanu *et al.* [4,3], for tree based algorithms, called *skin membrane broadcast* and *generalized broadcast*.

Example 4. We illustrate the algorithm for broadcasting to all descendants, for the hP system shown in Figure 2.

Step\Cell	σ_1	σ_2	σ_3	σ_4	σ_5	σ_6	σ_7	σ_8	σ_9
0	s_0a	s_0	s_0	s_0	s_0	s_0	s_0	s_0	s_0
1	s_1	s_0a	s_0a	s_0	s_0	s_0	s_0	s_0	s_0
2	s_1	s_1	s_1	s_0a	s_0aa	s_0a	s_0	s_0	s_0
3	s_1	s_1	s_1	s_1	s_1a	s_1	s_0a	s_0a	s_0
4	s_1	s_1	s_1	s_1	s_1	s_1	s_1	s_1a	s_0
5	s_1	s_1	s_1	s_1	s_1	s_1	s_1	s_1	s_0

Algorithm 2: Counting all paths from a given ancestor.

Precondition: Cells do not need any inbuilt knowledge about the network topology. All cells start in state s_0 and with the same rules. The initiating cell has an additional object a, not present in any other cell.

Postcondition: All descendant cells are eventually visited, enter state s_1 and will have a number of b's equal to the number of distinct paths from the initiating cell.

Rules:

1. $s_0a \rightarrow s_1ba_\downarrow$, with $\alpha = par$, $\beta = repl$.
2. $s_1a \rightarrow s_1ba_\downarrow$, with $\alpha = par$, $\beta = repl$.

Proof. This is a *deterministic* algorithm. Rule 1 is applied when the cell is in state s_0 and an a is available. This a is consumed, the cell enters state s_1, a b is generated and another a is sent to all its children, replicated as necessary. Additional a's may appear in a cell, because, in a dag structure, a cell may have more than one parent. Rule 2 is similar to rule 1. State s_1 is similar to state s_0 and is not essential here, it appears here only to mark visited cells. The number of generated b's is equal to the number of received a's, which eventually will be equal to the number of distinct paths from the initiating cell. All a's will eventually disappear from the system—however, cells themselves may never know that the algorithm has completed, that no other a's will come from their parents and all paths have been counted. A more rigorous proof will proceed by induction.   □

Remarks 5.

- This algorithm completes after $h + 1$ P-steps, where h is the *height* of the initiating node.
- State s_1 may be reached before the algorithm completes and cannot be used as a termination indicator.
- Several other path counting algorithms can be built in a similar manner, such as the number of *paths to a given descendant*.

Example 6. We illustrate the algorithm for counting all paths from a given ancestor, for the hP system shown in Figure 2.

Step\Cell	σ_1	σ_2	σ_3	σ_4	σ_5	σ_6	σ_7	σ_8	σ_9
0	$s_0 a$	s_0	s_0	s_0	s_0	s_0	s_0	s_0	s_0
1	$s_1 b$	$s_0 a$	$s_0 a$	s_0	s_0	s_0	s_0	s_0	s_0
2	$s_1 b$	$s_1 b$	$s_1 b$	$s_0 a$	$s_0 aa$	$s_0 a$	s_0	s_0	s_0
3	$s_1 b$	$s_1 b$	$s_1 b$	$s_1 b$	$s_1 bb$	$s_1 b$	$s_0 a$	$s_0 aa$	s_0
4	$s_1 b$	$s_1 b$	$s_1 b$	$s_1 b$	$s_1 bb$	$s_1 b$	$s_1 b$	$s_1 abb$	s_0
5	$s_1 b$	$s_1 b$	$s_1 b$	$s_1 b$	$s_1 bb$	$s_1 b$	$s_1 b$	$s_1 bbb$	s_0

Algorithm 3: Counting the children of a given cell.

Precondition: Cells do not need any inbuilt knowledge about the network topology. The initiating cell and its children start in state s_0 and with the same rules. The initiating cell has an additional object a, not present in any other cell.

Postcondition: The initiating cell ends in state s_1 and will contain a number of c's equal to its child count. The child cells end in state s_1. As a side effect, other parents (if any) of these children will receive superfluous c's—however, these c's can be discarded, if needed (rules not shown here).

Rules:

1. $s_0 a \rightarrow s_1 p_\downarrow$, with $\alpha = min$, $\beta = repl$.
2. $s_0 p \rightarrow s_1 c_\uparrow$, with $\alpha = min$, $\beta = repl$.

Proof. This is a *deterministic* algorithm with a straightforward proof, not given here.   □

Remarks 7.

- This algorithm completes after two P-steps.
- Several other algorithms that enumerate the immediate neighborhood can be built in a similar manner, such as *counting parents*, *counting siblings*, *counting neighbors*.

Algorithm 4: Broadcast for counting all children.

Precondition: Cells do not need any inbuilt knowledge about the network topology. All cells start in state s_0 and with the same rules. The initiating cell has an additional object a, not present in any other cell.

Postcondition: Each descendant cell enters state s_1 and, eventually, will contain a number of c's equal to its child count.

Rules:

0. For state s_0:
   1) $s_0 a \rightarrow s_1 p_\downarrow$, with $\alpha = min$, $\beta = repl$.
   2) $s_0 p \rightarrow s_1 p_\downarrow c_\uparrow$, with $\alpha = min$, $\beta = repl$.
1. For state s_1:
   1) $s_1 p \rightarrow s_1$, with $\alpha = par$.

Proof. This is a *deterministic* algorithm: the proof combines those from the broadcast algorithm (Algorithm 1) and the child counting algorithm (Algorithm 3). □

Remarks 8.

- This algorithm runs in $h + 1$ P-steps, where h is the *height* of the initiating cell.
- State s_1 may be reached before the algorithm completes its cleanup phase and cannot be used as a termination indicator.
- As a side effect, any parent of the visited children that is not a descendant of the initiating node will receive superfluous c's.
- Several other algorithms that broadcast a request to count the immediate neighborhood can be built in a similar manner, such as *broadcast for counting all parents, broadcast for counting all siblings, broadcast for counting all neighbors*.

Example 9. We illustrate the algorithm for counting all children via broadcasting, for the hP system shown in Figure 2.

Step\Cell	σ_1	σ_2	σ_3	σ_4	σ_5	σ_6	σ_7	σ_8	σ_9
0	$s_0 a$	s_0	s_0	s_0	s_0	s_0	s_0	s_0	s_0
1	s_1	$s_0 p$	$s_0 p$	s_0	s_0	s_0	s_0	s_0	s_0
2	$s_1 cc$	s_1	s_1	$s_0 p$	$s_0 pp$	$s_0 p$	s_0	s_0	s_0
3	$s_1 cc$	$s_1 cc$	$s_1 cc$	s_1	$s_1 p$	s_1	$s_0 p$	$s_0 p$	$s_0 c$
4	$s_1 cc$	$s_1 cc$	$s_1 cc$	s_1	$s_1 c$	$s_1 c$	$s_1 c$	$s_1 p$	$s_0 c$
5	$s_1 cc$	$s_1 cc$	$s_1 cc$	s_1	$s_1 c$	$s_1 c$	$s_1 c$	s_1	$s_0 c$

Algorithm 5: Counting heights by flooding.

Precondition: Cells do not need any inbuilt knowledge about the network topology. All cells start in state s_0, with the same rules and have no initial object.

Postcondition: All cells end in state s_2. The number of t's in each cell equals the distance from a furthest descendant.

Rules:

0. For state s_0:

    1) $s_0 \rightarrow s_1 a c_\uparrow$, $\alpha = min$, $\beta = repl$.

1. For state s_1, the rules will run under the following *priorities*, under the *weak interpretation*:

    1) $s_1 ac \rightarrow s_1 atc_\uparrow$, $\alpha = max$, $\beta = repl$.
    2) $s_1 c \rightarrow s_1$, $\alpha = max$.
    3) $s_1 a \rightarrow s_2$, $\alpha = min$.

Proof. Each cell emits a single object c to each of its parents in the first step. During successive active steps, a cell either: (a) uses rule 1.3 to enter the terminating state s_2 or (b) continues via rule 1.1 to forward one c up to each of its parents. In the latter case, since we have $\alpha = max$, and as enabled by the weak interpretation of priorities, rule 1.2 is further used to remove all remaining c's (if any), in the same step. The cell safely enters the end state s_2 when no more c's appear. Induction shows that the set of times that c's appear is consecutive: if a cell at $k > 1$ links away emitted a c, then there must be another cell at $k - 1$ links away emitting another c. Finally, the number of times rule 1.1 is applied is the number of times a cell receives at least one new c from below. These steps are tallied by occurrences of the object t.   □

Remarks 10.

- This algorithm, like other distributed flooding based algorithms, requires that all cells start at the same time. Achieving this synchronization is a nontrivial task—in Section 5, we suggest a simple and fast algorithm that achieves this synchronization.
- The time complexity of this quick algorithm is $h + 2$ P-steps, where h is the height of the dag. The two extra P-steps correspond to the initial step and the step to detect no more c's.
- This algorithm follows the approach by Ciobanu *et al.* [4,3], for the tree based algorithm called *convergecast*. Here we prefer to use the term *flooding*, and use the term *convergecast* for a result accumulation triggered by an initial broadcast.
- This algorithm makes critical use of the *weak interpretation* for *priorities*.

Example 11. We illustrate the algorithm for counting heights by flooding, for the hP system shown in Figure 2.

Step\Cell	σ_1	σ_2	σ_3	σ_4	σ_5	σ_6	σ_7	σ_8	σ_9
0	s_0	s_0	s_0	s_0	s_0	s_0	s_0	s_0	s_0
1	s_1acc	s_1acc	s_1acc	s_1a	s_1ac	s_1ac	s_1ac	s_1a	s_1ac
2	s_1acct	s_1act	s_1acct	s_2	s_1at	s_1act	s_1at	s_2	s_1act
3	s_1acctt	s_1att	s_1actt	s_2	s_2t	s_1att	s_2t	s_2	s_1actt
4	s_1act^3	s_2tt	s_1at^3	s_2	s_2t	s_2tt	s_2t	s_2	s_1at^3
5	s_1at^4	s_2tt	s_2t^3	s_2	s_2t	s_2tt	s_2t	s_2	s_2t^3
6	s_2t^4	s_2tt	s_2t^3	s_2	s_2t	s_2tt	s_2t	s_2	s_2t^3

Algorithm 6: Counting nodes in a single-source dag.

Precondition: Cells do not need any inbuilt knowledge about the network topology. All cells start in state s_0, with the same rules. The initiating cell is the source of a single-source dag and has an additional object a, not present in any other cell.

Postcondition: Eventually, the initiating cell will contain a number of c's equal to the number of all its descendants, including itself, which is also the required node count.

Rules:

0. For state s_0:
   1) $s_0a \rightarrow s_3p_\downarrow c$, with $\alpha = min$, $\beta = repl$.
   2) $s_0p \rightarrow s_1p_\downarrow$, with $\alpha = min$, $\beta = repl$.
1. For state s_1:
   1) $s_1 \rightarrow s_2c_\uparrow$, with $\alpha = min$, $\beta = one$.
2. For state s_2:
   1) $s_2c \rightarrow s_2c_\uparrow$, with $\alpha = max$, $\beta = one$.
   2) $s_2p \rightarrow s_2$, with $\alpha = max$.

Proof. We prove that the source will eventually contain k copies of object c, where k is the order of the single-source dag. The source cell will produce a copy of c following rule 0.1. A non-source cell σ_i will send one c to a parent σ_j, where $j \in \delta^{-1}(i)$, because a node is at state s_1 during at most one P-step, by rule 1.1. A cell σ_i will forward up, using rule 2.1, additional c's to one of its parents, which will eventually arrive at the source.   □

Remarks 12.

- This algorithm takes up to $2h$ P-steps, where h is the *height* of the initiating cell.
- The end state s_3 is not halting, may be reached before the algorithm completes and cannot be used as a termination indicator.

Example 13. We illustrate the algorithm for counting nodes in a single-source dag via convergecast, for the hP system shown in Figure 2, after removing node 9.

Step\Cell	σ_1	σ_2	σ_3	σ_4	σ_5	σ_6	σ_7	σ_8
0	s_0a	s_0	s_0	s_0	s_0	s_0	s_0	s_0
1	s_3c	s_0p	s_0p	s_0	s_0	s_0	s_0	s_0
2	s_3c	s_1	s_1	s_0p	s_0pp	s_0p	s_0	s_0
3	s_3c^3	s_2	s_2	s_1	s_1p	s_1	s_0p	s_0p
4	s_3c^3	s_2c	s_2cc	s_2	s_2p	s_2	s_1	s_1p
5	s_3c^6	s_2	s_2	s_2	s_2	s_2c	s_2c	s_2p
6	s_3c^6	s_2	s_2c	s_2	s_2	s_2c	s_2	s_2
7	s_3c^7	s_2	s_2c	s_2	s_2	s_2	s_2	s_2
8	s_3c^8	s_2	s_2	s_2	s_2	s_2	s_2	s_2

4   Basic Algorithms for Network Discovery–With IDs

In this section we assume each cell has an unique ID and the cells only know their own ID. Objects may be tagged with IDs to aid in communication.

Algorithm 7: Counting descendants by convergecast—with cell IDs.

Precondition: Cells do not need any inbuilt knowledge about the network topology. For each cell with index i, $1 \leq i \leq m$, the alphabet includes special ID objects c_i and $\bar{c}_i$. All cells start in state s_0 and have the same rules, except several similar, but custom specific, rules to process the IDs. The initiating cell has an additional object a, not present in any other cell.

Postcondition: All visited cells enter state s_1 and, eventually, each cell will contain exactly one $\bar{c}_i$ for each descendant cell with index i, including itself: the number of these objects is the descendant count.

Rules:

0. For state s_0 and cell σ_i (these are custom rules, specific for each cell):
   1) $s_0a \rightarrow s_1 p_\downarrow \bar{c}_i$, with $\alpha = min$, $\beta = repl$.
   2) $s_0p \rightarrow s_1 p_\downarrow c_{i\uparrow} \bar{c}_i$, with $\alpha = min$, $\beta = repl$.
1. For state s_1, the rules will run under the following *priorities*:
   1) $s_1 c_j \bar{c}_j \rightarrow s_1 \bar{c}_j$, for $1 \leq j \leq m$, with $\alpha = max$.
   2) $s_1 c_j \rightarrow s_1 c_{j\uparrow} \bar{c}_j$, for $1 \leq j \leq m$, with $\alpha = max$, $\beta = repl$.
   3) $s_1 p \rightarrow s_1$, with $\alpha = max$.

Proof. Assume that δ is the underlying dag relation. For each cell σ_i, consider the sets $C_i = \{c_j \mid j \in \delta^*(i)\}$, $\bar{C}_i = \{\bar{c}_j \mid j \in \delta^*(i)\}$, which consist of ID objects matching σ_i's children. By induction on the dag height, we prove that each visited cell σ_i will eventually contain the set $\bar{C}_i$, and, if it is not the initiating cell, will also send up all elements of the set C_i, possibly with some duplicates (up to all its parents). The base case, height $h = 0$, is satisfied by rule 0.1, if σ_i is the initiator, or by rule 0.2, otherwise. For cell σ_i at height $h + 1$, by induction, each child cell σ_k sends up C_k, possibly with some duplicates. By rules 0.1 and 0.2, cell σ_i further acquires one $\bar{c}_i$ and, if not the initiator, sends up one c_i. From its children, cell σ_i acquires the multiset C_i', consisting of all the elements of the set $\bigcup_{k \in \delta(i)} C_k = C_i \setminus c_i$, possibly with some duplications. Rule 1.3 sends up one copy of each element of multiset C_i' and records a barred copy of it. Rule 1.2 halves the number of duplicates in multiset C_i'. Rule 1.1 filters out duplicates in multiset C_i', if a barred copy already exists. Rule 1.4 clears all p's, which are not needed anymore. □

Remarks 14.

- Other counting algorithms can be built in a similar manner, such as *counting ancestors, counting siblings, counting sources* or *counting sinks*.
- The end state s_1 is not halting, it may be reached before the algorithm completes and cannot be used as a termination indicator.
- As a side effect, any parent of the visited children that is not a descendant of the initiating node may receive superfluous c_i's.
- This algorithm works under both *strong* and *weak* interpretation of *priorities*.

Example 15. We illustrate the algorithm for counting descendants via convergecast using cell IDs, for the hP system shown in Figure 2.

Step\Cell	σ_1	σ_2	σ_3	σ_4	σ_5	σ_6	σ_7	σ_8	σ_9
0	$s_0 a$	s_0	s_0	s_0	s_0	s_0	s_0	s_0	s_0
1	$s_1 c_1$	$s_0 p$	$s_0 p$	s_0	s_0	s_0	s_0	s_0	s_0
2	$s_1 c_2 c_3$	s_1	s_1	$s_0 p$	$s_0 pp$	$s_0 p$	s_0	s_0	s_0
	$\bar{c}_1$	$\bar{c}_2$	$\bar{c}_3$						
3	s_1	$s_1 c_4 c_5$	$s_1 c_5 c_6$	s_1	$s_1 p$	s_1	$s_0 p$	$s_0 p$	$s_0 c_6$
	$\bar{c}_1 \bar{c}_2 \bar{c}_3$	$\bar{c}_2$	$\bar{c}_3$	$\bar{c}_4$	$\bar{c}_5$	$\bar{c}_6$			
4	$s_1 c_4 c_5 c_5 c_6$	s_1	s_1	s_1	$s_1 c_8$	$s_1 c_7$	$s_1 c_8$	$s_1 p$	$s_0 c_6$
	$\bar{c}_1 \bar{c}_2 \bar{c}_3$	$\bar{c}_2 \bar{c}_4 \bar{c}_5$	$\bar{c}_3 \bar{c}_5 \bar{c}_6$	$\bar{c}_4$	$\bar{c}_5$	$\bar{c}_6$	$\bar{c}_7$	$\bar{c}_8$	
5	s_1	$s_1 c_8$	$s_1 c_7 c_8$	s_1	s_1	$s_1 c_8$	s_1	s_1	$s_0 c_6 c_7$
	$\bar{c}_1 \bar{c}_2 \bar{c}_3 \bar{c}_4 \bar{c}_5 \bar{c}_5 \bar{c}_6$	$\bar{c}_2 \bar{c}_4 \bar{c}_5$	$\bar{c}_3 \bar{c}_5 \bar{c}_6$	$\bar{c}_4$	$\bar{c}_5 \bar{c}_8$	$\bar{c}_6 \bar{c}_7$	$\bar{c}_7 \bar{c}_8$	$\bar{c}_8$	
6	$s_1 c_7 c_8 c_8$	s_1	$s_1 c_8$	s_1	s_1	s_1	s_1	s_1	$s_0 c_6 c_7 c_8$
	$\bar{c}_1 \bar{c}_2 \bar{c}_3 \bar{c}_4 \bar{c}_5 \bar{c}_6$	$\bar{c}_2 \bar{c}_4 \bar{c}_5 \bar{c}_8$	$\bar{c}_3 \bar{c}_5 \bar{c}_6 \bar{c}_7 \bar{c}_8$	$\bar{c}_4$	$\bar{c}_5 \bar{c}_8$	$\bar{c}_6 \bar{c}_7 \bar{c}_8$	$\bar{c}_7 \bar{c}_8$	$\bar{c}_8$	
7	s_1	s_1	s_1	s_1	s_1	s_1	s_1	s_1	$s_0 c_6 c_7 c_8$
	$\bar{c}_1 \bar{c}_2 \bar{c}_3 \bar{c}_4 \bar{c}_5 \bar{c}_6 \bar{c}_7 \bar{c}_8 \bar{c}_8$	$\bar{c}_2 \bar{c}_4 \bar{c}_5 \bar{c}_8$	$\bar{c}_3 \bar{c}_5 \bar{c}_6 \bar{c}_7 \bar{c}_8$	$\bar{c}_4$	$\bar{c}_5 \bar{c}_8$	$\bar{c}_6 \bar{c}_7 \bar{c}_8$	$\bar{c}_7 \bar{c}_8$	$\bar{c}_8$	
8	s_1	s_1	s_1	s_1	s_1	s_1	s_1	s_1	$s_0 c_6 c_7 c_8$
	$\bar{c}_1 \bar{c}_2 \bar{c}_3 \bar{c}_4 \bar{c}_5 \bar{c}_6 \bar{c}_7 \bar{c}_8$	$\bar{c}_2 \bar{c}_4 \bar{c}_5 \bar{c}_8$	$\bar{c}_3 \bar{c}_5 \bar{c}_6 \bar{c}_7 \bar{c}_8$	$\bar{c}_4$	$\bar{c}_5 \bar{c}_8$	$\bar{c}_6 \bar{c}_7 \bar{c}_8$	$\bar{c}_7 \bar{c}_8$	$\bar{c}_8$	

Algorithm 8: Shortest paths from a given cell.

Precondition: Cells do not need any inbuilt knowledge about the network topology. For each cells with indices i, j, $1 \leq i, j \leq m$, the alphabet includes special ID objects: p_i, $\bar{p}_i$, $\bar{c}_i$, x_{ij}. All cells start in state s_0 and have the same rules, except several similar but custom specific rules to process the IDs. The initiating cell has an additional object a, not present in any other cell.

Postcondition: This algorithm builds a *shortest paths* spanning tree, that is a breadth-first tree rooted at the initiating cell and preserving this dag's relation δ. Each visited cell σ_i, except the initiating cell, will contain one $\bar{p}_k$, indicating its parent σ_k in the spanning tree. Each visited cell σ_i will also contain one $\bar{c}_j$ for each σ_j that is a child of σ_i in the spanning tree, i.e., it will contain all elements of the set $\{\bar{c}_j \mid (i, j) \in \delta, \sigma_j \text{ contains } \bar{p}_i\}$.

Rules:

0. For state s_0 and cell σ_i (custom rules, specific for cell σ_i):
   1) $s_0 a \rightarrow s_1 p_{i\downarrow}$, with $\alpha = min$, $\beta = repl$.
   2) $s_0 p_j \rightarrow s_1 \bar{p}_j p_{i\downarrow} x_{ji\uparrow}$, for $1 \leq j \leq m$, with $\alpha = min$, $\beta = repl$.
   3) $s_0 x_{kj} \rightarrow s_0$, for $1 \leq k, j \leq m, k \neq i$, with $\alpha = max$.
1. For state s_1 and cell σ_i (custom rules, specific for cell σ_i):
   1) $s_1 x_{ij} \rightarrow s_1 \bar{c}_j$, for $1 \leq j \leq m$, with $\alpha = max$.
   2) $s_1 p_j \rightarrow s_1$, for $1 \leq j \leq m$, with $\alpha = max$.
   3) $s_1 x_{kj} \rightarrow s_1$, for $1 \leq k, j \leq m, k \neq i$, with $\alpha = max$.

Proof. It is clear that every visited cell σ_i, except the initiating cell, contains one $\bar{p}_k$ where $k \in \delta^{-1}(i)$ from rule 0.2. By a node's height, we prove that a cell σ_i will contain the set $C_i = \{\bar{c}_j \mid (i, j) \in \delta, \sigma_j \text{ contains } \bar{p}_i\}$. For height 0, $C_i = \emptyset$ is true since a sink σ_i does not have any children to receive an x_{ji}—see rule 0.2. For a cell σ_i of height greater than 0, first observe that rule 1.1 is only applied if rule 0.2 has been applied for a child cell σ_j. Thus, C_i contains all $\bar{c}_j$ such that (i, j) is in the spanning tree. Those x_{kj}'s are removed by rule 0.3, and x_{ij}'s that are not converted to $\bar{c}_j$ are removed by rule 1.3.   □

Remarks 16.

- For this algorithm, cells need additional symbols, see the precondition.
- This algorithm takes $h + 1$ P-steps, where h is the *height* of the initiating cell.
- The end state s_1 is not halting, it may be reached before the algorithm completes and cannot be used as a termination indicator.
- As a side effect, any parent of the visited children that is not a descendant of the initiating node will receive superfluous x_{ij}'s, but they are removed by rule 0.3.

- The rules for state s_0 make effective use of our rewriting mode refinement: rules 0.1 and 0.2 use $\alpha = min$, while rule 0.3 uses $\alpha = max$.
- Provided that arcs are associated with weights, this algorithm can be extended into a distributed version of the *Bellman-Ford algorithm* [7].

Example 17. We illustrate the algorithm for counting nodes in a single-source dag via convergecast, for the hP system shown in Figure 2. The thick arrows in Figure 3 show the resulting spanning tree.

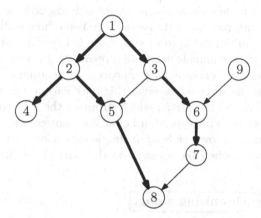

Fig. 3. A spanning tree created by the shortest paths algorithm (Algorithm 8)

Step\Cell	σ_1	σ_2	σ_3	σ_4	σ_5	σ_6	σ_7	σ_8	σ_9
0	$s_0 a$	s_0	s_0	s_0	s_0	s_0	s_0	s_0	s_0
1	s_1	$s_0 p_1$	$s_0 p_1$	s_0	s_0	s_0	s_0	s_0	s_0
2	$s_1 x_{12} x_{13}$	$s_1 \bar{p}_1$	$s_1 \bar{p}_1$		$s_0 p_2$	$s_0 p_2 p_3$	$s_0 p_3$	s_0	s_0
3	$s_1 \bar{c}_2 \bar{c}_3$	$s_1 \bar{p}_1 x_{24} x_{25}$	$s_1 \bar{p}_1 x_{25} x_{36}$	$s_1 \bar{p}_2$	$s_1 p_3 \bar{p}_2$	$s_1 \bar{p}_3$	$s_0 p_6$	$s_0 p_5$	$s_0 x_{36}$
4	$s_1 \bar{c}_2 \bar{c}_3$	$s_1 \bar{p}_1 \bar{c}_4 \bar{c}_5$	$s_1 \bar{p}_1 \bar{c}_6$	$s_1 \bar{p}_2$	$s_1 \bar{p}_2 x_{58}$	$s_1 \bar{p}_3 x_{67}$	$s_1 \bar{p}_6 x_{58}$	$s_1 p_7 \bar{p}_5$	s_0
5	$s_1 \bar{c}_2 \bar{c}_3$	$s_1 \bar{p}_1 \bar{c}_4 \bar{c}_5$	$s_1 \bar{p}_1 \bar{c}_6$	$s_1 \bar{p}_2$	$s_1 \bar{p}_2 \bar{c}_8$	$s_1 \bar{p}_3 \bar{c}_7$	$s_1 \bar{p}_6$	$s_1 \bar{p}_5$	s_0

5 The Firing-Squad-Synchronization-Problem (FSSP)

More sophisticated network algorithms can be built on the fundamental building blocks discussed in the previous sections.

For a given hP system, with cells $\sigma_1, \ldots, \sigma_m$, we now consider the problem of synchronizing a subset of cells $F \subseteq \{\sigma_1, \ldots, \sigma_m\}$, where all cells in the set F synchronize by entering a designated firing state, *simultaneously* and *for the first time*. The *commander* cell σ_c sends one or more orders, to one or more of its neighbors, to start and control the synchronization process; the commander itself may or may not be part of the firing squad. At startup, all cells start in the initial state s_0. The commander and the squad cells may contain specific objects, but all other cells are empty. Initially, all cells, except the commander, are idle, and will remain idle until they receive a message. Notifications may be further relayed to all cells, as necessary.

There are several ways to solve this problem. Here we assume that we can dynamically extend the dag structure of the initial hP system. Unlike the tree structures, which allow only limited extensions, the dag structures allow extensions that greatly simplify the solution to this problem and other similar problems, to the point that they may appear "trivial". We take this as an additional argument supporting the introduction of dag structures in the context of P systems. In our related paper [5], we propose a mechanism for dynamical extensions based on *mobile* channels. Here, we only describe a *partial* solution, which assumes that all required extensions have been "magically" completed.

Assume that the initial hP system was extended by an external cell, called *sergeant*, and additional channels from the sergeant to all cells in the set F. The commander initiates the synchronization process by sending a "notification" to the sergeant. When the sergeant receives this notification, the sergeant sends a "command" to all cells in the set F, which prompts the cells to synchronize by entering the firing state. The algorithm below does not consider the sergeant as part of the firing squad. However, with a simple extension (not shown here), we can also cover the case when the sergeant is also part of the firing squad.

Algorithm 9: Synchronizing a dag.

Precondition: We are given an hP system with m cells $\sigma_1, \ldots, \sigma_m$, a squad subset $F \subseteq \{\sigma_1, \ldots, \sigma_m\}$, and a commander cell $\sigma_c \in F$. We assume that the underlying dag structure was already extended with a new sergeant cell σ_{m+1} and additional channels from σ_{m+1}, as parent, to σ_i, as child, for each $i \in F \subseteq X$.

All cells start in state s_0 and have the same rules. State s_1 is here the designated firing state. Initially, the sergeant σ_{m+1} has an object c, the commander σ_c has an object a, and all other cells have no object.

Postcondition: All cells in the set F enter state s_1, simultaneously and for the first time, after three P-steps.

Rules:

0. For state s_0, the rules will run under the following *priorities* (either the weak or strong interpretation will work):
   1) $s_0 a \to s_0 b_\uparrow$, with $\alpha = min$, $\beta = repl$.
   2) $s_0 bc \to s_0 f_\downarrow$, with $\alpha = min$, $\beta = repl$.
   3) $s_0 b \to s_0$, with $\alpha = min$.
   4) $s_0 f \to s_1$ with $\alpha = min$.

Proof. At step 1, the commander sends a b notifier to all its parents, including the newly created sergeant, via rule 0.1. At step 2, the sergeant sends the firing command f to all squad cells, using rule 0.2. All other commander's parents clear their b notifiers at step 2, using rule 0.3. At step 3, all squad cells enter the firing state s_1, using rule 0.4.   □

Example 18. We illustrate the algorithm for synchronizing the hP system shown in Figure 4. This hP system consists of seven cells $\{\sigma_1, \ldots, \sigma_7\}$, $F = \{\sigma_1, \ldots, \sigma_5\}$ and σ_3 is the commander. The actual system structure is irrelevant in this case and was replaced by a blob that circumscribes the cells $\sigma_1, \ldots, \sigma_7$. In the diagram, this structure has already been extended by the sergeant cell σ_8 and the required channels.

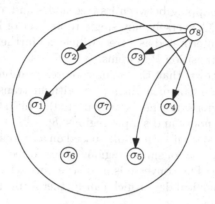

Fig. 4. An hP system for the synchronization algorithm (Algorithm 9), extended by the sergeant cell σ_8 and the required channels

Step\Cell	σ_1	σ_2	σ_3	σ_4	σ_5	σ_6	σ_7	σ_8
0	s_0	s_0	$s_0 a$	s_0	s_0	s_0	s_0	$s_0 c$
1	s_0	s_0	s_0	s_0	s_0	s_0	s_0	$s_0 bc$
2	$s_0 f$	$s_0 f$	$s_0 f$	$s_0 f$	$s_0 f$	s_0	s_0	s_0
3	s_1	s_1	s_1	s_1	s_1	s_0	s_0	s_0

In a related paper, [5], we propose a dynamic extension mechanism, which we believe is compatible with the existing P system framework, and will *complete* the whole algorithm, including the creation of all required extensions, in $e_c + 5$ P-steps, where e_c is the eccentricity of the commander in the underlying dag. In [5], we also provide a more constrained solution, which covers both hP and symmetric neural P systems, without requiring structural extensions. This solution applies traditional rules, under the weak priority scheme, and takes $6e_c + 7$ P-steps.

Previously known FSSP solutions only covered tree-based P systems. Bernardini *et al.* present a deterministic solution for tree-based P systems with polarizations and priorities [2], which works in time $4N + 2H$, where N and H are the number of tree nodes and tree height, respectively. Alhazov *et al.* present another deterministic solution for tree-based P systems with promoters and inhibitors [1], which works in time $3H$.

6   Planar Representation

We define a *simple region* as the interior of a simple closed curve (Jordan curve). By default, all our regions will be delimited by simple closed curves that are also smooth, with the possible exception of a finite number of points. This additional assumption is not strictly needed, but simplifies our arguments.

A simple region R_j is *directly contained* in a simple region R_i, if $R_j \subset R_i$ and there is no simple region R_k, such that $R_j \subset R_k \subset R_i$ (where $\subset$ denotes strict inclusion).

It is well known that any transition P system has a planar Venn-like representation, with a 1:1 mapping between its tree nodes and a set of hierarchically nested simple regions. Conversely, any single rooted set of hierarchically nested simple regions can be interpreted as a tree, which can further form the structural basis of a number of transition P systems.

We have already shown that this planar representation can be generalized for hP systems based on canonical dags (i.e., without transitive arcs) and arbitrary sets of simple regions (not necessarily nested), while still maintaining a 1:1 mapping between dag nodes and simple regions [8].

Specifically, any hP system structurally based on a canonical dag can be intensionally represented by a set of simple regions, where direct containment denotes a parent-child relation. The converse is also true, any set of simple regions can be interpreted as a canonical dag, which can further form the structural basis of a number of hP systems.

We will now provide several solutions to our open question [8]: How to represent the other dags, that do contain transitive arcs? First, we discuss a negative result. First, a counter-example that appeals to the intuition, and then a theorem with a brief proof.

Example 19. Consider the dag (a) of Figure 5, where nodes $1, 2, 3$ are to be represented by simple regions R_1, R_2, R_3, respectively. We consider the following three candidate representations: (e), (f) and (g). However, none of them properly match the dag (a), they only match dags obtained from (a) by removing one of its arcs:

(e) represents the dag (b), obtained from (a) by removing the arc $(1, 3)$;
(f) represents the dag (c), obtained from (a) by removing the arc $(1, 2)$;
(g) represents the dag (d), obtained from (a) by removing the arc $(2, 3)$.

Theorem 20. *Dags with transitive arcs cannot be planarly represented by simple regions, with a 1:1 mapping between nodes and regions.*

Proof. Consider again the counter-example in Example 19. The existence of arcs $(2, 3), (1, 2)$ requires that $R_3 \subset R_2 \subset R_1$. This means that R_3 cannot be directly contained in R_1, as required by the arc $(1, 3)$. □

It is clear, in view of this negative result, that we must somehow relax the requirements, if we want to obtain meaningful representations for general hP systems,

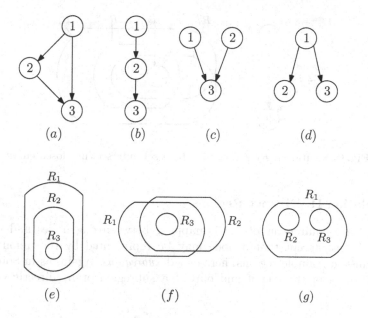

(a) $\qquad$ (b) $\qquad$ (c) $\qquad$ (d)

(e) $\qquad\qquad$ (f) $\qquad\qquad$ (g)

Fig. 5. A counter-example for planar representation of non-canonical dags

based on dag structure that may contain transitive arcs. We consider in turn five tentative solutions.

6.1   Solution I: Self-Intersecting Curves

We drop the requirement of mapping nodes to simple regions delimited by simple closed curves. We now allow self-intersecting closed curves with inward folds. A node can be represented as the union of *subregions*: first, a base simple region, and, next, zero, one or more other simple regions, which are delimited by inward folds of base region's contour (therefore included in the base region). For this solution, we say that there is an arc (i, j) in the dag if and only if a subregion of R_i directly contains region R_j, where regions R_i, R_j represent nodes i, j in the dag, respectively.

Example 21. The region R_1 in Figure 6 is delimited by a self-intersecting closed curve with an inward fold that defines the inner R_1'' subregion. Note the following relations:

- $R_1 = R_1' \cup R_1''$, thus R_1'' is a subregion of R_1;
- R_1 directly contains R_2, which indicates the arc $(1, 2)$;
- R_2 directly contains R_3, which indicates the arc $(2, 3)$;
- R_1'' directly contains R_3, which indicates the transitive arc $(1, 3)$, because R_1'' is a subregion of R_1.

Remark 22. It is difficult to visualize a cell that is modelled by a self-intersecting curve. Therefore, this approach does not seem adequate.

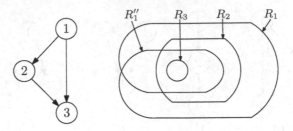

Fig. 6. Solution I: R_1 is delimited by a self-intersecting closed curve

6.2   Solution II: Distinct Regions

We drop the requirement of a 1:1 mapping between dag nodes and regions. Specifically, we accept that a node may be represented by the union of one or more distinct simple regions, here called *subregions*. Again, as in Solution I, an arc (i, j) is in the dag if and only if a subregion of R_i directly contains region R_j.

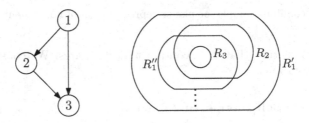

Fig. 7. Solution II: R_1 is the union of two simple regions, R_1' and R_1''

Example 23. In Figure 7, the simple region R_1 is the union of two simple regions, R_1' and R_1'', connected by a dotted line. Note the following relations:

- $R_1 = R_1' \cup R_1''$, thus R_1' and R_1'' are subregions of R_1;
- R_1' directly contains R_2, which indicates the arc $(1, 2)$, because R_1' is a subregion of R_1;
- R_2 directly contains R_3, which indicates the arc $(2, 3)$;
- R_1'' directly contains R_3, which indicates the transitive arc $(1, 3)$, because R_1'' is a subregion of R_1.

Remark 24. In Example 23, a dotted line connects two regions belonging to the same node. It is difficult to see the significance of such dotted lines in the world of cells. Widening these dotted lines could create self-intersecting curves— a solution which we have already rejected. Two distinct simple regions should represent two distinct cells, not just one. Therefore, this approach does not seem adequate either.

6.3   Solution III: Flaps

We again require simple regions, but we imagine that our representation is an infinitesimally thin "sandwich" of several superimposed layers, up to one distinct layer for each node (see Figure 8b). Initially, each region is a simple region that is conceptually partitioned into a *base subregion* (at some bottom layer) and zero, one or more other *flap subregions*, that appear as flaps attached to the base. These flaps are then folded, in the three-dimensional space, to other "sandwich" layers (see Figure 8c). The idea is that orthogonal projections of the regions corresponding to destinations of transitive arcs, which cannot be contained directly in the base region, will be directly contained in such subregions (or vice-versa). Because the thin tethered strip that was used for flapping is not relevant, it is represented by dots (see Figure 8d). As in the previous solutions, an arc (i, j) is in the dag if and only if a subregion S_k of R_i directly contains region R_j.

Superficially, this representation looks similar to Figure 7. However, its interpretation is totally different, it is now a flattened three-dimensional object. We can visualize this by imagining a living organism that has been totally flattened by a roller-compactor (apologies for the "gory" image).

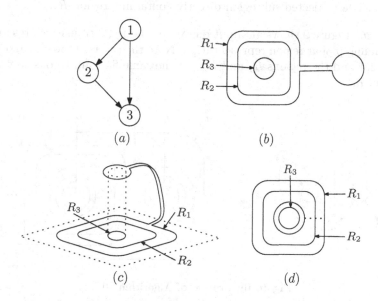

(a) (b)

(c) (d)

Fig. 8. The process described in Solution III

We next give a constructive algorithm that takes as input a dag (X, δ) and produces a set of overlapping regions $\{R_k \mid k \in X\}$, such that $(i, j) \in \delta$ if and only if a subregion of R_i directly contains R_j.

Algorithm 10: A dag to regions.

Input: dag (X, δ).
Output: flattened regions $\{R_k \mid k \in X\}$.

Step 1: Reorder the nodes of the dag (X, δ) to be in reverse topological order. (That is, sink nodes come before source nodes.)

Step 2: For each node i in δ ordered as in step 1 do:
    If i is a sink:
        Create a new region R_i disjoint from all previous regions.
    Otherwise:
        Create a base region of R_i by creating a simple closed region properly containing the union of all regions R_j such that $(i, j) \in \delta$. Further, for any transitive arc (i, j) create a flap subregion that directly contains R_j and attach it with a strip to the edge of the base region.

Remark 25. In the set constructed by this algorithm, if two or more transitive arcs are incident to a node j then the respective flaps (without tethers) may share the same projected subregion directly containing region R_j.

Example 26. Figure 9 shows an input dag with 6 nodes, 3 transitive arcs and its corresponding planar region representation. Note the reverse topological order is $6, 5, 4, 3, 2, 1$ and the regions R_1 and R_2 use the same flap subregions containing the region R_6.

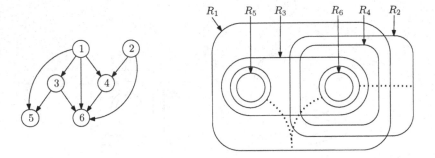

Fig. 9. Illustration of Algorithm 10

Theorem 27. *Every dag with transitive arcs can be represented by a set of regions with folded flaps, with a 1:1 mapping between nodes and regions.*

Proof. We show by induction on the order of the dags that we can always produce a corresponding planar representation. First, note that any dag can be recursively constructed by adding a new node i and arcs incident from i to existing nodes. Note that Algorithm 10 builds planar representations from sink nodes (induction base case) to source nodes (inductive case). Hence, any dag has at least one folded planar representation, depending on the topological order used. We omit the details of how to ensure non-arcs; this can be easily achieved by adding "spikes" to the regions—see our first paper for representing non-transitive dags [8].   □

Theorem 28. *Every set of regions with folded flaps can be represented by a dag with transitive arcs, with a 1:1 mapping between nodes and regions.*

Proof. We show how to produce a unique dag from a folded planar representation. The first step is to label each region R_k, which will correspond to node $k \in X$ of a dag (X, δ). We add an arc (i, j) to δ if an only if a subregion of R_i directly contains the region R_j.   □

Remark 29. One could imagine an additional constraint, that nodes, like cells, need to differentiate between its outside and inside or, in a planar representation, between up and down. We can relate this to membrane polarity, but we refrain from using this idea here, because it can conflict with the already accepted role of polarities in P systems. It is clear that, looking at our example, this solution does not take into account this *sense of direction*.

For example, considering the scenario of Figure (9), regions R_3, R_2 and R_1' (the base subregion of R_1) can be stacked "properly", i.e., with the bottom side of R_3 on the top side of R_2 and the bottom side of R_2 on the top side of R_1'. However, the top side of R_1'' (the flap of R_1) will improperly sit on the top side

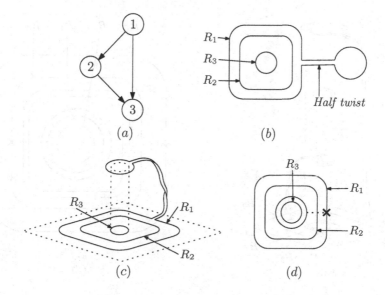

Fig. 10. The process described in Solution IV

of R_3, or, vice-versa, the bottom side of R_1'' will improperly sit on the bottom side of R_3.

Can we improve this? The answer follows.

6.4   Solution IV: Flaps with Half-Twists

This is a variation of Solution III, that additionally takes proper care of the outside/inside (or up/down) directions. We achieve this by introducing half-twists (as used to build Moebius strips), of which at most one half-twist is needed for each simple region.

Example 30. Figure 10 describes this process.

(a) a given dag with three nodes, $1, 2$ and 3;
(b) three simple regions, R_1, R_2 and R_3, still in the same plane;

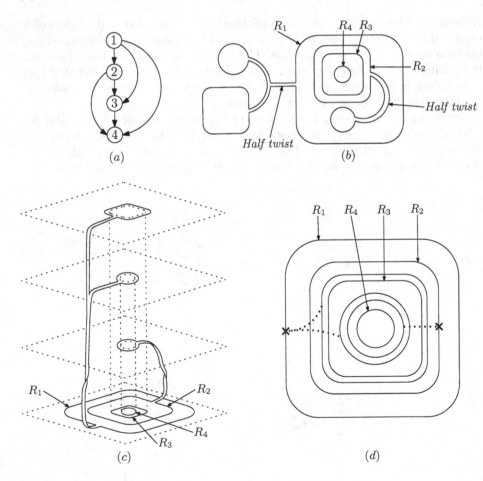

Fig. 11. The process described in Solution IV

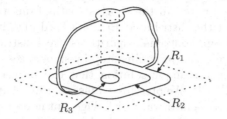

Fig. 12. The process described in Solution IV

(c) R_1 flapped and half-twisted in three-dimensional space;
(d) final "roller-compacted" representation, where dots represent the thin strip
    that was flapped, and the mark $\times$ a possible location of the half-twist.

Corollary 31. *Dags with transitive arcs can be represented by regions with half-twisted flaps, with a 1:1 mapping between nodes and regions.*

Proof. Since half-twisted flaps are folded flaps, the projection of the boundary of the base and flaps used for a region is the same region as given in the proof of Theorems 27 and 28, provided we always twist a fold above its base. $\square$

Remark 32. This solution solves all our concerns here and seems the best, taking into account the impossibility result (Theorem 20).

6.5   Solution V: Moebius Strips

To be complete, we mention another possible solution, which removes any distinction between up and down sides. This representation can be obtained by representing membranes by (connected) Moebius strips.

Perhaps interestingly, Solutions IV and V seem to suggest links (obviously superficial, but still links) to modern applications of topology (Moebius strips and ladders, knot theory) to molecular biology, for example, see [6].

7   Conclusions

In this paper we have presented several concrete examples of hP systems for the discovery of basic membrane structure. Our primary goal was to show that, with the correct model in terms of operational and transfer modes, we could present simple algorithms. Our secondary goal was to obtain reasonably efficient algorithms.

We first started with cases, where the cells could be anonymous, and showed, among other things, how an hP system could (a) broadcast to descendants, (b) count paths between cells, (c) count children and descendants, and (d) determine cell heights. We then provided examples where we allowed each cell to know its own ID and use it as a communication marker. This model is highlighted by our algorithm that computes all the shortest paths from a given source cell— a simplified version of the distributed Bellman-Ford algorithm, with all unity weights. For each of our nontrivial algorithms, we illustrated the hP system computations on a fixed dag, providing step-by-step traces.

We then moved onto a simple solution that can be used to synchronize a subset of (possibly all) cells. We presented a fast solution that requires structural extensions, which are straightforward with dags, but not applicable to trees. The solution given here assumes that the required extensions have already been built. In a related paper [5], we describe a natural way to dynamically extend a dag structure, which we believe is compatible with the P systems framework.

Finally, we focused on visualizing hP systems in the plane. We presented a natural model, using folded simple closed regions to model the membrane interconnections, including the transitive arcs, as specified by an arbitrary dag structure of an hP system.

As with most ongoing projects, there are several open problems regarding practical computing using P systems and their extended models. We end by mentioning just a few, closely related to the development of fundamental algorithms for discovery of membrane topology.

- In terms of using membrane computing as a model for realistic networking, is there a natural way to route a message between cells (not necessarily connected directly) using messages, tagged by addressing identifiers, in analogy to the way messages are routed on the internet, with dynamically created routing information?
- What are the system requirements to model fault tolerant computing? The tree structure seems to fail here, because a single node failure can disconnect the tree and make consensus impossible. Is the dag structure versatile enough?
- Do we have the correct mix of rewriting and transfer modes for membrane computing? For example, in which situations can we exploit parallelism and in which scenarios are we forced to sequentially apply rewriting rules?

Acknowledgements

The authors wish to thank John Morris and the three anonymous reviewers for detailed comments and feedback that helped us improve the paper.

References

1. Alhazov, A., Margenstern, M., Verlan, S.: Fast synchronization in P systems. In: Corne, D.W., Frisco, P., Paun, G., Rozenberg, G., Salomaa, A. (eds.) WMC 2008. LNCS, vol. 5391, pp. 118–128. Springer, Heidelberg (2009)
2. Bernardini, F., Gheorghe, M., Margenstern, M., Verlan, S.: How to synchronize the activity of all components of a P system? Int. J. Found. Comput. Sci. 19(5), 1183–1198 (2008)

3. Ciobanu, G., Desai, R., Kumar, A.: Membrane systems and distributed computing. In: Păun, G., Rozenberg, G., Salomaa, A., Zandron, C. (eds.) WMC 2002. LNCS, vol. 2597, pp. 187–202. Springer, Heidelberg (2003)
4. Ciobanu, G.: Distributed algorithms over communicating membrane systems. Biosystems 70(2), 123–133 (2003)
5. Dinneen, M.J., Kim, Y.-B., Nicolescu, R.: New solutions to the firing squad synchronization problem for neural and hyperdag P systems. EPTCS 15, 1–16 (2009)
6. Flapan, E.: When Topology Meets Chemistry: A Topological Look at Molecular Chirality. Cambridge University Press, Cambridge (2000)
7. Lynch, N.A.: Distributed Algorithms. Morgan Kaufmann Publishers Inc., San Francisco (1996)
8. Nicolescu, R., Dinneen, M.J., Kim, Y.-B.: Structured modelling with hyperdag P systems: Part A. In: Martínez del Amor, M.A., et al. (eds.) Seventh Brainstorming Week on Membrane Computing, vol. 2, pp. 85–107. Universidad de Sevilla (2009)
9. Păun, Gh.: Membrane Computing-An Introduction. Springer, Heidelberg (2002)
10. Păun, Gh.: Introduction to membrane computing. In: Ciobanu, G., Păun, Gh., Pérez-Jiménez, M.J. (eds.) Applications of Membrane Computing, pp. 1–42. Springer, Heidelberg (2006)

A Note on Small Universal
Spiking Neural P Systems

Linqiang Pan* and Xiangxiang Zeng

Key Laboratory of Image Processing and Intelligent Control
Department of Control Science and Engineering
Huazhong University of Science and Technology
Wuhan 430074, Hubei, People's Republic of China
lqpan@mail.hust.edu.cn, xzeng@foxmail.com

Abstract. In the "standard" way of simulating register machines by
spiking neural P systems (in short, SN P systems), one neuron is associated with each instruction of the register machine that we want to
simulate. In this note, a new way is introduced for simulating register
machines by SN P systems, where only one neuron is used for all instructions of a register machine; in this way, we can use less neurons to
construct universal SN P systems. Specifically, a universal system with
extended rules (without delay) having 10 neurons is constructed.

1   Introduction

The spiking neural P systems (in short, SN P systems) were introduced in [2], and
then investigated in a large number of papers. We refer to the respective chapter
of [7] for general information in this area, and to the membrane computing web
site from [11] for details.

Informally, an SN P system consists of a set of neurons placed in the nodes of
a directed graph, called the *synapse graph*. The content of each neuron consists
of a number of copies of a single object type, called the *spike*. The rules assigned
to neurons allow a neuron to send information to other neurons in the form of
electrical impulses (also called spikes). An output can be defined in the form of
the spike train produced by a specified output neuron.

Looking for small universal computing devices of various types is a well investigated issue in computer science, see, e.g. [3,8], and the references therein.
Recently, this issue was considered also in the case of SN P systems [5], where a
universal SN P system was obtained using 84 neurons for standard rules and 49
neurons for extended rules in the case of computing functions; used as generators
of sets of numbers, a universal system with standard rules (resp. extended rules)
having 76 neurons (resp. 50 neurons) was found. An improvement is presented in
[10] in the sense that less neurons are used to construct a universal SN P system.
Specifically, in the computing function mode, 68 neurons (resp. 43 neurons) are

---

* Corresponding author. Tel.: +86-27-87556070; Fax: +86-27-87543130.

G. Păun et al. (Eds.): WMC 2009, LNCS 5957, pp. 436–447, 2010.

used to construct a universal SN P system with standard rules (resp. extended rules); in the number generating mode, universal SN P systems are obtained with 64 neurons (resp. 43 neurons) using standard rules (resp. extended rules). All of the above universal SN P systems are obtained by simulating a register machine from [3], where a neuron is associated with each register of the register machine that we want to simulate; a neuron is associated with each instruction of the register machine; some auxiliary neurons are also used. If in the register machine that we want to simulate, there are m instructions and n registers, then the number of neurons in the universal SN P system obtained by this way is not less than $m + n$.

In this note, we present a new approach to simulate register machine, where one neuron (denoted by σ_{state}) is used for all instructions of the register machine. The function of neuron σ_{state} is similar with "the finite set of states" in a Turing machine. In this way, universal SN P systems with less neurons can be obtained. Specifically, a universal SN P system is constructed with extended rules (without delay) having 10 neurons.

The rest of this paper is organized as follows. In the next section, we introduce some necessary prerequisites. In Section 3, a small universal SN P system is constructed. Conclusions and remarks are presented in Section 4.

2   Prerequisites

We assume the reader to be familiar with (basic elements of) language theory [9], as well as basic membrane computing [6] (for more updated information about membrane computing, please refer to [11]), hence we directly introduce some basic notions and notations including register machines and SN P systems.

For an alphabet V, let V^* denotes the set of all finite strings over V, with the empty string denoted by λ. The set of all nonempty strings over V is denoted by V^+. When $V = \{a\}$ is a singleton, then we write simply a^* and a^+ instead of $\{a\}^*$, $\{a\}^+$.

A regular expression over an alphabet V is defined as follows: (i) λ and each $a \in V$ is a regular expression, (ii) if E_1, E_2 are regular expressions over V, then $(E_1)(E_2)$, $(E_1) \cup (E_2)$, and $(E_1)^+$ are regular expressions over V, and (iii) nothing else is a regular expression over V. With each expression E we associate a language $L(E)$, defined in the following way: (i) $L(\lambda) = \{\lambda\}$ and $L(a) = \{a\}$, for all $a \in V$, (ii) $L((E_1) \cup (E_2)) = L(E_1) \cup L(E_2)$, $L((E_1)(E_2)) = L(E_1)L(E_2)$, and $L((E_1)^+) = L(E_1^+)$, for all regular expressions E_1, E_2 over V. Non-necessary parentheses are omitted when writing a regular expression, and also $(E)^+ \cup \{\lambda\}$ can be written as E^*.

2.1   Register Machines

A register machine is a construct $M = (m, H, l_0, l_h, I)$, where m is the number of registers, H is the set of instruction labels, l_0 is the start label (labeling an ADD instruction), l_h is the halt label (assigned to instruction HALT), and I is the set of instructions; each label from H labels only one instruction from I, thus precisely identifying it. The instructions are of the following forms:

- $l_i : (\text{ADD}(r), l_j, l_k)$ (add 1 to register r and then go to one of the instructions with labels l_j, l_k non-deterministically chosen),
- $l_i : (\text{SUB}(r), l_j, l_k)$ (if register r is non-empty, then subtract 1 from it and go to the instruction with label l_j, otherwise go to the instruction with label l_k),
- $l_h : \text{HALT}$ (the halt instruction).

A register machine M generates a set $N(M)$ of numbers in the following way: we start with all registers being empty (i.e., storing the number zero), we apply the instruction with label l_0 and we continue to apply instructions as indicated by the labels (and made possible by the contents of registers); if we reach the halt instruction, then the number n present in specified register r_0 at that time is said to be generated by M. If the computation does not halt, then no number is generated. It is known (see, e.g., [4]) that register machines generate all sets of numbers which are Turing computable, even using register machines with only three registers as well as registers 1 and 2 being empty whenever the register machine halts, where we assume that the three registers are labeled with 0, 1, 2. Moreover, we may assume that the register where we place the computation result is not subject to subtraction operations [1].

Convention: when evaluating or comparing the power of two number generating/accepting devices, number zero is ignored.

2.2   Spiking Neural P Systems

We briefly recall the basic notions concerning spiking neural P systems (in short, SN P systems). For more details on such kind of systems, please refer to [2].

A *spiking neural P system* of degree $m \geq 1$ is a construct of the form

$$\Pi = (O, \sigma_1, \ldots, \sigma_m, syn, in, out),$$

where:

1. $O = \{a\}$ is the singleton alphabet (a is called spike);
2. $\sigma_1, \ldots, \sigma_m$ are neurons, of the form

$$\sigma_i = (n_i, R_i), 1 \leq i \leq m,$$

where:
   a) $n_i \geq 0$ is the initial number of spikes contained in σ_i;
   b) R_i is a finite set of rules of the following two forms:
      (1) $E/a^c \to a^p; d$, where E is a regular expression over a, and $c \geq 1$, $d \geq 0$, $p \geq 1$, with the restriction $c \geq p$;
      (2) $a^s \to \lambda$, for $s \geq 1$, with the restriction that for each rule $E/a^c \to a^p; d$ of type (1) from R_i, we have $a^s \notin L(E)$;

3. $syn \subseteq \{1, 2, \ldots, m\} \times \{1, 2, \ldots, m\}$ with $i \neq j$ for each $(i, j) \in syn, 1 \leq i, j \leq m$ (synapses between neurons);
4. $in, out \in \{1, 2, \ldots, m\}$ indicates the input and the output neurons, respectively.

If we always have $p = 1$ for all rules of the form $E/a^c \rightarrow a^p; d$, then the rules are said to be of the standard type, else they are called by extended rules.

The rules of type (1) are firing (we also say spiking) rules, and they are applied as follows. If the neuron σ_i contains k spikes, and $a^k \in L(E), k \geq c$, then the rule $E/a^c \rightarrow a^p; d \in R_i$ can be applied. This means consuming (removing) c spikes (thus only $k - c$ remain in σ_i), the neuron is fired, and it produces p spikes after d time units (as usual in membrane computing, a global clock is assumed, marking the time for the whole system, hence the functioning of the system is synchronized). If $d = 0$, then these spikes are emitted immediately, if $d = 1$, then these spikes are emitted in the next step, etc. If the rule is used in step t and $d \geq 1$, then in steps $t, t+1, \ldots, t+d-1$ the neuron is closed (this corresponds to the refractory period from neurobiology), so that it cannot receive new spikes (if a neuron has a synapse to a closed neuron and tries to send several spikes along it, then these particular spikes are lost). In the step $t + d$, the neuron spikes and becomes again open, so that it can receive spikes (which can be used starting with the step $t + d + 1$, when the neuron can again apply rules).

The rules of type (2) are forgetting rules; they are applied as follows: if the neuron σ_i contains exactly s spikes, then the rule $a^s \rightarrow \lambda$ from R_i can be used, meaning that all s spikes are removed from σ_i.

If a rule $E/a^c \rightarrow a; d$ has $E = a^c$, then we will write it in the simplified form $a^c \rightarrow a; d$.

If a rule $E/a^c \rightarrow a; d$ has $d = 0$, then we will write it in the simplified form $E/a^c \rightarrow a$.

In each time unit, if a neuron σ_i can use one of its rules, then a rule from R_i must be used. Since two firing rules, $E_1/a^{c_1} \rightarrow a^{p_1}; d_1$ and $E_2/a^{c_2} \rightarrow a^{p_2}; d_2$, can have $L(E_1) \cap L(E_2) \neq \emptyset$, it is possible that two or more rules can be applied in a neuron, and in that case, only one of them is chosen non-deterministically. Note however that, by definition, if a firing rule is applicable, then no forgetting rule is applicable, and vice versa.

Thus, the rules are used in the sequential manner in each neuron, at most one in each step, but neurons function in parallel with each other. It is important to notice that the applicability of a rule is established based on the total number of spikes contained in the neuron.

The initial configuration of the system is described by the numbers $n_1, n_2, \ldots, n_m$, of spikes present in each neuron, with all neurons being open. During the computation, a configuration of the system is described by both the number of spikes present in each neuron and by the state of the neuron, more precisely, by the number of steps to count down until it becomes open (this number is zero if the neuron is already open). Thus, $\langle r_1/t_1, \ldots, r_m/t_m \rangle$ is the configuration where neuron σ_i contains $r_i \geq 0$ spikes and it will be open after $t_i \geq 0$ steps, $i = 1, 2, \ldots, m$; with this notation, the initial configuration is $C_0 = \langle n_1/0, \ldots, n_m/0 \rangle$.

Using the rules as described above, one can define transitions among configurations. Any sequence of transitions starting in the initial configuration is called a computation. A computation halts if it reaches a configuration where all neurons are open and no rule can be used. In this note, we use SN P systems as number generating devices, we start from the initial configuration and we define the result of a computation as the number of steps between the first two spikes sent out by the output neuron.

In the next section, as usual, an SN P system is represented graphically, which may be easier to understand than in a symbolic way. We give an oval with rules inside to represent a neuron, and directed graph to represent the structure of SN P system: the neurons are placed in the nodes of a directed graph and the directed edges represent the synapses; the input neuron has an incoming arrow and the output neuron has an outgoing arrow, suggesting their communication with the environment.

3   A Small Universal SN P System

In this section we shall give a small universal SN P system (where extended rules, producing more than one spikes at a time, are used) by simulating a register machine.

Theorem 1. *There is a universal SN P system with extended rules (without delay) having 10 neurons.*

Proof. Let $M_u = (3, H, l_0, l_m, I)$ be a universal register machine with 3 registers labeled by 0, 1, 2, where $H = \{l_0, l_1, l_2, \ldots, l_m\}$ is the set of instruction labels, l_0 is the start label (labeling an ADD instruction) and l_m is the halt label (assigned to instruction HALT), I is the set of instructions. Without loss of generality, we assume that register 2 is not subject to subtraction operations.

We shall present an SN P system Π with 10 neurons to simulate register machine M_u. The structure of system Π is given in Figure 1, where spiking rules are omitted, which will be specified below. In system Π, neuron σ_{state} contains all spiking rules associated with all instructions of M_u (it is a point different with the "standard" way of simulating register machines by SN P systems, where one neuron is associated with each instruction of register machine that we want to simulate); neurons σ_i and σ_{a_i} ($i = 0, 1, 2$) are associated with registers 0, 1, 2; neuron σ_{out} is used to output the result of computation; auxiliary neurons σ_{b_1}, σ_{b_2} are used to send a fixed number of spikes to neuron σ_{state} at each step of computation. We point out that each neuron σ_i ($i = 0, 1$) has a synapse $(i, state)$ going to neuron σ_{state} except for neuron σ_2 (as you will see below, the difference originates from the fact that register 2 is not subject to substraction instructions); however, neuron σ_2 has a synapse $(2, b_2)$ going to neuron σ_{b_2}, which is used to stop the work of neurons σ_{b_1} and σ_{b_2} when the computation of system Π halts.

In system Π, each neuron is assigned with a set of rules, see Table 1, where $P(i) = 4(i+1)$, for $i = 0, 1, 2, \ldots, m$, and $T = 4(m+1)+1$. Neurons σ_i ($i = 0, 1$)

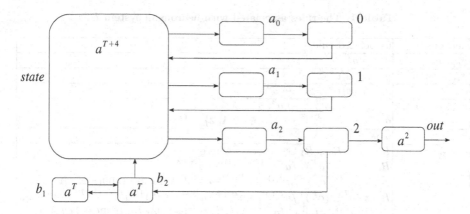

Fig. 1. The structure of system Π with the initial numbers of spikes

have the same set of rules except of neuron σ_2, the difference originates from the fact neuron σ_2 is not subject to subtraction instruction and it is related to output the result of computation. In neuron σ_{state}, there are $m + 1$ groups of rules $R_0, R_1, \ldots, R_m$, specifically, for each ADD instruction $l_i : (\text{ADD}(r), l_j, l_k)$, the set of rules $R_i = \{a^{P(i)}(a^T)^+/a^{P(i)+T-P(j)} \rightarrow a^{2r+3}, a^{P(i)}(a^T)^+/a^{P(i)+T-P(k)} \rightarrow a^{2r+3}\}$ is associated; for each SUB instruction $l_i : (\text{SUB}(r), l_j, l_k)$, the set of rules $R_i = \{a^{P(i)}(a^T)^+/a^{T+3} \rightarrow a^{2r+2}, a^{P(i)-1}(a^T)^+/a^{P(i)-1+T-P(j)} \rightarrow a, a^{P(i)-2}(a^T)^+/a^{P(i)-2+T-P(k)} \rightarrow a\}$ is associated; for instruction $l_m : \text{HALT}$, $R_m = \{a^{P(m)}(a^T)^+/a^{P(m)} \rightarrow a^6\}$ is associated. If the number of spikes in neuron σ_{state} is of the form $P(i) + sT$ for some $s \geq 1$ (that is, if the number of spikes is n, then $n \equiv P(i) \pmod{T}$; the value of multiplicity of T does not matter with the restriction that it should be greater than 0), then system Π starts to simulate instruction l_i. In particular, in the initial configuration of M_u, neuron σ_{state} has $T + 4$ spikes, which is the form $T + 4 = P(0) + T$, system Π starts to simulate the initial instruction l_0 of M_u; with $P(m) + sT = 4(m+1) + sT$ spikes in σ_{state}, system Π starts to output the result of computation; if the number of spikes in σ_{state} is of the form sT, then no rule in σ_{state} is enabled, which happens after the halt instruction is reached. That is why we use the label $state$ for this neuron, and the function of this neuron is somewhat similar with "the finite set of states" in Turing machine.

Initially, all neurons have no spike, with exception that each of neurons $\sigma_{b_1}, \sigma_{b_2}$ contains T spikes, neuron σ_{state} contains $P(0)+T = 4+T$ spikes, and neuron σ_{out} contains 2 spikes. As you will see, during the computation of M_u, the contents of registers r, $0 \leq r \leq 2$ are encoded by the number of spikes from neuron r in the following way: if the register r holds the number $n \geq 0$, then the associated neuron σ_r will contain $3n$ spikes; the increase (resp. decrease) of the number stored in register r is simulated by adding (resp. removing) three spikes.

With T spikes inside, neurons σ_{b_1} and σ_{b_2} fire by the rule $a^T \rightarrow a^T$, sending T spikes to each other; in this way, from step 1 until system Π starting to output

Table 1. The rules associated with neurons in system Π

neurons	associated rules
$\sigma_{b_1}, \sigma_{b_2}$	$a^T \to a^T$
$\sigma_i,\ i = 0, 1$	$a \to a,\ a(a^3)^+/a^4 \to a^2$
σ_2	$a \to a,\ a(a^3)^+/a^3 \to a^3$
$\sigma_{a_i},\ i = 0, 1, 2$	$a^{2i+2} \to a,\ a^{2i+3} \to a^3,\ a \to \lambda,$ $a^{2j+2} \to \lambda,\ a^{2j+3} \to \lambda,\ j \in \{0, 1, 2\} - \{i\}$
σ_{out}	$a \to a,\ a^3 \to \lambda,\ a^5 \to a$
σ_{state}	$R_{state} = R_0 \cup R_1 \cup \cdots \cup R_m,$ where: $R_i = \{a^{P(i)}(a^T)^+/a^{P(i)+T-P(j)} \to a^{2r+3},$ $\quad a^{P(i)}(a^T)^+/a^{P(i)+T-P(k)} \to a^{2r+3}\},$ for $l_i : (\text{ADD}(r), l_j, l_k);$ $R_i = \{a^{P(i)}(a^T)^+/a^{T+3} \to a^{2r+2},\ a^{P(i)-1}(a^T)^+/a^{P(i)-1+T-P(j)} \to a,$ $\quad a^{P(i)-2}(a^T)^+/a^{P(i)-2+T-P(k)} \to a\},$ for $l_i : (\text{SUB}(r), l_j, l_k);$ $R_m = \{a^{P(m)}(a^T)^+/a^{P(m)} \to a^6\},$ for $l_m : \textbf{HALT}$

the result of computation (that is, until a step when neuron σ_2 fires), at each step, neuron σ_{b_2} will send T spikes to σ_{state}.

In what follows, we check the simulation of register machine M_u by system Π, by decomposing system Π into three modules (i.e., modules ADD, SUB, and OUTPUT), and checking the work of each module.

Module ADD (Figure 2) – simulating an ADD instruction $l_i : (\text{ADD}(r), l_j, l_k)$

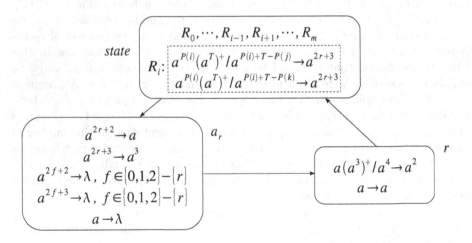

Fig. 2. Module ADD simulating $l_i : (\text{ADD}(r), l_j, l_k)$

The initial instruction of M_u, the one with label l_0, is an ADD instruction. Assume that we are in a step when we have to simulate an ADD instruction $l_i : (\text{ADD}(r), l_j, l_k)$, with the number of spikes being the form $P(i) + sT$ (for some $s \geq 1$) in neuron σ_{state} (in the initial configuration, neuron σ_{state} contains $P(0) + T$ spikes, and the simulation of the initial instruction with label l_0 is triggered). The rules $a^{P(i)}(a^T)^+/a^{P(i)+T-P(j)} \rightarrow a^{2r+3}$ and $a^{P(i)}(a^T)^+/a^{P(i)+T-P(k)} \rightarrow a^{2r+3}$ are enabled, non-deterministically choosing one of them to be applied.

If $a^{P(i)}(a^T)^+/a^{P(i)+T-P(j)} \rightarrow a^{2r+3}$ is applied, then neuron σ_{state} fires, sending out $2r+3$ spikes to neurons σ_{a_i} ($i = 0, 1, 2$). Neuron σ_{a_r} sends 3 spikes to neuron σ_r by rule $a^{2r+3} \rightarrow a^3$. In neurons σ_{a_t} ($t \in \{0, 1, 2\} - \{r\}$), these $2r+3$ spikes are forgotten by rule $a^{2r+3} \rightarrow \lambda$. Therefore, neuron σ_r increases its number of spikes by 3, and does not fire, which simulates the increase of the number stored in register r by 1. After consuming $P(i) + T - P(j)$ spikes by the rule $a^{P(i)}(a^T)^+/a^{P(i)+T-P(j)} \rightarrow a^{2r+3}$, the number of spikes in neuron σ_{state} is of the form $P(j) + sT$ (for some $s \geq 1$) (recalling that neuron σ_{state} receives T spikes from neuron σ_{b_2} at each step), hence system Π starts to simulate an instruction with label l_j.

Similarly, if $a^{P(i)}(a^T)^+/a^{P(i)+T-P(k)} \rightarrow a^{2r+3}$ is applied, then neuron σ_r increases its number of spikes by 3, and the number of spikes in neuron σ_{state} is of the form $P(k) + sT$ (for some $s \geq 1$). This implies that the number stored in register r is increased by 1, and system Π starts to simulate an instruction with label l_k.

The simulation of the ADD instruction is correct: we have increased the number of spikes in neuron σ_r by three, and we have passed to the simulation of one of the instructions l_j and l_k non-deterministically.

Remark: (1) The auxiliary neurons σ_{b_1} and σ_{b_2} are necessary for the function of system Π. They send T spikes to neuron σ_{state} at each step, which ensures that the number of spikes in neuron σ_{state} not less than 0.

(2) In the simulation of an ADD instruction, when neuron σ_{state} fires, it sends $2r+3$ spikes to all neurons σ_{a_i} ($i = 0, 1, 2$). Checking the rules in neurons σ_{a_i} ($i = 0, 1, 2$) (listed in Table 1), we can find that in neuron σ_{a_r} only rule $a^{2r+3} \rightarrow a$ is enabled and applied, sending three spikes to neuron σ_r; in neuron σ_{a_t} with $t \neq r$, only rule $a^{2r+3} \rightarrow \lambda$ is enabled and applied, these $2r+3$ spikes are forgotten, and neuron σ_t, $t \neq r$, receives no spike. In general, there is a bijection relation: neuron σ_r receives 3 spikes if and only if neuron σ_{state} sends out $2r+3$ spikes, where $r = 0, 1, 2$. So, the neurons σ_{a_i} ($i = 0, 1, 2$) work like a "sieve" such that only the register that the ADD instruction acts on can increase its number by 1.

(3) As you will see below, when a SUB instruction that acts on register r is simulated, neuron σ_{state} sends out $2r+2$ spikes. In this case, neurons σ_{a_i} ($i = 0, 1, 2$) also work like a "sieve", but with different bijection relation: neuron σ_r receives 1 spike if and only if neuron σ_{state} sends out $2r+2$ spikes, where $r = 0, 1, 2$.

Module SUB (Figure 3) – simulating a SUB instruction $l_i : (\text{SUB}(r), l_j, l_k)$.

The execution of instruction $l_i : (\text{SUB}(r), l_j, l_k)$ is simulated in Π in the following way. With the number of spikes in neurons σ_{state} having the form $P(i) + sT$ (for some $s \geq 1$), rule $a^{P(i)}(a^T)^+/a^{T+3} \rightarrow a^{2r+2}$ is enabled and applied, sending out $2r+2$ spikes; we suppose it is at step t. At step $t+1$, neuron σ_{a_r} spikes by the rule $a^{2r+2} \rightarrow a$, sending one spike to neuron σ_r; these $2r+2$ spikes in neuron σ_{a_t}, $t \neq r$, are forgotten by the rule $a^{2r+2} \rightarrow \lambda$ (that is, the "sieve" function of

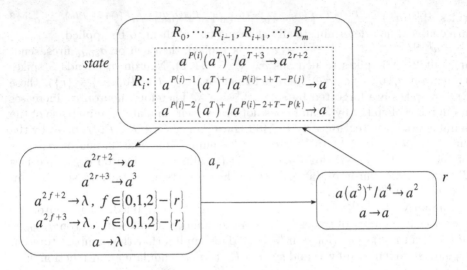

Fig. 3. Module SUB simulating $l_i : (\text{SUB}(r), l_j, l_k)$

neurons σ_{a_i}, $i = 0, 1, 2$, works again). For the number of spikes in neuron σ_r at step t, we consider the following two cases: (1) neuron σ_r contains at least three spikes (that is, register r is not empty); (2) neuron σ_r contains no spike (that is, register r is empty).

(1) If the number of spikes in neuron σ_r at step t is $3n$ with $n > 0$, then receiving one spike from neuron σ_{a_r} at step $t + 1$, neuron σ_r has $3n + 1$ spikes at step $t + 2$, and rule $a(a^3)^+/a^4 \to a^2$ is enabled and applied, consuming 4 spikes, sending 2 spike to neuron σ_{state}. In this way, the number of spikes in neuron σ_r is $3(n-1)$, simulating the number stored in register r is decreased by one. After receiving these 2 spikes, the number of spikes in neuron σ_{state} is of the from $P(i) - 1 + sT$ (for some $s \geq 1$), so rule $a^{P(i)-1}(a^T)^+/a^{P(i)+T-1-P(j)} \to a$ can be applied. Consuming $P(i) - 1 + T - P(j)$ at step $t + 3$ by rule $a^{P(i)-1}(a^T)^+/a^{P(i)+T-1-P(j)} \to a$, the number of spikes in neuron σ_{state} is of the form $P(j) + sT$ (for some $s \geq 1$), which means that the next simulated instruction will be l_j. Note that this one spike emitted by neuron σ_{state} will be immediately forgotten by all neurons $\sigma_{a_0}, \sigma_{a_1}, \sigma_{a_2}$ at the next step because of the rule $a \to \lambda$ in these neurons.

(2) If the number of spikes in neuron σ_r at step t is 0, then at step $t + 2$, neuron σ_r contains one spike (received from neuron σ_{a_r} at step $t + 1$), and the rule $a \to a$ is applied, consuming the single spike present in neuron σ_r and sending one spike to neuron σ_{state}. Neuron σ_{state} contains $P(i) - 2 + sT$ (for some $s \geq 1$) spikes at step $t + 3$, rule $a^{P(i)-2}(a^T)^+/a^{P(i)+T-2-P(k)} \to a$ is enabled and applied, consuming $P(i) - 2 + T - P(k)$ spikes. So, the number of spikes in neuron σ_{state} is of the form $P(k) + sT$ (for some $s \geq 1$), and system Π starts to simulate the instruction l_k.

The simulation of the SUB instruction is correct: starting from the simulation of the instruction l_i, we passed to simulate the instruction l_j if the register was non-empty and decreased by one, and to simulate instruction l_k if the register is empty.

Remark: In the set of rules R_i associated with a SUB instruction l_i, the regular expressions have numbers $P(i), P(i)-1, P(i)-2, P(i)-3$. Because $P(i) = 4(i+1)$ for each instruction l_i, which implies that $\{P(i_1), P(i_1) - 1, P(i_1) - 2, P(i_1) - 3\} \cap \{P(i_2), P(i_2) - 1, P(i_2) - 2, P(i_2) - 3\} = \emptyset$, for $i_1 \neq i_2$, the simulation of SUB instructions do not interfere with each other. On the other hand, in the set of rules R_i associated with an ADD instruction l_i, the regular expressions have number $P(i)$, it is not difficult to see that the simulations of an ADD instruction and a SUB instruction do not interfere with each other too. That is why we take $P(i)$ as a multiplicity of number 4.

Module OUTPUT (Figure 4) – outputting the result of computation.

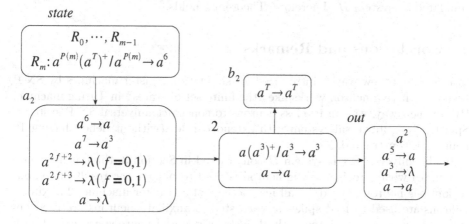

Fig. 4. Module OUTPUT

Assume now that the computation in M_u halts, which means that the halt instruction l_m is reached. For system Π, this means that neuron σ_{state} contains $P(m)+sT$ spikes (for some $s \geq 1$). At that moment, neuron σ_2 contains $3n$ spikes, for n being the content of register 2 of M_u. Having $P(m) + sT$ spikes inside, neuron σ_{state} gets fired and emits 6 spikes by the rule $a^{P(m)}(a^T)^+/a^{P(m)} \to a^6$. After that, the number of spikes in neuron σ_{state} is of the form sT (for some $s \geq 1$), no rule can be applied anymore in neuron σ_{state}.

At the next step, neurons $\sigma_{a_0}, \sigma_{a_1}$ forget these 6 spikes received from σ_{state} by the rule $a^6 \to \lambda$; only neuron σ_{a_2} sends one spike to neuron σ_2 by the rule $a^6 \to a$. In this way, neuron σ_2 has $3n + 1$ spikes, hence the rule $a(a^3)^+/a^3 \to a^3$ can be applied, sending three spikes to neuron σ_{out}. With five spikes inside (three spikes were received from neuron σ_2; two spikes were contained from the initial configuration), neuron σ_{out} fires by the rule $a^5 \to a$, which is the first spike sent out by system Π to the environment. Let t be the moment when neuron σ_{out} fires.

When neuron σ_2 spikes at step $t - 1$, neuron σ_{b_2} also receives 3 spikes from neuron σ_2, which gets "over flooded" and is blocked. So, neurons σ_{b_1} and σ_{b_2} stop their works.

Note that at step t, neuron σ_2 contains $3(n - 1) + 1$ spikes (three spikes were already consumed at step $t - 1$). From step t on, at each step, three spikes are consumed in neuron σ_2 by the rule $a(a^3)^+/a^3 \rightarrow a^3$, sending 3 spikes to neuron σ_{out}; these three spikes in neuron σ_{out} are forgotten by the rule $a^3 \rightarrow \lambda$. So, at step $t + (n - 1)$, neuron σ_2 contains one spike, and the rule $a \rightarrow a$ is enabled and applied, sending one spike to neuron σ_{out}. With one spike inside, neuron σ_{out} fires for the second (and last) time by the rule $a \rightarrow a$ at step $t + n$. The interval between these two spikes sent out to the environment by the system is $(t + n) - t = n$, which is exactly the number stored in register 2 of M_u at the moment when the computation of M_u halts.

From the above description, it is clear that the register machine M_u is correctly simulated by system Π. Therefore, Theorem 1 holds.

4   Conclusions and Remarks

In this note, a new way is introduced for simulating register machines by SN P systems, where a neuron works like "the finite set of states" in Turing machine. By this new way, we can use less neurons to construct universal SN P systems. Specifically, a universal system with extended rules (without delay) having 10 neurons is constructed.

In the universal SN P system Π constructed in Section 3, three neurons are associated with 3 registers; one neuron is used to output the result of computation; one neuron is used for all instructions of a register machine; 2 auxiliary neurons are used to feed spikes at each step; 3 auxiliary neurons are used between the neuron associated with all instructions and neurons associated with registers, which work as a "sieve". Can we remove these 4 auxiliary neurons to get smaller universal SN P systems? One possible way of removing these auxiliary neurons is to use more rules in the neuron associated with instructions realizing the function of "sieve".

In this note, we only considered SN P systems with extended rules without delay. Can we extend this way to the case of SN P systems with standard rules (it seems that a few more neurons are necessary), asynchronous SN P systems, or other variants and modes of SN P systems?

The universal SN P system constructed in this note is already quite small. If we start from universal register machines to construct universal SN P systems, then it may be not easy to get significant improvement. Of course, it is still possible to have smaller universal SN P systems, if we start construction from other small universal computational devices.

Acknowledgements

The comments from three anonymous referees are acknowledged. The comments from Artiom Alhazov, Enrique Ocejuel Piuedo, Rudolf Freund, Sergey Verlan during WMC10 are also greatly appreciated. The work was supported by National Natural Science Foundation of China (Grant Nos. 60674106, 30870826, 60703047, and 60533010), Program for New Century Excellent Talents in University (NCET-05-0612), Ph.D. Programs Foundation of Ministry of Education of China (20060487014), Chenguang Program of Wuhan (200750731262), HUST-SRF (2007Z015A), and Natural Science Foundation of Hubei Province (2008CDB113 and 2008CDB180).

References

1. Alhazov, A., Freund, R., Oswald, M., Slavkovik, M.: Extended spiking neural P systems. In: Hoogeboom, H.J., Păun, G., Rozenberg, G., Salomaa, A. (eds.) WMC 2006. LNCS, vol. 4361, pp. 123–134. Springer, Heidelberg (2006)
2. Ionescu, M., Păun, G., Yokomori, T.: Spiking neural P systems. Fundamenta Informaticae 71(2-3), 279–308 (2006)
3. Korec, I.: Small universal register machines. Theoretical Computer Science 168, 267–301 (1996)
4. Minsky, M.: Computation – Finite and Infinite Machines. Prentice Hall, New Jersey (1967)
5. Păun, A., Păun, G.: Small universal spiking neural P systems. BioSystems 90(1), 48–60 (2007)
6. Păun, G.: Membrane Computing – An Introduction. Springer, Berlin (2002)
7. Păun, G., Rozenberg, G., Salomaa, A. (eds.): Handbook of Membrane Computing. Oxford University Press, Oxford (2010)
8. Rogozhin, Y.: Small universal Turing machines. Theoretical Computer Science 168, 215–240 (1996)
9. Rozenberg, G., Salomaa, A. (eds.): Handbook of Formal Languages, 3 vols. Springer, Berlin (1997)
10. Zhang, X., Zeng, X., Pan, L.: Smaller universal spiking neural P systems. Fundamental Informaticae 87(1), 117–136 (2008)
11. The P System Web Page, http://ppage.psystems.eu

On the Power of Computing
with Proteins on Membranes

Petr Sosík[1,2], Andrei Păun[1,3,4,5], Alfonso Rodríguez-Patón[1], and David Pérez[1]

[1] Departamento de Inteligencia Artificial, Facultad de Informática
Universidad Politécnica de Madrid, Campus de Montegancedo s/n
Boadilla del Monte, 28660 Madrid, Spain
{psosik,apaun,arpaton,dperez}@fi.upm.es
[2] Institute of Computer Science, Silesian University, 74601 Opava, Czech Republic
[3] Department of Computer Science/IfM, Louisiana Tech University, P.O. Box 10348,
Ruston, LA 71272, USA
[4] Bioinformatics Department, National Institute of Research and Development for
Biological Sciences, Splaiul Independenţei, Nr. 96, Sector 6, Bucharest, Romania
[5] Faculty of Mathematics and Computer Science, Department of Computer Science,
University of Bucharest, Str. Academiei 14, 70109, Bucharest, Romania

Abstract. P systems with proteins on membranes are inspired closely
by switching protein channels. This model of membrane computing using
membrane division has been previously shown to solve an NP-complete
problem in polynomial time. In this paper we characterize the class of
problems solvable by these P systems in polynomial time and we show
that it equals **PSPACE**. Therefore, these P systems are computationally
equivalent (up to a polynomial time reduction) to the alternating Turing
machine or the PRAM computer. The proof technique we employ reveals
also some interesting trade-offs between certain P system properties,
as antiport rules, membrane labeling by polarization or the presence
of proteins.

1 Introduction

We continue the work on P systems with proteins on membranes, a model combining membrane systems and brane calculi as introduced in [6]. We consider a
rather restrictive case, where the "main" information to process is encoded in
the multisets from the regions of a P system, but these objects evolve under the
control of a bounded number of proteins placed on membranes. Also, the rules
we use are very restrictive: move objects across membranes, under the control
of membrane proteins, changing or not the objects and/or the proteins during
these operations. In some sense, we have an extension of symport/antiport rules
[5], with the mentioning that we always use minimal rules, dealing with only one
protein, one object inside the region and/or one object outside of it.

The motivation came from the observation by several authors recently that
the maximal parallelism way of processing different species of molecules in the
membrane structure is not very close to reality, thus we are considering a model

G. Păun et al. (Eds.): WMC 2009, LNCS 5957, pp. 448–460, 2010.

that is limiting the parallelism through the modeling of the trans-membrane proteins (protein channels) observed in nature. A second motivation comes from the brane calculi in which many rules act at the level of the membrane (unlike rules which act within the region enclosed by the membrane). In brane calculi introduced in [3], one works only with objects – called proteins – placed on membranes, while the evolution is based on membrane handling operations, such as exocytosis, phagocytosis, etc. In the membrane computing area we have rules associated with each region defined by a membrane, and in the recent years the rules in membrane computing have been considered mainly to work on symbol objects rather than other structures such as strings. The extension considered in [6] and in [7] was to have both types of rules (both at the level of the region delimited by membranes and also at the level of membrane controlled by a protein). The reason for considering both extensions was that in biology, many reactions taking place in the compartments of living cells are controlled/catalysed by the proteins embedded in the membranes bilayer. For instance, it is estimated that in the animal cells, the proteins constitute about 50% of the mass of the membranes, the rest being lipids and small amounts of carbohydrates. There are several types of such proteins embedded in the membrane of the cell; one simple classification places these proteins into two classes, that of integral proteins (these molecules can "work" in both inside the membrane as well as also in the region outside the membrane), and that of peripheral proteins (macromolecules that can only work in one region of the cell) – see [1].

In this paper we show that P systems with proteins on membranes can solve in polynomial time exactly the class of problems **PSPACE**. Mathematically, this property can be expressed as

$$M\text{-PTIME} = M\text{-NPTIME} = \textbf{PSPACE}, \tag{1}$$

where M-(N)PTIME is the class of problems solved in polynomial time by a (non-) deterministic machine M. (In our case, the machine M will be a P system with proteins on membranes.) This relation is also known as the *Parallel Computation Thesis* [12]. Computational devices with this property form the so-called *second machine class*. Another members of this class are the alternating Turing machine, SIMDAG (also known as SIMD PRAM) and other standard parallel computer models [12].

The rest of the paper is organized as follows: after introducing basic concepts used throughout the paper in Section 2, we show in Section 3 that the P systems with proteins on membranes can solve the problem QSAT in linear time. Then in Section 4 we show that such a P system can be simulated with a conventional computer (and hence also with Turing machine) in a polynomial space. Section 5 concludes the paper and mentions also some open problems.

2   Definitions

We will start by giving some preliminary notations and definitions which are standard in the area of membrane systems. The reader is referred to [4,8] for an introduction and overview of membrane systems, and to [13] for the most recent information. The membranes delimit *regions* precisely identified by the membranes. In these regions we place *objects* — elements of the set O. Several copies of the same object can be present in a region, so we work with *multisets* of objects. For a multiset M we denote by $|M|_a$ the multiplicity of objects a in M. A multiset M with the underlying set O can be represented by a string $x \in O^*$ (by O^* we denote the free monoid generated by O with respect to the concatenation and the identity λ) such that the number of occurrences of $a \in O$ in x represents the value $|M|_a$.

In the P systems which we consider below, we use two types of objects, *proteins* and usual *objects*; the former are placed **on** the membranes, the latter are placed **in** the regions delimited by membranes. The fact that a protein p is on a membrane (with label) i is written in the form $[_i p|\ ]_i$. Both the regions of a membrane structure and the membranes can contain multisets of objects and of proteins, respectively.

We consider the types of rules introduced in [6]. In all of these rules, a, b, c, d are objects, p is a protein, and i is a label ("res" stands for "restricted"):

Type	Rule	Effect
1res	$[_i p\|a]_i \to [_i p\|b]_i$	
	$a[_i p\|\ ]_i \to b[_i p\|\ ]_i$	modify an object, but not move
2res	$[_i p\|a]_i \to a[_i p\|\ ]_i$	
	$a[_i p\|\ ]_i \to [_i p\|a]_i$	move an object, but not modify
3res	$[_i p\|a]_i \to b[_i p\|\ ]_i$	
	$a[_i p\|\ ]_i \to [_i p\|b]_i$	modify and move one object
4res	$a[_i p\|b]_i \to b[_i p\|a]_i$	interchange two objects
5res	$a[_i p\|b]_i \to c[_i p\|d]_i$	interchange and modify two objects

In all cases above, the protein is not changed, it plays the role of a catalyst, just assisting the evolution of objects. A generalization is to allow rules of the forms below (now, "cp" means "change protein"):

Type	Rule	Effect (besides changing also the protein)
1cp	$[_i p\|a]_i \to [_i p'\|b]_i$	
	$a[_i p\|\ ]_i \to b[_i p'\|\ ]_i$	modify an object, but not move
2cp	$[_i p\|a]_i \to a[_i p'\|\ ]_i$	
	$a[_i p\|\ ]_i \to [_i p'\|a]_i$	move an object, but not modify
3cp	$[_i p\|a]_i \to b[_i p'\|\ ]_i$	
	$a[_i p\|\ ]_i \to [_i p'\|b]_i$	modify and move one object
4cp	$a[_i p\|b]_i \to b[_i p'\|a]_i$	interchange two objects
5cp	$a[_i p\|b]_i \to c[_i p'\|d]_i$	interchange and modify two objects

where p, p' are two proteins (possibly equal, and then we have rules of type *res*).

An intermediate case can be that of changing proteins, but in a restricted manner, by allowing at most two states for each protein, $p, \bar{p}$, and the rules either as in the first table (without changing the protein), or changing from p

to $\bar{p}$ and back (like in the case of bistable catalysts). Rules with such flip-flop proteins are denoted by $nff, n = 1, 2, 3, 4, 5$ (note that in this case we allow both rules which do not change the protein and rules which switch from p to $\bar{p}$ and back).

Both in the case of rules of type ff and of type cp we can ask that the proteins are always moved in their complementary state (from p into $\bar{p}$ and vice versa). Such rules are said to be of *pure ff* or *cp* type, and we indicate the use of pure ff or cp rules by writing ffp and cpp, respectively.

To *divide a membrane*, we use the following type of rule, where p, p', p'' are proteins (possible equal): $\qquad [_ip|\,]_i \rightarrow [_ip'|\,]_i[_ip''|\,]_i$

The membrane i can be non-elementary. The rule doesn't change the membrane label i and instead of one membrane, at next step, will have two membranes with the same label i and the same contents replicated from the original membrane: objects and/or other membranes (although the rule specifies only the proteins involved).

Definition 1. *A P system with proteins on membranes and membrane division (in the sequel simply P system, if not stated otherwise) is a system of the form*
$$\Pi = (O, P, \mu, w_1/z_1, \ldots, w_m/z_m, E, R_1, \ldots, R_m, i_o), \text{ where}$$

m *is the degree of the system (the number of membranes),*
O *is the set of objects, P is the set of proteins (with $O \cap P = \emptyset$),*
μ *is the membrane structure,*
$w_1, \ldots, w_m$ *are the (strings representing the) multisets of objects present in the
$\quad$ m regions of the membrane structure μ,*
$z_1, \ldots, z_m$ *are the multisets of proteins present on the m membranes of μ,*
$E \subseteq O$ *is the set of objects present in the environment (in an arbitrarily large
$\quad$ number of copies each),*
$R_1, \ldots, R_m$ *are finite sets of rules associated with the m membranes of μ, and*
i_o *is the label of the output membrane.*

The rules are used in the non-deterministic maximally parallel way: in each step, a maximal multiset of rules is used, that is, no rule is applicable to the objects and the proteins which remain unused by the chosen multiset. At each step we have the condition that each object and each protein can be involved in the application of at most one rule, but the membranes are not considered as involved in the rule applications except the division rules, hence the same membrane can appear in any number of rules of types 1–5 at the same time. By halting computation we understand a sequence of configurations that ends with a halting configuration (there is no rule that can be applied considering the objects and proteins present at that moment in the system). With a halting computation we associate a result, in the form of the multiplicity of objects present in region i_o at the moment when the system halts. We denote by $N(\Pi)$ the set of numbers computed in this way by a given system Π. We denote, in the usual way, by $NOP_m(pro_r; list\text{-}of\text{-}types\text{-}of\text{-}rules)$ the family of sets of numbers $N(\Pi)$ generated by systems Π with at most m membranes, using rules as specified in the list-of-types-of-rules, and with at most r proteins present on a membrane. When parameters m or r are not bounded, we use $*$ as a subscript.

Example: Consider the P system $\Pi = (O, P, \mu, w_0/z_0, w_1/z_1, E, R_0, R_1, i_0)$, where

$$O = \{a_1, \ldots, a_n\}$$
$$P = \{p, q\}$$
$$\mu = [_0[_1]_1]_0$$
$$w_0 = z_0 = E = \emptyset$$
$$w_1 = \{a_1\}$$
$$z_1 = \{p\}$$
$$R_0 = \emptyset$$
$$R_1 = \{[_1p|\;]_1 \to [_1q|\;]_1[_1q|\;]_1, [_1q|a_n]_1 \to a_n[_1q|\;]_1\}$$
$$\cup \{[_1q|a_i]_1 \to [_1p|a_{i+1}]_1 \mid 1 \le i \le n-1\}$$
$$i_0 = 0$$

In its initial configuration the system contains two membranes and one object. In every odd step all the membranes labelled 1 are divided and their membrane proteins are changed from p to q. In every even step the proteins change back from q to p, and objects a_i in the membranes evolve to a_{i+1}, for $1 \le i \le n-1$. Therefore, every two steps the number of membranes labelled 1 is doubled. In $2n$-th step the objects a_n are expelled to the membrane labelled 0, which is the output membrane, and the systems halts. The computation of the system is illustrated in Fig. 1. Therefore, we can write that $N(\Pi) = \{2^n \mid n \in \mathbb{N}\}$.

Several computational universality results are known to hold for P systems with proteins on membranes [7,6], from which we recall only two:

$$NOP_1(pro_2; 2cpp) = NRE,$$
$$NOP1(pro_*; 3ffp) = NRE,$$

where NRE is the class of all recursively enumerable sets of non-negative integers. In this paper, however, we focus on P systems working in accepting mode, described in the next section, which can solve decision problems.

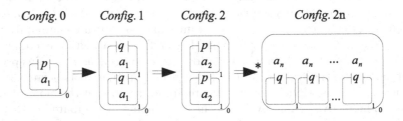

Fig. 1. An example of a P system with proteins on membranes

2.1 Families of Membrane Systems

Most of the membrane computing models are universal, i.e., they allow for a construction of a universal machine capable of solving any Turing-computable problem. However, when we try to employ the massive parallelism of P systems for effective solutions to intractable problems, the concept of one universal P systems solving all the instances of the problem is rather restrictive. The effective use of parallelism can be restricted by the particular structure of such a P system. For instance, the depth of the structure is fixed during the computation in most P system models, but for an effective parallel solution to various instances, various depths of the membrane structure might be needed.

Therefore, families of P systems are frequently used instead of a single P system. We start with formalizing a concept of *recognizer P system* for solving decision problems, which must comply with the following requirements: (a) the working alphabet contains two distinguished elements *yes* and *no*; (b) all computations halt; and (c) exactly one of the object *yes* (accepting computation) or *no* (rejecting computation) must be sent to the output region of the system, and only at the last step of each computation.

Consider a decision problem $X = (I_X, \theta_X)$ where I_X is a language over a finite alphabet (whose elements are called instances) and θ_X is a total boolean function over I_X.

Definition 2 ([9]). *A family* $\Pi = \{\Pi(w) : w \in I_X\}$ *of recognizer membrane systems without input membrane is* polynomially uniform by Turing machines *if there exists a deterministic Turing machine working in polynomial time which constructs the system* $\Pi(w)$ *from the instance* $w \in I_X$.

In this paper we deal with recognizer systems without input membrane, i.e., an instance w of a problem X is encoded into the structure of the P system $\Pi(w)$. The system $\Pi(w)$ is supposed to solve the instance w. Formally, [9] defines the conditions of *soundness* and *completeness* of Π with respect to X. A conjunction of these two conditions ensures that for every $w \in I_X$, if $\theta_X(w) = 1$, then every computation of $\Pi(w)$ is accepting, and if $\theta_X(w) = 0$, then every computation of $\Pi(w)$ is rejecting.

Note that the system $\Pi(w)$ can be generally nondeterministic, i.e, it may have different possible computations, but with the same result. Such a P system is also called *confluent*.

Definition 3 ([9]). *A decision problem X is solvable in polynomial time by a family of recognizer P systems belonging to a class $\mathcal{R}$ without input membrane* $\Pi = \{\Pi(w) : w \in I_X\}$, *denoted by* $X \in \mathbf{PMC}^*_{\mathcal{R}}$, *if the following holds:*

- *The family Π is polynomially uniform by Turing machines.*
- *The family Π is polynomially bounded; that is, there exists a natural number $k \in \mathbb{N}$ such that for each instance $w \in I_X$, every computation of $\Pi(w)$ performs at most $|w|^k$ steps.*
- *The family Π is sound and complete with respect to X.*

The family Π is said to provide a semi-uniform solution to the problem X. In this case, for each instance of X we have a special P system. Let us denote by $\mathcal{MP}$ the class of P systems with proteins on membranes. The following relation is proven in [7] for P systems with proteins on membranes:

$$NP \subseteq PMC^*_{\mathcal{MP}}.$$ (2)

3   Solving QSAT in Linear Time

In this section we show that P systems with proteins on membranes can solve in linear time the **PSPACE**-complete problem QSAT. More precisely, there exists a semi-uniform family of these P systems such that for each instance of QSAT, a proper P system solving that instance in a linear time can be constructed in a polynomial time w.r.t. the size of the instance. We also observe interesting trade-off between the use of certain elementary P systems operations.

The problem QSAT (satisfiability of quantified propositional formulas) is a standard **PSPACE**-complete problem. It asks whether or not a given quantified boolean formula in the conjunctive normal form assumes the value *true*. A formula as above is of the form

$$\gamma = Q_1 x_1 Q_2 x_2 \ldots Q_n x_n (C_1 \wedge C_2 \wedge \ldots \wedge C_m),$$ (3)

where each Q_i, $1 \leq i \leq n$, is either $\forall$ or $\exists$, and each C_j, $1 \leq j \leq m$, is a *clause* of the form of a disjunction

$$C_j = y_1 \vee y_2 \vee \ldots \vee y_r,$$

with each y_k being either a propositional variable, x_s, or its negation, $\neg x_s$. For example, let us consider the propositional formula

$$\beta = Q_1 x_1 Q_2 x_2 [(x_1 \vee x_2) \wedge (\neg x_1 \vee \neg x_2)]$$

It is easy to see that it is *true* when $Q_1 = \forall$ and $Q_2 = \exists$, but it is *false* when $Q_1 = \exists$ and $Q_2 = \forall$.

Theorem 1. PSPACE $\subseteq$ PMC$^*_{\mathcal{MP}}$.

Proof. Consider a propositional formula γ of the form (3) with

$$C_i = y_{i,1} \vee \ldots \vee y_{i,p_i},$$

for some $p_i \geq 1$, and $y_{i,j} \in \{x_k, \neg x_k \mid 1 \leq k \leq n\}$, for each $1 \leq i \leq m, 1 \leq j \leq p_i$. We construct the P system

$$\Pi = (O, P, \mu, w_0/z_0, w_1/z_1, \ldots, w_{n+2}/z_{n+2}, \emptyset, R_0, R_1, \ldots, R_{n+2}, 0)$$

with the components

$$O = \{a_i, t_i, f_i \mid 1 \leq i \leq n\} \cup \{r_i, \overline{r}_i \mid 1 \leq i \leq m\} \cup \{c_i \mid 1 \leq i \leq 3m + 5n\} \cup$$
$$\cup \{t, s, yes, no\},$$
$$P = \{p_0, p_+, p_-, p_x\},$$
$$\mu = [_0[_1 \cdots [_n[_{n+1}]_{n+1}[_{n+2}]_{n+2}]_n \cdots]_1]_0,$$
$$w_0 = c_1, \quad w_1 = a_1 no$$
$$w_i = a_i, \text{ for each } i = 2, \dots, n,$$
$$w_{n+1} = r_1 r_2 \dots r_m,$$
$$w_{n+2} = \lambda,$$
$$z_0 = p_0, \quad z_1 = p_x,$$
$$z_i = p_0, \text{ for all } i = 2, \dots, n + 2.$$

The rules contained in the sets R_i are defined below:

In R_i, $1 \leq i \leq n$:

$$[_ip_x|\,]_i \rightarrow [_ip_+|\,]_i [_ip_-|\,]_i, \quad [_ip_+|a_i]_i \rightarrow [_ip_+|t_i]_i, \quad [_ip_-|a_i]_i \rightarrow [_ip_-|f_i]_i \quad (4)$$

In R_i, $1 \leq i \leq n - 1$:

$$t_i[_{i+1}p_0|\,]_{i+1} \rightarrow [_{i+1}p_x|t_i]_{i+1}, \quad f_i[_{i+1}p_0|\,]_{i+1} \rightarrow [_{i+1}p_x|f_i]_{i+1} \quad (5)$$

In R_i, $3 \leq i \leq n$:

$$t_j[_ip_0|\,]_i \rightarrow [_ip_0|t_j]_i, \quad f_j[_ip_0|\,]_i \rightarrow [_ip_0|f_j]_i \quad \text{for all } j, \ 1 \leq j \leq i - 2 \quad (6)$$

In R_{n+1}:

$$t_i[_{n+1}p_0|r_j]_{n+1} \rightarrow r_j[_{n+1}p_0|t_i]_{n+1}, \quad [_{n+1}p_0|t_i]_{n+1} \rightarrow t_i[_{n+1}p_0|\,]_{n+1}$$
$$\text{for all } i, \ 1 \leq i \leq n \text{ and } j, \ 1 \leq j \leq m \text{ such that } C_j \text{ contains } x_i \quad (7)$$

$$f_i[_{n+1}p_0|r_j]_{n+1} \rightarrow r_j[_{n+1}p_0|f_i]_{n+1}, \quad [_{n+1}p_0|f_i]_{n+1} \rightarrow f_i[_{n+1}p_0|\,]_{n+1}$$
$$\text{for all } i, j, \ 1 \leq i \leq n, \ 1 \leq j \leq m \text{ such that } C_j \text{ contains } \neg x_i \quad (8)$$

In R_{n+2}:

$$r_1[_{n+2}p_0|\,]_{n+2} \rightarrow [_{n+2}p_0|\overline{r}_1]_{n+2} \quad (9)$$

$$r_{i+1}[_{n+2}p_0|\overline{r}_i]_{n+2} \rightarrow \overline{r}_i[_{n+2}p_0|\overline{r}_{i+1}]_{n+2} \quad \text{for all } i, \ 1 \leq i \leq n - 1 \quad (10)$$

$$[_{n+2}p_0|\overline{r}_m]_{n+2} \rightarrow t[_{n+2}p_0|\,]_{n+2} \quad (11)$$

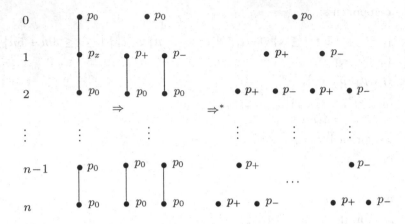

Fig. 2. Expansion of the initial membrane structure into a binary tree. The symbols at nodes indicate the proteins present on membranes.

In R_i, $1 \leq i \leq n$ such that $Q_i = \forall$:

$$[_i p_- | t]_i \to s[_i p_- |\ ]_i, \quad s[_i p_+ | t]_i \to t[_i p_+ | s]_i \tag{12}$$

In R_i, $1 \leq i \leq n$ such that $Q_i = \exists$:

$$[_i p_- | t]_i \to t[_i p_- |\ ]_i, \quad [_i p_+ | t]_i \to t[_i p_+ |\ ]_i \tag{13}$$

In R_1 :

$$t[_1 p_- | no]_1 \to s[_1 p_- | yes]_1 \tag{14}$$

In R_0 :

$$[_0 p_0 | c_i]_0 \to [_0 p_0 | c_{i+1}]_0 \quad \text{for all } i,\ 1 \leq i \leq 3m + 5n - 1 \tag{15}$$

In R_1 :

$$\begin{aligned} c_{3m+5n}[_1 p_- | yes]_1 &\to yes[_1 p_- | c_{3m+5n}]_1, \\ c_{3m+5n}[_1 p_- | no]_1 &\to no[_1 p_- | c_{3m+5n}]_1 \end{aligned} \tag{16}$$

It is easy to check that the size of the P system $\Pi(\gamma)$ (the number of objects, membranes, rules, the size of the initial configuration etc.) is $\mathcal{O}(nm)$, n being the number of variables and m the number of clauses. Also the system can be constructed in a polynomial (linear) time by a deterministic Turing machine whose input is the formula γ (a formal proof would be straightforward but cumbersome).

Initial phase of computation of the system $\Pi(\gamma)$ is illustrated in Fig. 2. In the first step the non-elementary membrane at level 1 is divided by the first rule in (4) into two parts with different membrane proteins. In the next step, symbols f_1 and t_1 are produced in the two resulting membranes, see the next rules in (4). In the third step, these symbols are moved one level lower, into the membranes labeled 2, see (5). The membrane protein on these membranes is changed to p_x.

This cycle is repeated n times and waves corresponding to the division by rules (4) descend the membrane tree towards its leaves. Simultaneously, the produced symbols t_i and f_i move towards the leaves of the tree thanks to the rules (6). This phase is finished after $3n - 1$ steps when the membrane structure forms a balanced binary tree, see Fig. 2. Each of its 2^n nodes at level n contains a set of objects $\{x_1, x_2, \ldots, x_n\}$, where $x_i \in \{f_i, t_i\}$, $1 \le i \le n$, such that all possible n-tuples are present.

Second phase consists of checking whether the formula without quantifiers is satisfied by the n-tuples of logical values $(x_1, x_2, \ldots, x_n)$. The checking is done for all the n-tuples in parallel. It starts by moving of those objects r_i, $1 \le i \le m$, corresponding to the clauses C_i which are satisfied by a particular n-tuple, from the membrane $[_{n+1}]_{n+1}$ to $[_n]_n$. Rules (7)–(8) are responsible for this process.

Whenever objects $r_1, \ldots, r_m$ appear in membrane $[_n]_n$, another process starts whose purpose is to check whether all r_i, $1 \le i \le m$, are present. This is done by their movements to-and-from membrane $[_{n+2}]_{n+2}$ driven by rules (9)–(11). Eventually, when all r_i's are present, meaning that all clauses are satisfied by the assignment $(x_1, x_2, \ldots, x_n)$ corresponding to a particular membrane $[_n]_n$, object t is released into the membrane $[_n]_n$.

The application of rules of the second phase can partially overlap with the initial phase: whenever first objects t_i or f_i arrive into the membrane $[_n]_n$, the second phase starts, while remaining t_i's and f_i's can arrive later. However, the application of the rules in the second phase described above is not altered.

Finally, third phase of computation checks whether the whole formula with quantifiers is satisfied. Objects t move upwards the membrane structure tree, checking at each level one quantifier $\forall$ or $\exists$. Observe that rules (12)–(13) allow for existence of more than one symbol t per membrane (in the case of $\exists$) which, however, do not alter the computation. Eventually, object t appears in membrane 0, signaling that the formula is satisfied. In this case, rules (14) guarantee that the object no is removed from membrane labelled 1 with protein p_- and replaced by the object yes.

The first phase of computation described above is finished in $3n - 1$ steps, the second phase takes up to $3m$ steps and the third phase up to $2n$ steps, hence, in total, less than $5n + 3m$ steps. Simultaneously with phases 1–3, symbols c_i, $1 \le i \le 5n + 3m$, together with rules (15)–(16) in membrane $[_0]_0$ implements a timer which, after $5n + 3m$ steps, moves either the symbol yes or no, whichever is present in membrane 1 with protein p_-, to the output membrane 0, and the computation halts. The object yes (no) is present if and only if the formula γ evaluates to $true$ ($false$, respectively), therefore the system $\Pi(\gamma)$ is sound and complete.

As shown above, the system always halts after $\mathcal{O}(n + m)$ steps and hence it is polynomially bounded. $\square$

The above proof is based on the technique employed by [10] when dealing with P systems with active membranes. However, since the function of membrane proteins is different, the technique was substantially adapted. Notice, e.g., that in the P systems with active membranes the division operation is driven by both membrane contents and polarization, while here it is controlled solely by membrane proteins. As a result, in [10] the membrane structure divides in the bottom-up manner while here the reverse top-down order must be employed.

Observe that rules (5) are the only rules of type $2cp$. All the rest are restricted (or division) rules. Furthermore, these $2cp$ rules are used only to control the membrane division process. The membrane division rules can be controlled solely by the presence of a specific membrane protein. Assume that we introduced division rules similar as in P systems with active membranes, i.e., of type $[_i p|a]_i \rightarrow [_i p|b]_i [_i p|c]_i$, controlled by the presence of certain object in a membrane. Then the rules (5) would not be needed and the whole P systems could use only restricted and division rules.

Hence, it turns out that the only necessary purpose of membrane proteins is the control of membrane division forced by the specific type of division rules. If we compare our proof with that in [10], we observe that the role played in [10] by the membrane polarization (which is in some sense generalized in the concept of membrane proteins) is in our proof frequently replaced by the use of antiport rules of types (4) and (5). Therefore, there is a trade-off between membrane labeling (polarization, proteins) and antiport rules.

This suggests that from the point of view of efficiency, there is no substantial difference between restricted and "change protein" rules. The paper [7] shows that the universality can be reached only with the restricted rules, too. However, there is another trade off between the number of membranes and the use of "change protein" rules in this case.

4   Simulation of a P System with Proteins on Membranes in Polynomial Space

In this section we demonstrate an algorithm for simulation of P systems with proteins on membranes which proves the relation reverse to that given in Theorem 1. Notice that the simulated P system is confluent (hence possibly nondeterministic), therefore the conditions of the Parallel Computation Thesis are satisfied. However, our simulation itself is deterministic – at each step we simulate only one chosen multiset of applicable rules. Hence we simulate one possible sequence of configurations of the P system. The algorithm of selection of the rules to be applied corresponds to introducing a weak priority between rules: (i) bottom-up priority between rules associated to different membranes, (ii) priority between rules in the same membrane, given by the order in which they are listed, including the priority between types 1–6, in this order. The confluency condition ensures that such a simulation leads always to a correct result.

We employ the technique of reverse-time simulation which is known from the general complexity theory when dealing with the second class machines. Instead of simulating a computation of a P system from its initial configuration onwards (which could require an exponential space for storing configurations), we create the recursive function State which returns the state of any membrane h after a given number of steps. The recursive calls evaluate contents of the membranes interacting with h in a reverse time order (towards the initial configuration). The key observation is that the state of the membrane is determined by its own state, states of the embedded membranes and its parent membrane at the previous computational step. In such a manner we do not need to store a state of any membrane, but instead we calculate it recursively whenever it is needed. The depth of the recursive calls is proportional to the number of steps of the simulated P system. Furthermore, at each level of the call stack we must store a state of a single membrane which can be done in a polynomial space. In this way a result of any $T(n)$-time-bounded computation of a recognizer P system with proteins on membranes can be found in a space polynomial to $T(n)$.

Theorem 2. PMC$^*_{\mathcal{MP}} \subseteq$ PSPACE.

The proof of the above theorem is not included for its extensive length. The interested reader can consult the technical report downloadable at the web address http://ui.fpf.slu.cz/~sos10um/TR_2009-01.pdf.

If we put together Theorems 2 and 1, we obtain the parallel computation thesis for semi-uniform families of confluent P systems with proteins on membranes:

Corollary 1. PMC$^*_{\mathcal{MP}} =$ PSPACE.

5   Discussion

We have shown that uniform families of recognizer P systems with proteins on membranes and without input can solve in polynomial time exactly the class of problems **PSPACE**. More precisely, they provide a semi-uniform solution to problems in this class. Therefore, they are computationally equivalent to other parallel computing model as PRAM or alternating Turing machine. We conjecture that the same result holds for uniform families of recognizer P systems with input providing uniform solutions (see [9] for definitions) but no formal proof is known yet. Possibly a construction similar to that in [2] could be used to solve this problem. Also the characterization of power of non-deterministic *non-confluent* P systems with proteins membranes remains open. The presented proof cannot be simply adapted to this case by using a non-deterministic Turing machine. The reason is that we cannot store non-deterministic choices of such a P system along a chosen trace of computation, as this would require an exponential space.

A similar result has been previously shown in [11] for the case of P systems with active membranes. Therefore, taking into the account another results of this kind related to other types of natural or molecular computing, one could suggest that the class **PSPACE** represents natural characterization of deterministic (or rather confluent) natural computations. It is important to note that certain operations used in P systems with proteins on membranes, as the division of non-elementary membranes, seem to have in practice very limited scalability, on one hand. On the other hand, some properties of biocomputing models, as the massive parallelism, minimal energy consumption, microscopic dimensions of computing elements etc. makes it very attractive to seek for ways how to harness the micro-biological machinery for algorithmic tasks.

Among further open problems we mention restricted variants of the P systems with proteins on membranes. How would the computational power of uniform families of such systems change if only certain types of rules were allowed?

Acknowledgements

Research was partially supported by the National Science Foundation Grant CCF-0523572, INBRE Program of the NCRR (a division of NIH), support from CNCSIS grant RP-13, support from CNMP grant 11-56 /2007, support from the Ministerio de Ciencia e Innovación (MICINN), Spain, under project TIN2006-15595 and the program I3, and by the Comunidad de Madrid (grant No. CCG06-UPM/TIC-0386 to the LIA research group).

References

1. Alberts, B., Johnson, A., Lewis, J., Raff, M., Roberts, K., Walter, P.: Molecular Biology of the Cell, 4th edn. Garland Science, New York (2002)
2. Alhazov, A., Martín-Vide, C., Pan, L.: Solving a PSPACE-complete problem by P systems with restricted active membranes. Fundamenta Informaticae 58(2), 67–77 (2003)
3. Cardelli, L.: Brane calculi – interactions of biological membranes. In: Danos, V., Schachter, V. (eds.) CMSB 2004. LNCS (LNBI), vol. 3082, pp. 257–280. Springer, Heidelberg (2005)
4. Frisco, P.: Computing with Cells. In: Advances in Membrane Computing. Oxford University Press, Oxford (2009)
5. Păun, A., Păun, G.: The power of communication: P systems with sym-port/antiport. New Generation Comput. 20(3), 295–306 (2002)
6. Păun, A., Popa, B.: P systems with proteins on membranes. Fundamenta Informaticae 72(4), 467–483 (2006)
7. Păun, A., Popa, B.: P systems with proteins on membranes and membrane division. In: Ibarra, O.H., Dang, Z. (eds.) DLT 2006. LNCS, vol. 4036, pp. 292–303. Springer, Heidelberg (2006)
8. Păun, G.: Membrane Computing – An Introduction. Springer, Berlin (2002)
9. Pérez-Jiménez, M.J.: A computational complexity theory in membrane computing. In: Păun, G., Pérez-Jiménez, M.J., Riscos-Núñez, A. (eds.) Tenth Workshop on Membrane Computing (WMC10), RGNC Report 3/2009, Sevilla, pp. 82–105. Universidad de Sevilla (2009)
10. Sosík, P.: The computational power of cell division in P systems: Beating down parallel computers? Natural Computing 2(3), 287–298 (2003)
11. Sosík, P., Rodríguez-Patón, A.: Membrane computing and complexity theory: A characterization of PSPACE. J. Comput. System Sci. 73(1), 137–152 (2007)
12. van Emde Boas, P.: Machine models and simulations. In: van Leeuwen, J. (ed.) Handbook of Theoretical Computer Science, vol. A, pp. 1–66. Elsevier, Amsterdam (1990)
13. The P systems web page, http://ppage.psystems.eu/

An Efficient Simulation
of Polynomial-Space Turing Machines
by P Systems with Active Membranes

Andrea Valsecchi, Antonio E. Porreca, Alberto Leporati,
Giancarlo Mauri, and Claudio Zandron

Dipartimento di Informatica, Sistemistica e Comunicazione
Università degli Studi di Milano-Bicocca
Viale Sarca 336/14, 20126 Milano, Italy
{valsecchi,porreca,leporati,mauri,zandron}@disco.unimib.it

Abstract. We show that a deterministic single-tape Turing machine,
operating in polynomial space with respect to the input length, can be
efficiently simulated (both in terms of time and space) by a semi-uniform
family of P systems with active membranes and three polarizations, using
only communication rules. Then, basing upon this simulation, we prove
that a result similar to the *space hierarchy theorem* can be obtained for
P systems with active membranes: the larger the amount of space we
can use during the computations, the harder the problems we are able
to solve.

1 Introduction

Membrane systems (also known as *P systems*) have been introduced in [11] as
a parallel, nondeterministic, synchronous and distributed model of computation
inspired by the structure and functioning of living cells. The basic model consists
of a hierarchical structure composed by several membranes, embedded into a
main membrane called the *skin*. Membranes divide the Euclidean space into
regions, that contain multisets of *objects* (represented by symbols of an alphabet)
and *evolution rules*. Using these rules, the objects may evolve and/or move from
a region to a neighboring one. Usually, the rules are applied in a nondeterministic
and maximally parallel way. A *computation* starts from an initial configuration
of the system and terminates when no evolution rule can be applied. The result
of a computation is the multiset of objects contained into an *output membrane*,
or emitted from the skin of the system. An interesting subclass of membrane
systems is constituted by *recognizer* P systems, in which: (1) all computations
halt, (2) only two possible outputs exist (usually named *yes* and *no*), and (3)
the result produced by the system only depends upon its input, and is not
influenced by the particular sequence of computation steps taken to produce it.
For a systematic introduction to P systems we refer the reader to [13], whereas
the latest information can be found in [22].

Since the introduction of membrane systems, many investigations have been
performed on their computational properties: in particular, many variants have

G. Păun et al. (Eds.): WMC 2009, LNCS 5957, pp. 461–478, 2010.

been proposed in order to study the contribution of various ingredients (associated with the membranes and/or with the rules of the system) to the achievement of the computational power of these systems. In this respect, it is known [14,20,6] that the class of all decision problems which can be solved in polynomial time by a family of recognizer P systems that use only basic rules, that is, evolution, communication and membrane dissolution, coincides with the complexity class **P**. Hence, in order to efficiently solve computationally difficult (for example, **NP**-complete) problems by means of P systems it seems necessary to be able to exponentially increase (in polynomial time) the number of membranes, that can be regarded as the size of the workspace.

In particular, two features have proven to be of paramount importance in establishing whether a membrane system is able to solve computationally difficult decision problems in polynomial time: membrane dissolution and division. Dissolution rules simply dissolve the surrounding membrane when a specified symbol occurs. Division rules are inspired from the biological process called *mitosis*: they allow to duplicate a given membrane that contains a specified symbol, possibly rewriting this symbol in a different way in each of the membranes produced by the process. All the other symbols, as well as the rules, which are contained in the original membrane are copied unaltered into each of the resulting regions. As for the membranes possibly contained in the original region (if any), we can consider the following situations. If no membrane occurs, then we say that the division is *elementary*; if one or more membranes occur, then we have to specify how they are affected by the division operation. If all the membranes are copied to each of the resulting regions, then we have a *weak* (non-elementary) division; if, instead, we can choose what membranes are copied into each of the resulting regions, then we have a *strong* (non-elementary) division.

Recognizer P systems with active membranes (using division rules and, possibly, polarizations associated to membranes) have been successfully used to efficiently solve **NP**-complete problems. The first solutions were given in the so called *semi-uniform* setting [12,20,9,10], which means that we assume the existence of a deterministic Turing machine that, for every instance of the problem, produces in polynomial time a description of the P system that solves such an instance. The solution is computed in a *confluent* manner, meaning that the instance given in input is positive (resp., negative) if and only if every computation of the P system associated with it is an accepting (resp., rejecting) computation.

Another way to solve **NP**-complete problems by means of P systems is by considering the *uniform* setting, in which any instance of the problem of a given length can be fed as input – encoded in an appropriate way – to a specific P system and then solved by it. Sometimes, a uniform solution to a decision problem Q is provided by defining a family $\{\Pi_Q(n)\}_{n \in \mathbb{N}}$ of P systems such that for every $n \in \mathbb{N}$ the system $\Pi_Q(n)$ reads in input an encoding of any possible instance of size n, and solves it. P systems with active membranes have thus been successfully used to design uniform polynomial-time solutions to some well-known **NP**-complete problems, such as SAT [15].

All the papers mentioned above deal with P systems having three polarizations, that use only division rules for elementary membranes (in [19] also division for non-elementary membranes is permitted, and in this way a semi-uniform solution to the **PSPACE**-complete problem QSAT is provided), and working in the *maximally parallel* way. As shown in [2], the number of polarizations can be decreased to two without loss of efficiency. On the other hand, in [5] the computational power of recognizer P systems with active membranes but *without* electrical charges and dissolution rules was investigated, establishing that they characterize the complexity class **P**. Finally, in [21] it was shown that polarizationless P systems with active membranes that use strong division for non-elementary membranes and dissolution rules, working in the maximally parallel way, are able to solve in polynomial time the **NP**-complete problem 3-SAT. This result establishes that neither evolution nor communication rules, and no electrical charges are needed to solve **NP**-complete problems, provided that we can use strong division rules for non-elementary membranes (as well as dissolution rules, otherwise we would fall in the case considered in [5]).

By looking at the literature one can see that, until now, the research on the complexity theoretic aspects of P systems with active membranes has mainly focused on the *time* resource. In particular, we can find several results that compare time complexity classes obtained by using various ingredients (such as, e.g., polarizations, dissolution, uniformity, etc.). Other works make a comparison between these classes and the usual complexity classes defined in terms of Turing machines, either from the point of view of time complexity [14,4,20], or space complexity [19,1,17]. A first definition of space complexity for P systems was given in [7], where the measure of space is given by the maximum number of objects occurring during the computation. The definition was then generalized to P systems with mutable membrane structure [16], in particular P systems with active membranes, thus formalizing the usual notion of exponential workspace generated through membrane division.

In this paper, basing upon the formal definitions given in [16], we present some results concerning the relations among space complexity classes defined in terms of P systems, under some specified constraints. In particular, we first show how to simulate a deterministic single-tape Turing machine by a semi-uniform family of P systems with active membranes and three polarizations. Then, by focusing our attention on computations occurring in polynomial space, we define a pseudo-hierarchy of space complexity classes. Such classes are inspired by the *space hierarchy theorem*, that we restate and prove (albeit in a slightly different form) for P systems with active membranes. Let us note that an analogous hierarchy for catalytic P systems with a fixed membrane structure has been introduced in [7].

The paper is organized as follows. In section 2 we recall the definition of recogniser P systems with active membranes, thus establishing our model of computation, and we recall some basic notions that will be used in the rest of the paper. In section 3 we show how to simulate a deterministic single-tape Turing machine by means of a semi-uniform family of P systems with active membranes. In section 4 we recall the space hierarchy theorem and, inspired by it, we define a pseudo-hierarchy of space complexity classes determined by P systems with active membranes. Finally, section 5 contains the conclusions and some directions for further research.

2   Definitions

We begin by recalling the definition of P systems with active membranes.

Definition 1. *A P system with active membranes of the initial degree $m \geq 1$ is a tuple*

$$\Pi = (\Gamma, \Lambda, \mu, w_1, \dots, w_m, R)$$

where:

- *Γ is a finite alphabet of symbols, also called* objects;
- *Λ is a finite set of labels for the membranes;*
- *μ is a membrane structure (i.e., a rooted unordered tree) consisting of m membranes enumerated by $1, \dots, m$; furthermore, each membrane is labeled by an element of Λ, not necessarily in a one-to-one way;*
- *$w_1, \dots, w_m$ are strings over Γ, describing the multisets of objects placed in the m initial regions of μ;*
- *R is a finite set of rules.*

Each membrane possesses a further attribute, named polarization *or* electrical charge, *which is either neutral (represented by 0), positive (+) or negative (−) and it is assumed to be initially neutral.*

The rules are of the following kinds:

- Object evolution rules, *of the form $[a \to w]_h^\alpha$*
  They can be applied inside a membrane labeled by h, having polarization α and containing an occurrence of the object a; the object a is rewritten into the multiset w (i.e., a is removed from the multiset in h and replaced by every object in w).
- Communication rules, *of the form $a[\ ]_h^\alpha \to [b]_h^\beta$*
  They can be applied to a membrane labeled by h, having polarization α and such that the external region contains an occurrence of the object a; the object a is sent into h becoming b and, simultaneously, the polarization of h is changed to β.
- Communication rules, *of the form $[a]_h^\alpha \to [\ ]_h^\beta b$*
  They can be applied to a membrane labeled by h, having polarization α and containing an occurrence of the object a; the object a is sent out from h to the outside region becoming b and, simultaneously, the polarization of h is changed to β.
- Dissolution rules, *of the form $[a]_h^\alpha \to b$*
  They can be applied to a membrane labeled by h, having polarization α and containing an occurrence of the object a; the membrane h is dissolved and its contents are left in the surrounding region unaltered, except that an occurrence of a becomes b.

- Elementary division rules, *of the form* $[a]_h^\alpha \to [b]_h^\beta [c]_h^\gamma$
  They can be applied to a membrane labeled by h, having polarization α, containing an occurrence of the object a but having no other membrane inside; the membrane is divided into two membranes having label h and polarizations β and γ; the object a is replaced, respectively, by b and c while the other objects in the initial multiset are copied to both membranes.
- Non-elementary division rules, *of the form*

$$[[\,]_{h_1}^+ \cdots [\,]_{h_k}^+ [\,]_{h_{k+1}}^- \cdots [\,]_{h_n}^-]_h^\alpha \to [[\,]_{h_1}^\delta \cdots [\,]_{h_k}^\delta]_h^\beta [[\,]_{h_{k+1}}^\varepsilon \cdots [\,]_{h_n}^\varepsilon]_h^\gamma$$

They can be applied to a membrane labeled by h, having polarization α, containing the positively charged membranes $h_1, \ldots, h_k$, the negatively charged membranes $h_{k+1}, \ldots, h_n$, and possibly some neutral membranes. The membrane h is divided into two copies having polarization β and γ, respectively; the positive children are placed inside the former, their polarizations changed to δ, while the negative ones are placed inside the latter, their polarizations changed to ε. Any neutral membrane inside h is duplicated and placed inside both copies.

A *configuration* in a P system with active membranes is described by its current membrane structure, together with its polarizations and the multisets of objects contained in its regions. The initial configuration is given by μ, all membranes having polarization 0 and the initial contents of the membranes being $w_1, \ldots, w_m$. A *computation step* changes the current configuration according to the following principles:

- Each object and membrane can be subject to only one rule during a computation step.
- The rules are applied in a *maximally parallel way*: each object which appears on the left-hand side of applicable evolution, communication, dissolution or elementary division rules must be subject to exactly one of them; the same holds for each membrane which can be involved in a communication, dissolution or division rule. The only objects and membranes which remain unchanged are those associated with no rule, or with unapplicable rules.
- When more than one rule can be applied to an object or membrane, the actual rule to be applied is chosen nondeterministically; hence, in general, multiple configurations can be reached from the current one.
- When dissolution or division rules are applied to a membrane, the multiset of objects to be released outside or copied is the one resulting *after* all evolution rules have been applied.
- The skin membrane cannot be divided, nor it can be dissolved. Furthermore, every object which is sent out from the skin membrane cannot be brought in again.

A (halting) *computation* $\mathcal{C}$ of a P system Π is a sequence of configurations $(\mathcal{C}_0, \ldots, \mathcal{C}_k)$, where $\mathcal{C}_0$ is the initial configuration of Π, every $\mathcal{C}_{i+1}$ can be reached from $\mathcal{C}_i$ according to the principles just described, and no further configuration can be reached from $\mathcal{C}_k$ (i.e., no rule can be applied).

We can use families of P systems with active membranes as language recognisers, thus allowing us to solve decision problems.

Definition 2. *A recogniser P system with active membranes Π has an alphabet containing two distinguished objects* yes *and* no, *used to signal acceptance and rejection respectively; every computation of Π is halting and exactly one object among* yes, no *is sent out from the skin membrane during each computation.*

If all computations starting from the initial configuration of Π agree on the result, then Π is said to be confluent; *if this is not necessarily the case, then it is said to be* non-confluent *(and the global result is acceptance iff an accepting computation exists).*

Definition 3. *Let $L \subseteq \Sigma^\star$ be a language and let $\mathbf{\Pi} = \{\Pi_x : x \in \Sigma^\star\}$ be a family of recogniser P systems. We say that $\mathbf{\Pi}$ decides L, in symbols $L(\mathbf{\Pi}) = L$, when for each $x \in \Sigma^\star$, the result of Π_x is acceptance iff $x \in L$.*

Usually, a condition of uniformity, inspired by those applied to families of Boolean circuits, is imposed on families of P systems.

Definition 4. *A family of P systems $\mathbf{\Pi} = \{\Pi_x : x \in \Sigma^\star\}$ is said to be* semi-uniform *when the mapping $x \mapsto \Pi_x$ can be computed in polynomial time, with respect to $|x|$, by a deterministic Turing machine.*

Time complexity classes for P systems are defined as usual, by restricting the amount of time available for deciding a language. By $\mathbf{MC}_{\mathcal{D}}^\star(f(n))$ we denote the class of languages which can be decided by semi-uniform families of confluent P systems of class $\mathcal{D}$ (e.g., $\mathcal{AM}$ denotes the class of P systems with active membranes) where each computation of $\Pi_x \in \mathbf{\Pi}$ halts within $f(|x|)$ steps. The class of languages decidable in polynomial time by the same families of P systems is denoted by $\mathbf{PMC}_{\mathcal{D}}^\star$.

Recently, a space complexity measure for P systems has been introduced [16]. We recall here the relevant definitions.

Definition 5. *Let $\mathcal{C}$ be a configuration of a P system Π. The* size $|\mathcal{C}|$ *of $\mathcal{C}$ is defined as the sum of the number of membranes in the current membrane structure and the total number of objects they contain[1]. If $\mathcal{C} = (\mathcal{C}_0, \ldots, \mathcal{C}_k)$ is a halting computation of Π, then the* space required by $\mathcal{C}$ *is defined as*

$$|\mathcal{C}| = \max\{|\mathcal{C}_0|, \ldots, |\mathcal{C}_k|\}.$$

The space required by Π *itself is then*

$$|\Pi| = \max\{|\mathcal{C}| : \mathcal{C} \text{ is a halting computation of } \Pi\}.$$

Finally, let $\mathbf{\Pi} = \{\Pi_x : x \in \Sigma^\star\}$ be a family of recogniser P systems; also let $f : \mathbb{N} \to \mathbb{N}$. We say that $\mathbf{\Pi}$ operates within space bound f iff $|\Pi_x| \leq f(|x|)$ for each $x \in \Sigma^\star$.

---

[1] An alternative definition, where the size of a configuration is given by the sum of the number of membranes and *the number of bits required to store the objects they contain*, has been considered in [16]. However, the choice between the two definitions is irrelevant as far as the results of this paper are concerned.

Next, we formally define the variant of Turing machines we use in the following sections.

Definition 6. *A single-tape deterministic Turing machine is a tuple*

$$M = (Q, \Sigma, \Gamma, \delta, q_0, A, R)$$

where:

- *Q is a finite and nonempty set of states;*
- *Σ is the finite input alphabet;*
- *Γ is the tape alphabet, a finite superset of Σ;*
- *the partial function $\delta\colon \Gamma \times Q \to \Gamma \times Q \times \{\leftarrow, -, \rightarrow\}$ is the transition function; we assume that δ is undefined on both accepting and rejecting states;*
- *$q_0 \in Q$ is the initial state;*
- *$A \subseteq Q$ is the set of accepting states;*
- *$R \subseteq Q$ is the set of rejecting states, disjoint from A.*

Finally, we recall the definition of constructible function (for further information on this topic see, for instance, [8,3,18]).

Definition 7. *A function $f\colon \mathbb{N} \to \mathbb{N}$ is said to be* time-constructible *iff the mapping $1^n \mapsto 1^{f(n)}$, i.e., from the unary representation of n to the unary representation of $f(n)$, can be computed by a deterministic Turing machine in $O(f(n))$ time.*

The function f is space-constructible *iff the mapping $1^n \mapsto 1^{f(n)}$ can be computed by a deterministic Turing machine in $O(f(n))$ space.*

3   Simulating Turing Machines

In this section we show that a single-tape Turing machine M having $\Sigma = \{0, 1\}$ as input alphabet and operating in polynomial space $f(n)$ and time $g(n)$ can be simulated efficiently (i.e., in $O(f(n))$ space and $O(g(n))$ time) by a semi-uniform family $\mathbf{\Pi}_M = \{\Pi_{M,x} : x \in \{0, 1\}^*\}$ of P systems with active membranes and three polarizations, where each $\Pi_{M,x}$ simulates the computation of M on input x. We also stress the fact that these P systems can be defined in such a way that communication is the only required kind of rule.

Turing machines operate by reading and writing symbols on a tape divided into cells: the main idea of our simulation is representing each cell by a membrane. In a Turing machine the tape cells are linearly ordered (we assume they are numbered by nonnegative integers); one way to organise the membranes without losing this piece of information is to nest them, i.e., to place one inside the other in a linear fashion. Either the innermost or the outermost membrane can be put into correspondence with the leftmost tape cell; without loss of generality, we choose the outermost one.

Each cell of the Turing machine contains a symbol taken from the tape alphabet, which we assume to be $\Gamma = \{0, 1, \_\}$, where $\_$ denotes the blank symbol. In the P system, the symbol written in a cell is stored *in the polarization* of the corresponding membrane. The default neutral polarization represents a blank cell, while the negative and positive polarizations represent 0 and 1, respectively.

A single object in the P system represents the state of the Turing machine (an element q of the finite set Q), and its location inside the membrane structure represents the position of the tape head: the object is located immediately inside the i-th membrane iff the tape head of the simulated machine is located on the i-th leftmost tape cell. The object is changed (via communication rules) both in form and location in order to reflect the change of state and position of the tape head of the Turing machine.

Finally, the transition function $\delta \colon \Gamma \times Q \to \Gamma \times Q \times \{\leftarrow, -, \to\}$ of the Turing machine is implemented by using a set of communication rules. The object representing the head position and state of the Turing machine is moved to the new position, while simultaneously changing the polarization of the current membrane in order to update the contents of the tape; it is also rewritten into a (possibly different) symbol, representing the new state of the machine. In order to execute these operations, the P system requires a constant number of steps for each computation step of the simulated Turing machine.

Let M be a single-tape deterministic Turing machine operating in space $f(n)$. Let $x = x_1 x_2 \cdots x_n \in \{0,1\}^n$ be an input for M. The membrane structure $\mu_{M,x}$ is made of $f(n)$ membranes labelled by h and nested one inside the other; this structure is surrounded by a further membrane h_0, which also contains a membrane labelled by w. The initial configuration of $\mu_{M,x}$ is as follows:

$$[[\hat{q}_0]_w^0 \overbrace{[x_1[x_2\cdots[x_n[\cdots[}^{f(n)\text{ membranes}} {}_h^0\cdots]_h^0 \overbrace{]_h^0\cdots]_h]_h^0}^{f(n)\text{ membranes}}]_{h_0}^0$$

$$\underbrace{}_{n\text{ membranes}} \qquad \underbrace{}_{n\text{ membranes}}$$

Each of the outermost n membranes labelled by h contains an object $x_i \in \{0,1\}$, representing the i-th input symbol of M; these objects are used to set up the initial contents of the tape (recall that, by definition, all membranes are initially required to be neutral). The following communication rules serve the purpose of changing the polarization of a membrane h according to the symbol contained in the corresponding tape cell:

$$[0]_h^0 \to [\,]_h^- \; \# \tag{1}$$

$$[1]_h^0 \to [\,]_h^+ \; \# \tag{2}$$

where $\#$ is a "junk" object, i.e., an object which does not appear on the left-hand side of any rule.

While the initial configuration of the simulated machine M is being set up (only one step is required to do so) the head/state object $\hat{q}_0$, where q_0 is the initial state of M, is sent out from w by means of the following rule:

$$[\hat{q}_0]_w^0 \to [\,]_w^0 \; \hat{q}_0 \tag{3}$$

After that, $\hat{q}_0$ enters the membrane corresponding to the leftmost tape cell, while simultaneously losing the "hat", by using one of the following communication rules:

$$\hat{q}_0 \, [\,]_h^\alpha \to [q_0]_h^\alpha \qquad\qquad \forall \alpha \in \{-, 0, +\} \qquad\qquad (4)$$

Object $\hat{q}_0$ is initially located inside w so that it requires two steps in order to reach the membrane corresponding to the initial cell of M, thus avoiding conflicts with the rules setting up the initial tape contents.

Now the real simulation begins. To each quintuple (a, q_1, b, q_2, d) describing a transition of M (i.e., denoting the fact that $\delta(a, q_1) = (b, q_2, d)$) corresponds a constant number of communication rules. If $\delta(a, q_1) = (b, q_2, \leftarrow)$ then there is a single rule

$$[q_1]_h^\alpha \to [\,]_h^\beta \, q_2 \qquad\qquad (5)$$

where α and β are $-$ or $+$ when a and b are 0 or 1 respectively. The rule moves the head/state object outwards (which corresponds to moving the tape head of M one position to the left) while changing it as the state of M does.

If the tape head does not move, as in $\delta(a, q_1) = (b, q_2, -)$, then two rules are needed:

$$[q_1]_h^\alpha \to [\,]_h^\beta \, q_2' \qquad\qquad (6)$$

$$q_2' \, [\,]_h^\beta \to [q_2]_h^\beta \qquad\qquad (7)$$

The first rule changes the symbol in the current cell, while the second one moves the (updated) head/state symbol back to that cell.

When the tape head moves right, i.e., $\delta(a, q_1) = (b, q_2, \rightarrow)$, five rules are needed:

$$[q_1]_h^\alpha \to [\,]_h^\beta \, q_2'' \qquad\qquad (8)$$

$$q_2'' \, [\,]_h^\beta \to [q_2']_h^\beta \qquad\qquad (9)$$

$$q_2' \, [\,]_h^\gamma \to [q_2]_h^\gamma \qquad\qquad \forall \gamma \in \{-, 0, +\} \qquad\qquad (10)$$

The three rules in (10) are used to move the head/state symbol one membrane deeper, thus completing the simulated movement of the tape head to the right.

Finally, the result of the computation of M is sent out of the membrane structure. If M enters an accepting state q, the head/state symbol is changed to *yes* and is expelled:

$$[q]_h^\alpha \to [\,]_h^\alpha \, yes \qquad\qquad \forall \alpha \in \{-, 0, +\} \qquad\qquad (11)$$

$$[yes]_h^\alpha \to [\,]_h^\alpha \, yes \qquad\qquad \forall \alpha \in \{-, 0, +\} \qquad\qquad (12)$$

$$[yes]_{h_0}^0 \to [\,]_{h_0}^0 \, yes \qquad\qquad (13)$$

An analogous situation occurs when q is a rejecting state:

$$[q]_h^\alpha \to [\,]_h^\alpha \, no \qquad\qquad \forall \alpha \in \{-, 0, +\} \qquad\qquad (14)$$

$$[no]_h^\alpha \to [\,]_h^\alpha \, no \qquad\qquad \forall \alpha \in \{-, 0, +\} \qquad\qquad (15)$$

$$[no]_{h_0}^0 \to [\,]_{h_0}^0 \, no \qquad\qquad (16)$$

Definition 8. *With a slight abuse of notation, we denote by $\mu_{M,x}$ the whole P system "module" consisting of both the membrane structure described above and the set of rules (1)–(16). This module will be used later as a part of a larger P system. We also denote by Π_M the family of P systems with active membranes $\{\Pi_{M,x} : x \in \{0,1\}^*\}$, where $\Pi_{M,x}$ consists of the module $\mu_{M,x}$ only.*

Theorem 1. *Let M be a single-tape deterministic Turing machine halting on every input and operating in space $f(n)$, where $f(n) = \Omega(n)$, $f(n) = O(n^k)$ for some fixed k and $f(n)$ is time-constructible. Also assume that M operates in time $g(n)$. Then Π_M is semi-uniform and decides the same language as M in $O(f(n))$ space and $O(g(n))$ time; furthermore, Π_M can be constructed in $O(f(n))$ time.*

Proof. Each P system $\Pi_{M,x}$ consists of $f(n) + 2$ membranes and contains $n + 1$ objects, where $n = |x|$; hence Π_M clearly uses $O(f(n))$ space.

Each transition of M on input x is simulated by $\Pi_{M,x}$ in at most three steps; another step is required to set up the initial contents of the tape. When the result object yes/no is produced, it is expelled from the system after a number of steps which equals the number of the tape cell where M enters the final state, plus a further step to exit the outermost membrane h_0. Hence, the total time is $O(g(n))$.

The mapping $x \mapsto \Pi_{M,x}$ can be computed in time $O(f(n))$, as

- the membrane structure consists of $f(|x|)$ identical membranes (and two further membranes w and h_0) and can be constructed in $O(f(n))$ time steps, as f is time-constructible by hypothesis;
- the initial configuration of the P system can be constructed in linear time from x, as exactly n symbols are to be placed inside the outermost membranes;
- the set of communication rules only depends on M, and not on x.

Finally, since $f(n)$ is bounded by a polynomial, the construction of Π_M is semi-uniform.   □

4   A Space Pseudo-Hierarchy

The space hierarchy theorem, a fundamental result in complexity theory, states that Turing machines are able to solve harder problems when given a larger amount of space to exploit. The proof [18] is constructive, as for every space bound $f(n)$ an explicit language is described which cannot be decided by using less space.

Definition 9. *Let $f: \mathbb{N} \to \mathbb{N}$. We denote by $L(f)$ the language of strings $x \in \{0,1\}^*$ of the form $\langle M \rangle 10^*$, where $\langle M \rangle$ is the binary description of a single-tape deterministic Turing machine that rejects x without using more than $f(|x|)$ space.*

Theorem 2 (Space hierarchy theorem). *Let f be a space-constructible function such that $f(n) = \Omega(n)$. Then $L(f)$ is decidable in space $O(f(n))$ but not in space $o(f(n))$.*

Proof (sketch). The language $L(f)$ can be decided by a deterministic Turing machine D which simulates M on x within a $f(|x|)$ space limit, flipping the result whenever the simulation completes successfully and rejecting if the non-blank portion of the tape of M becomes longer than the space limit. Such a simulation can be carried out in space $O(f(n))$.

If $L(f)$ could be decided in space $g(n) = o(f(n))$ by some deterministic Turing machine M, then D on input $\langle M \rangle 10^k$ (for large enough values of k) could complete the simulation within the space limit and give a different result from M, thus contradicting the hypothesis that $L(D) = L(M) = L(f)$.

The trailing k zeros in the description of $L(f)$ are a technical requirement: since $g(n)$ may be larger than $f(n)$ for small n even when $g(n) = o(f(n))$, for some inputs the simulation might not complete successfully; but D certainly answers the opposite of M on all strings $\langle M \rangle 10^k$ for large enough values of k, thus ensuring they decide different languages.   □

A related result can be proved in the setting of P systems with active membranes. The main idea is to modify the Turing machine D of the above proof in such a way that, instead of directly simulating the machine M it receives as input, it constructs a P system $\Pi''_{M,x,f}$ which carries out this task. $\Pi''_{M,x,f}$ is a variant of the P system $\Pi_{M,x}$ described in the previous section; notice that $\Pi_{M,x}$ is not suitable for the present task, as it is designed to simulate only halting Turing machines operating in polynomial space. The Turing machine D, instead, receives arbitrary machines M as input, which on some input x could try to use more space than we took into account when constructing $\Pi_{M,x}$; alternatively, they could also run forever, whereas we need to always give an answer.

We begin by modifying the P system module $\mu_{M,x}$ such that, when M exceeds the allocated space (i.e., when the tape head moves to the right of the rightmost cell), the simulation ends by rejecting. Furthermore, when the simulation is completed correctly, we return the opposite result of M.

The P system module $\mu'_{M,x,f}$, simulating M on x within a $f(|x|)$ space bound, has the following membrane structure and initial configuration:

$$[[\hat{q}_0]_w^0 \overbrace{[x_1 [x_2 \cdots [x_n}^{n \text{ membranes}} [\cdots [\ \overbrace{[\ ]_{h_1}^0]_h^0 \cdots]_h^0}^{f(n) \text{ membranes}} \overbrace{]_h^0 \cdots]_h^0]_h^0}^{f(n) \text{ membranes}}]_{h_0}^0$$

that is, the same structure of $\mu_{M,x}$ except for an additional membrane h_1 in the innermost position. Such a membrane is used to detect a space "overflow" and halt the simulation if this event occurs, according to the following rule:

$$[q]_{h_1}^0 \rightarrow [\ ]_{h_1}^0 \, yes \qquad\qquad \text{for all states } q \text{ of } M. \qquad (17)$$

Furthermore, the same rules (1)–(16) of definition 8 are used, except that rules (13) and (16), involving the outermost membrane, are changed in order to flip the result:

$$[yes]_{h_0}^0 \rightarrow [\ ]_{h_0}^0 \, no \qquad\qquad (13')$$

$$[no]_{h_0}^0 \rightarrow [\ ]_{h_0}^0 \, yes \qquad\qquad (16')$$

Definition 10. *The P system consisting only of module $\mu'_{M,x,f}$ is denoted by $\Pi'_{M,x,f}$; we also define the family $\Pi'_{M,f} = \{\Pi'_{M,x,f} : x \in \{0,1\}^*\}$.*

Lemma 1. *Let $f \colon \mathbb{N} \to \mathbb{N}$, with $f(n) = \Omega(n)$ and $f(n) = O(n^k)$ for some fixed k, be time-constructible; let M be a single-tape Turing machine which halts on every input. Then the family of P systems $\Pi'_{M,f}$ is constructible in $O(f(n))$ time, hence semi-uniform, and*

$$L(\Pi'_{M,f}) = \{x \in \{0,1\}^* : M \text{ rejects } x \text{ in } f(|x|) \text{ space}\}.$$

Proof. The family $\Pi'_{M,f}$ is obviously constructible in $O(f(n))$ time as in the proof of theorem 1, since there is only one extra membrane and the new rule (17) does not depend on x.

The language decided by $\Pi'_{M,f}$ is, by construction, the complement of that of M, except that strings x generating computations which require more that $f(|x|)$ cells are rejected. □

Another stumbling block we need to overcome is the fact that some Turing machines might operate within the space bound we fixed, but without halting. Fortunately, we know that a single-tape Turing machine, having tape alphabet $\{0, 1, \llcorner\}$ and operating in $f(n)$ space, either halts within $f(n) \cdot |Q| \cdot 3^{f(n)}$ steps (Q being its set of states), or does not halt at all. We can solve the problem by counting the number of simulated steps, and halting the simulation when such a time bound is exceeded. The usual solution, i.e., having an object which is successively rewritten into all values of the counter, does not work, as the counter may assume exponentially large values (with respect to n). Hence, a more sophisticated solution is needed.

Definition 11. *We define a P system module κ_n, having the following $(n+1)$-degree membrane structure and initial configuration:*

$$[\;\overbrace{[\cdots[}^{n-2}\,[d]^0_{c_0}]^0_{c_1}\cdots]^0_{c_{n-1}}]^0_{c_n}$$

The device is, essentially, an $(n+1)$-bit binary counter. Each membrane corresponds to one bit, c_0 and c_n being the least significant and most significant bits respectively. Neutral and positive polarizations represent 0 and 1, respectively. Thus, in the initial configuration, κ_n stores the value 0. By using communication rules, such value is incremented up to 2^n. Since all membranes $c_1, \ldots, c_{n-1}$ have identical behaviour, they can all be given the same label, thus simplifying the structure (and reducing the time required to construct it) as follows:

$$[\;\overbrace{[\cdots[}^{n-2}\,[d]^0_{c_0}]^0_c\cdots]^0_c]^0_{c_n}$$

Recall that incrementing a binary integer is performed by flipping its bits, one by one and starting from the least significant one, until a 0 is flipped into 1. The object d moves inside the membrane structure in order to perform this task. The following rules (which are identical for membranes labeled by c_0 and c) move d outwards, and change it into d' when the current increment operation has finished:

$$[d]_{c_0}^0 \rightarrow [\,]_{c_0}^+ d' \tag{18}$$

$$[d]_{c_0}^+ \rightarrow [\,]_{c_0}^0 d \tag{19}$$

$$[d]_c^0 \rightarrow [\,]_c^+ d' \tag{20}$$

$$[d]_c^+ \rightarrow [\,]_c^0 d \tag{21}$$

The next rules take d' back to the starting position; when d' re-enters the inner-most membrane c_0 it is rewritten into d, and the next increment operation may begin:

$$d' \,[\,]_c^\alpha \rightarrow [d']_c^\alpha \qquad\qquad \forall \alpha \in \{0, +\} \tag{22}$$

$$d' \,[\,]_{c_0}^\alpha \rightarrow [d]_{c_0}^\alpha \qquad\qquad \forall \alpha \in \{0, +\} \tag{23}$$

Finally, when d crosses the outermost membrane c_n it is left outside (i.e., there is no rule bringing it back inside), as a signal that the counter has reached the value 2^n:

$$[d]_{c_n}^0 \rightarrow [\,]_{c_n}^+ d \tag{24}$$

Lemma 2. *The P system module κ_n can be constructed from 1^n in $O(n)$ time; it sends out the object d after at least 2^n steps.*

Proof. The membrane structure is of linear size, and there is a constant number of communication rules, hence the construction can be performed in $O(n)$ time. Since incrementing the binary counter requires at least two applications of communication rules (and $2n$ in the worst case), the object d is not set out before 2^n time steps have passed. □

When the object d is sent out from κ_n, we can use it to stop the simulation of the Turing machine, as we know that if it has not halted yet, then it will never do (assuming we have chosen a suitable value for n). The obvious solution is to use d to dissolve the whole membrane structure $\mu'_{M,x,f}$; however, besides requiring the introduction of dissolution rules (recall that we have only used communication rules so far), there might exist computations during which d is not able to enter a membrane in $\mu'_{M,x,f}$ because it is blocked by the head/state object which continuously enters and exits from that membrane (e.g., if the head of the Turing machine is stuck on a single tape cell). Both problems can be solved by slightly changing the definition of module $\mu'_{M,x,f}$.

We define a P system module $\mu''_{M,x,f}$ with the following membrane structure and initial configuration:

$$[[\hat{q}_0]_w^0 \underbrace{[[x_1[[x_2 \cdots [[x_n[[\cdots [[[\,]_{h_1}^0]_{h}^0]_j^0 \cdots]_{h}^0]_j^0]_{h}^0]_j^0 \cdots]_{h}^0]_j^0]_{h}^0]_j^0]_{h_0}^0}$$

overbrace left: $2f(n)$ membranes   overbrace right: $2f(n)$ membranes
underbrace left: $2n$ membranes   underbrace right: $2n$ membranes

that is, each membrane labelled by h is surrounded by a further membrane labelled by j. The communication rules used to move the head/state object are changed so that, when it crosses a membrane h (in either direction), it also crosses the membrane j immediately outside *whenever it is neutrally charged*, without any further change. However, when the head/state object crosses a membrane labelled by j that is *positively* charged, it is changed into the object *yes*, so that it can be sent outside as if machine M has accepted.

The module $\mu''_{M,x,f}$ still simulates M on input x within space bound f; the double membrane structure, besides slowing down the simulation by a multiplicative constant, does not alter the simulated computation of M. However, we can combine this module with $\kappa_{\ell(n)}$ (where $\ell(n)$ is a value large enough to ensure that object d is not sent out prematurely) in such a way that when the object d is sent out, it traverses the membrane structure $\mu''_{M,x,f}$ and changes the polarization of all membranes labelled by j to positive, thus stopping the simulation if it has not already ended. Since the head/state object must always cross at least two membranes in order to simulate a transition of M, the situation in which it continuously crosses the same membrane forward and backward, thus blocking the object d, does never happen.

Definition 12. *The P system $\Pi''_{M,x,f}$ is defined as follows:*

$$[\mu''_{M,x,f} \; \kappa_{\ell(n)}]^0_s$$

that is, a skin membrane containing the P system modules $\mu''_{M,x,f}$ and $\kappa_{\ell(n)}$, with the initial configuration and rules given by those of the two modules together. A further pair of rules is used to send out the result from the skin membrane s:

$$[yes]^0_s \to [\;]^+_s \; yes \tag{25}$$

$$[no]^0_s \to [\;]^+_s \; no \tag{26}$$

As noticed above, the value of $\ell(n)$ must be large enough to ensure that object d is not sent out from $\kappa_{\ell(n)}$ before the result of a possibly halting computation of M is expelled from $\mu''_{M,x,f}$. Since each transition of M can be simulated by at most six steps of $\mu''_{M,x,f}$, and since M may accept when its head is on the rightmost position (the $f(n)$-th position) of the tape, thus requiring us to wait until the result object has travelled through the whole membrane structure, an appropriate value is

$$\ell(n) = \log\left(6 \cdot f(n) \cdot |Q| \cdot 3^{f(n)} + 2f(n) + 1\right).$$

The system is augmented with a set of communication rules which cause the object d, once it has been sent out from $\kappa_{\ell(n)}$, to traverse the nested membrane structure of $\mu''_{M,x,f}$ while changing the polarization of all membranes labelled by j to positive (without changing any other polarization), thus aborting any non-halting simulated computation.

We denote by $\boldsymbol{\Pi''_{M,f}}$ the family of P systems $\{\Pi''_{M,f,x} : x \in \{0,1\}^\star\}$.

From this definition, and lemmata 1 and 2, we can prove the following result.

Lemma 3. *Let* $f\colon \mathbb{N} \to \mathbb{N}$, *with* $f(n) = \Omega(n)$ *and* $f(n) = O(n^k)$ *for some fixed* k, *be time-constructible; let* M *be a single-tape Turing machine (which does not necessarily halt on every input). Then the family of P systems* $\mathbf{\Pi}''_{M,f}$ *is constructible in* $O(f(n))$ *time (hence semi-uniform), and*

$$L(\mathbf{\Pi}''_{M,f}) = \{x \in \{0,1\}^* : M \text{ rejects } x \text{ using at most } f(|x|) \text{ space}\}.$$

We are now finally able to prove that $L(f)$ can be recognised by a family of P systems in $O(f(n))$ space.

Theorem 3. *Let* $f\colon \mathbb{N} \to \mathbb{N}$, *with* $f(n) = \Omega(n)$ *and* $f(n) = O(n^k)$ *for some fixed* k, *be time-constructible. Then* $L(f)$ *can be decided by a semi-uniform family of P systems* $\mathbf{\Pi}_{L(f)}$ *using only communication rules, operating in space* $O(f(n))$ *and constructible in time* $O(f(n))$.

Proof. We only need to prove that the mapping $(\langle M \rangle, x) \mapsto \Pi''_{M,x,f}$ (i.e., we are given both M and its input, and not only x) can be computed in $O(f(n))$ time. The only feature of $\Pi''_{M,x,f}$ which depends on M (in contrast with other features depending on x) is the set of communication rules. The number of rules is linear with respect to the length of the encoding of M (due to the rules in (5)–(10) and (17)). Assuming a "reasonable" encoding of M, all the communication rules can be constructed in linear time, hence in $O(f(n))$ time.

The family $\mathbf{\Pi}_{L(f)}$ of P systems is then constructed in $O(f(n))$ time by the following Turing machine (here described informally):

> If the input x is not of the form $\langle M \rangle 10^*$, then construct a P system which rejects immediately. Otherwise, construct $\Pi''_{M,x,f}$.

The P systems constructed by this Turing machine work in $O(f(n))$ space, and the thesis follows. $\qquad\square$

In [17] a simulation algorithm for P systems with active membranes is described. Although the precise space requirements are not detailed (only an asymptotic upper bound is given), by looking at the description of the algorithm one can observe that, essentially, in order to simulate a P system Π we need to store its current configuration, step by step (some auxiliary space is needed; however, it does not exceed the space required by the configuration). Notice that a Turing machine storing the configuration of Π does *not* have the same space requirements as Π itself: indeed, a membrane structure of degree n may require up to $n \log n$ space, since the labels of the membranes (which do not contribute to the space required by Π) need to be stored in order to correctly apply the rules (especially non-elementary division rules); all the labels may be different in the worst case. Keeping in mind this detail, we can prove the following result.

Theorem 4. *Let* $f\colon \mathbb{N} \to \mathbb{N}$, *with* $f(n) = \Omega(n)$ *and* $f(n) = O(n^k)$ *for some fixed* k, *be time-constructible. Then no family* $\mathbf{\Pi}$ *of P systems with active membranes, constructible in* $o(f(n))$ *time and operating in* $o(f(n)/\log f(n))$ *space, decides* $L(f)$.

Proof. Suppose otherwise. Let M be the Turing machine constructing Π and consider a Turing machine M' implementing the following algorithm:

> On input x, simulate M on x thus obtaining a description of a P system Π_x deciding whether $x \in L(f)$. Then simulate Π_x and return the same result.

Then $L(M') = L(\Pi) = L(f)$, and M' has the following space requirements:

$$o\left(\overbrace{f(n)}^{\text{construction}} + \overbrace{\frac{f(n)}{\log f(n)} \log \frac{f(n)}{\log f(n)}}^{\text{simulation}} \right) = o(f(n))$$

This means that M' decides $L(f)$ in $o(f(n))$ space, thus contradicting the space hierarchy theorem.   □

Notice that there is no restriction on the kind of rules the family Π can use. By combining theorems 3 and 4 we obtain:

Theorem 5. *Let $f \colon \mathbb{N} \to \mathbb{N}$, with $f(n) = \Omega(n)$ and $f(n) = O(n^k)$ for some fixed k, be time-constructible. Then there exists a language L which can be decided by a semi-uniform family of P systems with active membranes (and using only communication rules) that can be built in $O(f(n))$ time and works in $O(f(n))$ space. On the other hand, L cannot be decided by any family of P systems constructible in $o(f(n))$ time and working in $o(f(n)/\log f(n))$ space.*

5 Conclusions

In this paper we showed that a deterministic single-tape Turing machine, which operates in polynomial space with respect to the input length, can be efficiently simulated (both in terms of time and space) by a semi-uniform family of P systems with active membranes and three polarizations. The proposed simulation contains, in our opinion, a very interesting construction which has never been considered before (to the best of our knowledge), and which is exploited to obtain the result: the contents of the cells of the simulated Turing machine are stored in the polarization of the membranes. This allowed us to use only communication rules to compute the result.

Basing upon the above simulation, we proved that a result similar to the *space hierarchy theorem* can be obtained for P systems with active membranes: the larger the amount of space we can use during the computations, the harder the problems we are able to solve.

Several open problems and research directions still remain to be investigated. First of all, the result related to the space (pseudo)-hierarchy for P systems contains a logarithmic factor, which arises from the simulation we proposed. Can we avoid such a factor, thus obtaining a theorem which exactly corresponds to the space hierarchy theorem related to Turing machines?

Following this direction, we could also consider different classes of P systems with active membranes (e.g., using different parallel semantics), and check whether the space (pseudo)-hierarchy theorem still holds for such classes.

As for the simulation of the single-tape deterministic Turing machine we presented in section 3, we conjecture that it can be extended to consider nondeterministic Turing machines, as well as multi-tape Turing machines, to obtain efficient simulations both in terms of time and space also in these cases. It would also be interesting to consider if such an efficient simulation can be performed for other different computational models.

Acknowledgements. This work has been partially supported by the Italian project FIAR 2007 "Modelli di Calcolo Naturale e Applicazioni alla Systems Biology".

References

1. Alhazov, A., Martín-Vide, C., Pan, L.: Solving a PSPACE-complete problem by recognizing P systems with restricted active membranes. Fundamenta Informaticae 58(2), 67–77 (2003)
2. Alhazov, A., Freund, R.: On the efficiency of P systems with active membranes and two polarizations. In: Mauri, G., et al. (eds.) Membrane Computing, pp. 81–94. Springer, Heidelberg (2005)
3. Balcázar, J.L., Díaz, J., Gabarró, J.: Structural Complexity I, 2nd edn. Springer, Heidelberg (1995)
4. Gutiérrez-Naranjo, M.A., Pérez-Jiménez, M.J., Riscos-Núñez, A., Romero-Campero, F.J.: P systems with active membranes, without polarizations and without dissolution: A characterization of P. In: Calude, C.S., Dinneen, M.J., Păun, G., Jesús Pérez-Jímenez, M., Rozenberg, G. (eds.) UC 2005. LNCS, vol. 3699, pp. 105–116. Springer, Heidelberg (2005)
5. Gutiérrez-Naranjo, M.-A., Pérez-Jiménez, M.J., Riscos-Núñez, A., Romero-Campero, F.J.: On the power of dissolution in P systems with active membranes. In: Freund, R., Păun, G., Rozenberg, G., Salomaa, A. (eds.) WMC 2005. LNCS, vol. 3850, pp. 224–240. Springer, Heidelberg (2006)
6. Gutiérrez-Naranjo, M.A., Pérez-Jiménez, M.J., Riscos-Núñez, A., Romero-Campero, F.J., Romero-Jiménez, A.: Characterizing tractability by cell-like membrane systems. In: Subramanian, K.G., Rangarajan, K., Mukund, M. (eds.) Formal Models, Languages and Applications. Machine Perception and Artificial Intelligence, vol. 66, pp. 137–154. World Scientific, Singapore (2006)
7. Ibarra, O.H.: On the computational complexity of membrane systems. Theoretical Computer Science 320, 89–104 (2004)
8. Kobayashi, K.: On proving time constructibility of functions. Theoretical Computer Science 35, 215–225 (1985)
9. Krishna, S.N., Rama, R.: A variant of P systems with active membranes: Solving NP-complete problems. Romanian Journal of Information Science and Technology 2(4), 357–367 (1999)
10. Obtulowicz, A.: Deterministic P systems for solving SAT problem. Romanian Journal of Information Science and Technology 4(1-2), 551–558 (2001)

11. Păun, G.: Computing with membranes. Journal of Computer and System Sciences 1(61), 108–143 (2000)
12. Păun, G.: P systems with active membranes: Attacking NP-complete problems. Journal of Automata, Languages and Combinatorics 6(1), 75–90 (2001)
13. Păun, G.: Membrane computing. An introduction. Springer, Berlin (2002)
14. Jesús Pérez-Jímenez, M., Romero-Jiménez, A., Sancho-Caparrini, F.: The P versus NP problem through cellular computing with membranes. In: Jonoska, N., Păun, G., Rozenberg, G. (eds.) Aspects of Molecular Computing. LNCS, vol. 2950, pp. 338–352. Springer, Heidelberg (2003)
15. Pérez-Jiménez, M.J., Romero-Jiménez, A., Sancho-Caparrini, F.: A polynomial complexity class in P systems using membrane division. In: Csuhaj-Varjú, E., et al. (eds.) Proc. Fifth Workshop on Descriptional Complexity of Formal Systems, DCFS 2003, Computer and Automation Research Institute of the Hungarian Academy of Sciences, Budapest, pp. 284–294 (2003)
16. Porreca, A.E., Leporati, A., Mauri, G., Zandron, C.: Introducing a space complexity measure for P systems. International Journal of Computers, Communications & Control 4(3), 301–310 (2009)
17. Porreca, A.E., Mauri, G., Zandron, C.: Complexity classes for membrane systems. RAIRO Theoretical Informatics and Applications 40(2), 141–162 (2006)
18. Sipser, M.: Introduction to the Theory of Computation, 2nd edn. Course Technology (2005)
19. Sosík, P.: The computational power of cell division in P systems: Beating down parallel computers? Natural Computing 2(3), 287–298 (2003)
20. Zandron, C., Ferretti, C., Mauri, G.: Solving NP-complete problems using P systems with active membranes. In: Antoniou, I., Calude, C.S., Dinneen, M.J. (eds.) Unconventional Models of Computation, pp. 289–301. Springer, Heidelberg (2000)
21. Zandron, C., Leporati, A., Ferretti, C., Mauri, G., Pérez-Jiménez, M.J.: On the computational efficiency of polarizationless recognizer P systems with strong division and dissolution. Fundamenta Informaticae 87(1), 79–91 (2008)
22. The P Systems Webpage, http://ppage.psystems.eu

Look-Ahead Evolution for P Systems

Sergey Verlan

LACL, Département Informatique
Université Paris Est
61, av. Général de Gaulle, 94010 Créteil, France
verlan@univ-paris12.fr

Abstract. This article introduces a new derivation mode for P systems. This mode permits to evaluate next possible configurations and to discard some of them according to forbidding conditions. The interesting point is that the software implementation of this mode needs very small modifications to the standard algorithm of rule assignment for maximal parallelism. The introduced mode has numerous advantages with respect to the maximally parallel mode, the most important one being that some non-deterministic proofs become deterministic. As an example we present a generalized communicating P system that accepts 2^n in n steps in a deterministic way. Another example shows that in the deterministic case this mode is strictly more powerful than the maximally parallel derivation mode. Finally, this mode gives a natural way to define P systems that may accept or reject a computation.

1 Introduction

P systems are defined as non-deterministic computational devices. However, for implementation reasons, it is better to limit the inherent non-determinism to a smaller degree and eventually have a deterministic evolution. One of such approaches is based on an examination of the next configuration(s) and cutting off non-deterministic computational branches that have some pre-defined properties. The notion of k-determinism [8,2] is closely related to such optimizations. For a system having the k-determinism property one can examine all possible future configurations for at most k steps and find a single evolution that is not forbidden. This gives an efficient procedure of the construction of the next configuration. However, it is not an easy task to prove that a P system has this property.

A derivation mode lies in the heart of the semantics of P systems as it permits to specify which multiset among different possible applicable multisets of rules can be applied. When P systems were introduced, only the maximally parallel derivation mode was considered which states that corresponding multisets should be maximal, i.e. non-extensible. With the apparition of the minimal parallel derivation mode [3] the concept of the derivation mode had to be precisely defined and [5] presents a framework that permits to easily define different derivation modes.

G. Păun et al. (Eds.): WMC 2009, LNCS 5957, pp. 479–485, 2010.

This article tries to express the notion of one-step look-ahead in terms of a derivation mode which gives a way to implement P systems in a more efficient way. The look-ahead is a forbidding condition formalized by a set of forbidden rules that should not be applicable after a maximally parallel multiset of normal (non-forbidding) rules was chosen. Such a formalization needs a small overhead and can be easily incorporated and efficiently implemented in already existing software simulators for P systems. In more general way, the look-ahead derivation can be considered as a further evolution of the notion of k-determinism (more precisely of 1-determinism), but without restricting to a deterministic evolution.

The look-ahead mode can give advantages in terms of deterministic evolution of the system and we show an example that demonstrates that the evolution in the look-ahead mode introduces more power into the system. Moreover, it is known that a deterministic evolution usually sequentializes the computation and it needs more steps. With the look-ahead derivation mode we show that deterministic computations can be efficient by giving an example of a P system with minimal interaction that can recognize 2^n in n steps. An interesting side effect of the definition permits to define computations that are accepted or rejected without introducing additional symbols.

2   Definitions

We do not present here standard definitions. We refer to [10] for all details.

We also assume that the reader is familiar with standard notions of P systems, which can be consulted in the book [9] or at the web page [13]. We shall only focus on the semantics of the evolution step. We will follow the approach given in [5], however we will not enter into deep details concerning the notation and the definition of derivation modes given there. Consider a P system Π of any type evolving in any derivation mode. The key point of the semantics of P systems is that according to the type of the system and the derivation mode δ for any configuration of the system C a set of multisets of applicable rules, denoted by $Appl(\Pi, C, \delta)$, is computed. After that, one of the elements from this set is chosen, non-deterministically, for the further evolution of the system. In order to define the look-ahead derivation mode we suppose that the set of rules of Π, denoted by R, is composed from two subsets: normal rules R_N and forbidden rules R_f, i.e. $R = R_N \cup R_f$. Then we define $Appl(\Pi, C, LA\delta)$ as follows:

$$Appl(\Pi, C, LA\delta) = \{R' \mid R' \in Appl(\Pi, C, \delta) \text{ and } R' \cap R_f = \emptyset\}.$$

This means that only those multisets of rules which do not contain any rule from the forbidden set R_f can be considered for further evolution of the system. The set R_f can be replaced by other checking conditions, we shall discuss them in Section 4. By convention, we skip δ if it is the maximally parallel derivation mode ($\delta = max$) and call the obtained mode simply look-ahead derivation mode or LA mode.

We remak that the look-ahead derivation can be considered for any derivation mode, however in this article we shall consider only look-ahead derivation for the maximally parallel derivation mode, which is the most commonly used.

Let us consider the particularities of the look-ahead derivation mode. In fact, the rule set R_f gives conditions that shall not be satisfied by the current configuration with the condition that a particular multiset of rules from $Appl(\Pi, C, \delta)$ will be applied. This differentiates the look-ahead derivation mode from permitting or forbidding conditions which are checked *before* the assignment of objects to rules is done (in order to see if the rule is applicable), while the conditions in LA mode are checked *after* all assignments of objects to rules are done. This gives a greater flexibility as such a procedure permits to evaluate next possible configurations and to cut off some of them according to R_f. In such a way the non-determinism of the system may be significantly decreased.

We remark that the overhead introduced by such a procedure is minimal and we discuss in Section 4 possible implementations of the look-ahead derivation mode.

Another interesting point is that it is possible that all multisets from the set $Appl(\Pi, C, \delta)$ contain rules from R_f. In this case, $Appl(\Pi, C, LA\delta)$ will be empty, hence a halting configuration is reached. It is possible to differentiate this halting case from the case when $Appl(\Pi, C, \delta)$ is also empty and naturally introduce rejecting and accepting computations. This is particulary interesting for decision P systems, because it gives a natural way to obtain an answer **yes** or **no** without the need for additional symbols.

3 Examples

In this section we give two examples that show the interest of the look-ahead derivation mode. The first example presents a deterministic recognition of 2^n in n steps using minimal symport/antiport and conditional uniport, while the second example shows how the initial number of symbols can be increased by minimal symport/antiport P systems in a deterministic way.

3.1 Deterministic Recognition of 2^n

In this subsection we consider P systems with minimal interaction which are a restricted variant of generalized communicating P systems [12]. We recall that the later systems are a purely communicating model defined on a graph and having rules of form $(A, i)(B, k) \rightarrow (A, j)(B, m)$, where A and B are two multisets of objects and i, j, k, m are labels of membranes (cells). This rule permits to move multisets A and B from cells i and k to cells j and m synchronously. We remark that symport, antiport and conditional uniport [11] rules are a particular case of these general communication rules.

The minimal interaction rules are obtained from the generalized communication rules by restricting multisets A and B to one symbol each. Minimal symport and minimal antiport rules are a particular case of minimal interaction rules.

Consider the following system $\Pi = (O, E, w_1, w_2, R)$, having 2 cells (0 denotes the environment) where $O = \{A, B, Z\}$, $E = \emptyset$, $w_1 = \{A^k\}$, $w_2 = \{B, Z\}$ and $R = R_N \cup R_f$ is defined as follows.

$R_N = \{1 : (A, 1)(A, 1) \rightarrow (A, 2)(A, 1),\ 2 : (A, 1)(B, 2) \rightarrow (A, 2)(B, 1)\}$ and
$R_f = \{3 : (A, 1)(Z, 2) \rightarrow (A, 0)(Z, 2)\}$.

We remark that the first rule is a conditional uniport rule that sends a copy of A from cell 1 to cell 2, providing that another A remains in cell 1. The second rule is an antiport rule exchanging A and B in cell 1 and 2. The third rule is in fact an uniport rule of A to the environment, but because of the definition an interaction of two symbols is required, hence a dummy symbol Z is present in cell 2.

Consider now the evolution of the system. Let C_0 be the initial configuration. If k is even then all three rules are applicable. Hence, $Appl(\Pi, C_0, max)$ contains two multisets of applicable rules: $\{1^{k/2}\}$ and $\{1^{k/2-1}, 2, 3\}$. By the definition of the LA mode, the second multiset is eliminated and only the first possibility remains. It is clear that a similar reasoning applies to all configurations C having an even number of symbols A in cell 1 and a symbol B in cell 2. If k is odd, then $Appl(\Pi, C_0, max)$ contains following multisets of rules: $\{1^{(k-1)/2}, 2\}$ and $\{1^{(k-1)/2}, 3\}$. By the definition of LA mode the second possibility is eliminated and only the first one remains. The same holds for all configurations having an odd number of A in cell 1 and a copy of symbol B in cell 2.

Now consider the first application of rule 2. It might happen only when the number of symbols A in cell 1 is odd. In all consequent configurations symbol B is present in cell 1. Consider a further configuration having m symbols A in the first cell. If m is even, then again two multisets of rules are applicable in max mode: $\{1^{m/2}\}$ and $\{1^{m/2-1}, 3\}$ and only the first one remains in the LA mode. If m is odd, then there is only one applicable multiset in max mode: $\{1^{(m-1)/2}, 3\}$ and there are no applicable rules in LA mode.

The recognition of 2^n is done as follows. It is known that if a number $k = 2^n$ is divided by 2 in a cycle, then at each step the quotient is always even, except at the end when it becomes 1. For a number $k \neq 2^n$, a similar process yields an odd number $t > 2$. Repeating this procedure for $t - 1$ yields another odd number $t' \geq 1$. Rule 1 permits to divide the number of A's in cell 1 by two at each step. Rule 2 permits to decrement one time the number of A's in cell 1. Hence, if initially $k = 2^n$, then at each step an even number of A's will be present in cell 2, except the last step where the rule 2 will be applied. Otherwise, when an odd number $t > 2$ of symbols A will be present in cell 1, both rules 1 and 2 will be applied. Further, an odd number of symbols A will appear in cell 1 and the computation will stop. In the first case we obtain an accepting computation, while in the second one the computation is rejecting. We remark that the acceptance of a computation may be done in other ways as it is shown in Section 4.

3.2   Deterministic Minimal Symport/Antiport on a Tree Structure

In this subsection we consider P systems (having a tree structure) with minimal symport and antiport rules. An antiport rule is denoted as $(u, in; v, out)$ and permits to exchange the multiset of objects v present in the membrane i where this rule is located with the multiset of objects u present in the parent membrane of i. A symport rule, denoted as (u, in) or (v, out), permits to send a multiset v to the parent membrane or the multiset u to one of inner membranes. In the case of minimal antiport, respectively symport, the size of the multisets u and v is equal to one, respectively two.

We start by the following remark.

Remark 1. *For any deterministic P system with minimal symport/antiport rules working in maximally parallel derivation mode, the number of objects initially present inside the system, i.e. not in the environment, cannot be increased.*

The proof of the above assertion may be done in a similar way as it was done for the case of one membrane in [1] and [7]. The main argument used in those articles remains valid: if the number of objects is increasing, then any rule that permits to bring an additional symbol from the environment will be used an arbitrary number of times because of the minimality of rules and determinism.

However, the situation changes if the look-ahead derivation mode is permitted. Then the following construction permits to bring one symbol from the environment, deterministically.

Let $\Pi = (\{p, A\}, \{A\}, [_1[_2]_2]_1, \{p\}, \emptyset, R_1, R_2 \cup R_2^f)$ be a P system with minimal symport rules having two membranes (the first membrane contains initially symbol p while the second one is empty). We define the sets of rules R_1 and R_2 as follows (by the superscript f we denote the forbidding set of rules).

$R_1 = \{1 : (p, out); \ 2 : (pA, in)\}$,

$R_2 = \{3 : (pA, in)\}$ and $R_2^f = \{4 : (A, in)\}$.

The system works as follows. Firstly the symbol p is sent to the environment by rule 1 and after that it brings a copy of symbol A by rule 2. Now, in the maximally parallel derivation mode there are two applicable multisets of rules: $\{3\}$ and $\{1, 4\}$. In LA mode the second multiset is eliminated, hence only rule 3 can be applied. In such a way, the number of symbols (A) is increased, deterministically.

Since the number of objects can be varied, we conjecture that a deterministic register machine can be simulated, i.e. we conjecture that deterministic P systems with minimal symport/antiport working in LA mode can recognize any recursively enumerable set of numbers.

4   Implementation Ideas

In this section we discuss some ideas about the practical implementation of the look-ahead maximally parallel derivation mode. We consider the classical implementation of the maximally parallel derivation mode which orders rules and applies the rules maximal number of times according to the order and after that uses backtracking to decrease the number of applications of rules of a higher order and increase the number of applications of rules of a lower order. In this setup it is enough to place rules from R_f after the rules from R_N and to use an additional condition that if a rule from R_f is chosen then the current multiset should be discarded and a new backtracking round should begin. Hence, only the last condition shall be additionally implemented, which is not so difficult.

Another possibility is to replace rules from R_f by an union of finite sets that check the presence of the symbols from the left-hand side of rules from R_f. In this case it is enough to check that these sets are not present in the configuration after all rules are chosen, supposing that the choice of rules marks or blocks in some way used symbols. We recall that the difference between this check and ordinary permitting/forbidding checks is that it should be done *after* an assignment of object to rules is done.

The rejecting condition may be replaced by an emptiness check of a particular cell, or, in a more general setup, by checking for some finite state conditions like it is done for P automata (see [4] for an overview).

5   Final Remarks

In this paper we introduced a new derivation mode for P systems: the look-ahead mode. In some sense, this mode is an extension of the maximally parallel derivation mode and all results formulated for the latter one are true for the look-ahead mode. We also think that in the non-deterministic case both modes have same computational properties. In a lot of cases forbidden rules can be replaced by trapping rules that will move corresponding symbols to a trap membrane or will transform them to trapping symbols and the computation will never stop. However, in the deterministic case the behavior of two modes is very different, as it is shown in Subsection 3.2.

We would like to mention some differences between the look-ahead mode and the concept of k-determinism introduced in [8]. The notion of k-determinism is a property of a P system that permits to examine all possible future configurations for at most k steps and find a single evolution that is not forbidden. This property cannot be easily checked for a P system. The look-ahead derivation mode is not a property but a procedure that permits to possibly limit the non-determinism of the system.

As further research topics we would mention the extension of the look-ahead for k steps ahead. However, it is not clear if the gain in power is justified as the computational overhead needed to compute further k configurations is quite big. Another interesting problem would be the study of the efficiency of the new mode. We think that a lot of existing proofs can be simplified using the look-ahead and, moreover, efficient deterministic or almost deterministic solutions for different computational problems may be constructed. In particular, it would be interesting to give a deterministic simulation of a register machine by deterministic minimal symport/antiport P systems.

Instead of forbidden rules one may consider sets or even multisets of rules that cannot be applied together. This is a generalization of the concept of forbidden rules, because set R_f corresponds to set of pairs $\{(r, r') \mid r \in R_N, r' \in R_f\}$. This permits a finer control of rules, in some sense similar to programmed grammars, and it can be implemented quite easily. A similar approach using applicability vectors is considered in [6].

Acknowledgements. The author would like to thank A. Alhazov, E. Csuhaj-Varjú and R. Freund for their precious comments related to the topic of the paper. The author also acknowledges the support by the Science and Technology Center in Ukraine, project 4032.

References

1. Alhazov, A., Rogozhin, Y., Verlan, S.: Symport/antiport tissue P systems with minimal cooperation. In: Proc. ESF Exploratory Workshop on Cellular Computing (Complexity Aspects), Sevilla, Spain, pp. 37–52
2. Binder, A., Freund, R., Lojka, G., Oswald, M.: Implementation of catalytic P systems. In: Domaratzki, M., Okhotin, A., Salomaa, K., Yu, S. (eds.) CIAA 2004. LNCS, vol. 3317, pp. 45–56. Springer, Heidelberg (2005)
3. Ciobanu, G., Pan, L., Păun, G., Pérez-Jiménez, M.J.: P systems with minimal parallelism. Theor. Comput. Sci. 378(1), 117–130 (2007)
4. Csuhaj-Varjú, E.: P automata. In: Mauri, G., Păun, G., Jesús Pérez-Jímenez, M., Rozenberg, G., Salomaa, A. (eds.) WMC 2004. LNCS, vol. 3365, pp. 19–35. Springer, Heidelberg (2005)
5. Freund, R., Verlan, S.: A formal framework for static (tissue) P systems. In: Eleftherakis, G., Kefalas, P., Păun, G., Rozenberg, G., Salomaa, A. (eds.) WMC 2007. LNCS, vol. 4860, pp. 271–284. Springer, Heidelberg (2007)
6. Freund, R., Kogler, M., Verlan, S.: P automata with controlled use of minimal communication rules. In: Bordihn, H., et al. (eds.) Proc. of Workshop on Non-Classical Models for Automata and Applications, Wroclaw, Poland, pp. 107–119 (2009)
7. Frisco, P., Hoogeboom, H.: P systems with symport/antiport simulating counter automata. Acta Informatica 41(2-3), 145–170 (2004)
8. Oswald, M.: P Automata. PhD thesis. Vienna Univ. of Technology (2003)
9. Păun, Gh.: Membrane Computing. An Introduction. Springer, Heidelberg (2002)
10. Rozenberg, G., Salomaa, A. (eds.): Handbook of Formal Languages. Springer, Berlin (1997)
11. Verlan, S., Bernardini, F., Gheorghe, M., Margenstern, M.: On communication in tissue P systems: conditional uniport. In: Hoogeboom, H.J., Păun, G., Rozenberg, G., Salomaa, A. (eds.) WMC 2006. LNCS, vol. 4361, pp. 521–535. Springer, Heidelberg (2006)
12. Verlan, S., Bernardini, F., Gheorghe, M., Margenstern, M.: Generalized communicating P systems. Theor. Comput. Sci. 404(1-2), 170–184 (2008)
13. The Membrane Computing Web Page, http://ppage.psystems.eu

Author Index